DEVELOPMENTS IN **SEDIMENTOLOGY** 56

INTRODUCTION TO THE PHYSICS OF COHESIVE SEDIMENT IN THE MARINE ENVIRONMENT

DEVELOPMENTS IN **SEDIMENTOLOGY** 56

INTRODUCTION TO THE PHYSICS OF COHESIVE SEDIMENT IN THE MARINE ENVIRONMENT

JOHAN C. WINTERWERP AND WALTHER G.M. VAN KESTEREN

WL / DELFT HYDRAULICS & DELFT UNIVERSITY OF TECHNOLOGY, DELFT, THE NETHERLANDS

SERIES EDITOR: T. VAN LOON

2004

ELSEVIER

Amsterdam – Boston – Heidelberg – London – New York – Oxford
Paris – San Diego – San Francisco – Singapore – Sydney – Tokyo

ELSEVIER B.V.	ELSEVIER Inc.	ELSEVIER Ltd	ELSEVIER Ltd
Sara Burgerhartstraat 25	525 B Street, Suite 1900	The Boulevard, Langford Lane	84 Theobalds Road
P.O. Box 211, 1000 AE	San Diego, CA 92101-4495	Kidlington, Oxford OX5 1GB	London WC1X 8RR
Amsterdam, The Netherlands	USA	UK	UK

First edition 2004

Library of Congress Cataloging in Publication Data
A catalog record is available from the Library of Congress.

British Library Cataloguing in Publication Data
A catalogue record is available from the British Library.

ISBN: 0-444-51553-4
ISSN: 0070-4571 (Series)

∞ The paper used in this publication meets the requirements of ANSI/NISO Z39.48-1992 (Permanence of Paper).
Printed in The Netherlands.

Mud! Mud! Glorious mud
Nothing quite like it for cooling the blood.
So, follow me, follow, down to the hollow,
And there let us wallow in glorious mud

the Hippo song

PREFACE

A comprehensive study of cohesive sediment requires a multi-disciplinary approach. The behaviour of cohesive sediment is determined by physical, biological and chemical aspects. Furthermore, there is a wide range of societal issues related to cohesive sediments, such as siltation in navigational channels, water quality problems, sustainable management of estuaries and wetlands, stability of continental slopes, etc.

In the last decade, substantial progress has been made in understanding the processes governing cohesive sediment behaviour. These developments have been reported in an overwhelming number of papers and congress proceedings, which are published in a wide variety of journals and books. However, there is no journal exclusively dedicated to cohesive sediment, nor is there a specialised book which summarises the recent work. The best introduction to the physical aspects of cohesive sediment is still found in the proceedings of the 1984-workshop on cohesive sediment edited by A.J. Mehta (1986). For engineering applications a handy introduction is provided by Whitehouse et al. (2000).

In the present book, we have therefore undertaken the task of summarising the recent progress made in understanding the physical processes of cohesive sediment in the marine environment. The book contains overviews of classical and recent literature and of new developments, which have not yet been published. We treat the physical processes in the water column, in the bed and at the water-bed interface. This is done from both a hydrodynamic and from a soil mechanical point of view, and we try to integrate these approaches where relevant. Moreover, we give some attention to biological and chemical effects, as they may (largely) affect the behaviour of cohesive sediment in the natural environment.

This book is written for all graduate students, scientists and engineers who want to study the physical behaviour of cohesive sediment in depth and at a fundamental level. As such the book does not contain, but a few, recipes for direct engineering applications. However, we do give examples and some numerical elaboration of the concepts and formulations derived in this book. Yet we feel that this book can contribute to sound managerial and engineering decisions through a better understanding of the underlying physics. We also hope that this book will stimulate and guide other scientists in further research on cohesive sediment, as many questions are still unsolved.

This book could not have been written without the financial and other support by WL | Delft Hydraulics, Delft University of Technology, the University of

Florida, the Netherlands Science Foundation, the U.S. Army Corps of Engineers and Nutech. We are very grateful to Prof. A.M. Mehta and Prof. J.A. Battjes for their review of almost the entire book, their constructive comments and suggestions and their encouragement. Individual chapters of the book were also reviewed by Dr. R.B. Kirby (Appendix C), Prof. F. Molenkamp (Chapter 8 & Appendix D), Mr. M. Sas (Appendix C), Dr. G.A. Sills (Chapter 5 & 7) and Dr. J. Widdows (Chapter 10). Isolated sections of the book were reviewed by Mr. G. de Boer, Mr. J.M. Cornelisse, Mr. W. Eysink, Mr. C. Kuijper, Dr. L. Merckelbach, Mr. J. Smit, dr. E.A. Toorman and Mr. A. Wijdeveld.

Delft, June 2004
Johan C. Winterwerp   Walther G.M. van Kesteren
WL | Delft Hydraulics   WL | Delft Hydraulics
Delft University of Technology

TABLE OF CONTENTS

1. Introduction   1

2. Boundary layer flow   5
   2.1 Hydrodynamics and mass balance   5
       2.1.1 The mean water movement   5
       2.1.2 The mass balance for suspended sediment   8
       2.1.3 The k-ε turbulence model   10
   2.2 The boundary layer   14
       2.2.1 The structure of the boundary layer   14
       2.2.2 Coherent structures in the boundary layer   20
   2.3 The effect of surface waves   24

3. The nature of cohesive sediment   29
   3.1 The composition of cohesive sediment   30
       3.1.1 Granular composition of cohesive sediment   30
       3.1.2 Mineral composition of cohesive sediment   35
       3.1.3 Organic composition of cohesive sediment   41
   3.2 Skeleton fabric composition of cohesive sediment   44
       3.2.1 Inter-particle forces   44
       3.2.2 Pore size distribution   50
   3.3 Geotechnical classification of cohesive sediment   52
       3.3.1 Geotechnical parameters   52
       3.3.2 Application to in-situ conditions   58
   3.4 Cohesive sediment in the marine environment   72
       3.4.1 Phenomenological classification of cohesive sediment   72
       3.4.2 Mud content of sand-mud mixtures   78
       3.4.3 Fluid mud   82

4. Flocculation processes   87
   4.1 Introduction   88
   4.2 Fractal structure of mud flocs   93
   4.3 Flocculation model   96
       4.3.1 Number equation for flocs   96
       4.3.2 Aggregation processes   99
       4.3.3 Floc break-up processes   100
       4.3.4 Eulerian flocculation model   103

4.3.5 Lagrangean flocculation model   106
4.3.6 Comparison with field data   109
4.4 Flocculation time   111
4.4.1 The gelling concentration   111
4.4.2 Non-equilibrium conditions   113

5. Settling and sedimentation   **121**
5.1 Introduction   122
5.2 Settling velocity and floc size   124
5.2.1 Settling velocity and floc size in still water   124
5.2.2 Hindered settling   127
5.3 Deposition and sedimentation   138
5.3.1 Deposition rate   138
5.3.2 Sedimentation in flowing water   151
5.3.3 Sedimentation in harbour basins   155

6. Sediment-fluid interaction   **161**
6.1 Introduction   163
6.2 Sediment-fluid interaction –literature overview   165
6.2.1 Non-cohesive sediment   165
6.2.2 Low-concentration mud suspensions   171
6.2.3 High-concentration mud suspensions   175
6.3 Sediment-induced buoyancy effects   180
6.3.1 The concept of saturation   180
6.3.2 HCMS in a turbidity maximum   193
6.3.3 HCMS and rapid siltation   200
6.4 Sediment-fluid interaction in the benthic boundary   207
      layer

7. Self-weight consolidation   **211**
7.1 Introduction – classical theory   211
7.2 The Gibson consolidation equation   214
7.3 Special cases of the Gibson equation   218
7.3.1 Terzaghi's consolidation theory   218
7.3.2 Kynch's sedimentation theory   219
7.4 Material functions for the Gibson equation   221
7.5 Application of Gibson's equation   225
7.6 Fractal description of bed structure   228
7.6.1 Effective stress   228
7.6.2 Permeability   231

7.7 Consolidation as an advection-diffusion process   233
     using fractal theory
7.8 Material functions for fractal approach   239
7.9 Application of fractal approach   243
7.10 Approximated solution of consolidation equation   246

8. Mechanical behaviour   **253**
     8.1 The seafloor as a multi-phase system   254
          8.1.1 Effective stress concept   255
          8.1.2 Drained and undrained behaviour   257
          8.1.3 Compressibility of gaseous water   261
     8.2 Stress-strain relations   264
          8.2.1 Elastic and plastic strains   264
          8.2.2 Elastic stress-strain relations   267
          8.2.3 Elasto-plastic stress-strain relations   270
          8.2.4 Cohesive sediment constitutive relations   273
     8.3 Failure mechanisms   284
          8.3.1 Type of failure   284
          8.3.2 Tensile failure   289
          8.3.3 Tensile failure parameters   298
          8.3.4 Shear failure   303
     8.4 Cyclical behaviour   310
          8.4.1 Cyclic stress-strain relations   311
          8.4.2 Examples of cyclical loading   319
     8.5 Strain-rate dependent behaviour   325
          8.5.1 Range of strain rates   325
          8.5.2 Rheological behaviour   327
          8.5.3 Creep   332
          8.5.4 Dynamic response   336

9. Erosion and entrainment   **343**
     9.1 Phenomenological description of erosion   343
     9.2 Literature on erosion   345
          9.2.1 Erosion formulae   345
          9.2.2 Effect of physico-chemical parameters   349
     9.3 Classification scheme for erosion   357
     9.4 Entrainment of fluid mud layers   362
     9.5 Erosion as a drained/undrained process   370
          9.5.1 Flow-induced erosion   370
          9.5.2 Wave-induced erosion   377

        9.5.3 Erosion by turbidity currents   379

10. **Biological effects**   **383**
    10.1 The role of vegetation   385
    10.2 Bio-deposition   386
    10.3 Bio-stabilisation   389
    10.4 Bio-destabilisation   393

11. **Gas in cohesive sediment**   **397**
    11.1 Introduction   397
    11.2 Gas-related processes in sediment   403
    11.3 Biogenic gas production   404
    11.4 Thermodynamic equilibrium of gas in water   407
        11.4.1 Solubility   407
        11.4.2 Liquid phase   410
        11.4.3 Gas hydrates   410
    11.5 Bubble mechanics   413
        11.5.1 Bubble nucleation   413
        11.5.2 Bubble growth   415
        11.5.3 Ebullition of bubbles   418
        11.5.4 Crack initiation and propagation   419
    11.6 Channel formation   423

References   **429**

Appendix A: Nomenclature   **A-1**

Appendix B: Definitions and useful relations   **B-1**

Appendix C: Measuring techniques   **C-1**
    C.1 Composition and properties of the sediment-water
         mixture   C-2
    C.2 Particle size distribution   C-6
        C.2.1 In-situ particle size distribution in the water column   C-6
        C.2.2 In-situ particle size distribution in the bed   C-7
        C.2.3 Particle size distribution in the laboratory   C-7
        C.2.4 Pore size distribution   C-8
    C.3 Sediment concentration and density   C-10
        C.3.1 In-situ sediment concentration in the water column   C-10

      C.3.2 In-situ sediment concentration (density) in the bed   C-11

      C.3.3 Sediment concentration in the laboratory   C-12

  C.4 Rheological parameters   C-14

  C.5 Soil mechanical parameters   C-17

      C.5.1 Atterberg limits   C-17

      C.5.2 Permeability   C-18

      C.5.3 Total and pore water pressure   C-19

      C.5.4 Consolidation parameters   C-20

      C.5.5 Stress-strain relations   C-22

  C.6 Settling velocity   C-23

      C.6.1 Settling velocity in the laboratory   C-23

      C.6.2 In-situ settling velocity   C-24

  C.7 Erodibility   C-26

      C.7.1 Erosion measurements in the laboratory   C-26

      C.7.2 Erosion measurements in the field   C-27

Appendix D: Tensor analysis   **D-1**

  D.1 The stress tensor   D-1

  D.2 The strain tensor   D-10

  D.3 The strain-rate tensor   D-15

  D.4 Momentum and energy equations   D-18

Appendix E: The 1DV POINT MODEL   **E-1**

  E.1 The 1DV-equations   E-1

  E.2 Numerical implementation of the 1DV-equations   E-14

  E.3 Requirements for numerical accuracy   E-15

Subject index   **I-1**

1.　INTRODUCTION

The earth's surface is almost entirely covered with larger or smaller amounts of cohesive sediment, or mud as it generally referred to, with the exception perhaps of some deserts and some parts of the ocean seabed. It is, and has been throughout time, both a blessing and a curse to mankind:

- Cohesive sediment is a valuable resource. Civilisation started in Egypt along the Nile River, in Mesopotamia along the Euphrates and Tigris, in India along the Indus, and in China along the Yellow River. All these areas are/were very fertile because of cohesive sediment deposits by the rivers on their flood plains. Also today, many river deltas belong to the most productive areas of the world, amongst which are the Yangtze delta, the Rhine-Meuse delta, and the Mississippi delta.
- At present, one realises that cohesive sediment also plays a key role in the functioning of healthy eco-systems. It is indispensable for the development of flora and fauna in estuaries, on the seabed and on flood plains of rivers, and in particular for the natural evolution of healthy wetlands.
- Cohesive sediment is also a valuable resource of building and construction material, such as plaster and bricks.
- Unfortunately, cohesive sediment is often contaminated these days, as organic (pcb's, etc.) and inorganic (heavy metals, etc.) pollutants adhere easily to the clay particles and organic material of the sediment. These contaminants can accumulate in the food web, sometimes to pathogenic and even lethal levels, endangering the entire eco-system.
- Because of the processes governing the transport and fate of cohesive sediment in open water systems, cohesive sediment tends to accumulate in still water regions, such as navigation channels and harbour basins. For instance, the Port of Rotterdam has to dredge about 10 Mm^3 of sediment on an annual basis to safeguard navigation, whereas this amounts to about 3 Mm^3 in the Port of Hamburg. Suffice it to say that the costs of maintaining these fairways and harbour basins can be very high, in particular when the sediments to be removed are contaminated.
- Thick layers of cohesive sediment are found on the slope of continental shelves. These layers can become unstable because of natural (earthquakes, tsunami's, gas releases, etc.) or human disturbance (pipelines), as a result of which mudflows and/or turbidity currents are generated which may damage cables, pipelines, and sometimes exploration platforms.
  Such mudflows are also found on land, in particular in mountainous areas, and their occurrence can be devastating.

As cohesive sediment, or mud, is such a familiar feature in our daily life, it is easily recognised as such. Yet, it is difficult to give a sound scientific definition of cohesive sediment. Mehta (2002), for instance, defines mud as a sediment-water mixture composed of grains that are predominantly less than 63 μm in size, exhibiting a rheological behaviour that is poro-elastic or visco-elastic when the matrix is particle-supported, and is highly viscous and non-Newtonian when it is in a fluid-like state.

One of the difficulties is that cohesive sediment varies so much in composition and can occur in so many appearances. It is therefore impossible to provide the reader with generally applicable rules or recipes to analyse and predict the behaviour, transport and fate of cohesive sediment in the natural environment. Hence, this book does not contain such general rules or recipes.

Yet, we present the ingredients and techniques to classify the sediment and quantify its properties and behaviour, and give examples of how these techniques can be implemented in simple and complicated tools (such as numerical models), and how they can be used to solve daily management and engineering problems. We believe that a thorough understanding of the underlying physical processes is a prerequisite for the proper use of these techniques and tools. Therefore, we present a rigorous analysis and derivation of the relevant physical-mathematical descriptions of the processes underlying the sediment behaviour. This will help the reader to appreciate the applicability and limitations of the various tools and techniques for his/her specific managerial or engineering problem.

Today, managerial and engineering studies are often carried out with numerical models. Therefore, we also present some examples of results of numerical studies, and in Chapter 2 we start with a detailed description of the relevant basic equations of the water movement in general, and of the boundary layer characteristics in particular, though these can be found in numerous specialised text books.

Chapter 3 provides a general description of our perception of the characteristics and properties of cohesive sediment and its classification, both from a microscopic and macroscopic point of view, and we give a more general classification of its appearances in the natural environment. We introduce a number of soil mechanical concepts to be used in the hydrodynamic framework of cohesive sediment, which covers a major part of this book. Next we present a detailed discussion of the various physical processes, and where relevant, of their mutual interactions. The Chapters 4 through 9 contain our views on what is commonly accepted as the key processes in cohesive sediment dynamics:

- flocculation, i.e. the formation and break-up of flocs of cohesive sediment – the flocculation process is the key in understanding the difference in behaviour of cohesive and non-cohesive sediment,

- settling and deposition of cohesive sediment, including a relation between settling velocity and floc size, and sedimentation in still and flowing water,
- interaction between suspended cohesive sediment and turbulent flow at high-concentrated sediment concentrations, such as in the case of fluid mud formation,
- consolidation, in particular self-weight consolidation, including the effects of minor amounts of fine sand, both in the classical material co-ordinates, and in a Eulerian frame of reference,
- mechanical behaviour of the bed under the influence of external stresses or deformations, such as fracturing and failure,
- non-Newtonian behaviour of sediment-water mixtures, i.e. the stress-strain relations when soft mud deposits flow, and
- erosion and entrainment of fresh and consolidated deposits, which are key processes in the large scale transport and fate of cohesive sediment in the marine environment.

In a number of chapters we discuss the interactions between the various processes, and show how these interactions affect the overall behaviour and appearance of cohesive sediment in nature. Amongst these is the behaviour of mixtures of cohesive and non-cohesive sediment. We recognise that our knowledge of these natural sediment occurrences is only very limited at present.

We focus mainly on the physical processes of cohesive sediment in the marine environment. Yet, non-physical aspects, such as biology and chemistry often affect the behaviour of cohesive sediment in natural marine environments. These effects are by no means of secondary importance: in-situ measurements have revealed, for instance, that the resistance to erosion of intertidal mud deposits can increase by an order of magnitude during algae bloom. We therefore have included a brief section on biological effects, e.g. Chapter 10.

The inclusion of (large quantities of) organic material may result in gas production in (thick) cohesive sediment deposits. As the permeability of the sediment is poor, gas may accumulate in bubbles, affecting the stability of such sediment deposits largely. At present, this is a major concern in exploration drilling activities on the continental slopes. Moreover, during maintenance dredging in harbours, and with sub-aqueous storage of contaminated mud, gas accumulation and gas release may affect the environment by mobilising contaminants. In Chapter 11 we give a concise description of the underlying processes, which are illustrated with a few case studies.

Where appropriate, we give practical examples on how the theory and/or methodology can be applied. We emphasise that we do not aim to present an exhaustive and in-depth treatment of all the work and disputes presented in literature on cohesive sediment behaviour. We merely present our own views, developed during numerous studies carried out in Delft. In particular, we feel that a thorough understanding of the various mutual interactions and the time effects related to the various processes is a prerequisite for a proper description of the transport and fate of cohesive sediments in the marine environment.

Appendix B contains a glossary of definitions and relations used in this book, which may be helpful to some readers to find their way in cohesive sediment dynamics. A summary of all symbols and parameters used in the book is found in Appendix A.

Appendix C contains a brief description of measuring techniques available to enable the quantification of the various parameters used in our physical-mathematical cohesive sediment descriptions. We do not aim to present in-depth discussions of the techniques available, but merely to make the reader aware of existing techniques and relevant instruments, often commercially available, and of some of the pitfalls in their use.

Appendix D gives a brief summary on tensor analysis, a technique required in studying and describing the mechanical processes in the seafloor, as treated in Chapter 8. This book also contains a series of numerical examples obtained with the 1DV POINT MODEL, which is obtained by stripping all horizontal gradients (except for the horizontal pressure gradient) from the full 3D equations described in Chapter 2 and from the various formulations derived in following chapters. This 1DV-model is described in Appendix E.

2. BOUNDARY LAYER FLOW

The majority of sediment transport processes treated in this book occur in what is commonly known as the benthic boundary layer. Also, water-bed exchange processes, which largely determine the transport and fate of cohesive sediment, are governed by the hydrodynamic conditions within this boundary layer. Therefore a brief overview of the relevant hydrodynamic processes within the boundary layer is presented. For further details and rigorous analyses, the reader is referred to one of the many excellent reference books on turbulence and/or boundary layer flow in general, and on the benthic boundary layer in particular (e.g. Pope, 2000; Tennekes and Lumley, 1994; Nezu and Nakagawa, 1993; or Boudreau and Jørgensen, 2001).

First, the three-dimensional continuity, momentum and mass balance equation are treated, together with the k-ε turbulence closure model – computational examples presented in this book have been obtained with this turbulence model. Next, the structure of the boundary layer is discussed, both for (quasi-) stationary flow and under short waves, together with relevant parameters and scales.

2.1 GOVERNING EQUATIONS

2.1.1 THE MEAN WATER MOVEMENT

This book focuses on the dynamics of cohesive sediment in shallow water, hence the hydrostatic pressure approximation is valid, and we assume that the bed is more or less horizontal. For flows in deeper water, the reader is referred to for instance Pedlosky (1987).

The mixture of water and cohesive sediment is treated as a single-phase fluid (see also Chapter 6). We assume that the fluid is incompressible and that the Boussinesq approximation is applicable, i.e. variations in the fluid density ρ can be neglected, except in the gravitational terms. Using Einstein's summation convention, the three-dimensional continuity equation reads:

$$\frac{\partial u_i}{\partial x_i} = 0 \tag{2.1}$$

and the three-dimensional momentum equation reads:

$$\frac{\partial u_i}{\partial t} + \frac{\partial u_i u_j}{\partial x_j} - e_{ijk} f_k u_j = \frac{1}{\rho} \frac{\partial \sigma_{ij}}{\partial x_j} - \delta_{i3} g \qquad (2.2)$$

The equation of state, $\rho = \rho(S,c)$, is specified in Section 2.2. In equ. (2.1), (2.2) and in the equation of state the following notation is used:

c = suspended sediment concentration by mass,

f = Coriolis parameter,

g = acceleration of gravity,

S = salinity,

t = time,

u_i = flow velocity, averaged over the turbulent time scale,

x_i = Cartesian co-ordinate (x_3 is positive upward - see sketch),

ρ = bulk density of water-sediment mixture, and

σ_{ij} = stress tensor (see also Chapter 8).

The Kronecker delta δ_{ij} ($\delta = 1$ for i = j, and $\delta = 0$ for i ≠ j) is used, together with the permutator e_{ijk} (e = 1 for cyclical i, j and k, e = -1 for anti-cyclical i, j and k; otherwise e = 0).

Further in this book, with emphasis on the processes in the vertical plane, x- and z- notation is preferred in stead of the tensor notation above:

x_1	x_3	u_1	u_3	σ_{13}
x	z	u	w	τ_{xz}

The stress tensor σ_{ij} consists of a part induced by fluid stresses and a part induced by inter-particle stresses. In the water column at suspended sediment concentrations up to the gelling point (structural density), the latter inter-particle stresses are negligible. Also stresses induced by particle-particle interaction are very small for cohesive sediment in the water column (see Chapter 6). The fluid and inter-particle stresses are superimposed according to the Kelvin-Voigt theory (see also Chapter 8):

$$\sigma_{ij} = \sigma_{ij}^w + \sigma_{ij}^{sk} \qquad (2.3)$$

where superscript $\bullet^w$ refers to the water phase and $\bullet^{sk}$ to the inter-particle stresses caused by the solids structure (skeleton) in the bed. Both stresses can be separated into a deviatoric part τ_{ij} and an isotropic part p:

$$\sigma_{ij} = \tau_{ij} - \delta_{ij} p, \quad \text{where } p = -\tfrac{1}{3}\sigma_{ii} \tag{2.4}$$

Note that tensile stresses are defined as positive, whereas in Chapter 8 we follow soil mechanical definitions and define compressive stresses as positive. According to (2.3), the isotropic pressure can be written as:

$$p = p^w + p^{sk} \tag{2.5}$$

In soil mechanics (2.5) is also known as the effective stress concept – this is further discussed in Chapter 7 and 8.

The decomposition given above allows the inclusion of contributions to the stress tensor resulting from:

- Molecular effects (viscosity), possibly affected by (weak) non-Newtonian effects,
- Turbulence - we assume that the eddy viscosity and eddy diffusivity are isotropic by applying the k-ε turbulence closure model (e.g. Section 2.3), and
- Interparticle stresses – this topic is elaborated in Chapter 7 and 8 where consolidation and strength evolution models are treated.

Thus, after applying the Reynolds averaging procedure for turbulent flow, the stress tensor becomes (Malvern, 1969):

$$\sigma_{ij} = \tau^w_{m,ij} - \delta_{ij} p^w_m + \sigma_{T,ij} + \tau^{sk}_{ij} - \delta_{ij} p^{sk} \tag{2.6}$$

in which $\tau^w_{m,ij}$ and p^w_m are the molecular stresses and $\sigma_{T,ij}$ the turbulent stresses, where we have omitted superscript $\bullet^w$ (see (2.8)). The molecular shear stress is modelled with Fick's law (Malvern, 1969; Hinze, 1975):

$$\tau^w_{m,ij} = \mu\left(\frac{\partial u_i}{\partial x_j} + \frac{\partial u_j}{\partial x_i}\right) \tag{2.7}$$

in which μ is the dynamic viscosity of the (Newtonian) fluid. The turbulent stress tensor is modelled with the Boussinesq eddy viscosity concept (e.g. Hinze, 1975; Rodi, 1984):

$$\sigma_{T,ij} = \rho v_T \left(\frac{\partial u_i}{\partial x_j} + \frac{\partial u_j}{\partial x_i} \right) - \frac{2}{3} \delta_{ij} \rho k \qquad (2.8)$$

in which v_T is a scalar quantity, representing the turbulent eddy viscosity and k is the turbulent kinetic energy, defined in Section 2.3.

In the remainder of this chapter the superscript $\bullet^w$ is dropped and the symbol p is used exclusively for the water pressure. Because of the hydrostatic pressure assumption, the momentum equation in the vertical direction x_3 simplifies to:

$$\frac{\partial p}{\partial x_3} = -\rho g \qquad (2.9)$$

We consider water systems with a free surface only, which may vary with time, though, as a result of which the pressure term is written as:

$$p(x_3) = p_{atm} + g \int_{x_3}^{Z_s} \rho dx_3' \qquad (2.10)$$

where p_{atm} is the atmospheric pressure at the water surface $x_3 = Z_s(t)$, i.e. $p(Z_s) = p_{atm}$. The inter-particle stress σ_{ij}^{sk} is elaborated in Chapter 7 and 8; it is zero, unless stated otherwise.

This set of equations is closed with proper initial and boundary conditions – these are not treated here, as these are site-specific. However, a no-slip condition at the bed is generally assumed (see also Section 2.2).

2.1.2 THE MASS BALANCE FOR SUSPENDED SEDIMENT

It is reasoned in Chapter 6 that suspensions of cohesive sediment can be treated as a single-phase fluid, the particles following the turbulent water movements, except for their settling velocity. The mass balance for fine-grained suspended sediment in three dimensions can therefore be described with the well-known advection-diffusion equation, which reads:

$$\frac{\partial c^{(i)}}{\partial t} + \frac{\partial}{\partial x_i}\left(\left(u_i - \delta_{i3}w_s^{(i)}\right)c^{(i)}\right) - \frac{\partial}{\partial x_i}\left(\left(D_s + \Gamma_T\right)\frac{\partial c^{(i)}}{\partial x_i}\right) +$$

$$-\frac{\partial}{\partial x_i}\left(\delta_{i3}\overline{w_s'c'}^{(i)}\right) = 0 \tag{2.11}$$

where $c^{(i)}$ is the ensemble averaged sediment concentration by mass of fraction $(i)^{1)}$, D_s the molecular diffusion coefficient, Γ_T the eddy diffusivity defined as $\Gamma_T = \nu_T / \sigma_T$ (see Section 2.3), with σ_T is the turbulent Prandtl-Schmidt number, and $w_s^{(i)}$ is the effective settling velocity of sediment fraction (i). It is often assumed that D_s and Γ_T are the same for all sediment fractions, which is correct for small particles ($w_s \ll u_3'$) in dilute suspensions (Felderhof and Ooms, 1990). The settling velocity is further elaborated upon in the Chapter 4 and 5. The last term in (2.11) stems from a correlation between the turbulent fluctuations in c and $w_s^{(i)}$. In general $w_s \ll w'$ (e.g. Chapter 6). Because the turbulent fluctuations of the settling velocity are much smaller than its mean value, the last term in (2.11) can be neglected with respect to the turbulent diffusion term, as this term represents the effect of the correlation $\overline{w'c'}$.

   The molecular diffusion term D_s is given by the Stokes-Einstein equation (Bird et al., 1960):

$$D_s = \frac{kT}{6\pi\mu D} \tag{2.12}$$

in which
D	= particle size,
k	= Boltzman constant (= $1.38 \cdot 10^{-23}$ J/K), and
T	= absolute water temperature.

The influence of the suspended sediment concentration on the bulk fluid density is given by the equation of state:

---

[1] Note that the superscript (i) is only an identifier for the sediment fraction and does not obey Einstein's summation convention.

$$\rho\left(S,T,c^{(i)}\right)=\rho_w\left(S,T\right)+\sum_i\left\{\left(1-\frac{\rho_w\left(S,T\right)}{\rho_s^{(i)}}\right)c^{(i)}\right\} \tag{2.13}$$

with $\rho_w(S,T)$ the density of the water as a function of salinity S and temperature T only.

Bed-boundary conditions for the mass balance are treated extensively in Chapter 5 and 9.

2.1.3 THE k-ε TURBULENCE MODEL

The flow in the majority of marine conditions treated in this book is characterised by high turbulence Reynolds numbers, i.e. $Re_T > 100$, where Re_T is defined as:

$$Re_T \equiv \frac{\sqrt{k}\,\ell}{\nu} \approx \frac{2\nu_T}{\nu} \tag{2.14}$$

in which k is the turbulent kinetic energy, defined in (2.15), ℓ is a typical length scale of the turbulence (mixing length) and ν is the kinematic molecular viscosity, i.e. $\nu = \mu/\rho$. For $Re_T < 100$ the kinematic viscosity becomes important at all turbulent length scales, and a so-called low-Reynolds-number turbulent closure model has to be applied. This may be the case at very high suspended sediment concentrations.

In shallow water, the turbulent flow field is not truly isotropic. Hinze (1975) and Nezu and Nakagawa (1993), amongst others, described experimental data in open channel flow showing that rms-values of turbulent fluctuations in the transverse velocity component are about 30 % smaller than those in the direction of the mean flow velocity, whereas the corresponding fluctuations in the vertical direction are even about 45 % smaller. However, at length scales of the order of the water depth, the differences in the turbulent velocity components are smaller, as was shown for instance by Nezu and Rodi (1986). Also, recent ADCP-measurements by Sukhodolov et al. (1998) in the Spree River (Germany) revealed fairly isotropic turbulence properties at length scales of the same order of and smaller than the water depth.

We are concerned with shallow water flows where the boundary layer approximation is valid. It has been shown that, in spite of the anisotropy in the turbulence field, isotropic turbulence models, such as the k-ε model, are then suited to establish the mixing properties perpendicular to the bed and the main shear flow. Such models have been applied with great success, as shown from comparisons with experimental data, e.g. Launder and Spaulding (1974); Rodi (1984); Uittenbogaard et al. (1992). We recognise that longitudinal and lateral mixing is underestimated by such isotropic models, but we focus on the processes in the vertical direction. Hence, we ignore anisotropic effects in the horizontal plane and apply the isotropic k-ε turbulence closure model. Rodi (1984) and Uittenbogaard (1995) have shown that this model even applies to fairly stratified flow conditions. For highly stratified flow conditions, however, the buoyancy effects are overestimated, resulting in too much damping of the vertical exchange of momentum (Simonin et al., 1989). This is attributed to internal waves, which may augment the vertical transport of momentum, the effects of which are, however, not included in the standard k-ε model. Uittenbogaard (1995) therefore advocates the inclusion of additional source and sink terms in the k-ε model, the descriptions of which are determined from internal wave properties. This approach is not further elaborated upon here.

The turbulent kinetic energy k and the turbulent dissipation rate ε per unit mass are defined by (see Tennekes and Lumley, 1994):

$$k \equiv \frac{1}{2}\overline{(u_i' u_i')} \quad \text{and} \quad \varepsilon = \frac{1}{2}\nu \overline{\left(\frac{\partial u_i'}{\partial x_j} + \frac{\partial u_j'}{\partial x_i}\right)^2} \tag{2.15}$$

where u_i' is the fluctuating part of the instantaneous velocity component, and the overbar represents averaging over the turbulent time scale.

The standard k-ε model with buoyancy destruction terms is used in a number of examples in this book. It reads:

$$\frac{\partial k}{\partial t} + \frac{\partial u_i k}{\partial x_i} - \frac{\partial}{\partial x_i}\left(\nu + \frac{\nu_T}{\sigma_k}\right)\frac{\partial k}{\partial x_i} = \tag{2.16a}$$

$$= \nu_T\left(\frac{\partial u_i}{\partial x_j} + \frac{\partial u_j}{\partial x_i}\right)\frac{\partial u_i}{\partial x_j} + \delta_{i3}\frac{g}{\rho}\frac{\nu_T}{\sigma_T}\frac{\partial \rho}{\partial x_i} - \varepsilon$$

$$\frac{\partial \varepsilon}{\partial t} + \frac{\partial u_i \varepsilon}{\partial x_i} - \frac{\partial}{\partial x_i}\left(v + \frac{v_T}{\sigma_\varepsilon}\right)\frac{\partial \varepsilon}{\partial x_i} = \tag{2.16b}$$

$$= c_{1\varepsilon} v_T \frac{\varepsilon}{k}\left(\frac{\partial u_i}{\partial x_j} + \frac{\partial u_j}{\partial x_i}\right)\frac{\partial u_i}{\partial x_j} + \delta_{i3}\left(1 - c_{3\varepsilon}\right)\frac{\varepsilon}{k}\frac{g}{\rho}\frac{v_T}{\sigma_T}\frac{\partial \rho}{\partial x_i} - c_{2\varepsilon}\frac{\varepsilon^2}{k}$$

The first term in (2.16) gives the rate of change of either k or ε, the second term represents advection, the third term represents diffusion, the fourth term is the turbulence production term, the fifth term represents destruction by buoyancy effects and the last term represents dissipation. Moreover, Reynolds' analogy is used, that is that also the diffusive transport of the turbulence parameters can be modelled as a gradient-type transport, using the turbulent Prandtl-Schmidt numbers σ_k and σ_ε. The values of most of the coefficients have been obtained from calibration of the model against grid-generated turbulence and matching with the logarithmic law of the wall, and they are well established in the literature (e.g. Spaulding and Launder, 1974; and Rodi, 1984).

The eddy viscosity in the turbulent stress tensor (2.8) and the eddy diffusivity in the advection-diffusion equation (2.11) are given by:

$$v_T = c_\mu \frac{k^2}{\varepsilon} \quad \text{and} \quad \Gamma_T = \frac{v_T}{\sigma_T} \tag{2.17}$$

The value of the Prandtl-Schmidt number σ_T and the coefficient $c_{3\varepsilon}$ are less well established. Here we follow Uittenbogaard (1995). He showed conclusively that in free turbulence, $\sigma_T = 0.7$, even under highly stratified conditions. Experimental data deviating from this value are explained in terms of the effects of internal waves, which do transfer momentum, but not mass. This effect is generally accounted for by a modification of σ_T, which is often modelled as a function of the Richardson number as well. Uittenbogaard (1995) instead, promotes the use of additional source and sink terms in the k-ε model through which the effects of internal waves can be described explicitly.

Uittenbogaard also argues why $\sigma_T < 1$. In turbulent flow, packages of fluid are deformed continuously by the turbulent stresses in the fluid. The deformation of these packages is restricted by the requirements of continuity: if the elongation in two directions is given at any instant, then the elongation in the third direction follows from continuity. In other words, if $\partial u_1'/\partial x_1$ and $\partial u_2'/\partial x_2$ are given, $\partial u_3'/\partial x_3$ is set. This affects the value of the correlation between the turbulent velocity components. This restriction does not apply to a solute, as a solute can diffuse freely through the fluid. Hence the correlation between c' and u_i' has more degrees of freedom than the correlation between the turbulent velocity components themselves. As we have concluded that the particles of fine-grained sediment can be treated as a passive tracer (apart from its settling velocity) in a single-phase description, the argument above is also valid for the turbulent diffusion of fine sediment treated in this book.

From an analysis of the experiments in stratified flow by Lienhard and Van Atta (1990), Uittenbogaard (1995) also concluded that for stable stratified flows, the buoyancy term in the ε-equation (2.16b) vanishes ($c_{3\varepsilon}$ = 1). For unstable stratified flow conditions $c_{3\varepsilon}$ = 0 is appropriate, because it represents ε-production, i.e. small-scale turbulence production by Rayleigh-Taylor instabilities.

The values of the various coefficients in the standard k-ε model are summarised in Table 2.1:

Table 2.1: Coefficients in standard k-ε turbulence model.

			Prandtl-Schmidt numbers				stable stratification	unstable stratification
c_μ	$c_{1\varepsilon}$	$c_{2\varepsilon}$	σ_k	σ_ε	σ_T	κ	$c_{3\varepsilon}$	$c_{3\varepsilon}$
0.09	1.44	1.92	1.0	1.3	0.7	0.41	1	0

With increasing suspended sediment concentration, viscous or so-called low-Reynolds number effects become important, the turbulent field becomes more and more anisotropic, and the standard k-ε model, tuned for high-Reynolds number flows, may not be applicable anymore. The literature describes a large number of low-Reynolds number models, often through more or less complex modifications of the standard k-ε model in the form of (Re_T-dependent) corrections to $c_{1\varepsilon}$, $c_{2\varepsilon}$ and c_μ. The reader is referred to Patel et al. (1984) or Goldberg and Apsley (1997) for a summary. Currently however, there is no consensus on which approach is most appropriate. Moreover, these models have rarely been applied to sediment-laden flow. However, suspended sediment affects the turbulent boundary layer shear flows often predominantly through buoyancy effects (e.g. Chapter 6). Therefore, low-Reynolds number effects are not treated further in this book.

2.2   THE BOUNDARY LAYER

2.2.1 THE STRUCTURE OF THE BOUNDARY LAYER

Hydraulic resistance in alluvial systems consists of two contributions:
- skin friction due to the no-slip wall condition, and
- form drag due to flow separation, stagnation effects, etc.

Skin friction is of relevance for the water-bed boundary condition, as friction forces are responsible for the erosion of sediment particles from the bed. Turbulence associated with form drag (separation-induced eddies, etc.) is important with respect to the vertical mixing of the sediment over the water depth. Many textbooks on non-cohesive sediment (e.g. Raudkivi, 1990; Vanoni, 1977; and Van Rijn, 1993) discuss methods to determine form drag as a function of bed topography (ripples, dunes, etc.). It is also possible to determine form drag through the calibration of bulk flow parameters, like tidal propagation, mean flow velocity, flow rate, etc. computed with numerical models. However, such a calibration often varies with the flow regime.

Skin friction follows from the structure of the shear flow near the bed. This structure is characterised by a boundary layer structure. Within this boundary layer, a number of regions can be distinguished through which the flow velocity reduces to the no-slip condition at the wall (e.g. seabed). This is shown in Fig. 2.1. Near the bed, viscous effects dominate in the so-called viscous sublayer:

$$\frac{u(z)}{u_*} = \frac{zu_*}{\nu} \tag{2.18}$$

where u_* is the shear velocity. Further away from the bed, in the lower $10 - 20$ % of the water column, the velocity profile becomes logarithmic:

$$\frac{u(z)}{u_*} = \frac{1}{\kappa}\ln\left(\frac{zu_*}{\nu}\right) + A \tag{2.19}$$

where κ is the von Kármàn constant and A has a value of about 5.6. The transition between the viscous sublayer and the logarithmic part is often said to occur at $z = \delta = 11.6v/u_*$, though in reality this transition occurs in a buffer layer bounded by $5 < zu_*/v < 30$ (e.g. Tennekes and Lumley, 1994).

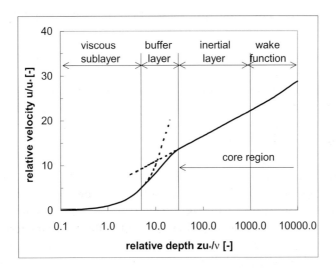

Fig. 2.1: The vertical velocity distribution under hydraulically smooth flow conditions.

In the upper 80 – 90 % of the water column the velocity profile follows the defect law (e.g. equ. (2.20a) from Coles, 1956 and equ. (2.20.b) from Tennekes and Lumley, 1994):

$$\frac{u(z)}{u_*} = \frac{1}{\kappa}\ln\left(\frac{zu_*}{v}\right) + A + \frac{2\Pi}{\kappa}\sin^2\left(\frac{\pi z}{2h}\right) \tag{2.20a}$$

$$\frac{u(z)}{u_*} = \frac{1}{\kappa}\ln\left(\frac{zu_*}{v}\right) + A + \frac{1}{2}\left[\sin\left(\pi\left(\frac{z}{h}-1\right)\right)+1\right] \tag{2.20b}$$

where Π has a value of about $\Pi \approx 0.2$ for stationary flow and large Reynolds numbers. The defect law is often written as (in Coles'-form):

$$\frac{u(h)-u(z)}{u_*} = -\frac{1}{\kappa}\ln\left(\frac{zu_*}{\nu}\right) + \frac{2\Pi}{\kappa}\cos^2\left(\frac{\pi z}{2h}\right) \tag{2.20c}$$

where $u(h)$ is the maximal velocity measured in the water column, commonly at the water surface. Nieuwstadt and Den Toonder (2001) argue that in open channel flow, the deviation from the logarithmic profile is only small. Therefore, the k-ε turbulence model is suitable, as this model predicts a logarithmic velocity profile throughout the entire water column for stationary homogeneous flow conditions, which corresponds to a parabolic eddy viscosity profile. This is substantiated by Lueck and Lu (1997) through a series of ADCP-measurements in the 30 m deep Cordova Channel (Canada). They found logarithmic velocity profiles in the major part of the water column over the major part of the tidal cycle.

The bed of alluvial channels is not smooth in general because of the presence of bed irregularities such as ripples, gravel, shell, etc. When these irregularities exceed the thickness of the viscous sublayer, as sketched in Fig. 2.2, the boundary layer flow is called hydraulically rough. It is expected that in cohesive sediment environments, where the bed structure has cohesive properties, the bed is often hydraulically smooth.

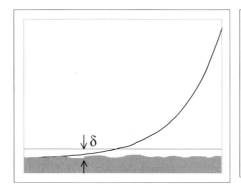
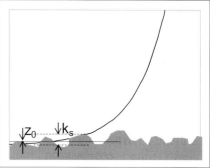

Fig. 2.2a: Hydraulically smooth conditions. *Fig. 2.2b: Hydraulically rough conditions.*

For hydraulically rough conditions, the viscous sublayer is absent (or exists as a thin layer around the bed irregularities). The velocity profile for smooth and rough conditions can be written as:

$$\frac{u(z)}{u_*} = \frac{1}{\kappa}\ln\left(\frac{z}{z_0}\right) \tag{2.21}$$

where z_0 is a roughness length, or "zero-velocity" level. This level is related to the equivalent sand roughness height (or Nikuradse height) k_s as shown in Table 2.2.

It may be convenient to eliminate u_* from (2.21), thus relating the vertical velocity profile to its depth-mean value $\bar{u}$:

$$\frac{u(z)}{\bar{u}} = \frac{\ln(z/z_0)}{\ln(\kappa h/z_0)} \tag{2.22}$$

Table 2.2: Roughness length for various flow conditions.

hydraulically smooth	$z_0 = 0.11\, v/u_*$	$u_* k_s/v \leq 5$
transitional	$z_0 = 0.11\, v/u_* + k_s/30$	$5 < u_* k_s/v < 70$ - 100
hydraulically rough	$z_0 = k_s/30$	$u_* k_s/v \geq 70$ - 100

Next to the mean velocity profile, the distribution of kinetic energy, Reynolds stresses etc. are of importance to understand the dynamics of cohesive sediment in turbulent flow (e.g. Fig. 2.3). As said, many of these turbulence parameters are not isotropic because of the presence of the bed in the shear flow. An excellent summary is given by Nezu and Nakagawa (1993), who advocate the following exponential relations in the so-called outer part of the boundary layer, e.g. $z/h > 5$ to 10 %:

$$\begin{aligned} u'_{rms}/u_* &= 2.30\exp(-z/h), \\ w'_{rms}/u_* &= 1.27\exp(-z/h) \text{ and} \\ k/u_*^2 &= 4.78\exp(-2z/h) \end{aligned} \tag{2.23}$$

Turbulence production P and dissipation rate ε are more or less in equilibrium in the lower 30 to 50 % of the boundary layer. Higher in the boundary layer diffusive effects become important, transporting turbulence from the wall region into the core region (Nieuwstadt and Den Toonder, 2001) and dissipation exceeds turbulence production. The vertical distribution of production and dissipation in the logarithmic region is approximated by:

$$\frac{Ph}{u_*^3} \approx \frac{\varepsilon h}{u_*^3} = \frac{1 - z/h}{\kappa \, z/h}$$

(2.24)

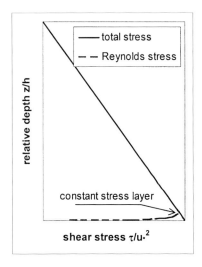

Fig. 2.3: Sketch of shear stress profile. The Reynolds stress is approximately constant in the inertial sublayer. Below this layer, viscous effects become important, and Reynolds stresses diminish rapidly.

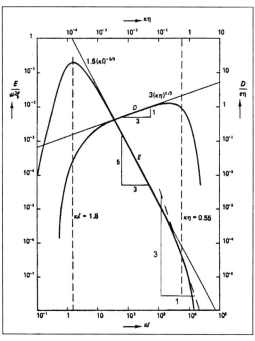

Fig. 2.4: Idealised 3D turbulent energy (E) and dissipation (D) spectra (after Tennekes and Lumley, 1994).

Turbulence and turbulent mixing are characterised by a Kolmogorov cascade of length and time scales. Relevant length scales are:

- the so-called mixing length is a scale for vertical mixing of suspended sediment, which scales linearly with distance from the bed in the lower 20 to 25 % of the water column,
- the water depth is a measure for the largest eddies in the vertical plane that can mix sediment over the water column,

- the macro or integral length scale is the largest spatial scale within which turbulent fluctuations are still mutually correlated,
- the micro or Taylor length scale is a measure for the smallest scales relevant for determining Reynolds stresses and eddy diffusivity, and
- the Kolmogorov scale is the smallest scale in the turbulent flow field at which dissipation takes place.

Tennekes and Lumley (1994) give an excellent summary of the scales of turbulence. These scales are not further treated here, except for the Kolmogorov scale λ_0. The latter is important, as it determines the size of flocs of cohesive sediment: the stresses that disrupt the flocs are in the viscous regime.

Per definition, $Re_{\lambda_0} \approx 1$, so eddies of scale $\lambda \approx \lambda_0$ have a velocity of about (e.g. Levich, 1962):

$$u_{\lambda_0} \approx \frac{\nu}{\lambda_0} \qquad (2.25)$$

where ν is the kinematic viscosity of the fluid. The time scale for eddies of the Kolmogorov scale can be approximated by:

$$T_{\lambda_0} \approx \frac{\lambda_0}{u_{\lambda_0}} \approx \frac{\lambda_0^2}{\nu} \equiv \frac{1}{G} \qquad (2.26)$$

G is the shear rate parameter that characterises the effects of turbulence on the evolution of floc size, as described in Chapter 4. According to Levich, turbulent eddy velocities scale with eddy size λ:

$$u_\lambda \approx \frac{u_{\lambda_0}}{\lambda_0} \lambda \qquad (2.27)$$

in which u_{λ_0} is the eddy velocity at scale $\lambda = \lambda_0$. This agrees with the normalised spectrum of the turbulent energy E at high wave numbers k around the Kolmogorov scale: E scales with k^{-3}, hence u with k^{-1}, thus with λ, as shown in Fig. 2.4, after Tennekes and Lumley (1994).

2.2.2 COHERENT STRUCTURES IN THE BOUNDARY LAYER

In the preceding sections mean values of flow velocity and turbulence parameters were discussed. However, it can be argued that not the mean values, but the peak values of flow velocity, shear stress, etc. govern the water-bed boundary conditions discussed in Chapter 5 and 9. Moreover, it has been shown that mean Reynolds stresses become small near the bed, in particular under hydraulically smooth conditions (e.g. Fig. 2.3).

   Therefore, some thought must be given to the variability in the stresses and the cause of this variability. This variability has been studied at length for hydraulically smooth conditions, e.g. Nezu and Nakagawa (1993); Pope (2000); and Nieuwstadt and Den Toonder (2001) for detailed summaries. The common view is that the flow in the viscous region near the bed ($z < 10\,v/u*$) becomes organised in slender longitudinal vortices, so-called streaks, which are advected at low velocity. At specific moments these streaks become unstable and form horseshoe or hairpin vortices, that are ejected into the boundary layer; these events are known as bursts or ejections. This is schematically shown in Fig. 2.5 (after Best, 1993). Mass conservation is maintained by flow towards the wall from locations higher in the water column - the so-called high-speed sweeps.

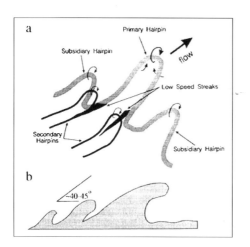
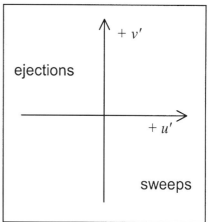

Fig. 2.5: Sketch of hairpin vortices showing streaks and ejections (after Best, 1993).

Fig. 2.6: Turbulent velocity quadrant.

Obi et al. (1996) have summarised statistical parameters of wall shear stresses measured in air, oil and water at a number of Reynolds numbers. Mean values, based on fourteen observations, are summarised in Table 2.3, where the standard deviation is normalised with the mean value of the shear stress.

Table 2.3: Bed shear stress statistics by Obi et al. (1996).

norm. standard deviation	coeff. of skewness	coefficient of kurtosis
0.4	1.0	4.5

The occurrence of ejections and sweeps is fairly random. However, from a statistical point of view, bursts occur more or less regularly in space and time, e.g. Nezu and Nakagawa (1993) for a summary.

Ejections ($u' < 0$ and $v' > 0$) and sweeps ($u' > 0$ and $v' < 0$) have an important contribution to turbulence production as the correlation between the turbulent velocity fluctuations is negative, i.e. they cause positive Reynolds stresses. This is depicted in the velocity quadrant of Fig. 2.6. Note that negative velocity correlations do not necessarily imply sweeps or ejections.

It is obvious that these streaks and ejections must have a large effect on the stability of granular beds: particles on the bed will be dragged with the ejections and mixed over the water column (Nezu and Nakagawa; 1993, Best, 1993). These coherent structures also give rise to a significantly skewed distribution of the bed shear stress, e.g. Table 2.3.

The statistics of the turbulent bed shear stress can be described with a probability density function (pdf). The following definitions are used, where $y(\tau)$ is the probability density distribution of the instantaneous bed shear stress τ.

mean:
$$\bar{\tau} = \int_{-\infty}^{\infty} \tau y(\tau) d\tau$$

normalised standard deviation:
$$\sigma = \frac{1}{\bar{\tau}} \sqrt{\int_{-\infty}^{\infty} (\tau - \bar{\tau})^2 y(\tau) d\tau}$$

(coefficient of) skewness:
$$m_3 = \frac{1}{\sigma^3} \int_{-\infty}^{\infty} (\tau - \bar{\tau})^3 y(\tau) d\tau$$

(coefficient of) kurtosis:
$$m_4 = \frac{1}{\sigma^4} \int_{-\infty}^{\infty} (\tau - \bar{\tau})^4 y(\tau) d\tau$$

Miyagi et al. (2000) conclude from a series of experiments in air that these statistical parameters are a function of the Reynolds number. However, the results of Obi et al. (1996) can be regarded as representative for high-Reynolds numbers.

No theory exists at present to predict the actual shape of the probability density function for bed shear stresses. However, it is known that the function should be skewed towards higher shear stresses because of the occurrence of turbulent bursts. Therefore the heuristic approach by Petit (1999) is followed. He proposes the following three-parameter density function:

$$y(\tau;\mu,m,s) = \int_{-\infty}^{\infty} \frac{1}{m}\exp\left\{-\frac{v}{m}\right\} n(\tau+m-\mu;v,s)\mathrm{d}v =$$

$$= \frac{1}{2m}\left(1+\mathrm{erf}\left\{\frac{m(\tau+m-\mu)-s^2}{ms\sqrt{2}}\right\}\right)\exp\left\{-\frac{2m(\tau+m-\mu)-s^2}{2m^2}\right\} \quad (2.28)$$

in which

$$n(\tau+m-\mu;v,s) = \frac{1}{s\sqrt{2\pi}}\exp\left\{-\frac{1}{2}\left(\frac{\tau+m-\mu-v}{s}\right)^2\right\} \quad (2.29)$$

Note that pdf (2.28) allows for negative bed stresses, i.e. stresses opposite to the mean flow direction, which are known to occur. The nth moment M_n of the density function reads:

$$M_n = \int_{-\infty}^{\infty} \tau^n y(\tau;\mu,m,s)\mathrm{d}\tau \quad (2.30)$$

and the parameters of the pdf can be expressed as:

$$\mu = M_1 \quad , \quad m = \frac{1}{2}\left(8M_1^3 - 12M_1M_2 + 4M_3\right)^{1/3}$$

$$\text{and} \quad s = \frac{1}{2}\sqrt{-2^{4/3}\left(2M_1^3 - 3M_1M_2 + M_3\right)^{2/3} - 4M_1^2 + 4M_2} \quad (2.31)$$

This density function is compared with the normalised data by Obi et al. in Fig. 2.7, in which $\tau_w^* = (\tau - \mu)/\sigma$. A reasonable agreement is shown, bearing in mind that pdf (2.28) does not contain a parameter for the flatness of the distribution.

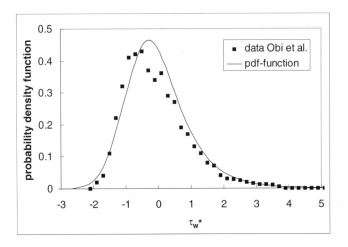

Fig.2.7: Comparison of pdf (2.28) with data by Obi et al. (1996).

Gust and Müller (1997) present a series of probability density functions of bed shear stresses measured in various devices to study the erosion of cohesive sediment, deployed both in the laboratory and in the field. They conclude that most erosion devices do not generate a skewed probability density function of the bed shear stress, which makes interpretation of experimental results obtained with these devices not straightforward.

All information on coherent structures and shear stress distribution has been obtained for hydraulically smooth flow over flat, impermeable walls. Alluvial beds are hydraulically rough in general, with the possible exception of muddy beds, and are characterised by bed irregularities due to bed forms, gravel and pebbles, bio-induced mounds and craters, etc. (e.g. Boudreau and Jørgensen, 2001). Moreover, alluvial beds are often permeable (sandy beds) and/or elastic (soft mud).

Nezu and Nakagawa (1993) reason that the structures of the turbulent boundary layer over a rough bed differ from that over a smooth bed. Under hydraulically rough conditions, streaks cannot develop. In this case, the local flow around the bed irregularities with flow separation, unstable wakes, etc. probably plays an important role in the production of turbulence and the generation of (coherent) structures. However, no literature exists with data on the (coherent) structure of the boundary layer under hydraulically rough conditions.

Description of sediment-laden flow becomes further complicated as suspended sediment induces smaller or larger vertical density gradients (e.g. Chapter 6), which can affect the efficiency of sweeps and ejections considerably.

From this analysis, it must be concluded that the non-stationary character of turbulence must play an important role in sediment transport processes in general, and on the bed boundary condition in particular. This is the case for flow-dominated environments, but probably even more in environments where waves play a role. Yet little quantitative information is available for the conditions where sediment transport processes are important, such as encountered in alluvial systems.

However, we need to understand the boundary layer structure and the shear stress distribution for hydraulically smooth conditions in sediment transport for two reasons:
1. flows over beds consisting of cohesive sediment often classify as hydraulically smooth,
2. numerous experiments are carried out in the laboratory, where hydraulically smooth conditions often prevail.

2.3   THE EFFECT OF SURFACE WAVES

Surface waves can affect the flow largely through a modification of the boundary layer and through the generation of stresses (so-called radiation stresses) throughout the water column. Modification of the flow field through the effect of waves is not treated in this book – the reader is referred to one of the many textbooks on this field (e.g. Fredsøe and Deigaard, 1992; Soulsby, 1997; or Van Rijn, 1990).

With respect to the common wave-paradigm in sediment transport, i.e. waves stir up sediment to be transported by the (tidal) flow, three wave-induced effects are important for cohesive sediment dynamics:
1. waves augment the bed shear stress (considerably), e.g. Chapter 9,
2. waves generate a stress field within the bed, e.g. Chapter 8, and
3. waves affect vertical mixing through turbulence generated within the wave-induced turbulent boundary layer.

Nielsen (1992) and Fredsøe and Deigaard (1992) give excellent reviews on the properties of the boundary layer in wave-affected environments. It is important to appreciate that the wave-induced boundary layer cannot develop fully because of the oscillating nature of the (orbital) flow. The thickness δ_w of the wave-induced boundary layer scales with the Stokes' length L_s:

$$\delta_w \approx 15 L_s \quad \text{where} \quad L_s = \sqrt{2 v_T / \omega} \tag{2.32}$$

where v_T is the eddy viscosity and ω is the angular frequency of the wave field. Hence, typical values for δ_w for short waves are of the order of several cm to a few dm at most.

The stresses generated within the bed (pressure gradients) are a function of the mechanical bed properties, and are treated in Chapter 8.

The rms-value of the instantaneous wave-induced bed shear stress vector $\tilde{\tau}_w$ and the corresponding shear velocity u_{*w} are defined by:

$$\tau_w = \left\langle \tilde{\tau}_w^2 \right\rangle^{1/2} = \rho u_{*w}^2 \quad ; \quad \tilde{\tau}_w = \tfrac{1}{2} \rho f_w u_{orb}^2 \quad ; \quad u_{*w}^2 = \tfrac{1}{4} f_w \hat{u}_{orb}^2 \tag{2.33}$$

in which the amplitude $\hat{u}_{orb}$ of the near-bed horizontal orbital velocity vector is determined from the rms wave height H_{rms}. The friction coefficient f_w for hydraulically rough conditions follows from Swart's formula (e.g. Soulsby et al., 1993):

$$\text{for} \ \frac{A}{k_s} > \frac{\pi}{2} : \ f_w = 0.00251 \cdot \exp \left\{ 5.21 \left(\frac{A}{k_s} \right)^{-0.19} \right\}, \ \text{and} \tag{2.34}$$

$$\text{for} \ \frac{A}{k_s} \le \frac{\pi}{2} : \ f_w = 0.3$$

where k_s is the Nikuradse roughness height related to the roughness length z_0 through $k_s \approx 30 z_0$ and A is the orbital amplitude defined by $A \equiv \hat{u}_{orb} / \omega$. Soulsby (1997) presents other formulae for f_w, with values for f_w varying within a factor two.

For hydraulically smooth beds, Whitehouse et al. (2000) propose:

$$f_w = 0.0251 Re_w^{-0.187} \quad \text{for} \quad Re_w > 5 \cdot 10^5 \tag{2.35}$$

where the wave-Reynolds number Re_w is defined as $Re_w \equiv \hat{u}_{orb} A / \nu$

It is noted that the mean bed shear stress may not be the most appropriate parameter to describe near-bed cohesive sediment dynamics, but that peak values are probably more significant, as discussed in Section 2.2.2.

The effect of waves on vertical mixing processes has been subject of many studies (e.g. Fredsøe, 1984). Here, the approach of Van Kesteren and Bakker (1984) is elaborated because of its simplicity and physical transparency. This method can be considered as a non-linear extension of the method by Grant and Madsen (1979). The effect of waves on vertical mixing is accounted for by an additional bed boundary condition to the flow, applying linear wave theory to relate wavelength and period, and orbital excursion and water particle velocity. This approach gives good results for large waves, but for smaller waves, the wave-effect is overestimated by about 20 % (e.g. Soulsby et al., 1993). A more accurate method, based on a parameterisation of a series of models and data is presented by Soulsby et al. (1993).

The wave-affected boundary layer thickness δ_w, as obtained by Van Kesteren and Bakker (1984) is given by:

$$\delta_w = 2\pi\kappa \frac{\hat{u}_{*w}}{\omega} = \kappa T \hat{u}_{*w} \tag{2.36}$$

where $\hat{u}_{*w}$ is the wave shear velocity related to the amplitude of the orbital velocity, κ is von Kármàn's constant and T is the wave period. Within the turbulent wave-boundary layer the (mean) eddy viscosity ν_T reads:

$$\text{for } \quad z_0 \leq (z - Z_b) \leq \delta_w \frac{16}{3\pi^3} \ln\left(\frac{8}{3\pi} \frac{\hat{u}_{*w}}{\xi \, u_{*f}}\right):$$

$$\nu_T = \frac{8}{3\pi \xi} \kappa \hat{u}_{*w} (z - Z_b) \cdot \exp\left\{-\frac{3\pi^3}{16} \frac{z - Z_b}{\delta_w}\right\} \tag{2.37a}$$

$$\text{for } \quad (z - Z_b) \geq \delta_w \frac{16}{3\pi^3} \ln\left(\frac{8}{3\pi} \frac{\hat{u}_{*w}}{\xi \, u_{*f}}\right):$$

$$\nu_T = \kappa u_{*f} (z - Z_b) \tag{2.37b}$$

where u_{*f} is the bottom shear velocity of the mean flow, Z_b is bed level and ξ is factor related to the angle between surface wave direction and current α_{wf} :

$$\xi = \frac{1}{2}\left(3 + \cos 2\alpha_{wf}\right) \tag{2.38}$$

For collinear wave-current conditions ξ = 2 and for perpendicular wave-current conditions, ξ = 1. Note that the near-bed mean flow shear stress $\tau_b = \rho_w u_{*f}^2$ is constant throughout the wave boundary layer.

The resulting vertical eddy viscosity profiles are shown in Fig. 2.8 for various combinations of u_{*f}/u_{*w} and α_{wf}.

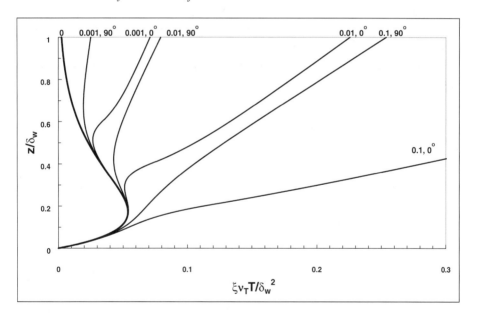

Fig.2.8: Eddy viscosity in turbulent wave boundary layer for various combinations of u_{*f}/u_{*w} and α_{wf} (Van Kesteren and Bakker, 1984).

Wave-induced turbulence contributes to the mean flow above the wave-boundary through an increase in effective roughness z_{wc}. For waves and currents together:

$$\delta_w \geq z_0 : \quad \frac{z_{wc}}{z_0} = \left(\frac{\delta_w}{z_0}\right)^{\beta} ; \quad \beta = 1 - \left(1 + \left(\frac{u_{*w}}{u_{*f}}\right)^2\right)^{-\frac{1}{2}} \tag{2.39}$$

The effective bed shear stress τ_b and shear velocity u_* follow from applying the logarithmic law of the wall using z_{wc} instead of z_0. For other formulations, which are in particular convenient for arbitrary angles between current and waves, the reader is referred to Soulsby et al. (1993), or Mellor (2002).

3.   THE NATURE OF COHESIVE SEDIMENT

The term "cohesive sediment" is generally associated with sticky, muddy, stinky and sometimes gassy sediment. The adverb "cohesive" relates to ductile behaviour when the sediment is remoulded. Actually all cohesive sediments consist of granular organic and mineral solids in a liquid phase. In general the liquid is water, but sometimes non-aqueous liquids can be present as well. Both the solids and the time scale of flow of the liquid phase govern the cohesive behaviour of the sediment. In this chapter, the description of cohesive sediment in terms of its origin, composition and classification is treated. A simple tool is provided for the assessment of cohesive mechanical behaviour based on standard geotechnical measurements.

In this book we refer to volumetric and mass concentration, to solids fraction and solids content, to sediment fractions, etc. The various definitions used are summarised in Table 3.1 – see also Appendix A and B.

Table 3.1: Nomenclature used in this book.

c	mass concentration	mass of solids / total wet volume
e	void ratio	volume of pores / volume of solids
n	porosity	volume of pores / total wet volume
W	water content	mass water / mass solids
ϕ_f	volume concentration of flocs	volume flocs / total wet volume
ϕ_s	volume concentration of solids	volume solids / total wet volume
ρ	bulk density	mass sediment / total wet volume
ρ_{dry}	dry bed density	$\equiv c$ = mass of solids / wet volume
ρ_s	mass density solids	mass of solids / volume of solids
ρ_w	mass density water	mass of water / volume of water
ξ^i	solids content	mass solid i / total mass solids
ψ^i	solids fraction	volume solid i / total solids volume

It is noted that solids fraction and solids content are identical when the specific density of all sediment fractions are identical, which is often more or less the case.

We use the superscripts $\bullet^{cl}$, $\bullet^{m}$, $\bullet^{sa}$ and $\bullet^{si}$ to refer to the sediment type, i.e. clay, mud (= clay + silt), sand and silt fraction, respectively, and the subscripts

$\bullet_s$, $\bullet_w$, $\bullet_g$ and $\bullet_f$ to refer to the sediment phase, i.e. solid, water, gas, and flocculated phase.

These definitions are further illustrated in Fig. 3.1.

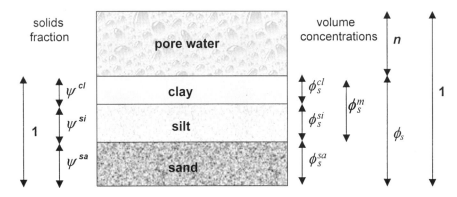

Fig. 3.1a: Definitions of solid fractions and volumetric concentrations of solids for sand, silt, clay and mud (mud = silt + clay).

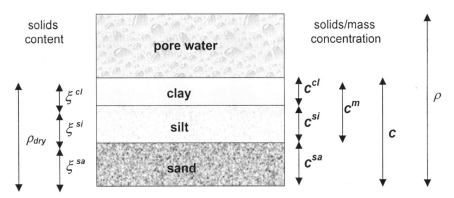

Fig. 3.1b: Definitions of solid content and mass concentrations of solids for sand, silt, clay and mud (mud = silt + clay).

3.1   THE COMPOSITION OF COHESIVE SEDIMENT

3.1.1 GRANULAR COMPOSITION OF COHESIVE SEDIMENT

By definition, sediment is formed of granular material that can settle in water by gravity. The granular size distribution and composition are of main importance for the mechanical behaviour. Cohesive sediment, or mud, as encountered in the marine environment, consists of a mixture of clay, silt, (fine) sand, organic material, water, and sometimes gas. Its composition, and thus its behaviour, varies in space and time and is governed by the availability of the sediment and its components, the meteo-hydrodynamic conditions, biological activity, history, etc. The clay particles and organic material, in relation to the chemical properties of the liquid phase, e.g. (pore) water, determine the cohesive behaviour of the sediment.

The solid phase is characterised by its particle size distribution. For practical purposes many classifications were defined and standardised, such as the US, British and Dutch standards, as given in Table 3.2. Fig. 3.2 shows a typical particle size distribution for sediments from the Port of IJmuiden, The Netherlands.

Table 3.2: Classification of particle size; diameter D in μm.

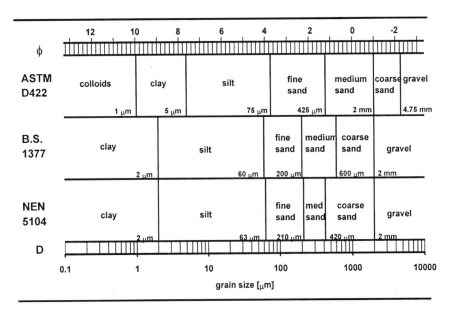

In sedimentology, the particle diameter is often given in terms of the fineness factor ϕ (Krumbein, 1941), defined by:

$$\phi = - {}^2\log D \quad \text{with } D \text{ in mm} \tag{3.1}$$

All size classifications distinguish between clay, silt, sand and gravel fractions. For cohesive sediments the gravel fraction is absent or very small. It is important to realise that fraction names refer to size, and do not distinguish the composition of the particles. Therefore, all fractions can contain mineral and organic solids. Important other fraction definitions in common use are: fines (< 45 μm) and mud fraction (< 63 μm).

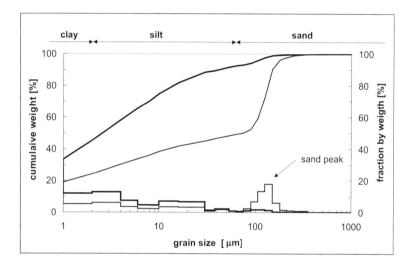

Fig. 3.2: Typical particle size distribution of harbour mud (Port of IJmuiden, The Netherlands) with variable sand content (Van Kesteren, 2004).

Within the clay fraction, a separate sub-fraction can be distinguished that does not settle in water due to Brownian motion: the colloidal fraction. The size of colloidal particles is in the order on 0.1 μm or less. Although present in water as a solid phase, it is often denoted as dissolved matter, because suspended solid concentration is measured as the solids fraction that does not pass through a 0.45 μm filter.

Organic colloids are indicated as DOC (Dissolved Organic Carbon). The colloidal fraction is only important for the behaviour of the clay minerals (formation of aggregates and flocs; see Chapter 4). This fraction also determines light extinction in the water column for more than 50 %. Therefore it plays a major role in the level of turbidity in marine ecosystems. The colloidal fraction is treated in this book as part of the clay fraction.

In sedimentology and soil mechanics it is common to classify soil with the so-called sand-silt-clay triangle. An example of such a classification is given in

Fig. 3.3. The percentage of sand, clay and silt are plotted in clockwise direction along the three sides of the triangle in the left panel.

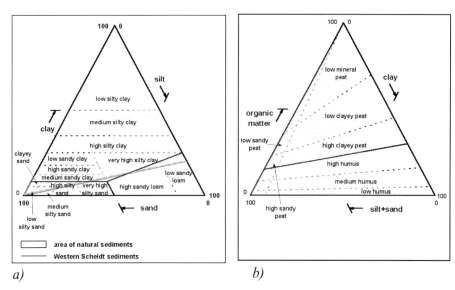

a)   b)

Fig. 3.3a/b: Triangular classification diagram according to NEN5104.

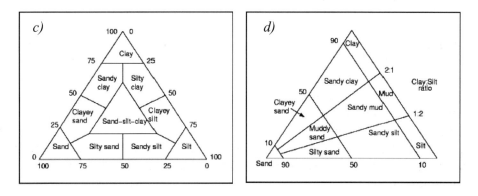

Fig. 3.3c/d: Triangular diagram for classification of sediment types by
Shepard and Folk (after Flemming, 2000).

Different soil types are shown in sub-zones of the triangle areas, according to the NEN5104 classification (Lubking, 1997; or NNI, 1989) in Fig. 3.3a/b and according to Flemming (2000) in Fig. 3.3c/d. The horizontal boundaries between soil types reflect the dominant role of the clay fraction in the soil behaviour. The solid lines separate the areas where sand, silt or clay dominate soil behaviour. Within the triangle of Fig. 3.3b, the classification is shown for organic rich soils. The solid line separates the two areas where organic matter is dominant (peat area) and where mineral solids dominate soil behaviour (humus area).

Fig. 3.3c/d show the classification diagram proposed by Flemming (2000), which is commonly used in sedimentology. The various sediment classes and their mutual distinction are not well defined, however.

The shaded areas in Fig. 3.3a/b cover most natural sediments. Furthermore it appears that for a given marine system, data on sediment composition from different locations and depth are located on one line in the triangular diagram. As an example, data for the Western Scheldt (The Netherlands) are plotted in Fig. 3.3a (Van Ledden, 2003; see also Section 3.3.2 and Chapter 8). This line corresponds to a constant ratio between the clay and silt fractions. Along this line, the sand content can be used as a variable. If the sand content is known at a certain location, together with the bulk density or porosity of the soil, the classification line in the triangular diagram enables us to determine the clay-water ratio, which is one of the key-parameters in soil mechanical behaviour.

However, also the skeleton formed by the different fractions determines the behaviour of mud. For instance, a pure granular skeleton can be formed when sand particles are in mutual contact. Such a skeleton differs largely in structure from the clay skeleton, and therefore differs in mechanical behaviour as well (see Section 3.3.2 and Chapter 8).

Organic matter is measured as part of the dry solids. The solids content is in general determined by drying the samples through heating at 110 ± 5 °C. The organic part is determined by burning of the dried sample at 500 °C. The loss of weight is a measure for the organic fraction (OM), and the weight of the ashes for the mineral part. However carbonates, organic matter adsorbed to clay minerals and free iron affect the loss on ignition and therefore corrections are necessary (see Appendix C).

Another parameter to quantify organic matter is TOC (total organic carbon), which is determined from the difference between the total carbon (TC) and inorganic carbon (IC). Both amounts are measured by oxidising the carbon to carbon-dioxide, the concentration of which is detected by non-dispersive infrared measurements. In case of hydrocarbons the ratio between TOC and OM is approximately 2.5. Note that in order to obtain OM from TOC, the chemical composition of organic matter must be known. Therefore it is difficult to provide a generic relation between OM and TOC.

The density of organic matter is much lower than the mineral solids density. It can be measured with a picnometer or assessed from empirical relations (f.i. Venmans et al., 1990; and Appendix C).

In general, the amount of organic matter is well correlated and proportional to the clay content, when the soil is dominated by the mineral part (lower part in Fig. 3.3b). This indicates a direct coupling between clay particles and organic matter. However in the "peat" region this correlation does not exist, because larger organic particles may be present in the silt and sand fractions. This is often the case in fresh water systems. By intensive remoulding, the larger organic silt and sand fraction can degrade to the clay fraction.

This granular classification enables us to assess the nature of soil behaviour, i.e. cohesive, granular or mixed, which has important implications for the mechanical and erosion properties of the sediment. This is further elaborated in Section 3.3.2 and Chapter 8.

3.1.2 MINERAL COMPOSITION OF COHESIVE SEDIMENT

Sillicates, with a variety in composition and structure, often form a major component of the mineral solids fraction. The most common silicate groups are quartz, feldspar and clay minerals. Non-silicate minerals are precipitates of salts (like carbonates), oxides and hydroxides, depending on chemical conditions.

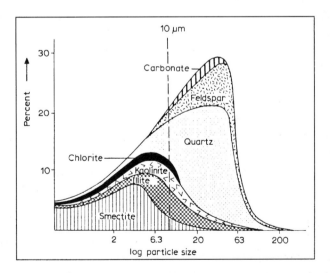

Fig. 3.4: Example of a mineral distribution as a function of grain size for Mississippi sediments (after Weaver, 1989).

An example of the distribution of different minerals as a function of grain size is given in Fig. 3.4 (Mississippi sediments, Weaver, 1989), showing that the clay fraction (< 2 μm) is mainly composed of clay minerals. The most common clay minerals in the marine environment are kaolinite, illite, smectite or montmorillonite, and chlorite. In the silt fraction (> 2 μm, < 63 μm) quartz, feldspar and carbonates are dominant, although clay minerals are still present. In the sand fraction (> 63 μm), clay minerals are absent.

Clay minerals are to a large extent responsible for cohesion. This is mainly because of the size and flat shape of the particles, yielding a very high specific surface area and an electrical charge distribution, which interacts with the ambient water (dipoles). Contrary to the shape of quartz and feldspar (so called tectosilicates), built up from three-dimensional silica (SiO_2) tetrahedra, clay minerals are so-called phyllo-silicates and consist largely of two-dimensional silica tetrahedra with aluminium-hydroxide octahedra (gibbsite) or magnesium-hydroxide octahedra (brucite) (Fig. 3.5). These sheets of silica tetrahedra and gibbsite or brucite can be combined in various ways to form different clay minerals. The basal layer structure of the four most common clay minerals and characteristic parameters are listed in Table 3.3.

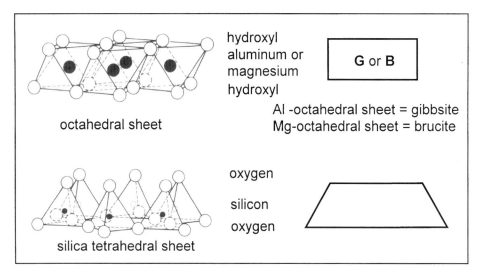

Fig. 3.5: Basic sheets in clay minerals; G = gibbsite, B = brucite, S = silicon.

Kaolinite is a 1:1 type clay mineral of the kaolin group with a basal layer of a silica tetrahedral and a gibbsite octahedral (Table 3.3). There is no charge deficiency over the basal layer. Multiple basal layers are attached by strong hydrogen bonds, forming thick crystalline flake-like particles with a thickness of about 100 nm and plate dimensions of 2 μm. Stacks of crystalline flakes can extend to large particle sizes up 100 μm or more (see Fig. 3.6a).

Table 3.3: Basal layers and properties in most common clay minerals (Mitchell, 1976).

clay mineral	basal layers	basal spacing	ρ_s kg/m^3	A_s m^2/g	CEC meq/100g	size μm
kaolinite		0.72 nm	2670 ± 10	15 ± 5	9 ± 6	6 sided flakes >0.1X> 2
montmorillonite exchangeable cations in G 1 of 6 Al is M		> 0.96 nm	2530 ± 170	prim: 85 ± 35 sec: 770 ± 70	115 ± 35	flakes >0.001 X <10
illite in G some Al⇒ Mg,Fe	in S some Si⇒ Al	1.00 nm	2800 ± 200	80 ± 20	25 ± 15	flakes >0.003 X <10
chlorite in B some Mg⇒ Al	in S some Si⇒ Al	1.40 nm	2800 ± 200		25 ± 15	flakes 1 X 1

Although the basal kaolinite layer has a very low negative charge, the large crystalline particles have a substantial negative charge on their flat face and a positive charge on their edges. This is due to substitution of some Si-ions by Al-ions in the outer silica tetrahedra of the crystalline particle. The positive charge at the edge is caused by dissociation of the gibbsite octahedral at low pH, and changes to a negative charge when pH exceeds 7. Because the attached layers cannot be separated easily by water molecules, kaolinite is known as a non-swelling clay mineral, with relative large crystalline particles resulting in a low specific surface area A_s and a low cation exchange capacity CEC (Table

3.3). CEC is defined as the number of cations that can be exchanged in the double layer around the crystalline particles. It is measured in milli-equivalents per 100 gram of dry clay particles.

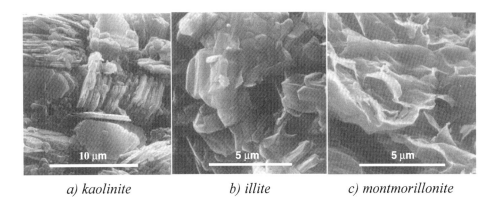

a) kaolinite   b) illite   c) montmorillonite

Fig.3.6: Scanning electron microscopic photos of typical clay minerals (from Tovey, 1971).

Both illite (belonging to the mica group, and also known as hydrous mica) and montmorillonite (smectite group) are 2:1 type clay minerals in which two silica tetrahedra are attached to a gibbsite octahedra (see Table 3.3). In illite, the negative charge caused by the substitution of less than one quarter of the Si-ions by Al-ions in the silica tetrahedra, is counterbalanced by K^+-ions between the basal layers. This results in strong ionic bonding, where K^+-ions are difficult to exchange with other cations, and polar water molecules are not able to enter between the basal layers. Only on the outside of the crystalline particles, negative charges are maintained on the particle's flat side due to the substitution of Si-ions by Al-ions in the silica tetrahedra. This limits the number of basal layers when potassium is not sufficiently available. Therefore the crystalline particles are much thinner than those of kaolinite (Fig. 3.6b), and their specific surface areas and CEC are higher.

In montmorillonite, Si-ions are not substituted in the silica tetrahedra. Therefore the negative charge is concentrated within the octahedral layer, where the ratio Mg/Al is about 1:5. Cations are present between the basal layers, but the low charge density enables water to enter between the basal layers, which results in swelling of the interlayer. These cations can be exchanged with cations in the ambient water. The large surface area of montmorillonite results in a CEC much larger than of kaolinite and illite. Due to the low bonding strength between the basal layers, the particles remain very thin, and curved flake-like structures are formed (Fig. 3.6c).

A synthetic analogue of montmorillonite is laponite in which Al-ions in the octahedral layer are replaced by lithium (Ramsey, 1986). This reduces the flake thickness below 20 nm, resulting in transparent cohesive sediment.

Chlorite is a 2:1:1 type clay mineral in which the basal structures of two silica tetrahedra and gibbsite octahedra are bonded by a brucite interlayer (see Table 3.3). As with illite, there is also substitution of Si-ions by Al-ions in the silica tetrahedra. In this case, substitution of some Mg-ions by Al-ions and Fe-ions in the brucite interlayer balances the charge deficiency of the basal 2:1 layers. Like kaolinite, large stacks can be formed of about 1 μm.

Kaolinite originates from weathering and hydrothermal alteration of granite. Deposits can be classified as primary at the location of formation and as secondary in alluvial sediments. Kaolinite tends to develop preferentially in the humid tropics.

Most smectite or montmorillonite clay deposits have formed in surface and subsurface environment by hydrolysis of extrusive volcanic activity. The strong global increase in extrusive volcanism in the Cretaceous period resulted in smectite deposits throughout the world, and are from that age or younger. When buried at greater depth (> 2000 m) smectites are altered to illite through the adsorption of potassium, or to chlorite when Mg is abundant.

The non-silicate minerals that are mainly found in cohesive sediments are salts, oxides and hydroxide precipitations:
- carbonates: calcite ($CaCO_3$) and siderite ($FeCO_3$),
- sulfates: gypsum ($CaSO_4$ $2H_2O$),
- sulfides: pyrite (FeS_2),
- phosphates: vivianite ($Fe_3(PO_4)_2$ $8H_2O$),
- oxides: hematite (Fe_2O_3) and polianite (MnO_2),
- hydroxides: goethite ($FeO(OH)$), manganite ($MnO(OH)$) and
                       gibbsite ($Al(OH)_3$).

Their occurrence depends on the local chemical conditions: availability of oxygen, iron, Ca, Mg and Mn, and their reduced state. Fig. 3.7 depicts the different stages of reduced equilibria as a function of p_e and pH (Stumm and Morgan, 1981). The relative electron activity p_e is proportional to the redox potential E_h (Bolt and Bruggenwert, 1976):

$$p_e = -\log[e^-] = \frac{F}{RT \ln 10} E_h \approx 17.9 E_h \qquad (3.2)$$

F is the Faraday constant (= 96,487 C/mol), R is the gas constant (= /mol), T is absolute temperature, and e^- is the electron concentration.

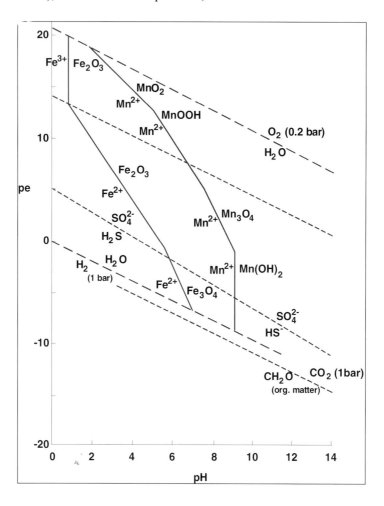

Fig. 3.7: Relative electron activity as a function of pH
(after Stumm and Morgan, 1981).

In shallow marine systems the pH is around 7.5 - 8, while in deeper ocean water, near the abyssal plane, pH can be much lower due to dissolved H_2S (e.g. the Black Sea). Positive E_h values are generally characteristic of bottom deposits that are well oxygenated, and those that consist of coarse sediments, or are poor in organic matter. Negative E_h values are generally associated with anoxic conditions, where anaerobic degradation of organic matter occurs. However, it should be noted that according to the equilibria in Fig. 3.7, there are also oxic conditions at negative E_h values related to pH. At a pH between 7.5 to 8 the E_h becomes negative when sulphates are not yet reduced, but manganese is. The final reduced state, where anaerobic degradation of organic matter occurs, is at $p_e = -8$ or $E_h = -0.447$ V.

3.1.3 ORGANIC COMPOSITION OF COHESIVE SEDIMENTS

Organic matter can have very large effects on the formation of mud flocs and the stability of the seabed, c.q. intertidal areas; the latter is further elaborated in Chapter 10. Organic matter consists primarily of organic polymers. Literature on the interactions between polymers and colloids is overwhelming, as this is an important issue in for instance process industry and sanitary engineering (e.g Birdi, 2003; and Hunter, 2001). However, literature on the role of organic polymers in the natural environment is scarce. Van Leussen (1988, 1994) gives a brief summary on the literature available at that time.

Organic matter in mud exists of particulate organic matter (POM) and dissolved organic matter (DOC). Organic matter can originate from outside a sedimentation area, or be generated within the sediment by biological processes. The allochtone organic matter is degraded during its transport and may consist mainly of resistant material like lignine. The autochtone organic material is used and produced in metabolic processes of organisms. The resulting organic material outside organisms is denoted as EPS (extra cellular substances).

The main organic substances encountered in marine mud are (Berner, 1980):
- Polysaccharides and proteins composed of peptides and amino acid,
- Lipides, hydrocarbons like cellulose, lignin composed of aliphatic and aromatic hydrocarbons,
- Humic acids.

The first group of organic substances can be regarded as flocculants, the second group is neutral, and the third are dispergents (deflocculants). Terrestial humus contains more lignine, while marine humus contains more amino acids, peptides and polysaccharides.

Polymers occur as charged, or as neutral particles. The charged particles are called poly-electrolytes and do not play a large role in natural environments. Non-ionic polymers on the other hand are very important. Polysaccharides, which are produced by a large variety of organisms (bacteria, algae, filter feeders, etc.), for instance, belong to the class of non-ionic polymers. These non-ionic polymers can adsorb to clay particles through three processes:

- Van der Waals forces,
- Bipolar forces, and
- Hydrogen bonding.

Bipolar forces are much stronger than the Van der Waals forces, and are quite effective, as the polymer-clay particle interaction is not affected by electrostatic repulsion (e.g. Section 3.2.1), because the polymer is overall electrically neutral. A long polymer string can adhere to a clay particle at a few locations, forming so-called loops and one or two tails, as sketched in Fig. 3.8 (e.g. Hunter, 2001). When another clay particle is adhered to the loops and/or tails of this polymer string, a strong particle pair is formed. This effect is known as bridging and may result in large inorganic/organic flocs with a strength 10 to 100 times larger than of pure inorganic material (Gregory, 1985). As water is bipolar, the clay-polymer structures can also bind large amounts of water.

Organic polymer occurrence in nature, however, is often hetero-dispersive. The smaller polymers adhere to the clay particles rapidly, and fewer possibilities exist for the longer strings to adhere. As a result, polymers cannot form large flocs in this case (Lyklema, 1988).

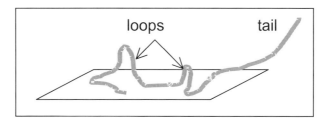

Fig. 3.8: Sketch of a polymer adsorbed to a clay particle (redrawn from Hunter, 2001).

Organic matter often contains humic acids with carboxyl groups (COOH) and phenol-hydroxyl groups with charge densities much higher than clay minerals (Hayes et al., 1990): 3-6 meq/g for humic acids and 0.01 to 0.2 meq/g for clay minerals (see Table 3.3). The carboxyl groups can easily form hydrogen bonds at pH = 8, which is typical for marine environments. However these bonds weaken rapidly when the concentration of mono-valence ions increases, as in seawater (Weaver, 1989). Hence, it may be expected that hydrogen bonds are not too important in estuaries and coastal seas.

Organic material in the water column is generally (partly) oxidised. However, after deposition, it will be reduced by, in sequential order, nitrate, manganese, iron and sulphate (Aller, 1982). If these reducers are depleted, anaerobic fermentation occurs (see Fig. 3.7), producing methane and carbon dioxide, resulting in gas formation in the bed (see Chapter 11). The order of reduction is determined by an energy cascade, known to as the "redox-tower". The reader is referred to Sigg and Stumm(1994) or Wiedemeyer and Schwamborn (1996) for further details. Note that the time scale of the chemical processes depends on both the sedimentation rate and the diffusion within the bed.

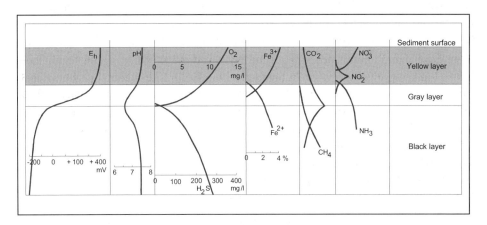

Fig. 3.9: Characteristic profiles of various chemical parameters in the upper part of the bed (after Fenchel and Riedl, 1970).

Reduction occurs in the upper part of the bed, i.e. a few mm's to about 1 cm at most. This is depicted in Fig. 3.9, after Fenchel and Riedl (1970). Close to the surface an oxidised layer develops in which free oxygen is present, although in decreasing amounts lower in the bed. This layer is usually yellow, yellow-brown, or yellow-ochre due to the presence of ferric iron. Other colours are found because of differences in crystalline structures and traces of other components. Deeper within the reduced layer, the sediment is black due to the presence of iron-sulphide (FeS). Other typical colours are: rust (Fe_2O_3) is dark brown, and bog-ore ($Fe(OH)_3$) is red brown.

The state of the organic material in the bed is characterised by the redox potential E_h and pH (see Fig. 3.9). Negative E_h values are characteristic of bottom deposits rich in organic matter and which consist largely of fine sediments. Thus it is expected that in cohesive sediment deposits, which in general contain (some) organic compounds, positive E_h values will only be found in a small layer near the bed surface.

Water-bed exchange processes in the marine environment are governed to a large extent by the erodibility of the upper few mm-cm of the bed (e.g. Chapter 9). Hence, the gradients in chemical-biological composition, depicted in Fig. 3.9, are very important, as the erodibility of the soil is largely affected by its chemical-biological composition (Chapter 9 and 10).

3.2   SKELETON FABRIC OF COHESIVE SEDIMENT

3.2.1 INTER-PARTICLE FORCES

The most characteristic property of cohesive sediment is that it can form flocs when the sediment is brought in contact with a fluid, like water. In fact, almost all cohesive sediment found in marine environments is flocculated. These flocs are characterised by very open structures formed by many (e.g. thousands to tens of thousand) clay particles and a high water content (up to 80 – 98 % by volume) – this is further elaborated in Chapter 4. The clay particles in these flocs are bound together by polymeric effects (e.g. Section 3.1.3) and/or the molecular Van der Waals attractive forces, which have to overcome the repulsive electrical forces induced by the negative charge on the particle's surface. This is possible when the negative charges on the clay particles are neutralised through cations in the ambient water, such as sodium ions in seawater (or organic polymers – see Section 3.1.3). Experimental studies (e.g. Mehta et al., 1982) showed that flocculation is practically complete at salinities between about 2 and 10 ppt.

Mixtures of charged clay particles in water with electrolytes are often referred to as lyophobic or hydrophobic colloid systems to distinguish from hydrophilic colloid systems, in which the particles contain groups that can dissociate in water, forming a stable colloidal solution (sol). The latter may occur in clay-water systems when humic and fulvic acids are bounded to the clay particles. In the following the hydrophobic colloid-clay-water-electrolyte system is discussed in more detail.

The cations in the water phase form a so-called diffusive double layer (Van Olphen, 1977; see also Fig. 3.13) around the clay particles. The cloud of particles neutralises the negative particle charge at some distance from the clay particle. As a result, the net force between two particles may become attractive at a specific cation concentration. This is schematically depicted in Fig. 3.10. The Van der Waals force, which attracts the particles, is independent of the water phase. For single molecules and spherical particle clusters, this force decays with d^{-7} and d^{-3}, respectively, where d is the distance from the particle. The electrostatic repulsive force depends on the cation concentration and decays exponentially. Hence, the net force between two particles can be attractive or repulsive, depending on cation concentration and particle distance, as sketched in Fig. 3.10. The net force may also exhibit a local minimum, the so-called energy barrier. The particles have to overcome this barrier before flocculation is possible. Under certain circumstances, two such barriers can exist at different distance from the particle (Birdi, 2003).

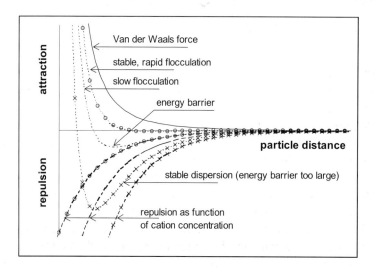

Fig. 3.10: Schematic diagram of net and gross attractive and repulsive forces between clay particles as a function of cation concentration.

The thickness of the diffusive double layer is a function of the valence of the clay particle, the valence of cations in the ambient water, and the cation concentration. According to the Gooy-Chapman theory (Everett, 1988), the thickness of the diffusive double layer is given by:

$$\lambda = \sqrt{\frac{\mu k T}{8\pi c_0 N e^2 \upsilon^2}} \approx 3\cdot 10^{-9}\, \frac{1}{\upsilon\sqrt{c_0}} \qquad (3.3)$$

in which λ is the double layer thickness, μ is the dielectrical constant of the pore water (= $8\cdot 10^{-10}$ C^2/Jm), k is the Boltzmann constant (= $1.38\cdot 10^{-23}$ J/°K), N is the Avogadro number (= $6\cdot 10^{23}$ mol^{-1}), e is the unit charge (= $1.6\cdot 10^{-19}$ C), T is absolute temperature, c_0 is the cation concentration in surrounding water [mol/m^3], and υ is the cation valence. The double layer thickness can vary between about 5 and 1,000 Å (Van Olphen, 1977).

The equilibrium between the concentrations of adsorbed cations in the double layer and the surrounding water is given by Gapon's equation (Mitchell, 1976):

$$\frac{[Na^+]_s}{[Ca^{++}]_s + [Mg^{++}]_s} = k \frac{[Na^+]_e}{\sqrt{0.5([Ca^{++}]_e + [Mg^{++}]_e)}} = k \times SAR \qquad (3.4)$$

in which the ion-capacity is measured in meq/l, subscript $_s$ refers to adsorbed cations, subsrcipt $_e$ refers to cations in the surrounding fluid, and the coefficient k (= 0.017 (meq/l)$^{-0.5}$) is a constant. Note that ion-capacity is the product of the ion-concentration in [mol/m^3] and valence v.

The right hand side term in equ. (3.4) is called the Sodium Adsorption Ratio (SAR), which is a useful parameter to quantify the effect of cations in the surrounding water on mechanical properties of a sediment like shear strength, permeability and erodibility. In the marine environment the ratio of monovalent and bivalent cations is almost constant, and therefore SAR and salinity are uniquely related (Fig. 3.11 for low salinity, after Ariathurai, 1974).

Note that equ. (3.4) is applicable to the chemistry in the water column only. The binding of clay particles in the bed is affected by the cation concentration of Fe^{++} and Mn^{++} in the pore water as well (see also Chapter 9).

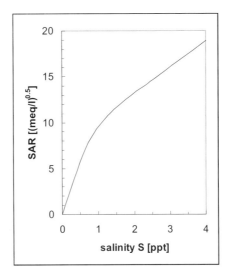

Fig. 3.11: Relation between SAR and salinity
(redrawn from Ariathurai, 1974).

Because SAR is only related to the cation concentrations and cation valence, it only reflects the double layer thickness (see equ. (3.6)): a low SAR means a thin double layer, while a high SAR means a thick double layer. The actual effect on mechanical properties such as strength, permeability and erodibility also depends on the repulsive force between particles, and hence on the charge density on the clay surfaces. The charge density σ [C/m^2] is expressed as a function of the cation exchange capacity and the specific surface area:

$$\sigma = \frac{\text{CEC}\, F}{100\, A_s} \qquad\qquad (3.5)$$

in which F is Faraday's constant (= 96.5 C/meq), CEC is the cation exchange capacity [meq/100g] and A_s is the specific surface [m^2/g].

This explains why, at low SAR, an increase in CEC results in higher shear strength, while at a high SAR, an increase in CEC results in lower shear strength, as depicted in Fig. 3.12 (see also Chapter 9).

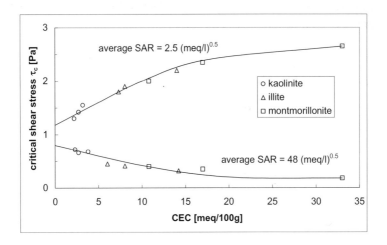

Fig. 3.12: Critical shear strength for erosion as function of SAR and CEC (redrawn from Kandiah, 1974).

The net result of simultaneous repulsion and attraction between particles can be determined from the double layer potential at which mobility is possible, known as the ζ–potential, as shown in Fig. 3.13 (after Mitchel, 1976).

According to The Gooy-Chapman theory the ζ–potential [V] is related to the charge density and double layer thickness:

$$\zeta \propto -4\pi\sigma\,\frac{\lambda}{\mu} \tag{3.6}$$

in which μ is the di-electrical constant of (pore) water (= $8\cdot10^{-10}$ C^2/Jm), λ is the double layer thickness (see equ. (3.3)) and σ is the charge density.

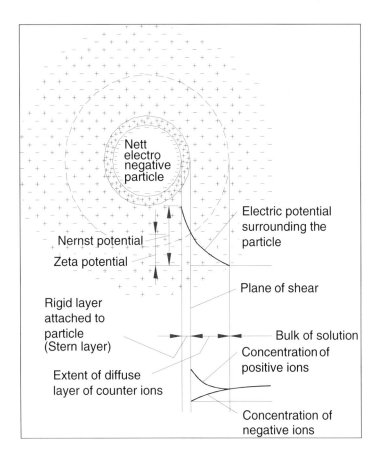

Fig. 3.13: Definition of ζ-potential (after Mitchel, 1976).

The ζ–potential is a measure of the repulsive forces between particles and can be determined as a characteristic function of pH, cation concentration and valence. An example of ζ–potential as function of pH is shown in Fig. 3.14 for kaolinite and montmorillonite (Vane et al., 1997). In general increasing pH (i.e. a decrease in H^+-ion concentration) will yield a decreasing ζ–potential, starting at a positive value at low pH and becoming negative at high pH. When the ζ–potential becomes zero at the so-called iso-electrical point, the net repulsive force is zero, and particles can stick together in face-to-face contact (coagulate) very rapidly (e.g. Chapter 4).

At a positive or negative ζ–potential, the coagulation process is much slower and particle-particle contacts will be different. Especially at negative ζ–potential, edge-to-face contact occurs, resulting in a very open floc structure. Given the fact that the flat side of the clay minerals is negatively charged, the particles will be completely dispersed when the distance between the clay particles is more than the thickness of the double layer, and the ζ–potential reaches its maximum possible negative value.

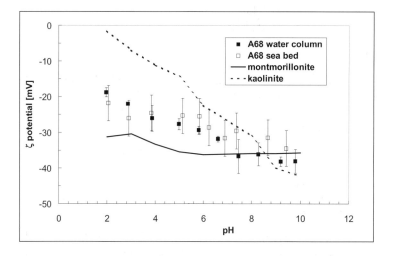

Fig. 3.14: Relation ζ-potential as function of pH (Vane et al., 1997) and data from IJmuiden mud.

The pH-dependency of the ζ–potential in Fig. 3.14 for kaolinite shows an iso-electrical point at pH $\approx$ 2, while the ζ–potential of montmorillonite is always negative and is almost unaffected by pH. In Fig. 3.14, the ζ–potential characteristics for suspended and bed sediment from the Port of IJmuiden are also shown (samples A68, recent measurements by authors). This ζ–potential has no iso-electrical point and resembles the one for montmorillonite, but is less negative with a larger dependency of pH. This indicates the dominant presence of montmorillonite. The difference with the reference characteristics for montmorillonite is due to the presence of other clay minerals, such as kaolinite, and salts, especially calcium-carbonate. The bed material shows the same characteristic, but at a potential less negative. This can be related to a more flocculated skeleton as a result of consolidation of the seabed.

Although at present the ζ–potential is not a common parameter in cohesive sediments studies, it is very useful in determining flocculation characteristics of sediment. Determination of the ζ–potential requires small amounts of sediment and therefore can be measured in dilute water samples. Besides an indicator for mineral composition of the suspended solids, the ζ–potential characteristics can be used to determine flocculation conditions in the water column, especially when water chemistry or seabed pore water chemistry is changing (f.i. salt to brackish or fresh water environment).

For further background and information on the behaviour of clay particles at micro-scale, the reader is referred to Kruijt (1952); Van Olphen (1977); or to the more recent books by Birdi (2003) or Hunter (2001).

3.2.2 PORE SIZE DISTRIBUTION

In Section 3.1.1 we discussed the grain size distribution of cohesive sediments. This characterisation gives only information on the size of the basic granular components, but not on the skeleton formed by the particles. Complementary to the grain size distribution, the pore size distribution contains information on the type of skeleton structure of the sediment.

In general the pore size distribution is represented in a cumulative graph as the amount of pore water in pores larger than a given diameter (see Fig. 3.15). The distribution of specific surface area is associated with the pore size distribution, and therefore the pore size distribution can be used to characterise the type of skeleton structure, such as granular or flocculated. Especially the flocs formed during settling create a very open particles network in the sediment (e.g. Chapter 4). When consolidation occurs, this network is squeezed, and therefore the larger pores in the size distribution disappear.

The pore size distribution can be used to determine the structure of the sediment network, which can be quantified through a fractal dimension. The application of fractal modelling is treated in Chapter 4. In this section, we use the pore size distribution as an experimental tool for classifying sediment structure.

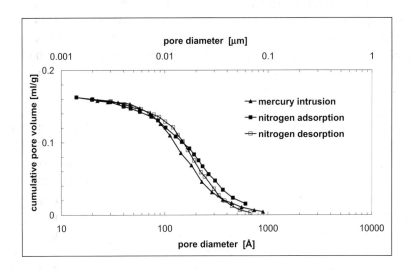

Fig. 3.15a: Pore size distributions in Westerwald kaolin clay (Van Kesteren, 2004).

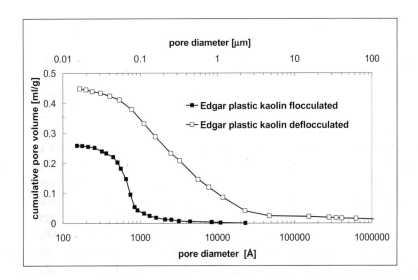

Fig. 3.15b: Pore size distributions of Edgar plastic kaolin (Diamond, 1970).

Because the cohesive sediment fabric is determined predominantly by the clay size fraction (< 2 μm), the pore sizes and related specific surface areas of interest are in the range of 10 nm up to 10 μm. In this range two methods can be used to measure the pore size distribution: nitrogen adsorption/desorption (capillary condensation) and mercury intrusion. These methods, and other methods for coarse porous systems, are discussed in more detail in Appendix C. An example is shown in Fig. 3.15a, where both methods are compared for a kaoline clay from Westerwald (Van Kesteren, 2004). Both the adsorption and desorption curves for nitrogen correspond well with the curve determined with mercury intrusion. As these methods are based on different physical properties, they yield a good confirmation of the measured pore size distribution. Fig. 3.15b presents an example of how the pore size distribution reflects the skeletal fabric. The distributions for a flocculated and a deflocculated kaolin are given, showing a large change in the characteristic of the pore size distribution (Diamond, 1970), indicating that the larger pores disappear in flocculated sediment.

The process of capillary condensation, as used in the nitrogen adsorption/desorption technique, is not only important for determining pore size distributions, but also for the mechanical behaviour of cohesive sediment. This is of particular relevance in failure mechanisms, where pore water and local cavitation are crucial factors, as discussed in Chapter 8.

3.3 GEOTECHNICAL CLASSIFICATION OF COHESIVE SEDIMENT

3.3.1 GEOTECHNICAL PARAMETERS

In this section we examine how the various components presented in the preceding sections determine the mechanical behaviour of cohesive sediment. For this purpose, we use simple, readily available soil mechanical tools. The most important parameters are the so-called Atterberg or plasticity limits: the plastic limit PL and the liquid limit LL. These parameters yield the water content of a cohesive sediment sample from two standardised soil mechanical tests. The water content W in soil mechanics is defined as the ratio of the weight of water to the weight of solids in a sample. To determine the liquid limit LL, a groove of specified dimensions is made in remoulded samples of different water content. At the liquid limit LL, this groove is to be closed within a certain number of blows on the sample through (local) fluidisation. To determine the plastic limit

PL, cylinder shaped samples (sausages) at different water content are prepared. These cylinders begin to crumble, when deformed, at the plastic limit.

These tests do not seem to have a sound scientific basis. However, they do have a solid geotechnical engineering basis, and have been applied in geotechnical practice for a long time, and are still part of standards applied throughout the world. As a result, a large empirical database exists, relating these limits to various soil mechanical parameters.

The tests to measure LL and PL can in fact be considered as undrained shear strength tests on remoulded samples, and therefore reflect the mechanical behaviour as a function of water content or porosity. Currently, the Atterberg limits are also determined with more sophisticated tests, such as the Swedish fall-cone test (Head, 1986). In general it can be stated that the LL corresponds to an undrained shear strength on the order of 1 kPa, while the PL corresponds to an undrained shear strength on the order of 100 kPa. It is important to appreciate that the ratio between these two shear strengths classifies the behaviour of the sediment, more than their absolute values. More information about test procedures etc. can be found in many geotechnical textbooks, such as Head (1986), Mitchell (1976) and Lambe and Whitman (1979).

The difference in water content between the liquid and plastic limit is called the plasticity index PI (= LL - PL). It is a measure of the amount of water bounded within the sediment at specific stress or strength levels. This is shown in the so-called plasticity-plot in Fig. 3.16, where the PI is plotted as function of the LL. This plot can be used to classify soil; for instance, distinction can be made between inorganic and organic-rich soils, different levels of plasticity, and cohesive or non-cohesive behaviour. The so-called A-line distinguishes between inorganic clays and sediments rich in organic matter and silt, whereas the so-called B-line envelops sediments found in the natural environment.

The PI-LL relation of soils from one specific sedimentation area, with varying silt and sand content, but the same type of clay composition, should form a straight line, such as the samples form the Port of IJmuiden, shown in Fig. 3.16. Note that such a plasticity plot does not include information on the grain size distribution, or on the amount of clay-sized particles (< 2 μm).

A further application of the Atterberg limits is obtained when the PI is related to the clay fraction in the so-called activity plot. When the clay fraction determines the mechanical behaviour of the sediment, PI is linearly related to this clay fraction. The slope of this linear relation is a measure for the type of clay mineral involved, and is defined as the activity A (Skempton, 1965):

$$A = \frac{\text{PI}}{\xi^{cl} - \xi_0}$$
(3.7)

where ξ^{cl} is the clay content [% w/w] and ξ_0 is the intercept [% w/w]. Note that ξ_0 can be regarded as the clay content beyond which the sediment acquires cohesive properties (e.g. Section 3.2.2 and Chapter 8).

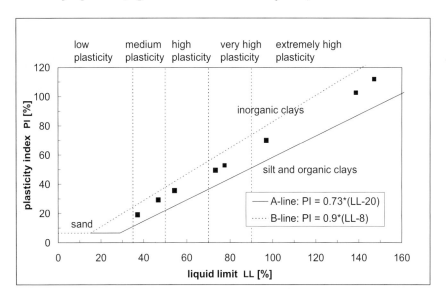

Fig. 3.16: Plasticity-plot for sediment with little organic matter ($\xi^{org} < 15\%$) (after Head, 1986) and data from IJmuiden mud.

For kaolinite, the acitivity is low and around 0.4, for illite it is around 0.9, and for montmorillonite it can go up to 7. Although the activity itself cannot be used to determine mineral composition, it does reflect the type of mineral involved. As an example, samples from various locations in and around IJmuiden Harbour (The Netherlands) are plotted in Fig. 3.17, showing a straight line, which implies that the mineral composition is the same at all locations. From the slope, an activity $A = 2.76$ is found, indicating that montmorillonite is present to bind water (this finding is consistent with our analysis of the ζ-potential in Section 3.2.1).

An important consequence of this classification by activity is the applicability of mud density as a parameter for mechanical behaviour. Two mud samples from different locations and with different dominant clay minerals may have identical densities, while for instance the undrained shear strength, the permeability and the erodibility may differ considerably. This is further elaborated in Chapter 8. Therefore, density is not used in geotechnical engineering to correlate mechanical properties, but a dimensionless water content instead, the so-called liquidity index LI:

$$LI = \frac{W - PL}{PI} = \frac{W - PL}{LL - PL} \tag{3.8}$$

where W is the actual water content [% w/w]. Examples are given in Fig. 3.18 with the undrained shear strength c_u of remoulded samples and in Fig. 3.19 with the permeability k as a function of LI. The LI enables us to correlate different samples with different clay contents and clay minerals. Though the accuracy of this approach is not high, it can be used for a first assessment of c_u. The undrained shear strength c_u is defined as the residual stress in the bed after failure of a sample assessed with a vane shear test, e.g. Fig. 3.20. Note that in soil mechanics the undrained shear strength is currently often denoted by s_u.

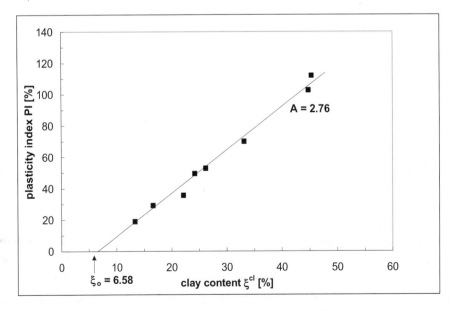

Fig. 3.17: Activity-plot for IJmuiden mud (Van Kesteren, 2004).

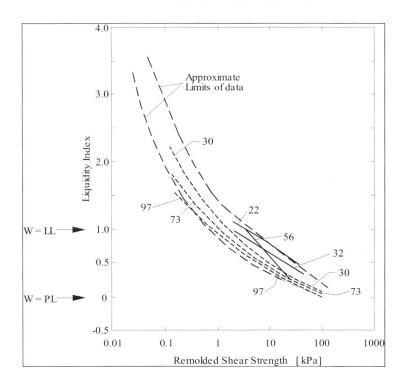

Fig. 3.18: Undrained shear strength as function of LI (after Mitchell, 1976).

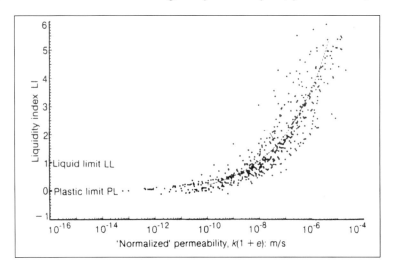

Fig. 3.19: Permeability as a function of LI (Van Kesteren, 2004 and Lubking, 1997).

Fig. 3.18 shows that there are two regions with different slopes: one for LI > 2 and one for LI < 2. Only the lower part covers the water content between LL (LI = 1) and PL (LI = 0). Extrapolation to high LI-values is not accurate. Yet, this is typically the region for cohesive sediments: undrained shear strengths less than 10 kPa down to several Pa's.

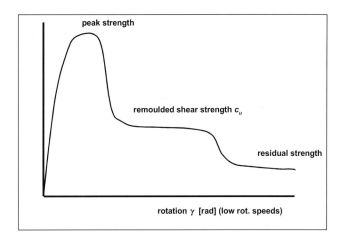

Fig. 3.20: Definition of undrained shear strength c_u obtained from shear vane tests.

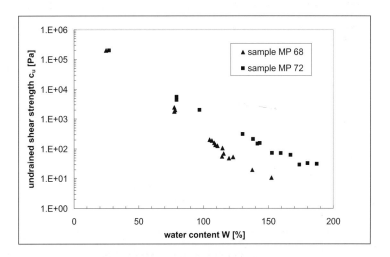

Fig. 3.21a: Undrained shear strength as function of water content for IJmuiden mud (Van Kesteren, 2004).

For high water contents (i.e. soft soils / cohesive sediments) it is therefore more accurate to use the ratio of water content to plasticity index as shown in Fig. 3.21. In Fig. 3.21a, the undrained shear strength c_u is plotted against water content W for two different samples, and in Fig. 3.21b the same samples are plotted, but now the water content is normalised with PI. In the latter case, the data collapse onto one straight line, and this representation is particularly useful for soft soils. Because PI is linearly related to the clay content by equ. (3.7), the W/PI-curve reflects the water content of the clay fraction. The same approach is applicable to other mechanical properties such as permeability, viscosity, compressibility, etc.

These bulk parameters, classifying the sediment properties, form the basic tools to quantify the erodibility and stability of cohesive sediment deposits, as treated in Chapter 8 and 9.

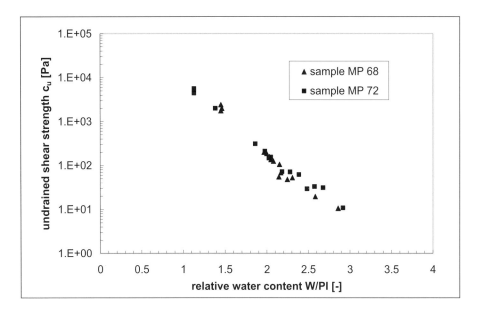

Fig. 3.21b: Undrained shear strength as function of W/PI for IJmuiden mud (Van Kesteren, 2004).

3.3.2 APPLICATION TO IN-SITU CONDITIONS

In Section 3.1.1 and 3.3.1 cohesive sediments were classified by common geotechnical standards for grain size distribution and Atterberg limits. This classification only describes the granular composition of the sediment in a remoulded state, and does not account for the actual fabric (skeleton) in-situ. In Section 3.2 the physics behind particle interactions in a cohesive sediment skeleton and its effect on skeleton structure, reflected by the pore size distribution, is discussed. Although the pore size distribution is an indicator for the in-situ fabric of cohesive sediment, no classification is currently available reflecting the actual, in-situ condition. In this section we therefore propose a classification that reflects such in-situ conditions in combination with the above-mentioned geotechnical standards.

Classification of in-situ condition for sand and silts

For non-cohesive soils, such as sand, in-situ conditions are usually classified by means of the in-situ density at zero stress, relative to the density at minimum and maximum pore volume. For instance the most dense packing of spherical grains of one diameter is a tetrahedral packing (see Fig. 3.22a) with a porosity $n = (1 - \pi / 3\sqrt{2}) = 25.6$ %. The most loose packing of such grains is a cubic-centred packing (see Fig. 3.22b) with a porosity $n = (1 - \pi / 6) = 47.6$ %. The state of packing has a large effect on the mechanical behaviour of the soil: loosely packed soil will tend to decrease in volume (compaction) when grains are displaced with respect to each other, while densely packed soil must increase in volume upon displacement (dilation). The actual *in-situ* packing can be related to the most dense and loose packing with the density index I_D [1] defined by:

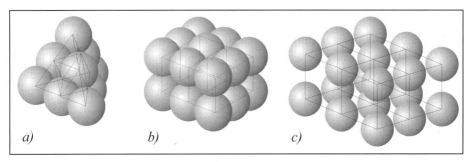

a)   *b)*   *c)*

Fig. 3.22: Network structure of granular material.

---

[1] In geotechnical literature I_D is referred to as the relative density D_r; however I_D is recommended by the Int. Soc. of Soil Mech. and Found. Eng. (ISSMFE).

$$I_D = \frac{e_{max} - e}{e_{max} - e_{min}} 100 \ \% \ \left(= \frac{\phi_{s,max} - \phi_s}{\phi_{s,max} - \phi_{s,min}} \frac{\phi_{s,min}}{\phi_s} \right) \tag{3.9}$$

in which e is the actual void ratio, e_{max} the void ratio at the most loose packing and e_{min} the void ratio at the most dense packing.

The void ratio e, defined as the ratio of pore to solids volume, is related to the porosity n, defined as the ratio of pore to total volume, or to the solid concentration ϕ_s as (see also Appendix B):

$$e = \frac{n}{1-n} \quad \Leftrightarrow \quad n = \frac{e}{1+e} \ ; \ \phi_s = 1 - n = \frac{1}{1+e} \tag{3.10}$$

Many correlations between the density index I_D and mechanical properties of sand are described in the literature (Lubking, 1997). In this section we focus on in-situ conditions, where sand-like skeletons determine the mechanical behaviour of the soil.

Particle interactions within a clay skeleton (see Fig. 3.6) differ largely from those with a sand-like skeleton (see Fig. 3.22):
- the stiffness of a sand-like skeleton is orders of magnitude higher than that of a clay skeleton, and
- relative particle movement in a sand-like skeleton is geometrically more constrained than in a clay skeleton; this is due to the shape and stiffness of the basic particles.

Therefore an important discriminator for in-situ conditions is whether the sand grains are able to form a skeleton or not. It is obvious that when sand grains are not in mutual contact (see Fig. 3.22c), but isolated in a clay-water mixture, the sand skeleton stiffness reduces to zero, and the mobility of sand grains is not constrained anymore. This implies that we can use the maximum void ratio or porosity of the sand skeleton as a discriminator between sand-skeleton-dominated sediment and clay-skeleton-dominated sediment.

The overall porosity of the sand skeleton n^{sa} can be expressed in the actual in-situ sediment porosity by means of the volume balance (see Fig. 3.1):

$$n^{sa} \equiv \frac{n + \phi^{cl} + \phi^{si}}{n + \phi^{cl} + \phi^{si} + \phi^{sa}} = 1 - \psi^{sa}(1-n) \qquad (3.11)$$

where ψ^{sa} is the sand fraction defined as the ratio of sand to total solids volume:

$$\psi^{sa} \equiv \frac{\phi^{sa}}{\phi_s} \qquad (3.12)$$

The various definitions have been given in Fig. 3.1. If the solid densities of all fractions are equal, ψ^{sa} can be read directly from the grain size distribution.

The same approach can be applied when the silt fraction is so large that it determines the skeleton, and sand particles are only filling space within the silt skeleton. However, as non-quartz minerals are present in the silt fraction, the stiffness and mobility of the silt skeleton are less constrained than those of a sand skeleton. As with the behaviour of a sand skeleton, the behaviour of a silt skeleton is determined by its porosity with respect to the silt fraction. This porosity equals the volume of the pore water and clay solids divided by the volume of the silt-clay-water mixture, which equals the total volume minus the sand volume. The volume balance of this mixture yields:

$$n^{si} \equiv \frac{n + \phi^{cl}}{n + \phi^{cl} + \phi^{si}} = \frac{n + \psi^{cl}(1-n)}{1 - \psi^{sa}(1-n)} \qquad (3.13)$$

in which ψ^{cl} is the clay fraction.

The maximum porosity can be used as a discriminator for sand- and/or silt-dominated skeletons and clay-dominated skeletons. This is illustrated in Fig. 3.23, where the measured maximum and minimum porosity of sand-silt mixtures as a function of sand and silt fraction is shown (Van Kesteren, 1996). The measured porosities show a minimum at ψ^{sa} = 78 %. These measured porosities agree with the results of other studies on silt-sand mixtures and sand-gravel mixtures (e.g. Kuerbis et al., 1988; and Floss, 1970).

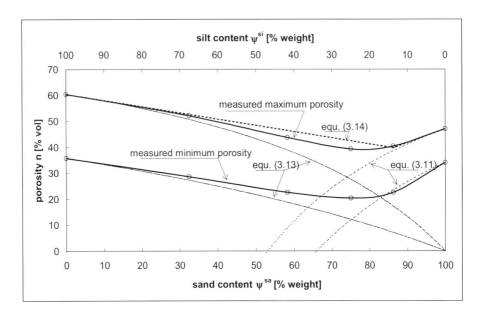

*Fig. 3.23: Minimum and maximum porosity for sand-silt mixtures
(Van Kesteren, 1996).*

Comparison of the measured porosities with equ. (3.11), assuming constant n^{sa}, indicates that the maximum and minimum porosity of a sand skeleton is not affected by silt particles up to $\psi^{sa} = 15$ %: silt particles are only occupying pore space within the sand skeleton. Beyond $\psi^{sa} = 15$ %, the sand skeleton is affected and its maximum and minimum porosity are higher than predicted by equ. (3.11). Similarly, comparison of the measured porosities with equ. (3.13), assuming constant n^{si} indicates that sand particles do not affect the silt skeleton up to $\psi^{sa} = 30$ %. In this analysis we assume $\psi^{cl} = 0$ for simplicity.

The maximum porosity can be approximated with the bold dashed line in Fig. 3.23, given by:

$$
n_{\max} = \frac{n_{100}^{si}(100 - (100 - n_{100}^{si})\psi^{sa}) - \psi^{cl}}{100 - \psi^{cl}} \quad \text{for} \quad 0 < \psi^{sa} < 86 \ \%
$$

$$
n_{\max} = n_{100}^{sa} - (100 - n_{100}^{sa})(100 - \psi^{sa}) \quad\quad \text{for} \quad 86 < \psi^{sa} < 100 \ \%
$$

(3.14)

in which n_{100}^{sa} and n_{100}^{si} are the maximum porosity of the 100 % sand and 100 % silt skeleton.

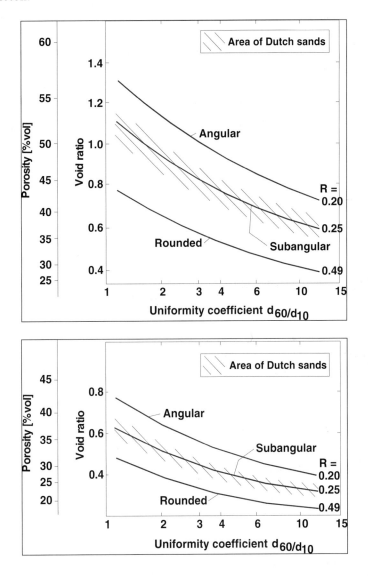

Fig. 3.24: Minimum and maximum porosity as function of D_{60}/D_{10} (redrawn from Youd, 1973).

The minimum in the porosity curves at ψ^{sa} = 78 % in Fig. 3.23 corresponds well with Fuller's grain size distribution for optimum packing conditions at

ψ^{sa} = 82 % (Fuller et al., 1907). Optimum packing is obtained for well-graded distributions, to be characterised with the ratio of D_{60}/D_{10}, which is also known as the uniformity coefficient. A high uniformity coefficient implies low packing. By definition the Fuller distribution has a uniformity coefficient of 36.

Fig. 3.24 shows relations between n_{min} and n_{max} and D_{60}/D_{10}, including the effects of particle shape and roughness (Youd, 1973). The roundness R refers to the sharpness of the corners and edges of a grain and is defined as the ratio of the average radius of curvature of the corners to the radius of the largest inscribed circle. Typical values for marine sands range from R = 0.22 to 0.3.

The uniformity coefficient D_{60}/D_{10} can be obtained from the grain size distribution, and can be used to assess minimum and maximum porosity levels from Fig. 3.23 and 3.24 at in-situ sand content.

Classification of in-situ condition for cohesive sediments

The next step in classifying in-situ cohesive sediments is combining the discriminators for sand- or silt-dominated skeletons with the cohesive behaviour of the clay-water system. This can be done with the sand-silt-clay triangle that was presented in Fig. 3.2 to classify soils on the basis of grain size distribution. We combine this diagram with in-situ density or in-situ porosity. For this purpose, we have to convert solids fraction into solids content. If the specific density for all size fractions is the same, $\xi^i = \psi^i$, as argued in the beginning of this chapter.

Given a certain in-situ porosity, the discriminator between sand-/silt-dominated skeletons and clay-dominated skeletons is given by equ. (3.14). These discriminators are drawn in the sand-silt-clay diagram in Fig. 3.25 in the form of dashed lines for an in-situ porosity n = 36, 40, 45 and 50 % (ρ_{dry} = 1696, 1590, 1457 and 1325 kg/m^3, respectively).

In the zone beyond the discriminator for a certain in-situ porosity (for instance n = 40 % given by bold dashed lines in Fig. 3.25), n^{sa} or n^{si} exceed their critical value, and the skeleton is dominated by the clay-water-system. Below this discriminator, the sediment structure is dominated by the sand or silt skeleton. These discriminators agree well with the empirical classification of soils in Fig. 3.2, confirming the applicability of equ. (3.14) to classify the structure of cohesive sediments through their sand-, slit- or clay-skeleton in a quantitative way.

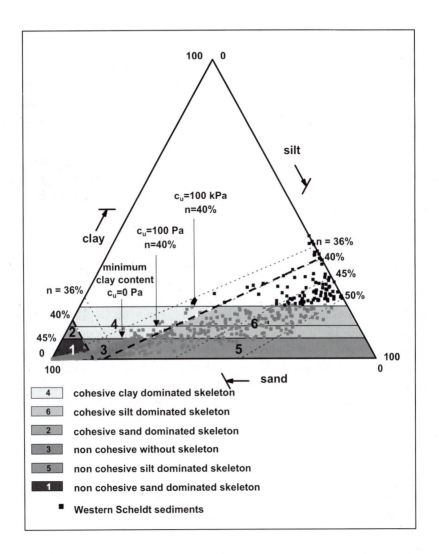

Fig. 3.25: Classification in sand-silt-clay diagram, $c_u = 0$ at $\xi^{cl} = \xi_0$.

For a certain in-situ density (or porosity) and a given grain size distribution, the ratio of ϕ^{si} to ϕ^{cl} is often constant (e.g. Fig. 3.1). We have shown in Section 3.3.2 that the mechanical properties of a clay-water system determine the cohesive behaviour of the sediment. Properties such as the undrained shear strength and viscosity are a function of W/PI, or similarly (e.g. equ. (3.7)) a function of $W/(\xi^{cl} - \xi_0)$. Using these relations, lines of constant undrained shear strength can be plotted in the sand-silt-clay triangle of Fig. 3.25 as well, and appear as horizontal lines. As an example results are shown for IJmuiden harbour mud for the undrainded shear strength relation given in Fig. 3.18. Lines are drawn for an undrained shear strength $c_u = 100$ Pa and $c_u = 100$ kPa at $n = 40$ and 50%. It is shown that c_u increases rapidly beyond ξ_0 (at which $c_u = 0$) and therefore the critical clay content ξ_0 can be used as a discriminator between non-cohesive and cohesive behaviour.

The various discriminators divide the sediment triangle in six sub-zones, distinguishing six modes of sand-mud behaviour:
1. non-cohesive sediment dominated by sand skeleton,
2. cohesive sediment dominated by sand skeleton,
3. non-cohesive sediment with unstable skeleton,
4. cohesive sediment dominated by clay skeleton,
5. non-cohesive sediment dominated by silt skeleton, and
6. cohesive sediment dominated by silt skeleton.

These six sub-zones are indicated in Fig. 3.25 for $n = 40$ %. Mode 1, 5 and 3 are characterised by a non-cohesive behaviour with respectively a sand-dominated network, a silt-dominated network, and a mixed sand-silt network with a very loose skeleton. Mode 3 is characterised by its sensitivity to liquefaction (see Chapter 8 for details). Mode 4 is characterised by a cohesive behaviour with a clay network, and modes 2 and 6 are characterised by a cohesive behaviour with a sand and silt network, respectively.

In Fig. 3.25, we have also drawn data on the granular composition of sediments from Western Scheldt, The Netherlands. The data more or less fall within a narrow band around a straight line, indicating a constant ratio between clay and silt content of about 1:4, as discussed in Section 3.1. Moreover, these data suggest that not all six modes defined above are relevant for the Western Scheldt, as elaborated below.

The clay-silt ratio appears to be constant in many other marine environments (e.g. Flemming, 2000), suggesting that silt and clay particles are transported together. Probably, silt particles are entrapped within cohesive sediment flocs, which are transported with the (tidal) flow. This implies that the sediment composition of an in-situ sample can be given by the sand content ξ^{sa} alone (in the phase diagram to be discussed below, we refer to in-situ dry densities of the soil – hence we prefer the use of solids content instead of solids fraction).

Thus we can reduce the general sand-silt-clay triangle of Fig. 3.25 to a site-specific diagram if the in-situ porosity n or dry bed density ρ_{dry} is known. This results in the ξ^{sa} - ρ_{dry} diagram of Fig. 3.26 that is constructed for Western

Scheldt sediments as an example. It can be regarded as a phase-diagram for bed behaviour in natural environments. To establish this diagram, we have used an empirical relation drawn up by Allersma (1988) on the basis of samples from Dutch estuaries and a few reservoirs elsewhere:

$$\rho_{dry} = 480\alpha_c + (1300 - 280\alpha_c)\xi^{sa\,0.8} \qquad\qquad (3.15)$$

where ρ_{dry} is the dry bed density, ξ^{sa} is the sand content of the sediment sample and α_c a consolidation coefficient, ranging from $\alpha_c = 0$ for fresh deposits to $\alpha_c = 2.4$ for old deposits, resulting respectively in a lower and upper bound for the dry density.

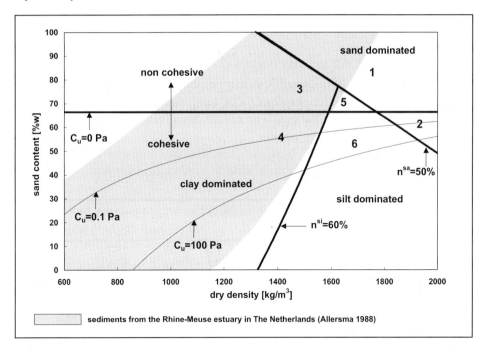

Fig. 3.26: Sediment-phase diagram for Western Scheldt sediments.

The area between the lower and upper bound of this relation is also depicted in Fig. 3.26 (grey shaded area). From this figure it can be deduced that three modes of sand-mud behaviour are expected to be encountered in the Western Scheldt:

mode 1: non-cohesive sand dominated behaviour,
mode 3: non-cohesive very loose (supercritical) sand/silt skeleton, and
mode 4: cohesive clay dominated behaviour.

Construction and application of phase diagram

In the following an example is elaborated how a phase diagram can be constructed from available geotechnical data and how it can be applied to assess sediment behaviour. The identified modes of sediment behaviour are also compared with observations.

Our case study concerns the Orange River fan on the West Coast of Africa (Van Kesteren, 1996). Grain size distributions (2 μm up to 2 mm) and dry densities were obtained from ten cores. The following steps are required to construct the phase diagram:

Step1:

Determine the sand, silt and clay content ξ^{sa}, ξ^{si} and ξ^{cl} (which are identical to ψ^{sa}, ψ^{si} and ψ^{cl}, if the specific density of the solids fractions are identical) and the uniformity coefficient D_{60}/D_{10} of sand and silt solids only from the measured grain size distributions. The results are given in Table 3.4, together with the measured in-situ dry density.

Table 3.4: Geotechnical data from Orange River fan.

Sample	clay < 2 μm	silt 2–63 μm	sand > 63 μm	D_{60}/D_{10}	n_{max} [%vol]	ρ_{dry} [kg/m³]	PL [%w]	LL [%w]
1	2.0	10.0	88.0	1.8	49.0	1674		
2	3.2	11.5	85.4	2.0	48.4	1521		
3	3.0	19.0	78.0	1.7	49.3	1413		
4	5.9	26.7	67.5	5.4	42.2	1476		
5	10.3	32.3	57.4	9.4	38.6	1413	26.4	34.3
6	7.0	30.0	63.0	10.0	38.2	1476		
7	19.9	46.5	33.6	11.4	37.4	1423	25.9	36.4
8	7.6	37.2	55.2	10.9	37.7	1472		
9	34.8	64.3	0.9	2.2	47.9	1108	29.2	74.8
10	39.3	57.3	3.4	2.5	47.0	1259	22.7	60.7

Step 2:
Plot the clay, silt and sand content in the triangular sand-silt-clay diagram (see Fig. 3.27). The samples form a single, curved line, indicating that the clay and silt fractions are correlated, thus that the samples only differ in sand content.

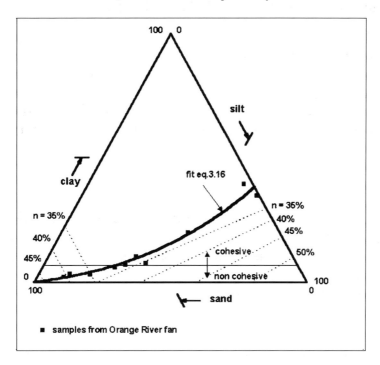

Fig. 3.27: Sand-silt-clay diagram for Orange River fan sediments.

Step 3:
Determine the conditions for sand- and silt-dominated skeleton on the basis of equ. (3.14) as a function of sand content and dry density along the sediment line in Fig. 3.27. Eliminate the clay content in equ. (3.14) with a fit between the measured clay and sand content of the various samples, which is given by:

$$\xi^{cl} = 0.27(1-\xi^{sa})^2 + 0.112(1-\xi^{sa}) \tag{3.16}$$

The resulting discriminators for sand-, silt- and clay-dominated skeletons are depicted in the phase diagram Fig. 3.27, using the measured dry densities given in Table 3.4.

Step 4:

Determine the maximum porosity of the sand and silt fraction with Fig. 3.24 for the D_{60}/D_{10} ratio from Table 3.4 and plot the results in a figure like Fig. 3.23, yielding Fig. 3.28. Choose the maximum porosity for 100 % silt and 100 % sand such that equ. 3.14 fits the maximum porosity data obtained from Fig. 3.24 using the measured D_{60}/D_{10} ratio. In our example of Fig. 3.28, the maximum porosity for 100 % silt and 100 % sand are set at 55 % and 50 %, respectively.

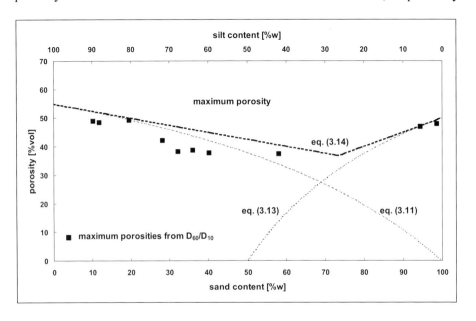

Fig. 3.28: Assessment of maximum porosity.

Step 5:

Next, we determine the discriminator for cohesion. The Atterberg limits were determined for some samples (see Table 3.4). In the activity plot, Fig. 3.29, the plasticity index as a function of clay content is shown, indicating that the clay has an activity of $A = 1.3$. The offset with the horizontal axis gives the critical clay content for cohesive behaviour $\xi_0 = 6.8$ %. This value is close to the average value of 7 %, which is recommended to be used if no Atterberg data are available. The cohesion discriminator appears as a horizontal line in the phase diagram Fig. 3.30.

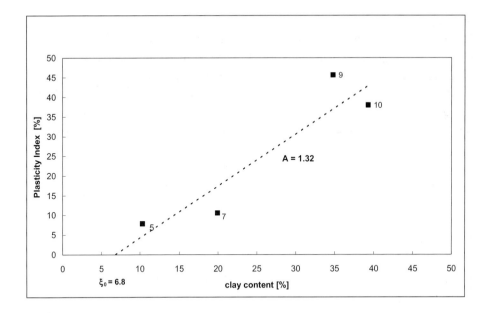

Fig. 3.29: Activity plot for Orange River fan sediments.

Step 6:
Finally, plot the measured data of dry density and sand content in the phase diagram Fig. 3.30. Equ. 3.16 appears as an S-shaped curve, fitting the data.

The phase diagram is now complete. It shows that the sediments of the Orange River fan are characterised by Mode 1 (non-cohesive, sand-dominated skeleton), Mode 3 (non-cohesive, unstable skeleton) and Mode 4 (cohesive clay-dominated skeleton). Mode 5 (non-cohesive silt-dominated skeleton) has almost vanished. Samples 1 and 2 can be classified as low to medium dense sands. Sample 2 is close to critical porosity, where shear deformation results in compaction.

Samples 3 and 4 are characterised by poor stability conditions. Their porosity is beyond the maximum value and therefore compaction is expected upon shear deformation. However, the sediment's permeability is low because of the clay content, and pore water pressures can easily build up. It is expected that these samples are sensitive to liquefaction.

The clay-water system within the pores of the samples 5, 6 and 8 builds up strength and can therefore sustain a certain load. When strength information is available, for instance when Atterberg limits are determined, lines of constant undrained shear strength can be plotted in the sediment phase diagram. As an example, the line of constant $c_u = 100$ Pa is drawn, which is a typical undrained shear strength encountered in the top of a muddy seafloor.

Van Kesteren (2003) describes further tests on the samples. In particular triaxial tests showed the sensitivity of the samples 3 and 4 to liquefaction.

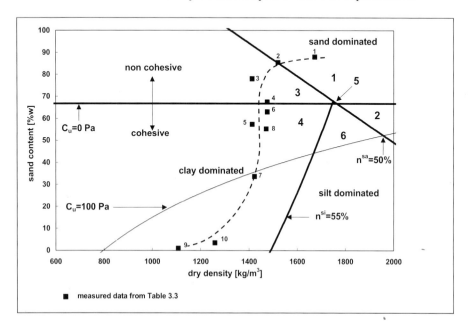

Fig. 3.30: Sediment-phase diagram for Orange River fan sediments.

3.4   COHESIVE SEDIMENT IN THE MARINE ENVIRONMENT

3.4.1   PHENOMENOLOGICAL CLASSIFICATION OF COHESIVE SEDIMENT

The classification in the preceding sections helps us to understand and quantify the behaviour and properties of cohesive sediment deposits at a microscopic scale. However, the behaviour and appearance of cohesive sediment on a larger scale is not only dependent on the physico-chemical properties of the sediment and its pore water, but also on a number of environmental parameters.

As a result, cohesive sediment is encountered in a large variety of modes, amongst which wash load in rivers, estuaries and seas, layers of fluid mud in estuaries and navigational channels, thick deposits on the slope of continental shelves, etc. It is important to recognise that the mode of appearance is dependent on the sediment and pore water properties (as discussed in the preceding sections), on the meteo-hydrodynamic conditions, the (stress) history of the sediment, and last, but not least, the availability of sediment. With its appearance, its behaviour and overall properties change, and such changes may occur on very small to very large spatial and temporal scales: cohesive sediment may be regarded as wash load in one part of a system, whereas it may form thick layers of fluid mud a few hundred metres further downstream. Because of the important role of organic material in general and of the effects of chemistry and biology in particular, significant seasonal effects on the sediment behaviour occur as well, in particular in the intertidal areas of estuaries and coastal seas.

In the literature a variety of terminology is used, and one does not agree even on apparently obvious definitions (e.g. Migniot, 1968; Parker and Hooper, 1994). To minimise confusion in the present book, we define a number of cohesive sediment appearances in this section. This classification is based on the work of Ross and Mehta (1989) and the more recent classification proposed by Bruens (2003). The classification by Ross and Mehta (1989) follows the schematic vertical profile of sediment concentration and velocity, shown in Fig. 3.31.

The terminology used in this book is presented in Table 3.5, together with a number of bulk properties and the dimensionless scaling parameters, which govern the relevant processes.

The forms of cohesive sediment in the first column of Table 3.5 have been ordered by increasing sediment concentration c. However, no absolute c-values can be assigned to this column, as the mud appearances are determined by sediment and pore water properties and the stress history of the sediment. LCMS is our abbreviation for Low-Concentration Mud Suspensions and HCMS for High-Concentration Mud Suspensions; in the latter case sediment-fluid interaction plays an important role. Turbidity currents are also referred to as mudflows (Coussot, 1997) and can develop into debris flow when carrying (large amounts of) debris (such as gravel, boulders, etc.). However, the term debris flow is often reserved for flows on land, i.e. above water.

It is illustrative to give a few examples of occurrences of the various mud appearances defined in Table 3.5:

- LCMS is found in the majority of natural systems, i.e. rivers, large parts of estuaries and coastal areas, etc. when the concentrations are too low to affect the flow field.
- HCMS is for instance found in estuaries, in particular near their turbidity maxima, and above mud banks in coastal waters, when the turbulent flow field is (largely) affected by the suspended sediment (e.g. Chapter 6).

- Turbidity currents may be found on the slopes of deep seas, such as the slope of continental shelves; some further observations are presented in Section 9.4.3.

- Fluid mud is found in many navigational channels and harbour basins throughout the world. Fluid mud flow may also occur on the slopes of continental shelves. Section 3.4.3 contains some further discussions on fluid mud.

- Consolidating and consolidated beds are found everywhere where cohesive sediments are found. Their behaviour is discussed extensively in Chapter 7 and 8.

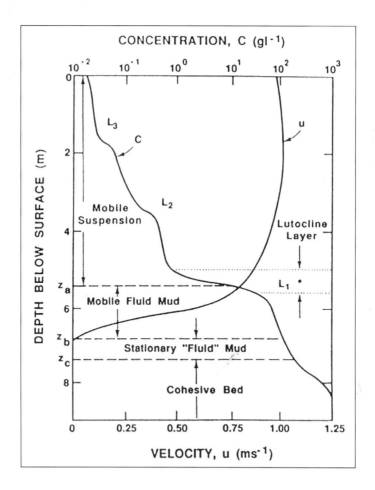

Fig. 3.31: Typical vertical profiles of suspended sediment concentration and velocity for high-concentration conditions (after Ross and Mehta, 1989).

Table 3.5: Classification of cohesive sediment modes in the marine environment.

	concentr.	flow characteristics		governing scale numbers				
LCMS	$c \ll c_{gel}$	turbulent	Newtonian	Re	β			
HCMS	$c < c_{gel}$	turbulent	Newtonian	Re	Ri			
turbidity current	$c \approx c_{gel}$	turbulent	non-Newt.	Re_e	Ri			
mobile fluid mud	$c \approx c_{gel}$	trans./lam.	non-Newt.	Re_e	Ri		Pe	
stationary fluid mud	$c \approx c_{gel}$	trans./creep	non-Newt.			p^w/p_e	Pe	
consolidating bed	$c > c_{gel}$	creep	non-Newt.			p^w/p_e	Pe	σ/τ_v
consolidated bed	$c \gg c_{gel}$	stationary	non-Newt.					σ/τ_v

where the following symbols have been used:

c = suspended sediment concentration

c_{gel} = gelling concentration

Re = Reynolds number

Re_e = effective Reynolds number

β = Rouse number

Ri = Richardson number

p^w = total pore water pressure

p_e = hydrostatic pore water stress

Pe = Peclet number

σ = externally applied stress

τ_v = yield strength

Notice that the various appearances may or may not occur in one system. Wash load may remain wash load more or less indefinitely, and thick layers of mud on the continental shelf slope may become unstable and generate turbidity currents, which may freeze at the foot of the slope; these occurrences will hardly affect the system. On the other hand, in estuaries, layers of fluid mud may be formed during neap tide, being remobilized during accelerating flow, in which case the majority of the appearances given in Table 3.5 are found in one tidal cycle.

In this book, we define the gelling concentration c_{gel} as the concentration at which a network structure exists, i.e. when flocs are in direct contact with each other, and yield strength is developing (e.g. Chapter 4 and 7). In soil mechanics, c_{gel} is also known as the structural density, e.g. Chapter 4 and 7.

The Rouse number β determines the vertical suspended sediment concentration profile (e.g. Chapter 5), and is defined as:

$$\beta = \sigma_T W_s / \kappa u_* \tag{3.17}$$

where σ_T is the turbulent Prandtl-Schmidt number, W_s is settling velocity, κ is the von Kármàn constant and u_* the shear velocity.

The Reynolds number Re defines whether the flow is laminar or turbulent. For non-Newtonian, Bingham plastic flow (e.g. Chapter 8), the effective Reynolds number Re_e is defined as:

$$\frac{1}{Re_e} = \frac{1}{Re} + \frac{1}{Re_y} \quad \text{where} \quad Re = \frac{4U_m \delta_m}{v_m} \quad \text{and} \quad Re_y = \frac{8\rho_m U_m^2}{\tau_B} \tag{3.18}$$

where U_m is the mean velocity in the fluid mud layer with thickness δ_m, τ_B is the Bingham strength of fluid mud (see also Chapter 8 and 9), and v_m and ρ_m are the viscosity and density of fluid mud. According to Liu and Mei (1989) the critical effective Reynolds number amounts to about $Re_{e,c} = 2\text{-}3 \cdot 10^3$. This criterion was validated by Van Kessel (1997) against experiments on laminar and turbulent mud flows over a sloping bed.

The Richardson number Ri determines whether (sediment-induced) buoyancy effects (stratification) on the turbulent properties of the flow are important. These effects are treated extensively in Chapter 6.

The difference between the total pore water stress p^w and the hydrostatic pore water stress p_e is called the water over-pressure or excess pore water pressure. It is non-zero in a consolidating bed (see also Chapter 7).

The Peclet number Pe is a measure to determine whether a deformation process should be regarded as drained or undrained, and is defined as:

$$Pe = V\ell / c_v \tag{3.19}$$

where V is the velocity of the deformation process, ℓ is a length scale of the deformation process, and c_v is the consolidation coefficient of the soil (e.g. Chapter 7, 8 and 9).

The ratio of the (externally) applied stresses σ and the yield strength τ_y determines whether the soil may flow under the influence of said stresses.

The vertical concentration profile, such as sketched in Fig. 3.31, is determined by a number of processes, which are treated subsequently in this book:

- Flocculation: because of the cohesive nature of mud, flocs are formed, affecting the settling velocity and bed structure,
- Settling and mixing: mud particles fall through the water column due to gravity, opposed by mixing processes generated by the turbulent water movement,
- Deposition: settling mud particles may become part of the bed,
- Resuspension: during accelerating flow, particles freshly deposited on the bed may be re-entrained into the water column by the turbulent flow,
- Entrainment: turbulent flow over or underneath a less turbulent fluid entrains water and matter from this less turbulent layer,
- Gelling: deposited mud particles, when left still for sufficient time, will form a structure, causing the build-up of strength that can resist re-entrainment,
- Consolidation: another step in bed formation is self-weight consolidation, when pore water is squeezed out of the bed, and the strength of the bed increases further,
- Liquefaction: when subject to cyclical loading, bonds between particles can be broken gradually, reversing the consolidation process, weakening the bed, and
- Erosion: even when the bed has achieved a considerable strength, it can still be eroded by turbulent flow or waves.

These processes may act simultaneously or successively. Often however, only some of these play a role, depending on the prevailing conditions. In the case of concentrated mud suspensions, however, several of these processes become a function of the mud concentration itself, and a complicated interactive picture emerges (e.g. Chapter 6).

Also accelerations and deceleration of sediment-water mixtures in horizontal direction can be important. For instance, the deceleration of a sediment-laden flow near its transport capacity may lead to the formation of fluid mud, whereas mobile mud on a slope may accelerate to become a turbidity current, or even mix entirely with the ambient fluid by entrainment processes.

3.4.2 MUD CONTENT OF SAND-MUD MIXTURES

In the previous sections we have analysed the behaviour of sand-mud mixtures in the marine environment. We have argued that in practice often only a few of the six possible modes of sediment behaviour are expected to occur, which we have illustrated with examples from the Western Scheldt and Orange River fan. This is of course the result of the hydrodynamic conditions prevailing and the properties and availability of the sediment. The influence of local hydrodynamic conditions and sediment supply on the mud content ξ^m (again, mud contains clay and silt particles) was elaborated by Van Ledden (2002, 2003), and we follow his reasoning.

Let us analyse the sediment behaviour under tidal conditions in a small domain, such as a (part of a) intertidal flat. Both the water column and bed in the domain contain sand and mud. If the domain is small enough, the hydro-sedimentilogical conditions within the domain will be homogeneous. Thus, the mass balance for mud can be written as:

$$\frac{dhC^m}{dt} = E^m - D^m + k_e\left(C_\infty^m - C^m\right) \tag{3.20}$$

in which C^m is the depth-averaged mud concentration in the domain, C_∞^m is the mud concentration outside this domain, k_e is a positive exchange coefficient, and E^m and D^m are the erosion and deposition rate within the domain, respectively.

For the deposition rate[2] we use the sediment flux $W_s C^m$ (e.g. Section 5.3.1). The erosion rate is a function of the mud content in the bed, and is modelled with a simple erosion formula (e.g. Section 9.2.1, equ. (9.3)):

$$E^m = \xi^m M \left(\tau_b / \tau_e - 1 \right) H \left(\tau_b / \tau_e - 1 \right)$$ (3.21)

where M is an erosion rate coefficient, τ_b is the actual bed shear stress, τ_e is the critical shear stress for erosion, and H is the Heavyside step function. We assume to first order that M nor τ_e are affected by the sand-mud composition of the bed.

Next, we consider tidal flow, assuming a sinusoidal velocity variation with a maximum bed shear stress $\tau_{b,0}$ and integrate (3.20) and (3.21) over the tide. At equilibrium, C^m and the bed level and bed composition are constant, hence $C^m = C_\infty^m$ (e.g. (3.20)), and we find:

$$\xi_{equ}^m = \frac{W_s C_\infty^m}{M \alpha_E} = \frac{F_s}{\alpha_E}$$ (3.22)

where F_s is a relative settling flux, i.e. the ratio between potential deposition to erosion rates, and α_E is given by Van Ledden (2003):

$$\alpha_E = 0 \qquad\qquad \text{for } \tau_{b,0} < \tau_e$$

$$\alpha_E = \left(\frac{\tau_{b,0}}{2\tau_e} - 1 \right) \left(1 - \frac{2}{\pi} \arcsin \left\{ \sqrt{\frac{\tau_e}{\tau_{b,0}}} \right\} \right) + \tag{3.23}$$

$$+ \frac{1}{2\pi} \sqrt{\frac{\tau_{b,0}}{\tau_e} - 1} \qquad\qquad \text{for } \tau_{b,0} > \tau_e$$

---

[2] Note that Van Ledden (2002, 2003) used Krone's deposition formula (e.g. Chapter 5). As this is not the case in our analysis, our results differ quantitatively from those of Van Ledden; qualitatively our results and conclusions are identical.

We have implicitly assumed that also the transport of and bed level changes due to sand are in equilibrium, as a result of which we do not have to take the balance for the sand fraction into account.

Fig. 3.32 shows the variation of the equilibrium mud content ξ^m_{equ}, described by equ. (3.22), as a function of the relative bed shear stress for various values of the relative settling flux $F_s = W_s C^m_\infty / M$. This diagram suggests that for small values of F_s, a sharp transition is to be expected between mud-rich and mud-poor sediments, whereas for large values of F_s, this transition is more gradual.

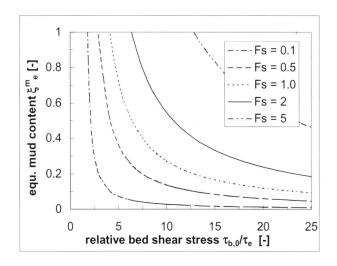

Fig. 3.32: Equilibrium mud content as a function of bed shear stress for various values of the relative settling flux F_s.

This picture is further elaborated in Fig. 3.33, comparing values of the mud content ξ^m measured on an intertidal mudflat in the Western Scheldt, The Netherlands ("de Molenplaat", e.g. Van Ledden, 2002, 2003) with its equilibrium value of equ. (3.22). The latter was obtained using characteristic values of the model parameters for intertidal mudflats in The Netherlands: $W_s = 0.25$ mm/s, $C^m_\infty = 200$ mg/l, $M = 5 \cdot 10^{-4}$ kg/m^{2}s, and $\tau_e = 0.5$ Pa.

We observe that the equilibrium curve envelops most data, and that the observed sharp transition between mud-rich and mud-poor sediments is properly predicted. However, the observed mud content varies largely within the equilibrium curve. This may be due to:

- The field data do not reflect equilibrium conditions; Van Ledden (2003) estimates that equilibrium is only achieved after a few months. Within this period the hydrodynamic and sedimentological conditions may vary largely,
- The erosion properties do vary with mud content, contrary to our assumption,
- In our derivations, we have implicitly assumed that the water depth is constant. This of course is not true under tidal conditions, in particular not on intertidal flats.

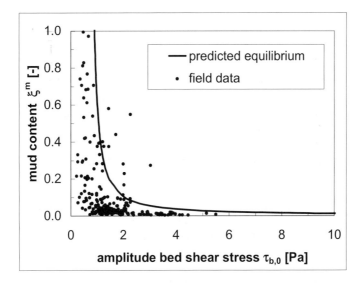

Fig. 3.33: Comparison of mud content observed on an intertidal mud flat with the predicted equilibrium mud content.

However, we may conclude that a sharp transition between mud-rich and mud-poor sediment deposits can be expected when the relative settling flux F_s is small. This will be often the case in marine environments characterised by LCMS. Under HCMS-conditions, the relative settling flux F_s is large, and a gradual transition between mud-rich and mud-poor deposits is to be expected. The latter conclusion is enforced by the fact that hydrodynamic sorting at HCMS-conditions is less developed than at LCMS-conditions.

3.4.3 FLUID MUD

Fluid mud is a suspension of cohesive sediment at a concentration at or beyond the gelling point (e.g. Chapter 5), i.e. of the order of several 10 to 100 g/l, as shown in Table 3.6. This suspension exhibits profound non-Newtonian behaviour, and it is either stationary or moving. In the latter case the fluid mud flow is laminar (by definition) and its dynamics will be fairly independent of the flow in the water column above. The thickness of fluid mud layers ranges from a few dm to many m's. Fluid mud is generated through rapid siltation, or by liquefaction of mud deposits.

Table 3.6: Typical fluid mud concentrations in estuaries;
the large concentrations in the Amazon may be attributed to high shear rates.

estuary	reference	fluid mud conc. [g/l]	remarks
Severn	Crickmore ('82)	appr. 10	
	Kirby & Parker ('83)		
Parret	Odd et al. ('93)	40 - 80	above cons. bed
Ems Estuary	Van Leussen ('94)	appr. 40	0.35 m from bed
Amazon	Kineke et al. ('95)	40 - 250	0 - 2 m from bed
	Kineke et al. ('96)		
Loire	Le Hir ('97)	appr. 40	1 m thick layers
Jiaojiang River	Guan et al. ('98)	appr. 40	0.35 m from bed

Fluid mud is in a transient state as no mechanism exists to keep the particles in suspension: the mud layer is subject to consolidation. The consolidation process for stationary fluid mud is treated in Chapter 7. The consolidation rate for mobile fluid mud is larger than that of stationary fluid mud (e.g. Wolanski et al., 1990). This is explained by the shearing of the mud layer, as a result of which pore water can be expelled more easily. No formulations for the effect of shearing on the permeability of the mud are known to the authors. However, the effect of shear failure on consolidation is treated in Chapter 8.

Fluid mud is a general plastic, shear thinning material, characterised by a yield strength τ_y and viscosity decreasing with increasing shear rate $\dot{\gamma}$; a typical flow curve is sketched in Fig. 3.34. Coussot (1997) advocates the Herschel-Bulkley model to describe the mud's stress-strain relations. This model reads in one-dimensional form:

$$\tau = \tau_y + K\dot{\gamma}^n \tag{3.24}$$

The model parameters in (3.24) vary largely with the solids concentration in the mud (e.g. Coussot, 1997): τ_y varies from about 0.01 to 0.1 Pa at concentrations of about 10 g/l to 1 to 10 kPa for concentrations up to 1000 g/l. In that range of concentrations K varies between 0.01 Pa·s^n and 100 to 1000 Pa·s^n and $n < 1$ has typical values of about 1/3 to 1/2. For $n = 1$, (3.24) simplifies to the well-known Bingham model, often used to model the behaviour of fluid mud, in which case K is the fluid mud's viscosity (see also Section 8.5).

Coussot (1997) elaborates on a large number of flow conditions for fluid mud, and derives backwater curves for laminar mudflows.

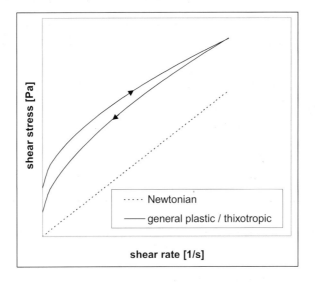

Fig. 3.34: Schematised flow curve of fluid mud, including hysteresis as a result of shear thinning.

Upon shearing, (the) bonds between the mud particles break. As it takes time to restore these bonds, the stresses in the mud during a second deformation are smaller, which results in the hysteresis in the flow curve sketched in Fig. 3.34. This effect is known as thixotropy, which is elaborated further upon in Chapter 8.

Fluid mud is important from a navigational point of view, as rapid siltation hinders navigation. However, fluid mud deposits are so soft, in particular directly after their formation, that ships may sail through the mud. This has led to the concept of Navigable Depth (Kirby and Parker, 1983), distinguishing between the hard sediment bed and the soft mud.

The vertical sediment concentration profile, i.e. the mud density profile, often exhibits a number of distinct steps, which affect its acoustic impedance. As a result, these steps are revealed from dual-, or multi-echo sounding, used to measure the channel depth. An example of the occurrence of fluid mud is presented in Fig. 3.35, showing longitudinal profiles of the reflections of the 210 and 15 kHz acoustic signal in the access channel to Rotterdam Port on November 13, 1998 (Fig. 3.35a) and December 29, 1998 (Fig. 3.35b), indicating layers of fluid mud with a thickness of 1 to 3 m[3].

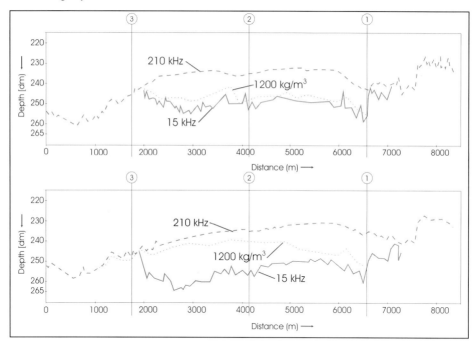

Fig. 3.35: Fluid mud occurrence in access channel to Rotterdam Port, measured by dual-frequency echo sounding on November 13, 1998 (upper panel) and December 29 (lower panel). Cross sections 1 – 3 refer to the vertical profiles presented in Fig. 3.36.

---

[3] These figures have kindly been made available by the Netherlands Ministry of Transport and Public Works, Rijkswaterstaat, Directorate North Sea.

Simultaneously, vertical density profiles were measured, some of which are presented in Fig. 3.36. It is remarkable that on Nov. 13 the 1200 kg/m^3 contour is found close to the 15 kHz reflection, whereas its location varies on Dec. 29. This is further elaborated in Fig. 3.36, showing a change from a gradual density profile to a very steep profile. The latter is representative for a very thin fluid mud layer.

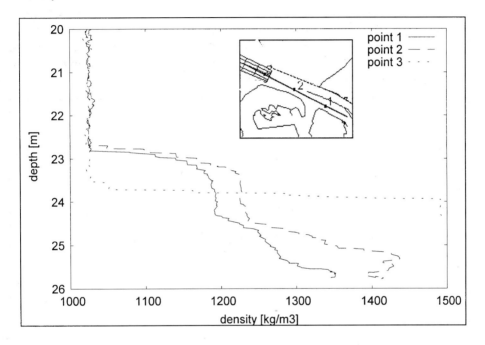

Fig. 3.36: Vertical density profiles measured on December 29, 1998 in the mud deposits in access channel to Rotterdam Port, measured by gamma-densitometer.

4.   FLOCCULATION PROCESSES

The physical properties of particulate clusters of cohesive sediment, or flocs, which are formed by the flocculation processes described in Chapter 3, differ strongly from the properties of the particles from which these flocs are formed. This is due to the large water content of the flocs, as a result of which flocs tend to have very open structures with densities only slightly larger than that of water. This is depicted in Fig. 4.1, showing the excess density of mud flocs as a function of floc size, as presented by a number of authors.

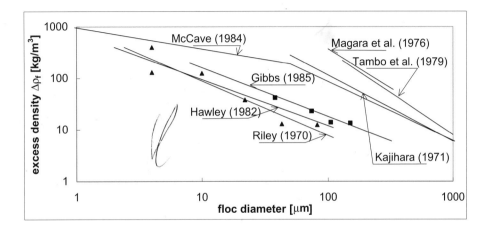

Fig. 4.1: Excess floc density $\Delta \rho_f$ as a function of floc size D_f (redrawn from Van Leussen, 1994).

Krone (1986) was one of the first investigators who studied the physical properties of mud flocs as a function of their structure. Krone hypothesised that flocs are built in a hierarchical way, yielding an ordered structure. This is schematically depicted in Fig. 4.2, showing three orders of aggregation. The primary particles, which form the building stones of the structure, are formed by coagulation, and may contain hundreds to thousands of clay particles. In this definition we follow Van Olphen (1977) who distinguishes between flocculation and coagulation:

> *flocculation is a reversible process, and is the result of simultaneously occurring aggregation and floc break-up processes,*

coagulation *is an irreversible process through which the primary mud
particles (the smallest aggregates in a natural environment, also known
as flocculi) are formed – note that irreversibility in the present context
is related to the maximum shear stresses in the marine environment.*

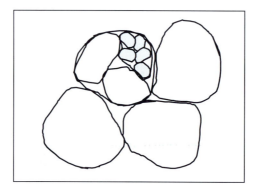

Fig. 4.2: Sketch of hierarchical floc structure (redrawn from Krone, 1986).

In the literature coagulation sometimes is defined as "aggregation" and is then
related to the face-to-face bonding of clay particles.

In this chapter the structure of mud flocs in the water column, and the processes
that govern their size and properties are discussed. A simple model is proposed
to quantify the evolution of mud flocs under turbulent flow conditions.

4.1   INTRODUCTION

Aggregation is the result of mutual collisions of, and subsequent adherence
between, particles of cohesive sediment, and break-up is caused by turbulent
stresses and mutual collisions. The effects of the latter are small for cohesive
sediment. Flocculation is governed by three processes:
1. Brownian motions cause the particles to collide, resulting in the formation of
   aggregates,

2. Particles with a large settling velocity will overtake those with a smaller
   settling velocity: collisions between these particles may result in aggregation
   (generally referred to as flocculation by differential settling), and
3. Turbulent motion will cause particles, carried by turbulent eddies, to collide
   and form flocs; turbulent shear on the other hand may disrupt the flocs,
   causing floc break-up (generally referred to as flocculation by shear effects).

Many studies have been reported on the effects of flocculation on particle size distributions of suspended matter, e.g. Winterwerp (1999, 2002) or McAnally (1999) for a review. Mathematical-physical formulations for the collision frequency by the three agents mentioned above are well known and are given in many papers and reviews, e.g. McCave (1984) and Van Leussen (1994). From the studies by Krone (1962); O'Melia (1980); McCave (1984);and Van Leussen (1994) it can be concluded that aggregation due to Brownian motion is negligible in estuarine and coastal environments.

Stolzenbach and Elimelich (1994) studied the effect of differential settling in detail by analysing theoretical results presented in the literature (Wacholder and Sather, 1974), and by carrying out experiments in a settling column. They showed that the chance that a large, but less dense, rapidly falling particle will actually collide with a small, but denser, slowly falling particle is very small, because the trajectory of the smaller particle is deflected around that of the larger one. They concluded that the effect of differential settling on aggregation in marine environments is much smaller than predicted by the classical collision functions, and might even be entirely absent in many situations.

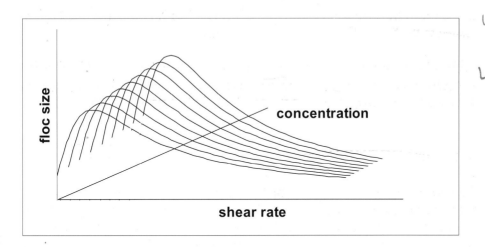

Fig. 4.3: Conceptual flocculation diagram (redrawn from Dyer, 1989).

From these results we conclude that the effects of Brownian motion and differential settling are probably small in estuarine and coastal environments. Therefore, we focus on the effect of turbulence. A qualitative picture of this effect, shown in Fig. 4.3, was published by Dyer (1989). It shows that the mean floc size first increases with shear rate, followed by a decrease, and that the floc size increases with increasing suspended sediment concentration. These trends have been quantified through a number of measurements, amongst which recent detailed floc size measurements on the Amazon Continental Shelf as part of the AMASEDS project (Berhane et al., 1997). The qualitative picture of Fig. 4.3 is further elaborated upon in Section 4.4.

A heuristic formula of this picture was advocated by Van Leussen (1994), based on the work by Argaman and Kaufman (1970), which is discussed below:

$$w_s = w_{s,r} \frac{1+aG}{1+bG^2} \tag{4.1}$$

where w_s and $w_{s,r}$ are the actual and reference settling velocities, respectively[1], a and b are empirical coefficients, and G is a shear rate parameter, defined in Chapter 2 as:

$$G = \sqrt{\frac{\varepsilon}{\nu}} = \frac{\nu}{\lambda_0^2} \tag{4.2}$$

in which ε is the turbulent dissipation rate per unit mass, ν is the kinematic viscosity of the suspension, and λ_0 is the Kolmogorov micro-scale of turbulence: $\lambda_0 = (\nu/\varepsilon)^{1/4}$. Typical values of λ_0 in estuaries range from 100 to 1000 μm, depending on the water depth and flow velocity. For isotropic turbulence, the dissipation rate per unit mass ε is given by (e.g. Tennekes and Lumley, 1994; see also equ. (2.15)):

$$\varepsilon = 2\nu \overline{s_{ij} s_{ij}} \approx 15\nu \overline{\left(\partial u_1'/\partial x_1\right)^2} \tag{4.3}$$

in which s_{ij} is the turbulent rate of strain, u_1' is the turbulent velocity fluctuation in x_1-direction, and the coefficient 15 stems from all the components that contribute to the rate of strain. It follows that G is a measure of the turbulent shear rate of the flow at the smallest turbulent length scales.

---

[1] We will use capital letters for a constant settling velocity W_s, and small letters for w_s when the settling velocity varies because of flocculation and/or hindered settling effects.

It is illustrative to visualise the variation of $G(U,h,z)$ over the water depth h for various values of h and depth-averaged flow velocity U. For this purpose we apply the relation for ε given by Nezu and Nakagawa (1993):

$$\varepsilon \approx \frac{u_*^3}{\kappa h}\frac{1-\zeta}{\zeta} \qquad\qquad (4.4)$$

where u_* is the shear velocity, κ the Von Kármàn constant, and $\zeta = z/h$ the relative elevation above the bed. Substitution of (4.4) into (4.2) yields the variation of $G(U,h,z)$ for clear water ($v = 10^{-6}$ m^2/s), the results of which are plotted in Fig. 4.4, showing that G varies between 0.1 and 10 s^{-1} in general. Only for U of about 1 m/s and larger, G attains values beyond 10 s^{-1} near the bed.

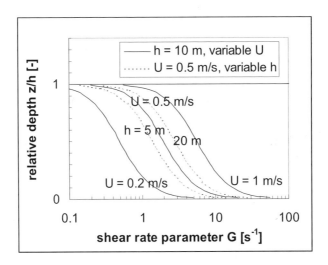

Fig. 4.4: Typical values for the shear rate parameter G in open channel flow (equ. (4.4)).

Formula (4.1) was applied by Malcherek (1995) with a three-dimensional model in a numerical study on the suspended sediment concentrations in the Weser estuary. That author reported that he was able to simulate the observations properly only when using this formulation.

The effects of turbulence on flocculation are widely addressed in the field of sanitary engineering. The two classical studies generally referred to are those by Argaman and Kaufman (1970) and by Parker et al. (1972). These authors studied the simultaneous effects of aggregation and floc break-up; in the latter paper the processes of floc break-up were evaluated in somewhat greater detail. The various parameters in their models, however, are difficult to measure and they proposed a formulation similar to (4.1), describing the variation of the number of particles in a turbulent environment as a function of G and their residence time in that turbulent environment.

Ayesa et al. (1991) developed an algorithm to calibrate the parameters of the model by Argaman and Kaufman against measured variations of the number of particles from experiments carried out in a mixing tank, obtaining an equilibrium concentration for large residence times. Lee et al. (1994) evaluated a slightly different model, and applied it to flocculation processes in a mixing tank for waste water treatment.

Boadway (1978) derived a model for the time evolution of floc size, including the effects of aggregation and floc break-up for massive, spherical (Euclidian) particles. Aggregation and break-up were assumed to be driven by the velocity gradient. A simplified form of the Smoluchowski collision formula (see Section 4.3) and a heuristic floc break-up description were used. The various parameters in his model were calibrated against experiments with alum flocs in a roto-viscometer-like instrument with a gap width of 6.75 mm. The velocity difference over the gap determines the velocity gradient. For equilibrium conditions, the floc diameter appeared to scale with the suspended sediment concentration.

Akers et al. (1987) tried to validate Parker's model for the maximum floc size $D_{f,max}$ through experiments with a dilute suspension in a mixing tank. In agreement with theory, they found that $D_{f,max} \propto \lambda_0$. On the basis of physical reasoning and experiments in a mixing tank with a paddle flocculator (impeller), Tambo and Hozimu (1979) found $D_{f,max} \propto \lambda_0^n$ with $\frac{1}{2} < n < \frac{3}{4}$.

In summary it can be concluded that many studies have been carried out on the effects of turbulence on the flocculation process. Some of these studies are rather descriptive, others more fundamental, and some entirely empirical. They all describe the variation in the number of particles or estimate the maximum floc size that can be achieved. Only the model of Boadway (1978) explicitly describes the important variation in floc size and settling velocity with time.

In the following sections, an alternative approach is proposed, based on a simple flocculation model. We implicitly include both clay and silt particles in the flocculation process. The effects of the neutral silt particles on flocculation rates will be incorporated in the efficiency coefficients of the flocculation process. We do not have to make these effects explicit, as the clay-silt ratio at a specific marine site appears to be more or less constant (e.g. Chapter 3).

For a detailed literature review, and further details of the proposed flocculation model, the reader is referred to Winterwerp (1998, 1999, 2002).

4.2   FRACTAL STRUCTURE OF MUD FLOCS

We treat the mud flocs as self-similar fractal entities, which is a well-accepted approximation at present, e.g. Tambo and Watanabe (1979); Krone (1984); Kranenburg (1994); Huang (1994); Logan and Kilps (1995); and Chen and Eisma (1995). This approximation allows us to establish relations between size and settling velocity of the flocs. It is noted that the conceptual order-of-aggregate description of mud flocs by Krone (1984), as described in the beginning of this chapter is similar to this fractal picture. We distinguish between individual flocs in the water column, where each floc contains a large, but limited number of primary particles, and space-filling network structures in the fluid mud layer where all particles are more or less connected with each other.

Fractal theory implies that various physical processes obey power-law behaviour at all scales. However, in reality, structures are bounded within certain spatial ranges, and often are not self-similar at all scales. Yet, from many physical experiments, it can be concluded that several processes governing cohesive sediment obey such power-law behaviour, amongst which are viscosity, yield strength and permeability, see Kranenburg (1994). This concept is further applied in Chapter 5, 7 and 8.

The fractal dimension n_f is obtained from the description of a growing object with linear size La and volume $V(La)$, where a is the linear size of the nucleus or primary particle (e.g. Viscek, 1992). It is assumed that the linear size of the primary objects has unit dimension, $V(La) \propto N(La)$, where N is the number of primary objects (particles). Then, the fractal dimension n_f is defined as:

$$n_f = \lim_{L\uparrow\infty} \frac{\ln(N(L))}{\ln(L)} \qquad (4.5)$$

From this approach it follows that the differential (or excess) density $\Delta\rho_f$ of mud flocs is given by (Kranenburg, 1994):

$$\Delta\rho_f = \rho_f - \rho_w = \left(\rho_s - \rho_w\right)\left[\frac{D_p}{D_f}\right]^{3-n_f} \tag{4.6}$$

where ρ_f, ρ_w and ρ_s are the densities of the mud flocs, the (interstitial) water and the sediment (primary particles), and D_f and D_p are the diameters of the flocs and of the primary particles. From (4.6) it follows that the relation between the solids concentration c [kg/m^3] and the volumetric concentration of flocs ϕ_f reads:

$$\phi_f = \left(\frac{\rho_s - \rho_w}{\rho_f - \rho_w}\right)\frac{c}{\rho_s} = \frac{c}{\rho_s}\left[\frac{D_f}{D_p}\right]^{3-n_f} = f_s N D_f^3 \tag{4.7}$$

where f_s is a shape factor and N the number concentration of mud flocs (see Section 4.3).

When the volumetric concentration of the flocs ϕ_f becomes unity, the flocs form a space-filling network. This condition is known as gelling in colloid chemistry (Van Olphen, 1977). We therefore refer to the gelling concentration c_{gel} of the mud suspension when $\phi_f = 1$; c_{gel} can be obtained from (4.7):

$$c_{gel} = \rho_s\left[\frac{D_p}{D_f}\right]^{3-n_f} \tag{4.8}$$

Note that due to consolidation effects, the ratio c/c_{gel} can exceed unity within a fluid mud layer and a consolidating bed.

The primary particles of size D_p (or flocculi), which form the building stones of the fractal floc structure, have been generated by coagulation processes, as discussed in Chapter 3 and the beginning of this chapter. Hence, the size of D_p should be a function of the clay mineral and the chemistry of the pore water. However, such a quantitative function is not known at present, and D_p will have to be assessed from measurements. Note that the density of the primary flocs is not necessarily equal to the specific solids density; most probably it is (much) lower.

Measurements of the fractal dimension of flocs of cohesive sediment in the water column reveal values from about $n_f \approx 1.4$ for very fragile flocs, like marine snow, to $n_f \approx 2.2$ for strong estuarine flocs. Typical values within estuaries and coastal waters range from $n_f \approx 1.7$ to 2.2, with an average value of $n_f \approx 2.0$. Within the bed much larger fractal dimensions are found, i.e. $2.6 < n_f < 2.8$ (e.g. Chapter 7, and Kranenburg, 1994). This difference in the fractal dimension has not yet been explained, but must be related to the fracture of flocs under self-weight consolidation. However, we need a working hypothesis to accommodate for this transition and for that purpose we follow Vicsek (1992) in the following paragraphs.

The aggregation process of particles moving in a fluid falls in either of two classes. In the class of Diffusion Limited Aggregation (DLA), the aggregation process starts from a single particle, the seed (or primary particle in cohesive sediment dynamics). Other, identical particles are released one by one to collide with the aggregate by diffusion processes, thus forming flocs. It is clear that in sediment-water mixtures particles are not released one by one, but that many particles move through the fluid simultaneously, approaching each other by diffusion processes. In this case the aggregation process falls within the so-called class of Cluster-Cluster Aggregation (CCA).

This CCA-class is commonly further subdivided in two sub-classes (e.g. Weitz et al., 1985). When the sticking probability of two particles approaching each other is approximately unity, the diffusion process is the limiting agent in the particle growth. In this case the aggregation process falls within the sub-class of Diffusion Limited Cluster-Cluster Aggregation (DLCCA). Both from theoretical studies and from numerical and physical experiments the fractal dimension for DLCCA is established at about $1.7 < n_f < 1.8$ (e.g. Meakin, 1986).

If the sticking probability is much smaller than unity, for instance when the particles have to overcome a repulsive barrier, sticking becomes the limiting factor, and the aggregation process falls within the sub-class of Reaction Limited Cluster-Cluster Aggregation (RLCCA). From theoretical studies and numerical and physical experiments the fractal dimension for RLCCA is established at about $1.9 < n_f < 2.1$ (e.g. Brown and Ball, 1985).

When clay particles prevail in the suspension, RLCCA is likely, as the electrical charges on the clay particles generate short range repulsive forces (e.g. Section 3.2.1). Indeed, the fractal dimensions of suspended cohesive sediment flocs, consisting mainly of mineral material, are of the order of $n_f \approx 2$, and the aggregation efficiency is small in general (e.g. Section 4.1). This also follows from values for the aggregation efficiency e_c which describes the number of effective collisions. O'Melia (1985), for instance, found values of e_c of 10^{-3} to

10^{-1}. Such a low efficiency value suggests that the aggregation process of cohesive sediment in a turbulent environment falls indeed within the class of Reaction Limited Cluster-Cluster Aggregation.

The presence of certain types of organic matter increases the sticking probability of approaching particles considerably through sticking filaments, polysaccharides etc. (e.g. Section 3.1.3). In that case DLCCA is more likely to occur, resulting in a decrease in fractal dimension (e.g. Weitz et al., 1985). It is postulated that it is the augmented sticking probability of organic compounds in a DLCCA process, that causes low excess density, porous flocs of small strength in sediment suspensions rich of organic matter. Indeed Chen and Eisma (1995) found fractal dimensions of $1.7 < n_f < 1.8$ for mud aggregates with a fairly high organic content.

All theories and experiments, leading to the n_f-values given above, were derived for low-concentrated suspensions, i.e. $\phi_f \ll 1$. For High-Concentration Mud Suspensions near the bed this is no longer the case, particularly in the hindered settling regime around slack water. Kolb and Herrmann (1987) argue that a very abrupt transition from low-concentration CCA to high-concentration CCA (HCCCA) occurs when the average mutual distance between the clusters becomes of the order of the cluster dimensions themselves. When ϕ_f approaches unity, very compact clusters are formed with large fractal dimensions up to $n_f \approx 3$. This implies that on a micro-scale loose aggregates with $n_f \approx 2$ become densely packed at $n_f \approx 3$. On a macro-scale the overall fractal dimension attains a value somewhere in between, that is $n_f \approx 2.6 - 2.8$. Similar conclusions are drawn by Haw et al. (1997) from a theoretical study on cluster-cluster aggregation near the gelling point (see also Chapter 7).

4.3   FLOCCULATION MODEL

4.3.1 NUMBER EQUATION FOR FLOCS

In this section we derive an equation for the evaluation of floc size to determine the time and spatial variations of settling velocity. We restrict ourselves to the effect of turbulent shear on the flocculation process of cohesive sediment. All secondary effects, such as the influence of the particles on the turbulence structure itself, are omitted. Only the influence of the suspension on the buoyancy effects in the fluid is accounted for. It is assumed that the particles are advected with the large scale turbulent movements of the fluid, which is correct even for fairly large sand particles with a diameter up to $D_f = 200$ µm, as discussed in Chapter 6.

Let us consider a control volume V at a fixed location in a Eulerian co-ordinate system which is enveloped by a closed surface A. At time t this control volume contains fluid with N particles per unit volume. The total number of particles within this volume is given by $\iiint N dV$. The total number of particles entering or leaving this control volume is given by $\oiint N \vec{v} \cdot \vec{n} dA$, where $\vec{n}$ is the unit vector normal to A, positive outward, and $\vec{v}$ the velocity of the particles.

Next, we introduce a flocculation function F_N describing the rate of change of the number of particles per unit volume as a result of turbulence-induced aggregation and floc break-up processes; F_N is specified in the following sections. Then the rate of change of the number of particles in the control volume V is given by:

$$\frac{d}{dt} \iiint N dV = -\oiint N \vec{v} \cdot \vec{n} dA + \iiint F_N dV \qquad (4.9)$$

Using Gauss' theorem, the first integral on the right hand side of (4.9) can be transformed into a volume integral. As V is fixed in space we obtain:

$$\iiint \left(\frac{\partial N}{\partial t} + \nabla \cdot \vec{v} N - F_N \right) dV = 0 \qquad (4.10)$$

This relation should be valid for all choices of the control volume V, which is only possible if the integrand is identically zero everywhere in the fluid (Batchelor, 1983). The convective velocity $\vec{v}$ consists of a fluid-induced part and the settling velocity of the particles induced by gravity, including the effects of hindered settling (e.g. Section 5.3). It is convenient for our further analyses to rewrite the result in Einstein notation:

$$\frac{\partial N}{\partial t} + \frac{\partial}{\partial x_i} \left((u_i - \delta_{i3} w_s) N \right) = F_N \qquad (4.11)$$

in which δ_{i3} is the Kronecker delta. This equation differs from the dynamic equation (for the number concentration of dust particles) given by Friedlander (1977) only in the settling and flocculation terms.

From (4.7) we find a relation between N and D_f:

$$N = \frac{1}{f_s} \frac{c}{\rho_s} D_p^{n_f - 3} D_f^{-n_f}$$

(4.12)

where f_s is a shape factor (for spherical particles $f_s = \pi/6$), and c is the suspended sediment concentration by mass. Substitution of (4.12) into (4.11) yields:

$$\frac{\partial}{\partial t}\left(cD_f^{-n_f}\right) + \frac{\partial}{\partial x_i}\left((u_i - \delta_{i3}w_s)cD_f^{-n_f}\right) = f_s\rho_s D_p^{3-n_f} F_N$$

(4.13)

In case we have no flocculation ($F_N = 0$, hence D_f = constant), (4.13) is identical to the mass balance for suspended sediment (2.11) prior to averaging. If we multiply (4.13) with $D_f^{n_f+1}$, and subtract the mass balance equation, we obtain an equation for the evolution of the floc size D_f:

$$\frac{\partial D_f}{\partial t} + \left(u_i - \delta_{i3}w_s\right)\frac{\partial D_f}{\partial x_i} = -f_s \frac{\rho_s}{c} D_p^{3-n_f} D_f^{n_f+1} F_N = -\frac{F_N}{N}D_f$$

(4.14)

in which N/F_N can be considered as a time scale for flocculation (see also equ. (4.11)). Again, considering the case without flocculation ($F_N = 0$), we observe that (4.14) then indicates that the material derivative of D_f is identically zero, i.e. the particle size remains constant along particle trajectories.

Next we apply the classical Reynolds averaging procedure to the turbulent quantities in (4.11). We define $N = \overline{N} + N'$, $c = \overline{c} + c'$, $D_f = \overline{D}_f + D'_f$, $u_i = \overline{u}_i + u'_i$, and $w_s = \overline{w}_s + w'_s$. The last decomposition is necessary, as w_s is a function of D_f, hence of N. The description of the flocculation function F_N is highly empirical and therefore we omit its turbulent contribution. Substitution and averaging over the turbulent time scale, applying Fick's law, and assuming isotropic turbulence, yields an equation for the number of mud flocs in a turbulent fluid:

$$\frac{\partial N}{\partial t} + \frac{\partial}{\partial x_i}\left((u_i - \delta_{i3}w_s)N\right) - \frac{\partial}{\partial x_i}\left((D_s + \Gamma_T)\frac{\partial N}{\partial x_i}\right) +$$
$$+ \frac{\partial}{\partial x_i}\left(\delta_{i3}\overline{w_s'N'}\right) = F_N \tag{4.15}$$

in which Γ_T is the turbulent eddy diffusivity, defined earlier in Section 2.3. We have omitted the overbars for convenience. Note that in our derivation we have obtained the additional turbulent flux term $\overline{w_s'N'}$, which stems from the turbulent fluctuations in the settling velocity. A similar term is present in Chapter 2 in the mass balance, where we have reasoned that it can be omitted as $w_s' \ll w'$.

As a result of the turbulent random motions, the particles will collide to form flocs. It was shown by Smoluchowski (1917) that these collisions can be described as a diffusion process. This conclusion is generally accepted, and all aggregation models presented in the literature are based on this concept. It is therefore only briefly summarised in Section 4.3.2.

Turbulence generates large deformations of the fluid; the resulting turbulent stresses can disrupt the flocs (Hinze, 1975). Several studies have been published on floc break-up processes, e.g. Argamam and Kaufman (1970); Parker et al. (1972); Tambo and Huzumi (1979); Sontag and Russel (1987); and West et al. (1994). Most of the formulations given by these authors are not very practical and/or contain a number of unmeasurable coefficients; neither is there agreement on which of these formulations is most suitable. Therefore, a new, simple break-up function is proposed in Section 4.3.3. In our derivation of the flocculation function we will follow small, homogeneous clouds of particles on their trajectory through the fluid.

4.3.2 AGGREGATION PROCESSES

Levich (1962) discussed the aggregation process in turbulent flow of particles smaller than the Kolmogorov scale λ_0. At this scale viscous effects are dominant. In his derivation, Levich assumed that small eddies with size λ $(\lambda \geq \lambda_0)$ bring these particles together. Integrating the Smoluchowski diffusion equation over a finite volume $\gg \lambda_0$ yields the rate of coagulation between the particles in a turbulent fluid (e.g. McCave, 1984):

$$\frac{dN}{dt} = -\frac{3}{2}(1-\phi_*)e_c \pi e_d GD_f^3 N^2 \tag{4.16}$$

where t is time, e_c an efficiency parameter accounting for the fact that not all collisions will result in coagulation, and e_d is an (efficiency) parameter for diffusion. The efficiency parameter e_c is a function of the physico-chemical properties of the sediment and the water, of the organic compounds (coatings, polysaccharides, etc.) in the sediment, the silt-clay ratio, and of the form and structure of the flocs (Gregory, 1997). Hence e_c is therefore basically an empirical parameter, see Van Leussen (1994) and McAnally (1999), for example.

We have heuristically inserted the factor $(1-\phi_*)$ to guarantee that the aggregation processes will never result in volumetric concentrations beyond unity, in which we defined $\phi_* = \min\{1, \phi_f\}$. From a conceptual point of view, the insertion of this factor is consistent with the observations that at high concentrations the free path length of the particles is reduced, resulting in a decrease in effective diffusion at the scale of the floc forming eddies.

Equ. (4.16), except for the $(1-\phi_*)$-factor, is used in most studies on flocculation mentioned in the literature. Moreover, Vicsek (1992) shows that (4.16) is valid for the Reaction Limited Cluster-Cluster Aggregation processes, described in Section 4.2, which are believed to govern the aggregation of cohesive sediment in marine environments.

McAnally (1999) also evaluated the contribution of multiple collisions. He found that the frequency of three-body collisions in turbulent flow is not larger than 2 % of the frequency of two-body collisions for floc diameters up to 1 mm and suspended sediment concentrations up to 10 g/l. For smaller flocs at lower concentrations, this frequency drops rapidly. The frequency of four-body is even much lower. More-body collisions are therefore not considered here.

4.3.3 FLOC BREAK-UP PROCESSES

The effects of floc break-up processes by inter-particle collisions are omitted as these effects will be small for flocs with a small excess density, as dealt with here. Only break-up by turbulence-induced stresses is analysed; this break-up process acts continuously in a turbulent shear flow, contrary to for instance break-up by collisions. Turbulence exerts fluctuating shear stresses on the flocs. It is therefore likely that the break-up rate of a floc will be a function of the ratio of the turbulence induced stresses τ_T and the strength of the floc τ_y. The flocs are considered as self-similar fractal entities, the smallest units being formed by the primary particles (flocculi). The number of sub-flocs that can be generated by the break-up processes is therefore dependent on the size of the flocs D_f in relation to the size of the primary particles D_p. Finally, the break-up rate will be a function of the time scale T_ε of the disrupting turbulent eddies. On the basis of dimensional considerations, the break-up process should read:

$$\frac{1}{N}\frac{dN}{dt} \propto \frac{1}{T_\varepsilon}F\left(\frac{D_f}{D_p},\frac{\tau_T}{\tau_y}\right) \tag{4.17}$$

For simplicity we assume that the function F can be written as a power law:

$$F = a\left(\frac{D_f - D_p}{D_p}\right)^p\left(\frac{\tau_T}{\tau_y}\right)^q \tag{4.18}$$

in which the coefficient a and the exponents p and q are to be determined from experimental data.

We are considering flocs with the dimensions of the Kolmogorov length scale or smaller. The disruptive stresses that need to be considered are therefore in the viscous regime. As $Re_{\lambda_0} \approx 1$, eddies of scale $\lambda \approx \lambda_0$ have a velocity of about (e.g. Levich, 1962):

$$u_{\lambda_0} \approx \frac{\nu}{\lambda_0} \tag{4.19}$$

where ν is the kinematic viscosity of the fluid. The time scale for eddies of the Kolmogorov scale can be approximated by:

$$T_{\lambda_0} \approx \frac{\lambda_0}{u_{\lambda_0}} \approx \frac{\lambda_0^2}{v} \equiv \frac{1}{G} \tag{4.20}$$

According to Levich, turbulent eddy velocities scale with the eddy size λ:

$$u_\lambda \approx \frac{u_{\lambda_0}}{\lambda_0} \lambda \tag{4.21}$$

in which u_{λ_0} is the eddy velocity at scale $\lambda = \lambda_0$. This agrees with the normalised spectrum of the turbulent energy E at high wave numbers k around the Kolmogorov scale: E scales with k^{-3}, hence u with k^{-1}, thus with λ, as shown in Fig. 2.4. As a consequence, turbulent stresses in the viscous regime at scale $\lambda = D_f$ are given by:

$$\tau_T = \mu \frac{\partial u}{\partial z} \approx \mu \frac{u_{\lambda_0} D_f / \lambda_0}{D_f} \approx \mu G \tag{4.22}$$

in which μ is the dynamic viscosity of the fluid.

The yield strength of flocs F_y is determined by the number of bonds in a plane of failure, which, because of the fractal structure of the flocs, is independent of the size of the aggregate (e.g. Kranenburg, 1994). Hence F_y is constant, as a result of which τ_y scales with $1/D_f^2$. The actual value of F_y is determined by the physico-chemical properties of the sediment and the water, and can be related to the ζ-potential of the water-sediment suspension; such relations are highly empirical though, and no explicit formulations are known at present.

Finally, the time scale for the eddies at $\lambda = D_f$, at which scale the relation $T_\varepsilon = T_D$ is valid, maintains the value of equ. (4.20), hence, $T_\varepsilon = 1/G$. By substitution into equ. (4.17) we obtain the growth rate of particle number concentration N by floc break-up:

$$\frac{dN}{dt} = ae_b G \left(\frac{D_f - D_p}{D_p} \right)^p \left(\frac{\mu G}{F_y/D_f^2} \right)^q N \qquad (4.23)$$

in which e_b is an efficiency parameter for the break-up process.

4.3.4 EULERIAN FLOCCULATION MODEL

A complete flocculation model should include both aggregation and floc break-up processes. These two processes are the result of two different manifestations of the turbulent flow field: aggregation is the result of diffusion induced by the turbulence, whereas break-up is the result of turbulence-induced shear. However, the formulations of these aggregation and floc break-up processes still contain uncertainties, and the empirical parameters involved are poorly known. Therefore, a simple linear combination of (4.16) and (4.23) is proposed, thus assuming simultaneous aggregation and floc break-up but without mutual interaction. The differential equation for the flocculation of cohesive sediment under the influence of turbulent shear then becomes:

$$\frac{\partial N}{\partial t} + \frac{\partial}{\partial x_i}\left((u_i - \delta_{i3} w_s) N \right) - \frac{\partial}{\partial x_i}\left((D_s + \Gamma_T) \frac{\partial N}{\partial x_i} \right) =$$
$$= -k_A'(1 - \phi_*) G D_f^3 N^2 + k_B G^{q+1} (D_f - D_p)^p D_f^{2q} N \qquad (4.24)$$

in which we define the parameters k_A' and k_B as follows:

$$k_A' = \frac{3}{2} e_c \pi e_d , \quad \text{and}$$
$$k_B = ae_b D_p^{-p} \left(\frac{\mu}{F_y} \right)^q \qquad (4.25)$$

When solving (4.24) we assume that no particles leave or enter the water column through the water surface. At the bed, flocs may enter or leave the computational domain through erosion or deposition processes. These processes are discussed in detail in the Chapters 5 and 9. Yet, for convenience, we present the bed boundary conditions for the flocculation model here.

We assume that the flocs that leave the domain through deposition have the local floc size described by (4.24). We further assume that the particles that enter the domain will have a grain size equal to the local equilibrium conditions determined by the local bed shear stress (i.e. the local value of the shear rate parameter G at that time). This implies that the mass of sediment brought into the water column through erosion is determined by the number of particles N, their volume V_f and density ρ_f, in which V_f and ρ_f are determined by equilibrium conditions[2].

From (4.7) we find:

$$V_{f,e} = f_s D_{f,e}^3 , \quad \text{and}$$

$$\rho_{f,e} = \rho_w + (\rho_s - \rho_w) \left[\frac{D_p}{D_{f,e}} \right]^{3-n_f} \qquad (4.26)$$

where $D_{f,e}$ is the equilibrium floc size (e.g. Section 4.3.5). The aggregation and floc break-up terms are set to zero at the water surface and at the horizontal water-bed interface. Hence the boundary conditions read:

$$\left\{ (u_3 - w_s)N - (D_s + \Gamma_T)\frac{\partial N}{\partial x_3} \right\} \Bigg|_{x_3 = Z_s} = 0$$

$$\left\{ (u_3 - w_s)N - (D_s + \Gamma_T)\frac{\partial N}{\partial x_3} \right\} \Bigg|_{x_3 = Z_b} = \theta_{b,N} \qquad (4.27)$$

The source-sink term $\theta_{b,N}$ represents the exchange with the bed and is modelled with a simple erosion formulae, so that the water-bed exchange formulation is consistent with the one for the mass balance:

---

[2] We use equilibrium parameters for the special case $n_f = 2$ for convenience, see equ. (4.32).

$$\theta_{b,N} = -w_s N + \frac{E_b}{f_s D_e^3 \rho_{f,e}} \qquad (4.28)$$

in which the bed erosion rate E_b [kg/m^2/s] is elaborated in Chapter 9.

Our flocculation model consists of the coupled set of equations (2.11) for the mass balance, (4.24) and (4.25) for the number of particles, the geometrical relation (4.12), and the boundary conditions (4.27) and (4.28), together with the relations (4.26) and (4.32 - see below). A one-dimensional form of this flocculation model is obtained by neglecting the horizontal gradients in the various equations, which is implemented in a one-dimensional vertical (1DV) numerical model - see Winterwerp (2001, 2002) for details.

For the initial conditions in numerical simulations we propose to apply the conditions for equilibrium for $n_f = 2$. Given an initial flow field and sediment concentration (by mass), the equilibrium floc size can be established with (4.32) below. Formula (4.12) then gives the initial number of particles N_0.

The proposed model has been derived for mono-dispersive cohesive sediment suspensions; the predicted floc size can be regarded to represent the median floc size. In reality, cohesive sediment suspensions consist of a distribution of floc sizes (e.g. Migniot, 1968). A number of conceptual models for floc populations have been presented in literature (e.g. Jeffrey, 1982; McCave, 1984; and Spicer and Pratsinis, 1996). Only McAnally (McAnally, 1999 and McAnally and Mehta 2002) implemented a population model to account for flocculation between various classes. However, the kinetics of multi-dispersive suspensions are by no means trivial. For instance, which flocs can aggregate with each other – do only flocs of similar size join? And do flocs break-up into primary particles, in more or less identical sub-flocs, as suggested by Krone's order of aggregation theory, or completely at random?

Another uncertainty is the flocculation process under short waves. Turbulence under wave action is limited to the wave-induced boundary layer, as discussed in Section 2.6. However, this layer is very thin, i.e. a several cm to a few dm at most. So, are short waves important in the floc-forming processes? Hill et al. (2001) and Agrawal and Traykovski (2001), for instance, measured floc break up during wave activity on North Atlantic Continental Shelf off Massachusetts, and found that the floc size decreased by a factor 3.

At present, we cannot answer these questions, and further studies are required. In particular, we will need data on the kinetics of multi-dispersive suspensions, and on flocculation processes under wave action.

4.3.5 LAGRANGEAN FLOCCULATION MODEL

First, we study the behaviour of a simplified version of the flocculation model. We assume small volumetric concentration ($\phi_f \ll 1$) and apply (4.12). In the case where we have no net advective or turbulent transport (i.e. homogeneous turbulence), as for instance in a suspension in still water, the sediment concentration is constant, and (4.24) reduces to:

$$\frac{\mathrm{d}D_f}{\mathrm{d}t} = k_A c G D_f^{4-n_f} - k_B G^{q+1}\left(D_f - D_p\right)^p D_f^{2q+1} \tag{4.29}$$

in which the dimensional aggregation parameter k_A [m^2/kg] and floc break-up parameter k_B [s$^{1/2}$/m^2] are defined as:

$$k_A = k_A' \frac{D_p^{n_f-3}}{n_f f_s \rho_s} \tag{4.30a}$$

$$k_B = \frac{ae_b}{n_f} D_p^{-p} \left(\frac{\mu}{F_y}\right)^q = k_B' \frac{D_p^{-p}}{n_f} \left(\frac{\mu}{F_y}\right)^q \tag{4.30b}$$

where k_A' and k_B' are non-dimensional coefficients. This is the Lagrangean flocculation model elaborated in Winterwerp (1998). For equilibrium conditions, i.e. $\mathrm{d}D_f/\mathrm{d}t = 0$, an equilibrium floc size $D_{f,e}$ can be defined. The exponents p and q can be obtained from empirical data available in literature. In many experimental studies it was found that D_e is proportional to λ_0 (e.g. Bratby (1980); Akers et al. (1987); Leentvaar and Rebhun (1983); Van Leussen (1994); and more recently Berhane et al., 1997)). This can be explained from the turbulent energy spectrum: at length scales exceeding λ_0, the turbulent stresses increase rapidly, hence floc break-up becomes dominant.

For our analysis we need to introduce a relation between the settling velocity of mud flocs and their size. Because of the legibility of the text, we will present this relation only in Section 5.2 of the next chapter. Here we will use the results, and the reader is referred to that section for details and rationale.

Fig. 5.3 shows the variation of the settling velocity measured by Owen as published by Thorn (1981) and Mehta (1986), by Ross (1988) and by Wolanski et al. (1992). Here we concentrate on the left branch of the curve, representing data affected by flocculation effects. The scatter is large, and it is difficult to establish an accurate relation from these data. For the present analysis we assume that the data given represent variations around equilibrium values of $w_{s,e}$, and that $w_{s,e} \propto c$ (e.g. equ. (5.3)). We note that the measured w_s - c relation presented in Fig. 5.3 may be affected by spurious correlations: large values of c in general occur at large current velocity. At such high velocities larger sediment particles (with a larger settling velocity) may be mobilised. This effect will particularly take place in shallow and intertidal areas with a pronounced spring-neap tidal cycle.

By substituting these additional assumptions into (4.40) we find $q = 1/2$ and $p = 3 - n_f$, see Winterwerp (1998) for details. If we further set the fractal dimension at the average value for mud flocs in the water column at $n_f = 2$, we obtain a simple flocculation equation that can be used to study the behaviour of the flocculation model analytically:

$$\frac{dD_f}{dt} = k_A c G D_f^2 - k_B G^{3/2} D_f^2 \left(D_f - D_p \right) \tag{4.31}$$

For small values of D_f, the first term on the right hand side of (4.31), e.g. the aggregation term dominates, whereas for large D_f the second term, e.g. the break-up term dominates. From this flocculation equation an equilibrium floc size $D_{f,e}$ is obtained for $dD_f/dt = 0$. A mathematically trivial, but physically unsound and unstable solution is $D_{f,e} = 0$. In that case small particles always grow. The other equilibrium solution for the simplified model reads:

$$D_{f,e} = D_p + \frac{k_A c}{k_B \sqrt{G}} \quad \left(= D_p + \frac{k_A c}{k_B} \frac{\lambda_0}{\sqrt{v}} \right) \tag{4.32}$$

which is a stable equilibrium, as for $D_f < D_{f,e}$ the flocs grow and for $D_f > D_{f,e}$ the flocs break up.

For the chosen fractal dimension $n_f = 2$, the differential equation (4.31) can easily be solved analytically, if the sediment concentration by mass c is constant. First a time scale parameter T' is defined:

$$T' = \left(k_B G^{3/2} D_e^2\right)^{-1} \tag{4.33}$$

The solution of equ. (4.31) then is of the following implicit form:

$$t = T'\left[\ln\left(\frac{D_{f,e} - D_0}{D_{f,e} - D_f}\frac{D_f}{D_0}\right) + \frac{D_{f,e}}{D_0} - \frac{D_{f,e}}{D_f}\right] \tag{4.34}$$

in which D_0 is the floc size at $t = 0$. This solution describes the aggregation/floc-break-up process for flocs initially either smaller or larger than the flocs at equilibrium size.

From (4.34) a time constant for flocculation T_f can be defined in the case that the initial floc size D_0 is much smaller or much larger than the equilibrium value, yielding the maximum time scales of the aggregation and floc break-up processes:

$$\text{for } D_{f,e} \gg D_0 \quad T_f \approx T' D_{f,e}/D_0 \approx \frac{1}{k_A c G D_0}$$

$$\text{for } D_{f,e} \ll D_0 \quad T_f \approx 2T' \approx \frac{2k_B}{k_A^2 c^2 \sqrt{G}} \tag{4.35}$$

$$\text{hence:} \quad T_f \propto \left(\frac{U^3}{h}\right)^{-n} c^{-m} \quad \text{with} \quad \frac{1}{4} \le n \le \frac{1}{2}, \quad 1 \le m \le 2$$

The effects of the flocculation time on the flocculation process are further elaborated in Section 4.4.

The flocculation model contains a flocculation parameter k_A and a floc break-up parameter k_B, which have to be determined empirically. This has been done at present for one set of data only, i.e. on the basis of flocculation experiments carried out in the settling column of Delft Hydraulics (see Van Leussen, 1994). This column is a 4.25 m high perspex cylinder with an inner diameter of 0.29 m. Within the column a grid can be oscillated at various frequencies and amplitudes to generate a homogeneous turbulence field. For details on the calibration, the reader is referred to Winterwerp (1998, 1999, 2002).

Table 4.1: Parameters of flocculation model with $n_f = 2$,
after Winterwerp (1998).

α'	k_A	k_A'	k_B	ae_b	$\sqrt{\mu/F_y}$
[-]	[m^2/kg]	[-]	[s$^{1/2}$/m^2]	[-]	[s$^{1/2}$/m]
1/18	14.6	0.31	14.0 10^3	2 10^{-5}	5.6 10^3

From this analysis the parameter values presented in Table 4.1 were found. It is stressed that these parameters have been obtained for cohesive sediment from the Ems estuary in The Netherlands, and that no other data are available at present. On the basis of our analysis in Chapter 3, one would expect though, that these parameters are not universally applicable. However, as no other data are available, their values will be used anyway in the remainder of this chapter as an illustration without claiming general applicability, when the flocculation model is applied.

From the parameter values obtained from the column tests, the non-dimensional coefficient k_A' can be determined, and k_B' can be estimated. For $\rho_s = 2650$ kg/m^3 and $D_p = 4$ μm, we find $k_A' \approx 0.31$ and $e_c = O\{0.01\}$, as $e_d \approx 0.5 - 1$ (e.g. Levich, 1962). This agrees fairly well with the values reported by O'Melia (1985), given in Section 4.2.

For the assessment of k_B', we need to know the floc strength F_y. Very little information about this strength is available, but from some literature (e.g. Van Leussen, 1994 and Matsuo and Unno, 1981) F_y is estimated at $F_y \approx O\{10^{-10}\}$ N (though this can vary by several orders of magnitude), as a result of which k_B' is estimated at $k_B' = O\{10^{-5}\}$. Apparently, the efficiencies of both the flocculation and floc break-up processes are very small, the latter being even much smaller than the former. This was also concluded by Van Leussen (1994) on the basis of his estimates of the time scale for floc break-up.

4.3.6 COMPARISON WITH FIELD DATA

We have applied the flocculation model to compute the floc size evolution in the Ems-Dollard estuary, using the model parameters obtained in the settling column, as presented in Table 4.1. For details of these computations, the reader is referred to Section 6.3.2 or Winterwerp (1999, 2002). The results are presented in Fig. 4.5, showing considerable variation of the floc size over the tidal period and over the water column, with lower values of about 100 μm and maximum values of about 800 μm. This is the result of the variation of G and c

over the tidal period and water depth, which is accounted for in the proposed aggregation and floc break-up formulations. The computed floc sizes presented in Fig. 4.5 should be considered as a characteristic floc size of the suspension, something like the median floc size. A further analysis of these results is given in Section 5.3.2.

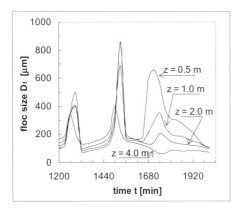 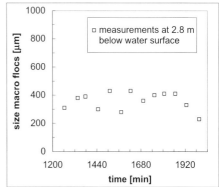

Fig. 4.5: Computed evolution of floc size distribution as a function of time for the Ems-Dolard estuary.

Fig. 4.6: Measured size of macro flocs in Ems-Dollard estuary (after Van Leussen, 1986).

Van Leussen (1986) measured the floc size distribution in the Ems-Dollard estuary by means of an optical system. He concentrated on the larger flocs, referred to as macro flocs. It appeared that the size of the macro flocs did not vary much over the tidal period, as depicted in Fig. 4.6, showing sizes between about 200 and 400 μm. Note that these values are in the range of size of the larger flocs computed with the flocculation model.

Floc size distributions measured around High Water Slack and Maximum Ebb Velocity are presented in Fig. 4.7 (see also Fig. 6.17). These results suggest a median floc size of about 100 to 150 μm at the measuring times, which corresponds well with the computed results (e.g. Fig. 4.5). The data also suggest a small increase in median floc size from HWS to MEV, which is again consistent with the computational results.

From this brief analysis we can conclude that the spatial and time variation of floc size can be significant. This is reflected directly in the settling velocity distribution, and has important implications for the transport and fate of cohesive sediment in the marine environment, as discussed in Section 6.3.2.

This conclusion also implies that the flocculation time becomes an important parameter, relating the time scale of the flocculation process to the time scale of the driving forces. This is further elaborated in the next sections.

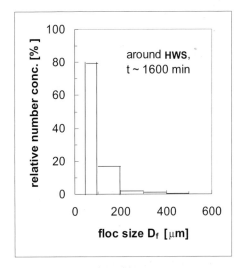 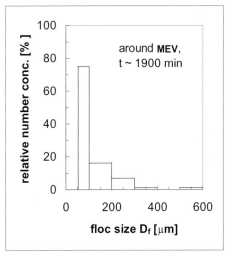

Fig. 4.7: Measured floc size distribution in Ems-Dollard estuary (after Van Leussen, 1986).

4.4   FLOCCULATION TIME

4.4.1   THE GELLING CONCENTRATION

When mud flocs settle on the bed, they may form a space-filling network structure, called a gel, and a measurable strength builds up (e.g. Chapter 8). The concentration at which this happens is referred to in this book as the gelling point. The volumetric concentration ϕ_f is then assumed to become unity, see equ. (4.8). If we assume that the flocculation model is also valid for these conditions, we can establish the gelling concentration c_{gel} for equilibrium conditions ($dD_f/dt = 0$) as a function of the shear rate parameter G for various values of the fractal dimension n_f from the geometrical relation (4.8) and the simplified flocculation model (4.29) proposed in this chapter. We assume $D_{f,e} \propto \lambda_0$ and $w_{s,e} \propto c$, hence $q = 1/2$ and $p = n_f - 3$. We further assume that $D_p \ll D_f$ (e.g. equ. (4.32)) and that the sediment concentration is constant, yielding:

$$c_{gel} = \left(\frac{ae_b}{k_A'} \sqrt{\frac{\mu}{F_y}} \right)^{\frac{3-n_f}{4-n_f}} D_p^{\frac{3-n_f}{4-n_f}} \rho_s G^{\left(\frac{3-n_f}{2(4-n_f)}\right)} \qquad\qquad (4.36)$$

Using the various model parameters assessed in Section 4.3, $c_{gel}(G,n_f)$ can be established for equilibrium conditions. The results are presented in Fig. 4.8, showing the high sensitivity of c_{gel} to n_f at larger values of n_f. For $n_f = 3$, we obtain the trivial result that $c_{gel} = \rho_s$. We observe that for $n_f = 2.6$ to 2.7, we find $c_{gel} \approx 100$ g/l, which is a characteristic concentration for fluid mud occurrences (e.g. Section 3.4.3) - see also Fig. 4.4 for typical values of G in open-channel flow.

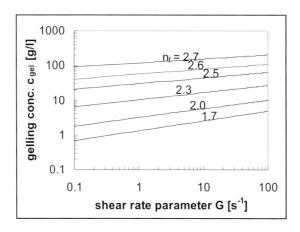

Fig. 4.8: Variation of gelling concentration c_{gel} with n_f and G for equilibrium conditions.

It appears not possible to obtain c_{gel}-values of the order of 100 g/l by modifying the other parameters of equ. (4.36) within reasonable limits. Hence, by applying the flocculation model for equilibrium conditions, fluid-mud concentrations of about 100 g/l can only be obtained with fractal dimensions substantially larger than the average n_f-values encountered in the mud flocs in the water column. This is an important result, which is caused by the response time of the flocs to a variation in hydrodynamic conditions. This is further elaborated in the next section, and through a computational example in Section 6.3.2.

It is also noted that (4.36) yields unrealistically large values for c_{gel} at small G, because $D_{f,e}$ becomes unrealistically large. Such large flocs would collapse, as their yield strength decreases with D_f^2, a process not accounted for in the present model.

4.4.2 NON-EQUILIBRIUM CONDITIONS

From our analysis of the Lagrangean flocculation model in Section 4.3.5, we obtained the time scales for flocculation (e.g. equ. (4.35)). The flocculation time can vary between a few minutes to many days, and even more, depending on the hydrodynamic conditions and the suspended sediment concentration. As a result, flocs of cohesive sediment, observed in marine environments, are often not in equilibrium with the local hydrodynamic conditions.

Let us first compute the time evolution of floc size, as predicted with the Lagrangean flocculation model (4.31). The results are shown in Fig. 4.9, together with data obtained by Van Leussen (1994) in a settling column (for details, see Winterwerp, 1998). We observe that for two cases, equilibrium conditions are obtained within a few hundred seconds. However, for one situation, the flocculation time exceeds one hour.

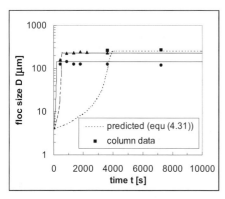

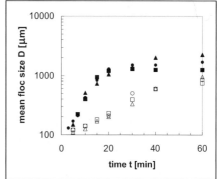

Fig. 4.9: Floc growth as a function of time (after Winterwerp, 1998).

Fig. 4.10: Floc growth measured by Hogg et al. (1985).

Fig. 4.10 presents the time evolution of flocs, as measured by Hogg et al. (1985) on a suspension of kaolinite clay, peptised with a polymeric flocculant. The clay concentration was about 80 g/l, and flocculation conditions are extremely favourable through the addition of the flocculant. Yet, the flocculation time was still about one hour.

Also McAnally (1999) found flocculation times of many hours, depending on the turbulence intensity and sediment concentration.

These observations have important implications for the flocculation in marine conditions. Flocs settle continuously due to gravity and are remixed by turbulence. Hence, they experience varying hydrodynamic conditions (shear rates) during their journey through the water column. This implies that the ratio between flocculation time and residence time in a specific turbulent environment becomes an important parameter.

This can be quantified by considering a situation where the turbulence field is homogeneous over the water depth, as can be realised in a settling column, for example. The mean residence time T_r for all particles in the water column with initial height Z_0 above the bottom of the water column can be obtained from:

$$\int_0^{T_r} w_s \, dt = \alpha'' \int_0^{T_r} D_f \, dt = Z_0 \tag{4.37}$$

in which w_s is the settling velocity and $\alpha'' = \alpha' D_p g \Delta / v \, (\approx 3 \text{ s}^{-1} \text{ for } D_p = 4 \text{ } \mu\text{m})$ (e.g. Section 5.2). The residence time T_r can be solved directly from this equation using the simplified flocculation model (4.31) to yield:

$$T_r = \frac{Z_0}{\alpha'' D_{f,e}} + \frac{D_f - D_0}{k_A c G D_f D_0} = \frac{Z_0}{w_{s,e}} + T' \left[\frac{D_{f,e}}{D_0} - \frac{D_{f,e}}{D_f} \right] \tag{4.38}$$

in which $w_{s,e}$ is the equilibrium settling velocity. The maximum floc size, as a function of a limited residence time, follows from (4.37) and (4.38) ($t = T_r$):

$$D_{f,max} = \frac{D_{f,e}}{\left(\dfrac{D_{f,e}}{D_o} - 1 \right) \exp\left\{ -\dfrac{k_A c G Z_0}{\alpha''} \right\} + 1} \tag{4.39}$$

The maximum floc size predicted with equ. (4.39) is plotted in Fig. 4.11 for the model parameters of Table 4.2, a suspended sediment concentration of 1 g/l, and

Z_0 = 1, 2, 5 and 10 m, respectively. It is shown that for small values of G, the flocs cannot attain equilibrium conditions because of the limited residence time. If the concentration would decrease by a factor 10, the shear rate at which equilibrium would be possible, would increase by the same factor.

It is noted that this picture is qualitatively similar to the heuristic picture presented in Fig. 4.3. Moreover, similar results have been observed during the AMASEDS-campaign, as shown in Fig. 4.12. However, one has to be careful with a quantitative interpretation of Fig. 4.12, as the various data points were obtained at different levels in the water column (so different residence times), and at different suspended sediment concentration.

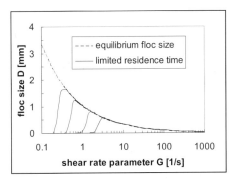 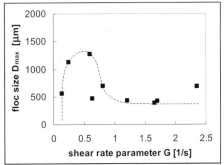

Fig. 4.11: Effects of limited residence time on the relation between floc size and shear rate (after Winterwerp, 1998).

Fig. 4.12: Relation between floc size and shear rate, measured by Berhane et al. (1997).

We can substantiate these observations further, and investigate when the classical picture of flocculation processes in the water column is correct. This picture was drawn by Van Leussen (1986) and is sketched in Fig. 4.13. It shows larger flocs higher in the water column, where turbulent shear is relatively small, and smaller flocs near the bed, where turbulent shear is high.

This picture can of course only hold when the flocculation time is small in comparison to the mixing and settling time of sediment. As the settling velocity of mud flocs is generally of the order of a few 0.1 mm/s, or smaller, the vertical mixing time is generally much smaller than the settling time. Moreover, the time scale for floc break-up is almost always smaller than the time scale for aggregation, as follows from equ. (4.35). Hence, we can compare aggregation time with settling time only, i.e. the time necessary to form larger flocs with size D_h higher in the water column through the aggregation of smaller flocs with size D_l lower in the water column, which are (assumed to be) in equilibrium with the hydrodynamic conditions near the bed. Also, we assume that the vertical gradient in suspended sediment concentration is small.

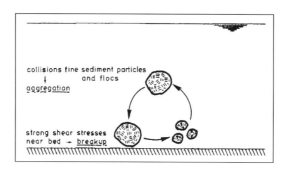

Fig. 4.13: Schematic picture of flocculation processes in the water column (after Van Leussen, 1986).

The flocculation time T_f for $D_h > D_l$ reads:

$$T_f \approx \frac{1}{k_A c G_u D_l} \tag{4.40}$$

where G_u is the mean value of the shear rate in the upper 25 % of the water column. We assume that D_l is the floc size in equilibrium with the shear rate G_l in the lower 25 % of the water column. G_u and G_l are found from averaging equ. (4.4):

$$G_u = \frac{1}{0.25} \int_{0.75}^{1} G \mathrm{d}\zeta = 0.36 \sqrt{\frac{u_*^3}{\kappa v h}} \quad \text{and} \tag{4.41a}$$

$$G_l = \frac{1}{0.25} \int_{0}^{0.25} G \mathrm{d}\zeta = 3.82 \sqrt{\frac{u_*^3}{\kappa v h}} \tag{4.41b}$$

and the equilibrium floc size in the lower part of the water column D_l follows from:

$$D_l = \frac{k_A c}{k_B \sqrt{G_l}} \tag{4.42}$$

The settling time of large flocs in the upper part of the water column T_s reads:

$$T_s = \frac{h}{W_s} = \frac{h}{\alpha'' D_u} \tag{4.43}$$

where D_u is the floc size in equilibrium with G_u, and α'' is defined in Section 5.2. If the relative flocculation time $T_f/T_s < 1$, we expect that aggregation can take place and that flocs higher in the water column are larger than those near the bed. Substitution from equ. (4.40) through (4.43) yields T_f/T_s:

$$\frac{T_f}{T_s} = \frac{\alpha''}{k_A ch} \frac{1}{G_u} \sqrt{\frac{G_l}{G_u}} \approx 0.0012 \sqrt{\frac{1}{hc^2 u_*^3}} \tag{4.44}$$

where we use the various parameter values obtained for Ems mud, e.g. Section 4.3.5. The relative flocculation time T_f/T_s is depicted in Fig. 4.14 for a variety of suspended sediment concentrations, showing under which conditions vertical gradients in floc size (i.e. $D_u \gg D_l$) are expected to occur.

This graph suggests that for estuarine conditions, with shear velocities of a few cm/s, vertical gradients in floc size do only occur for suspended sediment concentrations beyond a few 100 mg/l. Around slack water, suspended sediment concentrations should even be much larger for vertical gradients in floc size to occur (e.g. Winterwerp, 2002). It is noted that these results are obtained for the parameter settings obtained from laboratory experiments with Ems mud, and therefore cannot be considered to be universally valid. However, this analysis clearly indicates that vertical gradients in floc size, as depicted in Fig. 4.13, can only occur if the residence time of the flocs in the upper part of the water column is large enough.

If the flocculation time is large, the mean floc size throughout the water column can be estimated, assuming that the floc size is in equilibrium with near-bed hydrodynamic conditions (i.e. local G and c).

A limited residence time will also have implications on the floc size distribution. Fig. 4.15 presents a picture of the initial floc size distribution, i.e. prior to flocculation, and the floc size distribution for equilibrium conditions, as measured by Kranck (Kranck and Milligan, 1992). It shows a shift towards lager flocs in the equilibrium situation, as expected. However, if equilibrium conditions are not met, more or less any floc size distribution between the two curves of Fig. 4.15 may occur. Of course, this observation also holds for the median floc size. It is clear that this has major implications for the interpretation of floc size data obtained in the marine environment.

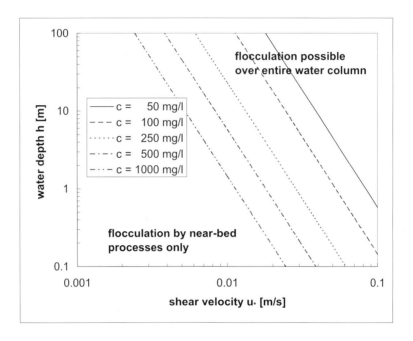

Fig. 4.14: Relative flocculation time of mud flocs in water column; the diagonal lines represent the condition $T_f = T_s$.

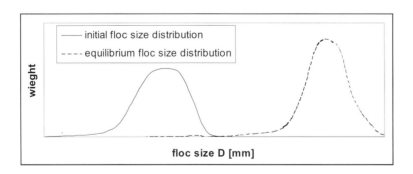

Fig. 4.15: Particle size distribution for initial and equilibrium conditions (after Kranck and Milligan, 1992).

A final remark on the implications of non-equilibrium conditions concerns the structure of the flocs. Fig. 4.16 presents the fractal dimension of flocs found in the Tamar estuary, as derived by Manning (e.g. Winterwerp et al., 2002). It is shown that the fractal dimension of flocs can be clustered in two groups. In group 1, prior to High Water, $n_f = 2 - 2.5$, whereas in group 2, after High Water, $n_f = 2.6 - 3.0$. It is hypothesised that the group 2 sediments contain many particles that were eroded from the bed, hence have a different structure than the particles of group 1 (e.g. Chapter 7), which have been residing in the water column for a longer time. The particles of group 2 apparently had not sufficient time to become adapted to their new flow environment.

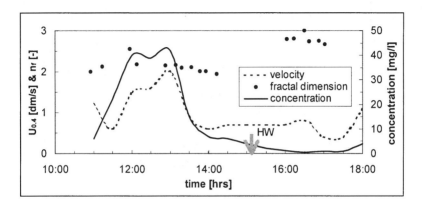

Fig. 4.16: Variation in fractal dimension measured by Manning in Tamar estuary (after Winterwerp et al., 2002).

5.   SETTLING AND SEDIMENTATION

Settling of mud flocs is one of the most important aspects in assessing the transport and fate of cohesive sediment suspensions in the marine environment. Yet, this settling velocity is difficult to determine. One cannot take samples from the suspension to bring to the laboratory measuring the settling velocity of the material under well-controlled conditions, as the flocs do not survive the sampling and/or transport. Therefore, the settling velocity has to be measured in-situ, for which purpose a variety of instruments has been developed. Eisma (1996) gives an excellent review of the instruments available at that time, their deployment, accuracy and shortcomings – see also Appendix C.2.

The literature contains an overwhelming amount of papers on settling velocity measurements (e.g. Whitehouse et al., 2000;Manning, 2001: and McAnally, 1999 for an overview). More recently, the commercially available LISST–system (e.g. Appendix C.2.1) is often used, and the literature already contains a large number of publications on the results obtained with this instrument.

The development in instrumentation is made possible through the developments in monitoring and data processing techniques, in particular optical methods as Particle Tracking and Particle Image Velocimetry. On the other hand, theoretical studies on the settling velocity are scarce, and a rigorous picture is still incomplete.

This chapter presents a relation between settling velocity and particle size for individual flocs in still water, and for flocs in high-concentrated suspensions, where hindered settling plays a role. Next, the sedimentation and the deposition rate of cohesive sediment in still and flowing water are treated. We distinguish between sedimentation and deposition as follows:

> *deposition* is defined as the gross flux of cohesive sediment flocs on the seabed.
> *sedimentation* is defined as the net increase in bed level (accretion or shoaling), i.e. the sedimentation rate is the deposition rate minus the erosion rate.

5.1   INTRODUCTION

Prior to diving into detailed theories of settling and deposition, one should appreciate that a uniquely defined settling velocity for cohesive sediment does not exist:

1. In general, suspended sediment is characterised by a distribution in particle size, hence settling velocity. In particular when the sediment is poorly sorted, the deposition rate is not simply the product of median settling velocity and suspended sediment concentration. In that case, the suspension should be treated in different fractions.

2. The particle size distribution, and its median size, may vary largely (sometimes by orders of magnitude!) in time and space as a result of flocculation (e.g. Chapter 4) and sorting processes.

3. The settling velocity is defined under the condition of equilibrium between the gravitational and drag forces on the settling particle. However, in the marine environment, the particles may not be able to follow the turbulent fluid motions in the eddies, violating the equilibrium assumption. This is the case for large particles only, as elaborated in Chapter 6.

4. Yet, it is often possible to define a characteristic settling velocity to describe the transport and fate of a cohesive sediment suspension satisfactorily. However, this characteristic settling velocity is a function of the scale of the problem to be addressed (near field, far field, seasonal variations, etc.).

5. Different practical (engineering) problems require different approaches, studies and/or modelling techniques, which often lead to different settling velocities. For instance, for water quality problems one is mainly interested in the finer fraction of the sediment to which contaminants adhere, whereas for siltation studies, the coarser, bulk fraction is of greater importance.

These considerations imply that no generally applicable recipe can be given to determine settling velocities. Even direct, in-situ measurement of the settling velocity may not necessarily yield the characteristic settling velocity suited to a particular practical problem, because:

- in-situ measurements are often biased towards the larger size class, as the larger particles are easier to detect,

- in-situ measurements are necessarily limited in their extent, and the measuring location and time frame may not be characteristic for the system.

These drawbacks do not imply that we are not in favour of in-situ measurements. On the contrary, we feel that such measurements are essential for learning and understanding the behaviour of cohesive sediment at a particular site.

A final remark concerns spurious correlations that may mislead our perception of the sediment behaviour, and the settling velocities deduced therefrom:

6. The relation between settling velocity and suspended sediment concentration. Most data sets show an increase in settling velocity with increasing sediment concentration. This is often caused by an increase in aggregation rate (see Section 4.7). However, suspended sediment concentrations tend to increase with increasing flow velocity because of erosion of the seabed. Such a larger flow velocity, however, can mobilise and suspend larger sediment particles, as a result of which the composition of the sediment in suspension changes. This, of course, also implies an increase in mean settling velocity w_s, but not as a result of flocculation processes. A proper modelling of this effect requires a description of the sediment in a number of fractions.
   At even higher suspended sediment concentrations the effective settling velocity decreases again because of hindered settling effects. These are treated extensively in this chapter.

7. Sediment-induced density currents and cloud formation. A cloud of sediment is not only subject to convective settling, but also to self-induced density currents. Because of continuity, advective transport of the cloud's front causes an equivalent lowering of the cloud's interface, which can wrongly be interpreted as the settling velocity of the mud flocs forming this cloud: the "settling velocity" established in this way may be an order of magnitude larger than the actual particle settling velocity. This process plays a role for instance during dumping of dredged sludge. Proper simulation of such activities requires a full description of the processes in the vertical (e.g. 2DV or 3D models), accounting for sediment-induced density currents.

8. Variations in measured vertical concentration profiles are caused both by local processes, i.e. settling and vertical mixing, and by horizontal advection, even in the case that measurements are carried out with a floating device. This implies that the determination of the settling velocity from vertical concentration profiles may be inaccurate, as horizontal flow velocities are always at least an order of magnitude larger than the vertical settling velocities.

In this chapter we try to circumvent these difficulties and restrict ourselves to the following definitions:

W_s	a characteristic settling velocity without further specification, constant in time and space
$w_{s,r}$	settling velocity of a single mud floc in still water; $w_{s,r}$ may vary in time and space as a result of flocculation processes
w_s	effective settling velocity of a particle in a suspension of cohesive sediment

Where appropriate, the reader should interpret these settling velocities as representative of one class of a specific floc size distribution in a cohesive sediment suspension.

5.2   SETTLING VELOCITY AND FLOC SIZE

5.2.1   SETTLING VELOCITY AND FLOC SIZE IN STILL WATER

The settling velocity of mud flocs is a function of their shape, size D_f and differential density $\Delta\rho_f$, i.e. the excess density relative to water. For estuarine sediment this is discussed by Dyer (1989), Gibbs (1985), Hawley (1982) and Van Leussen (1994), amongst others. Due to aggregation effects, flocs form with relatively small $\Delta\rho_f$; typical values for $\Delta\rho_f$ being in the order of 50 to 300 kg/m^3 (and sometimes even smaller), as discussed in Chapter 4.

The settling velocity $w_{s,r}$ of individual mud flocs can be obtained from a balance between the gravitational and drag forces F_g and F_d for a single floc in a homogeneous fluid (e.g. Batchelor, 1983):

$$F_g = \alpha \frac{\pi}{6} D_f^3 g \Delta\rho_f \quad \text{and} \quad F_d = \beta c_D \frac{1}{2} \rho_w \frac{\pi}{4} D_f^2 w_{s,r}^2 \qquad (5.1)$$

in which α and β are coefficients depending on the shape (sphericity) of the particles. According to Vanoni (1977), Raudkivi (1976) and Graf (1977) the formulation for the drag coefficient c_D, that adequately matches most empirical data on non-cohesive sediment for $Re_p < {\sim}800$, reads:

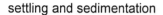

$$c_D = \frac{24}{Re_p}\left(1 + 0.15 Re_p^{0.687}\right) \qquad (5.2)$$

in which $Re_p = w_{s,r}D_p/\nu$ is the particle Reynolds number. We assume that this relation is also valid for flocs of cohesive sediment, as no data on the drag coefficient for falling flocs are available.

Implicitly we assume that the fluid flows around, and not through the flocs. This is, however, not a trivial assumption. Johnson et al. (1996) for instance studied the flow through flocs. They treated the flocs as self-similar fractal entities and reasoned that the size of the largest pores within the floc scales with the overall floc size. The absolute permeability of the flocs, hence the flow through the falling flocs, would therefore increase with increasing floc size. The effect of pore size was studied through a series of settling experiments with latex micro-spheres and numerical simulations. From these experiments the settling velocity and floc density, the latter obtained through counting the number of primary particles within each floc, were determined as a function of the floc diameter. The measured settling velocities were consistently larger by an order of magnitude than the settling velocity computed with Stokes' law of impermeable flocs of the same size and mass. Next the settling velocity of flocs of similar size and mass were computed, allowing flow through these flocs; their permeability was assumed to be homogeneously distributed within each floc. The settling velocity of the permeable flocs appeared slightly larger than that of the impermeable flocs. From these studies Johnson et al. concluded that the settling velocity of porous flocs is affected by the flow through the pores of the flocs, and that the pore size distribution is an important parameter. It is not clear, however, why the settling velocity would increase if fluid can flow through the pores. One would expect the opposite, as the effective friction-affected floc surface would increase with permeability.

Gregory (1997) pointed out that fractal aggregates are hardly permeable at fractal dimensions beyond $n_f = 2$. However, the permeability rapidly increases with decreasing n_f, affecting largely the collision frequency between the particles. This might have implications for the flocculation efficiency in the case of Diffusion Limited Cluster-Cluster Aggregation, which is hypothesised to be the governing aggregation process when the active organic content of the mud is high (see Section 4.2).

Moudgil and Vasudevan (1988) performed settling experiments with clay flocs in a settling column under normal conditions, the results of which were compared with the settling velocity of frozen flocs of the same size and mass, and under the same conditions. No significant differences in settling velocity were found, from which these researchers concluded that, during settling, no fluid flows through the flocs.

Also from the experiments by Stolzenbach and Elimelich (1994), who studied the effects of differential settling, one can conclude indirectly that flow through flocs will be small. If not, larger and smaller particles would collide more often, and differential settling would not be negligible (e.g. Gregory, 1997).

From these arguments we conclude that we may treat flocs as porous, but effectively impermeable entities as a fair approximation to determine their settling velocity in marine conditions.

For mud flocs with a fractal structure, relations (4.6) and (5.2) can be substituted into equ. (5.1), to yield an implicit formula for the settling velocity of single mud flocs in still water:

$$w_{s,r} = \frac{\alpha}{18\beta} \frac{(\rho_s - \rho_w)g}{\mu} D_p^{3-n_f} \frac{D_f^{n_f-1}}{1+0.15Re_f^{0.687}} \qquad (5.3)$$

For spherical ($\alpha = \beta = 1$), Euclidean ($n_f = 3$) particles in the Stokes' regime, for which $Re_f \ll 1$, the well-known Stokes' formula for a stationary settling particle is obtained:

$$w_{s,r} = \frac{(\rho_s - \rho_w)gD_f^2}{18\mu} \qquad (5.4)$$

It is emphasised that the water temperature has a significant effect on the viscosity, hence on the settling velocity of the mud flocs. However, we will not elaborate on this, as this effect is extensively treated in the literature.

Fig. 5.1 presents the settling velocity as established with equ. (5.3) with $\alpha = \beta = 1$, $D_p = 4$ μm (which is a typical value found by Van Leussen (1994) for Ems mud), $\rho_s = 2650$ kg/m^3, $\rho_w = 1020$ kg/m^3 and $\mu = 10^{-3}$ Pas, for three values of n_f, i.e. $n_f = 1.7$, 2 and 2.3, together with observations. These observations consist of data on settling velocity and floc size measured in the Delft Hydraulics settling column, in the Ems estuary and in the North Sea (Van Leussen, 1994; VIS in Fig. 5.1 means size and settling velocity measurements with a Video In Situ camera system), in Chesapeake Bay (Gibbs, 1985), and in the Tamar estuary (Fennessy et al., 1994). It is observed that equ. (5.3) matches the data adequately. This is also the case for data presented by Hawley (1982), who summarises the work of

Japanese researchers in the ocean (not shown here). The overall trend of the data points seems slightly steeper than the fit with $n_f = 2$, because no data are available in the lower right corner of the graph, i.e. small W_s and large D_f. However, when the individual data sets are studied, the slopes agree better with $n_f = 2$.

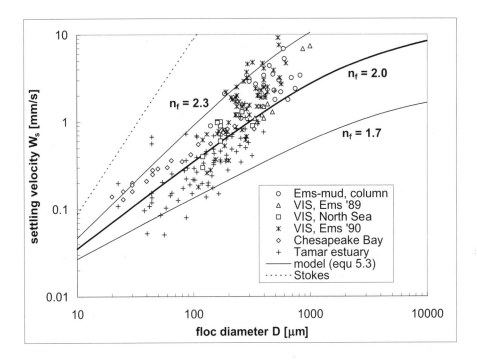

Fig. 5.1: Relation between settling velocity and floc size.

Figure 5.1 shows that for particles with a diameter up to a few 100 μm, the data can be properly represented with the fit $W_{s,r} \propto D_f$, i.e. with an average fractal dimension $n_f = 2$. At floc diameters beyond $D_f = 1$ mm, equ. (5.3) predicts a rapid deviation from a simple power law behaviour because of the increasing role of the particle Reynolds number. We can expect that the validity of (5.3) becomes limited at floc sizes beyond a few mm.

5.2.2 HINDERED SETTLING

In Section 5.2.1 we have treated the settling velocity of individual mud flocs in still water. That theory is applicable to cohesive sediment suspensions at low (volumetric) concentration. When this concentration increases, the flocs start to hinder each other in their settling. This effect is called hindered settling. When the volumetric concentration increases further and becomes unity, a space-filling network develops and the flocs cannot settle further, except as a result of self-weight consolidation (e.g. Chapter 7).

An extensive and detailed review on hindered settling was presented by Scott (1984), part of which was published in a paper by Mandersloot et al. (1986), summarising many theoretical and empirical studies on hindered settling in chronological order. That review is focused on the behaviour of Euclidean particles (sand), however. As the flocs of cohesive sediment are not Euclidean, we cannot follow Scott in detail in our analysis of the hindered settling behaviour of mud flocs, as will be explained in this section. However, we use his analysis as a guide to ours.

From the literature and from physical reasoning, we can identify the following seven processes that affect the settling velocity of individual particles in a suspension:

1. **Return flow and wake formation**. A falling particle generates a return flow and a wake. When other particles in the vicinity of this falling particle are located within this return flow, their effective settling velocity (with respect to a fixed reference frame) will be affected, and the overall effective settling velocity of the suspension will decrease by a factor $(1 - \phi_f)$. We note that for Euclidean particles $\phi_f = \phi_s$, where ϕ_s is the volumetric concentration of the primary particles ($\phi_s = c/\rho_s$), whereas for cohesive sediment the volume fraction of the flocs is relevant. If, however, a second particle is caught in the wake of the falling particle, its settling velocity is increased, as is pointed out by Reed and Anderson (1980). The latter effect is often ignored but can be important (see process 7.).

2. **Dynamic or flow effect**. Several authors (e.g. Smith, 1998 and Darcovich et al., 1996) stress the role of neighbouring particles on the velocity gradients around a falling particle, as a result of which the pressure distribution around this particle, its hydrodynamic drag and added mass are affected. Some of the theoretical and numerical studies on hindered settling are focused on evaluating one or more of these processes (Reed and Anderson, 1980; Smith, 1998; and Darcovich et al., 1996). These effects are often lumped in an effective viscosity coefficient. It is noted that the influences of the return flow and wake formation discussed under process 1 are of course also flow effects.

3. **Particle-particle collisions**. Collisions between particles cause additional stresses in the suspension. Since the pioneering work by Bagnold (1954), it is common to incorporate also this effect in an effective viscosity. As a consequence, the settling of individual particles is hindered, so that the

effective settling velocity of the suspension decreases. This effect of course becomes more pronounced at higher concentrations (e.g. Buscall, 1990). It is noted that this effect is unimportant for mud suspensions, because cohesive sediment flocs do not exhibit elastic behaviour. Collisions between flocs result in plastic deformations and/or break-up of these flocs, consuming kinetic energy.

4. **Particle-particle interactions**. Batchelor (1982) discussed the role of mutual attraction and repulsion of particles by electrical charges or otherwise. Also Batchelor's analysis for small concentrations led to an augmented effective viscosity. In the case of cohesive sediment attractive and repulsive forces will result in floc formation, or in a stable dispersion in the absence of a floc promoting solute.

5. **Viscosity**. Einstein was the first to realise that the presence of particles in a fluid increases the strain rate within that fluid, resulting in an increase in apparent or effective viscosity of the suspension. Since then, many studies have been carried out and published on this effect. A concise summary is given by Scott (1984). Most formulae are of the form $\mu_{eff} = \mu_0(1 - a\phi_s)^{-b}$, where a (≥ 1) and b (≥ 2.5) are coefficients; b is often assumed to depend on ϕ_s. Each individual particle within a suspension is then considered to be falling in the remainder of that suspension, which has an increased viscosity. This would then decrease the effective settling velocity of all particles. The formulations in the literature for μ_{eff} differ widely, which is probably caused by implicitly accounting for the various effects discussed under 2., 3. and 4. through an (augmented) effective viscosity.
Note that in principle, Einstein's approach, and the formulae of his followers, is valid for Euclidean spheres only. Cohesive sediment flocs may deform under the influence of the fluid stresses, dissipating part of the energy, as a result of which Einstein's approach may not be accurate for cohesive sediment.

6. **Buoyancy or reduced gravity**. By the same argument, that is that an individual particle settles in the remainder of the suspension with an increased bulk density, the effective settling velocity of the suspension decreases by a factor $(1 - \phi_s)$.

7. **Cloud formation or settling convection**. Darcovich et al. (1996) and Tacker and Lavelle (1997) stress the role of cloud formation. Particles in the wake of another particle (see 1.) will be dragged by this particle. The wake around this group of particles increases, catching more particles, and a cloud of settling particles is formed. Such a cloud may behave as a settling entity by itself (referred to as a thermal in the literature), as a result of which the effective settling velocity of, or within, the suspension may increase. This

effect is probably important during the dumping of dredge sludge, for instance, but less so in the more or less homogeneous suspensions encountered in estuarine and coastal environments.

Combining the processes mentioned under 1., 5. and 6. yields a formula for the effective settling velocity w_s of the form:

$$w_s = W_{s,r}\left(1 - k\phi_s\right)^n \tag{5.5}$$

which was the rationale of Scott (1984) in advocating the semi-empirical hindered settling formula by Richardson and Zaki (1954), in which $k \approx 1$ and n is a function of the particle Reynolds number: $2.5 < n < 5.5$. Richardson and Zaki derived this formula from a dimensional analysis, yielding the relevant parameters for hindered settling, and an extensive series of sedimentation and fluidization experiments with particles of a large variety in shape and Reynolds numbers.

The larger particle Reynolds numbers in these experiments were obtained by using fairly large particles with diameters up to about 0.5 to 1 cm. The mutual distance between the particles then becomes so large, that the suspension can no longer be regarded as a continuum, and the picture that each particle settles in a suspension with higher density and viscosity is no longer valid. Only the effects of return flow, uneven flow distribution and particle-particle collisions are relevant. As a result the exponent attains a small value of about $n \approx 2.5$ for high Re_p.

The basic form of Richardson and Zaki's formula is confirmed by several experimental studies (e.g. Landman and White, 1992 and Davis and Birdsell, 1988) and by theoretical and numerical studies (e.g. Reed and Anderson, 1980 and Darcovich et al., 1996), amongst which studies with two-phase models are prominent (Tacker and Lavelle, 1980; Buscall, 1990; Ingber and Womble, 1994; and Smith, 1998). We note that hindered settling effects are automatically accounted for by means of a proper modelling of the sediment-fluid interactions in a two-phase formulation, as in the two-phase models by Teisson et al. (1992) and Le Hir (1997), for instance.

Recently, Cheng (1997) proposed an alternative approach for sand. He explicitly included the effects of return flow, augmented viscosity and density. The effective settling velocity of an individual particle in a suspension is obtained by assuming again that this particle falls in the remainder of the suspension with larger viscosity and density. Cheng defined the grain parameters d_* and $d_{*,m}$ as

$$d_* = \left(\frac{(\rho_s - \rho_w)g}{\rho v^2}\right)^{\frac{1}{3}} D, \quad \text{and} \quad d_{*,m} = \left(\frac{(\rho_s - \rho_m)g}{\rho_m v_m^2}\right)^{\frac{1}{3}} D \qquad (5.6)$$

where ρ_w is the clear water density, v is the clear water kinematic viscosity and subscript $*_m$ represents mixture characteristics. Cheng described the increase in viscosity by the sediment particles by:

$$\frac{v_m}{v} = \frac{2}{2 - 3\phi_s}, \quad \text{assuming} \quad \max\{\phi_s\} = \frac{2}{3} \qquad (5.7)$$

Substitution into Stokes' expression yields the following hindered settling formula:

$$w_s = W_{s,r} \frac{2 - 2\phi_s}{2 - 3\phi_s} \left(\frac{\sqrt{25 + 1.2d_{*,m}^2} - 5}{\sqrt{25 + 1.2d_*^2} - 5}\right)^{1.5} \qquad (5.8)$$

This formula can also be written in the "classical" Richardson and Zaki form. The resulting variation of the exponent n in (5.5) with particle Reynolds number Re_p for various concentrations (ϕ_s = 0.05, 0.1 and 0.4) is presented in Fig. 5.2, showing a good agreement between the two approaches.

It is common to apply equ. (5.5) for cohesive sediment as well, using the overall volumetric concentration of the flocs for ϕ_f (e.g. Mehta, 1986). This is probably not correct because:

- The viscosity of mud suspensions at concentrations below the gelling point does not scale with $(1 - a\phi_s)^{-b}$. Such a scaling law would lead to unrealistic large viscosity coefficients. From a large series of roto-viscometer measurements with natural mud from various sites in The Netherlands at concentrations of 50 and 100 g/l (Delft Hydraulics, 1992), it can be concluded that at the high shear rates occurring in turbulent flow, the viscosity of these cohesive sediments does not increase by more than a factor of 2 to 3 with respect to the clear water viscosity. As we are not aware of any formulation for the viscosity of a suspension of cohesive sediment flocs in this range of concentrations, we therefore propose to apply the classical formula of Einstein:

$$v_m = v\left(1 + 2.5\phi_f\right)$$ (5.9)

Equ. (5.9) agrees reasonably with the Delft Hydraulics' (1992) data.

• The buoyancy effect does not scale with $(1 - \phi_f)$, but with $(1 - \phi_s)$, where ϕ_s is the volumetric concentration of the primary particles $(\phi_s \equiv c/\rho_s)$, i.e. with the bulk density of the sediment.

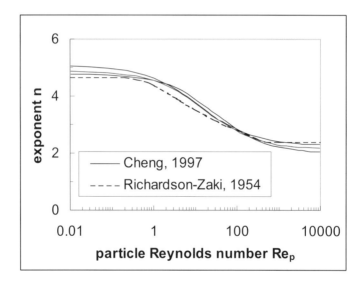

Fig. 5.2: Variation of n in (5.5) with Re_p according to Richardson and Zaki (1954) ($\phi_s = 0.05$, 0.1 and 0.4) and Cheng (1997).

These observations provide us with the ingredients to derive a hindered settling formula for cohesive sediment. Next, we include the effects of small amounts of fine sand, as sediment suspension in many marine environments consist of a mixture of cohesive sediment and fine sand. The reference settling velocity of fine sand can be obtained with Stokes' formula, and is not further elaborated here. In our derivation we use the superscript $\bullet^m$ to refer to the mud fraction (cohesive sediment, i.e. clay and silt, e.g. Section 3.1) and the superscript $\bullet^{sa}$ to refer to the sand fraction (e.g. Chapter 3).

We assume that only the mud fraction will affect the viscosity of the mixture, but that its increased buoyancy is dependent on both fractions. The hindered settling regime is characterised by the fact that the settling velocity of the sand and mud particles are not equal, i.e. that segregation of the fractions occurs. First, the general hindered settling formula for sand-mud mixtures is derived, after which the formulation for cohesive sediment only is compared with some data; no data are currently available on hindered settling of sand-mud mixtures.

First the effect of the return flow by the entire sand-mud mixture on the mud fraction only is derived. The effective settling velocity W_s^m of a single mud floc with respect to a fixed reference frame reads:

$$W_s^m = W_{s,r}^m - v_f \tag{5.10}$$

where $W_{s,r}^m$ is the reference settling velocity of an individual floc in still water and v_f is the return flow induced by the rest of the mixture. Because of continuity we have the following relationship:

$$W_s^m \phi_f^m + W_s^{sa} \phi_s^{sa} - \left(1 - \phi_f^m - \phi_s^{sa}\right) v_f = 0 \tag{5.11}$$

from which it follows that:

$$v_f = \frac{W_s^m \phi_f^m + W_s^{sa} \phi_s^{sa}}{1 - \phi_f^m - \phi_s^{sa}} \tag{5.12}$$

and substitution into (5.10) yields:

$$W_s^m \left(1 - \phi_f^m - \phi_s^{sa}\right) = \left(1 - \phi_f^m - \phi_s^{sa}\right) W_{s,r}^m - W_s^m \phi_f^m - W_s^{sa} \phi_s^{sa} \tag{5.13}$$

from which we find for $\phi_s^{sa} \ll \phi_f^m$:

$$W_s^m = \frac{\left(1 - \phi_f^m - \phi_s^{sa}\right)W_{s,r}^m - W_s^{sa}\phi_s^{sa}}{1 - \phi_s^{sa}} \approx \left(1 - \phi_f^m - \phi_s^{sa}\right)W_{s,r}^m \tag{5.14}$$

Similarly we find for $W_{s,r}^{sa} > W_{s,r}^m$ the effective settling velocity W_s^{sa} of a single sand particle with respect to a fixed reference frame, using the approximation in equ. (5.14):

$$W_s^{sa} = \frac{1 - \phi_f^m - \phi_s^{sa}}{1 - \phi_f^m}\left(W_{s,r}^{sa} - \phi_f^m W_{s,r}^m\right) \tag{5.15}$$

Including the buoyancy and viscosity effects yields a hindered settling formula for the fractions of cohesive and non-cohesive sediment (for $\phi_s^{sa} \ll \phi_f^m$):

$$W_s^m = \frac{\left(1 - \phi_f^m - \phi_s^{sa}\right)\left(1 - \phi_s^m - \phi_s^{sa}\right)}{1 + 2.5\phi_f^m}W_{s,r}^m \tag{5.16a}$$

$$W_s^{sa} = \frac{\left(1 - \phi_f^m - \phi_s^{sa}\right)}{\left(1 - \phi_f^m\right)}\frac{\left(1 - \phi_s^m - \phi_s^{sa}\right)}{1 + 2.5\phi_f^m}\left(W_{s,r}^{sa} - \phi_f^m W_{s,r}^m\right) \tag{5.16b}$$

with $\phi_f^m = c^m/c_{gel}$, $\phi_s^m = c^m/\rho_s$ and $\phi_s^{sa} = c^{sa}/\rho_s$; the gelling concentration c_{gel} is computed with the flocculation model of Chapter 4, or is obtained from measurements. The hindered settling regime is bounded by $\phi_f^m + \phi_s^{sa} < 1$; for $\phi_f^m + \phi_s^{sa} \geq 1$ we are in the consolidation regime (e.g. Chapter 7).

If we omit the sand fraction, we obtain a hindered settling formula for cohesive sediment with a mono-dispersive floc distribution:

$$W_s^m = \frac{\left(1 - \phi_f^m\right)\left(1 - \phi_s^m\right)}{1 + 2.5\phi_f^m}W_{s,r}^m \tag{5.17}$$

This hindered settling formula is compared with data of Thorn (1981), Ross (1988) and Wolanski et al. (1992) in Fig. 5.3, right hand side data only (e.g. $c >$ 2 to 3 g/l). It is shown that the data can be described properly with (5.17), using reasonable values for the gelling concentration c_{gel} , which were obtained by trial and error. Equ. (5.17) matches the data favourably, and is certainly comparable to the fit that can be obtained with equ. (5.5), e.g. Mehta (1986).

The left-hand side of Fig. 5.3 shows an increase in settling velocity with increasing sediment concentration, as used in the assessment of the flocculation model parameters in Section 4.7 (see also Winterwerp, 1998).

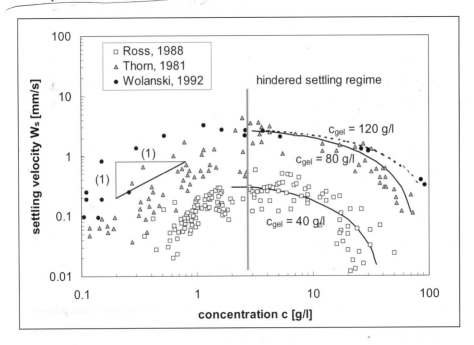

Fig. 5.3: *Variation of settling velocity with data – verification of hindered settling formula (5.17).*

It is noted that we have used the gelling concentration c_{gel} in our derivation of the hindered settling formula for fine sand. This implicitly implies that we assume that the strength of the space-filling network of the cohesive sediment - water mixture is sufficient to suspend the fine sand particles. This remains correct in still water for sand fraction up to 70% of the solids (Van Kesteren, 1998) However, experimental evidence shows that stable sand-mud suspensions segregate when the mixture starts to flow (Jin et al., 1994). This effect can be included by coupling the strength of the space-filling network to the strain rate of the flow (see Chapter 8).

In still water, such as a harbour basin, a settling column or a depot, the initial sedimentation phase of cohesive sediment suspensions is governed by the hindered settling process described above. In that case, the mass-balance (2.11) simplifies into the simple wave equation in z-direction:

$$\frac{\partial c}{\partial t} - \frac{\partial}{\partial z}\left(W_{s,r}\,\mathrm{f}(c)c\right) = 0 \qquad (5.18)$$

in which $\mathrm{f}(c)$ is a hindered settling function, such as (5.5) or (5.17)[1]. Equ. (5.18) is also known as Kynch' equation, in honour of Kynch (1952), who developed a kinematic theory of sedimentation. Equation (5.18) describes the first phase of the settling and consolidation process of a cohesive sediment suspension in still water, also referred to as the hindered settling phase. Note that (5.18) is no longer valid in the consolidation regime (e.g. Concha and Bustos, 1985, and Chapter 7).

It is important to appreciate that the hyperbolic equation (5.18) can yield one or two interfaces (shock waves), depending on the initial conditions of the suspension, as was shown by Kynch (1952) using the method of characteristics. Kranenburg (1992) elaborated further on Kynch' analysis, and we follow his reasoning. Kranenburg applied (5.18) to the volumetric concentration (either flocs, primary particles, or sand grains) and rewrote it as follows:

$$\frac{\partial \phi}{\partial t} - W_{s,r}\,\mathrm{F}(\phi)\frac{\partial \phi}{\partial z} = 0 \qquad (5.19)$$

in which $\mathrm{F}(\phi) = \mathrm{d}\big(\phi\mathrm{f}(\phi)\big)/\mathrm{d}\phi$. Equ. (5.19) can be solved by integrating along the characteristic lines $\mathrm{d}z/\mathrm{d}t = W_{s,r}\mathrm{F}(\phi)$, and it can be shown that concentration gradients increase when the characteristic lines converge. This implies that a stable interface (shock wave) will develop if $\mathrm{dF}/\mathrm{d}\phi < 0$. Fig. 5.4 presents the variation of the hindered settling function F with ϕ, and it is shown for equ. (5.5) that F has a minimum at $\phi_m = 2/(n+1)$. For $\phi < \phi_m$, $\mathrm{dF}/\mathrm{d}\phi < 0$, and two interfaces develop. For $\phi > \phi_m$, $\mathrm{dF}/\mathrm{d}\phi > 0$, and only one upper interface develops.

---

[1] Note that $\phi_s = \phi_f\, c_{gel}/\rho_s$.

This is schematically sketched in Fig. 5.5, showing the direction of the characteristic lines (after Kranenburg, 1992). Fig. 5.5a shows the case of two interfaces, an upper interface lowering in time, and a lower interface (the bed) rising in time. Fig. 5.5b shows the situation of only one interface; the lower interface has vanished and a gradual transition from bed to suspension is observed. The distinct kink in the upper interface in Fig. 5.5a is known as the point of contraction. Beyond this point, a space-filling network develops, and the hindered settling formulae are no longer valid, as effective stresses build up in the consolidation bed, e.g. Chapter 7. However, in the first phase of the consolidation process, permeability effects dominate, so the consolidation equation has the form of (5.19), and the analysis presented here can still be used (e.g. Chapter 7).

Note that the hindered settling function F for equ. (5.17) has no minimum (i.e. $dF/d\phi < 0$ for all ϕ). Hence it can be concluded that the use of (5.17) will always result in the occurrence of two interfaces in a settling suspension.

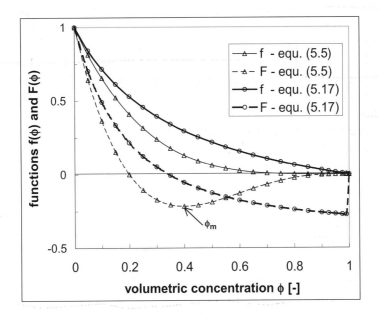

Fig. 5.4: Variation of hindered settling functions f and F with ϕ for equ. (5.5) (k = 1, n = 4) and equ. (5.17), with n = 1, after Kranenburg (1992).

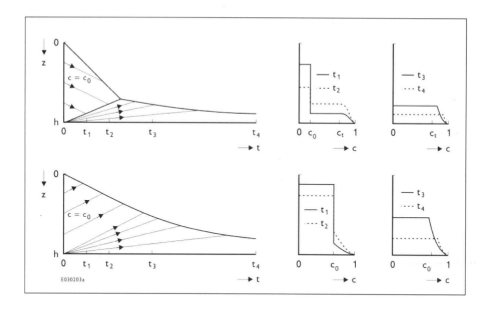

Fig. 5.5: a) Upper panel: formation of two interfaces: $\phi_0 < \phi_m$ and b) lower panel: formation of one interface: $\phi_0 > \phi_m$.

5.3 DEPOSITION AND SEDIMENTATION

5.3.1 DEPOSITION RATE

The settling velocity of cohesive sediment flocs does not directly yield the deposition rate from the suspension, as required as a bed boundary condition to the mass balance equation (2.11). As the deposition rate is at the heart of quantifying cohesive sediment dynamics in the marine environment, it is discussed in detail in this section.

Formulations for the deposition rate of cohesive sediment used world-wide are based on two series of sedimentation experiments described in the literature. The first series consists of experiments in a straight flume. These were first carried out by Krone (1962; see also Einstein and Krone, 1962) in a 33 m long flume, 1 m wide with a water depth of 0.33 m.

A suspension of fine sediments from San Francisco Bay was mixed with fresh water, salted with sodium-chloride and circulated through the flume at high flow velocity. After thorough mixing, the flow velocity was reduced to allow the sediment to settle. The floc size of the sediment was governed by the return flow system and hardly any flocculation occurred in the flume itself; the settling velocity of the sediment amounted to $W_s = 6.6 \cdot 10^{-6}$ m/s. The decrease in suspended sediment concentration c, as measured during three experiments in the flume, is presented in Fig. 5.6, showing a logarithmic decay at concentrations below about 300 mg/l. The data for $c < 300$ mg/l were fitted by equ. (5.20):

$$c = c_0 \exp\{-(1-p)W_s t/h\} \qquad (5.20)$$

where c_0 is the initial concentration, t is time and h is the water depth. The coefficient p represents the overall probability of resuspension of deposited material (Krone, 1962, 1993). At present, p is often referred to as the fraction of flocs that are too weak to survive the large shear stresses near the bed, and therefore will be broken and resuspended (Partheniades et al., 1968; see also Section 4.4.2).

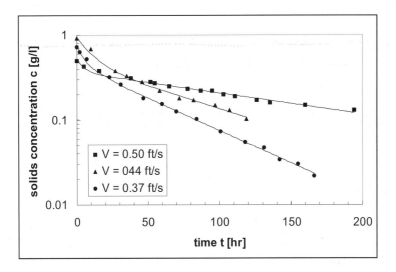

Fig. 5.6: Decay with time of suspended sediment concentration as measured by Krone (1961, 1962).

Equ. (5.20) forms the basis for Krone's worldwide used deposition formula (e.g. equ. 2.11):

$$\frac{dh\overline{c}}{dt} = -D = -W_s c_b \left(1 - \frac{\tau_b}{\tau_d}\right) \quad \text{for} \quad \tau_b < \tau_d \qquad (5.21)$$

where D is the deposition rate, $\bar{c}$ is the depth-averaged concentration, c_b is the near-bed concentration (often c_b is set equal to $\bar{c}$, e.g. Section 5.6), τ_b the bed shear stress and τ_d the so-called critical shear stress for deposition. Note that (5.21) predicts that eventually all sediment will be deposited at a specific flow velocity for which $\tau_b < \tau_d$. Krone (1962, 1993) also presented a fit for the deposition rate at larger concentrations; this analysis is not presented here however. Typical values of τ_d are 0.05 - 0.1 Pa, e.g. Self et al. (1989).

Later, similar tests in a straight flume were performed by Partheniades (1965), Mehta et al. (1982) and Kuijper et al. (1989). Neither Mehta nor Kuijper found full deposition, which Kuijper et al. attributed to the limited length of the flume.

The second series of sedimentation experiments were carried out in rotating annular flumes, for the first time by Partheniades and co-workers (1968, 1986), later by Mehta (1973, 1975), and repeated by Kuijper et al. (1989), amongst others. The rotating annular flume deployed by Mehta and Partheniades consisted of a circular channel, 0.2 m wide and 0.45 m deep, and a mean diameter of 1.5 m. The water-sediment suspension is driven by a rotating upper lid. The flume itself can rotate in opposite direction to minimise secondary currents. Sedimentation experiments by Mehta were carried out with San Francisco Bay mud, Maracaibo mud and processed kaolinite clay. A suspension was made with distilled water and salted tap water by mixing the sediment thoroughly in the flume at high flow velocity. Then the rotation of the flume was set at the required speed, and the sediment was allowed to settle. Contrary to Krone's experiments, not all sediment deposited when $\tau_b < \tau_d$, and an equilibrium concentration was found, as shown in Fig. 5.7.

It appeared that the equilibrium concentration c_{eq} scaled with the initial concentration C_0, and that the ratio c_{eq}/C_0 could be described with a log-normal distribution of a function of the bed shear stress (see also Partheniades, 1986). The behaviour shown in Fig. 5.7 is often explained from a distribution in floc size, hence settling velocity and floc strength. This hypothesis is substantiated by the measured decrease in median diameter with time (Mehta and Lott, 1987). Such a distribution is implicitly accounted for by the log-normal distribution mentioned above. Note that suspensions at equilibrium conditions may also contain a colloidal fraction that will not settle anyway. In that case the chemistry of the water (pH, salinity, etc.) starts to play a role in the behaviour of these colloidal fractions.

Mehta and Lott (1987) reanalysed the experiments assuming a floc size distribution in Krone's equ. (5.20). This (skewed) distribution contained 14 discrete classes, each with a characteristic W_s and τ_d. Mehta and Lott were indeed able to reproduce the observed concentration evolution. Later, Verbeek et al. (1993) were successful with a similar analysis, using a continuous distribution for W_s and τ_d however. Also Teeter (2001) applied a multiple fraction model, distinguishing between cohesive particles (with a critical shear stress for deposition) and non-cohesive silt particles, and was able to reproduce Mehta's measurements.

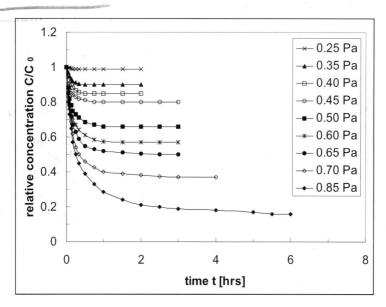

Fig. 5.7: Decay with time of suspended sediment concentration as measured by Mehta (1973, 1975).

Self et al. (1989) carried out erosion and deposition experiments in a "radial flow-cell chamber" with non-cohesive silt particles of varying diameter under neutral and non-neutral chemical conditions. They observed a critical shear stress for deposition, different from a critical shear stress for the onset of movement, beyond which particles could not settle. However, the experiments were conducted under laminar flow conditions, and it is not clear how the results should be interpreted with respect to full-scale field conditions with turbulent flow characteristics.

 Partheniades (1965) carried out three experiments in a rotating annular flume, which form the basis of the current cohesive sediment paradigm, that sedimentation and erosion cannot occur simultaneously in cohesive sediment dynamics. The first was a deposition experiment at a specific rotational speed of the annular flume, in which a mud suspension was allowed to settle partly. After equilibrium was obtained, the remaining suspension was carefully exchanged with clear water, maintaining the speed of the flume. It was observed that the exchanged water remained clear, from which Partheniades concluded that erosion does not occur during deposition.

Partheniades et al. (1968) carried out a second experiment, similar to the one described above. However, in this case, the suspension was only partly exchanged with clear water when equilibrium was obtained. Then, the deposited sediment was remixed over the water depth, after which the deposition experiment was repeated at the same hydrodynamic conditions, until a new equilibrium was achieved. It appeared that the new relative equilibrium concentration was lower than at the end of the first phase of this experiment, but not zero. This implies that some of the fine material, that could not settle during this second phase, was entrained into the bed during the first phase of the experiment. This behaviour was attributed to a change in flocculation properties of the sediment by removing part of the finer fraction.

In the third experiment, a mud bed was eroded at another rotational speed of the flume. Also this experiment was continued until equilibrium was achieved, and again the suspension, now containing eroded bed material, was carefully exchanged with clear water, maintaining the speed of the flume. Again, the water remained clear, from which Partheniades concluded that deposition does not occur during erosion.

It is noted that this paradigm does not explain an observation by Krone (1962, 1993) during another deposition experiment in his straight flume. In this experiment, sediment particles were labelled with radioactive gold, and allowed to settle fully to form a bed. Then, a suspension of the same sediment, but not labelled, was brought into the flume under such hydrodynamic conditions that the sediment could settle slowly. It appeared that the total amount of sediment in the flume decreased exponentially with time as described by (5.20), but that initially some gold-labelled particles were entrained from the bed. After a few hours however, the absolute amount of labelled sediment particles did not grow further, its fraction, with respect to the total amount of sediment that remained in suspension in the flume, kept increasing though. This observation can only be explained if a part of the labelled sediment particles in/on the bed is mixed within the suspension with a relative low fraction of labelled sediment. This of course requires erosion of the bed.

Sanford and Halka (1993) analysed a series of field measurements under tidal conditions in Chesapeake Bay. They observed that the suspended sediment concentration started to decrease when the flow started to decelerate. This

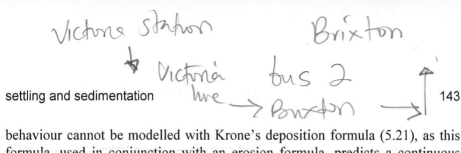
behaviour cannot be modelled with Krone's deposition formula (5.21), as this formula, used in conjunction with an erosion formula, predicts a continuous increase in suspended sediment concentration, until the flow velocity (bed shear stress) decreases below its critical value for deposition ($\tau_b < \tau_d$). Sandford and Halka (1993) were able to simulate the observed concentration pattern only when they applied a continuous deposition formula, i.e. $D = W_s c$, as shown in Fig. 5.8.

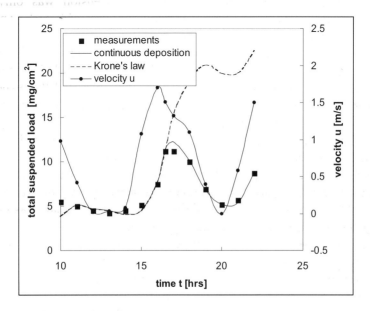

Fig. 5.8: Measured and predicted total sediment load (redrawn from Sandford and Halka, 1993).

Sanford and Halka (1993) also presented an overview of (field) data with the same behaviour, i.e. a decrease in concentration when the flow decelerates, including observations in Long Island Sound, San Francisco Bay, the Eastern Scheldt, and a number of British estuaries. From these observations and their analysis, they concluded that the paradigm of mutually exclusive erosion and deposition of mud is apparently valid under laboratory conditions only, but cannot explain many field observations.

This conclusion is of course rather unsatisfactory, and relies heavily on a selective set of laboratory experiments. Moreover, we feel that the paradigm of mutually exclusive erosion and deposition is not supported by a sound explanation of the underlying physical processes. We therefore propose an alternative description of the hydro-sedimentological conditions during deposition. This description consists of the following four elements:

1. Simultaneous erosion and deposition

Contrary to the classic cohesive sediment paradigm, it is assumed that erosion and deposition can occur simultaneously. The deposition rate D is given by the sediment flux at the bed, thus:

$$D = W_{s,b} c_b \tag{5.22}$$

where c_b and $W_{s,b}$ are the suspended sediment concentration and settling velocity of the sediment at the bed, accounting for vertical concentration gradients on c_b and possible flocculation effects on $W_{s,b}$.

2. Erosion rate

For the present analysis it is assumed that the erosion of sediment from the bed occurs when the bed shear stress exceeds a critical value for erosion τ_e. Because of physico-chemical effects (consolidation, gelling, restructuring), the bed strength increases with time, hence the critical shear stress for erosion (Chapter 9) of sediment deposits increases with time. The erosion rate E is assumed to scale with the excess bed shear stress and reads:

$$E(t) = F\left(\frac{\tau_b - \tau_e(t)}{\tau_e(t)}\right) S(\tau_b - \tau_e(t)) \tag{5.23a}$$

where F is an excess shear stress function, and $S(x)$ is a step function: $S = 0$ for $x < 0$, and $S = 1$ for $x > 0$. Of course, τ_b may also vary with time, but not necessarily at the same time scale as τ_e. Note that τ_e may also decrease in time because of swell, liquefaction and other processes when the sediment is subjected to turbulent shear or wave induced stresses. This is further elaborated in Chapter 8.

3. Bed shear stress

Further to Partheniades (Partheniades and Paaswell, 1970, Partheniades, 1965, 1986), it is assumed that at any time t the bed shear stress τ_b can be described by a probability density function $y(\tau)$. However, we do not use a Gaussian distribution, as at present it is known (e.g. Christensen, 1965, amongst others) that such a density function over a rigid bed should be skewed, as has been

elaborated in Section 2.5. If it is assumed that F in (5.23a) is linear pro$_l$
to τ_b, (5.23a) becomes:

$$E(t) = \frac{M}{\tau_e(t)} \int_{\tau_e(t)}^{\infty} y(\tau)\tau \, d\tau$$

(5.23b)

This co-called stochastic erosion process is schematically depicted in Fig. 5.9,
showing that erosion can also occur when the mean bed shear stress $\overline{\tau}_b$ is
smaller than τ_e.

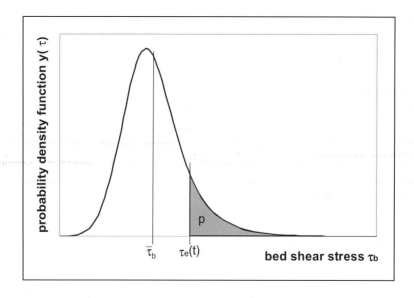

Fig. 5.9: Conceptual picture of stochastic erosion process.

The critical value for erosion τ_e may also have a probability density function.
No information exists on such a function, though Partheniades (1965) reasons
that this function may be quite narrow. However, the precise form of the pdf of
τ_e is not important for the arguments in the next sections, and it is assumed that
τ_e can be represented by a single value only.

4. Flocculation
It was shown in Section 4.4.2 that at low concentrations the flocculation time of
cohesive sediment particles higher in the water column is so large, that the floc
size is mainly determined by the near-bed turbulent flow conditions. In tidal
conditions, the high turbulent stresses around maximum flow and ebb velocities
form the dominant conditions. Hence, at low concentrations, variation in

flocculation over the water depth or with time does not play an important role (see also Hill et al., 2001). This agrees with Krone's remark that he did not observe ongoing flocculation processes within his flume.

At high-concentration conditions, flocculation is important and results in variations in floc size over the water depth and the tide (e.g. Section 5.6). However, in these cases, deposition results in the formation of fluid mud, and water-bed exchange processes are governed by other processes than those treated in this section.

Application of the physical picture, described above, to the experiments by Krone (Krone, 1962; Einstein and Krone, 1962) implies elaboration of the following zero-dimensional mass balance equation, assuming spatially homogeneous conditions:

$$\frac{\mathrm{d}h\overline{c}}{\mathrm{d}t} = E - D = \frac{M}{\tau_e(t)}\int_{\tau_e(t)}^{\infty} y(\tau)\tau\,\mathrm{d}\tau - W_s c_b \tag{5.24}$$

As the settling velocity of the sediment in the flume is very small, virtually no vertical concentration gradient is to be expected. Note that the solution of (5.24) does not yield an exponential decay.

Exponential decay, however, was observed by Krone at concentrations below about 0.3 g/l (see Fig. 5.6). Below this concentration, the deposition flux is very small (i.e. $0.3 \cdot 10^{-6}$ kg/m^2/s), in fact smaller than the potential erosion flux. As the deposited material can also increase in strength sufficiently during the 100 to 200 hours duration of the depositional tests, the erosion flux is limited by the availability of erodible material. This explains why the settling curve is not linear in the initial phase of the deposition experiment, when $c > 300$ mg/l. Hence, the mass balance for Krone's experiments should read:

$$\frac{\mathrm{d}h\overline{c}}{\mathrm{d}t} = -D + \alpha E = -W_s c_b + \alpha \frac{M}{\tau_e(t)}\int_{\tau_e(t)}^{\infty} y(\tau)\tau\,\mathrm{d}\tau \tag{5.25}$$

where E is to be interpreted as the *potential* erosion rate. The actual erosion rate is smaller than the potential erosion rate, when the amount of erodible sediment is limited. This is modelled with the coefficient α, which is the fraction of sediment that can be eroded: $\alpha = 1$ for $D \geq E$, $\alpha = W_s c_b / M$ for $D < E$. Note that in the latter case, equ. (5.25) is very similar to Krone's deposition formula, in which p then represents the probability that the bed shear stress exceeds the critical shear stress for erosion of recently deposited material (e.g. Fig. 5.9).

The probability density function, based on equ. (2.28) and the parameters derived in Section 2.2.2, for three of Krone's sedimentation experiments ($V = 0.50$ ft/s, $\tau_b = 0.0515$ Pa; $V = 0.44$ ft/s, $\tau_b = 0.0415$ Pa; $V = 0.37$ ft/s, $\tau_b = 0.0305$ Pa) is shown in Fig. 5.10.

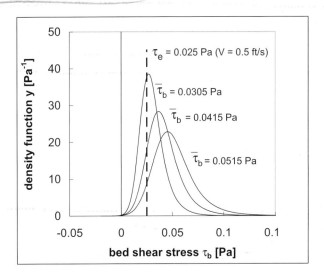

Fig. 5.10: Bed shear stress pdf for Krone's experiments (1961, 1962) on the basis of equ. (2.28).

The mass balance equation (5.25) still contains the two unknowns M and τ_e. These can be obtained through calibration of (5.25) against one of Krone's experiments. For this, his $V = 0.50$ ft/sec experiment is used, yielding the values in Table 5.1, obtained after some trial and error. Note that the variation of τ_e with time does not have to be known precisely in this case, because of the large time scales of the processes involved.

Table 5.1: Erosion/deposition parameters from calibration of (5.25) against Krone's $V = 0.5$ ft/s experiment (1962).

c_b	W_s	$\tau_e (t = 0)$	M
$\overline{c}$	$6.6 \cdot 10^{-6}$ m/s	0.025 Pa	$5.5 \cdot 10^{-8}$ kg/m^2/s

The resulting hindcast with equ. (5.25) of Krone's $V = 0.50$ ft/s test is shown in Fig. 5.11, showing a good agreement with the original data. With the parameters of Table 5.1 and the pdf of Fig. 5.10, also Krone's $V = 0.44$ ft/s and $V = 0.37$ ft/s experiments have been simulated. The results are also presented in Fig. 5.11, showing good agreement with the data for the $V = 0.44$ ft/s experiment, and somewhat lesser agreement with the data for the $V = 0.37$ ft/s experiment.

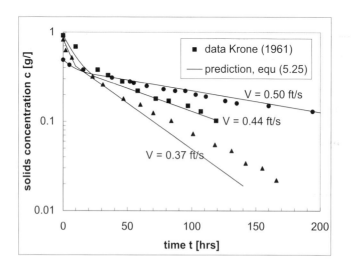

Fig. 5.11: Decay with time of suspended sediment concentration during deposition; comparison of measurements by Krone (1961, 1962) and prediction by equ. (5.25); model parameters calibrated with V = 0.5 ft/s test.

It is certainly possible to tune the various parameters further to obtain an even better overall agreement. This is not done however, as the purpose of this section is to analyse Krone's experiments in a qualitative way, explaining the physical meaning of the "critical shear stress for deposition" and the difference in deposition rate at lower and higher concentrations. Moreover, no information is available on the accuracy of Krone's data; in particular the bed shear stress must have been difficult to measure in the smooth flume used by Krone. A small error in τ_b will have a substantial influence on the area below the pdf-function beyond τ_e of Fig. 5.10.

From our analysis, we conclude that the so-called probability of deposition (hence the critical shear stress for deposition) does not exist, but that this phenomenon is to be interpreted as a probability of resuspension. It is argued that the water-bed exchange is then affected by two mechanisms:

1. The bed shear stress is stochastic in nature, which can be described through a skewed probability density function, and
2. The erodibility of freshly deposited cohesive sediment particles decreases with time, as the critical shear stress of erosion increases with time because of consolidation and physico-chemical effects.

In Section 4.2.2 it was shown that at low suspended sediment concentration, flocculation processes are so slow that the floc size, hence settling velocity, is fully determined by the near-bed high energy conditions, and does not vary over the water depth. So, floc formation and floc break-up processes will only play a role in the sedimentation process at high-concentration conditions (Winterwerp, 2002). Note that this does not imply that cohesive sediment is not flocculated at low sediment concentrations – our analysis merely states that floc size is constant over the water depth.

These conclusions imply that the common engineering practice, in which the water-bed exchange processes are described with a combination of Krone's deposition formula and Partheniades' erosion formula (e.g. Ariathurai and Arulanandan, 1978), does not describe the physics correctly.

Thus it is proposed to model the sedimentation flux for applications at *low-concentration* cohesive sediment suspensions for engineering applications simply by the flux itself:

$$D = W_s c_b \tag{5.26}$$

in which the settling velocity W_s may be a function of time, but does not vary over the water depth. It is further proposed not to use the pdf-distribution described here, or otherwise, as the local bed shear stress and the erosion parameters τ_e and M are poorly known in general (see also Chapter 9).

This approach agrees with Sanford and Halka's (1993) analysis and the data they present. The increase of τ_e with time should be modelled through a consolidation model when evolution in time is important, for instance over a spring-neap cycle, or in case of seasonal effects. The increase of τ_e during short periods around slack water in a semi-diurnal tidal cycle is often too small to justify its modelling.

It is noted that τ_e may decrease in time as well because of swell, liquefaction, etc. as a result of turbulent or wave-induced stresses. Ultimately, τ_e may vanish, if time scales are long enough, as argued in Chapter 9.

This sedimentation concept sheds another light on the paradigm of mutually exclusive erosion and deposition. The observed mutual exclusion of erosion and deposition should be attributed to an increase of τ_e with time as a result of which only recently deposited sediment may be re-entrained from the bed.

The arguments above can easily be implemented in the approach by Mehta and Lott (1987) and Verbeek et al. (1993) to describe the sedimentation behaviour of graded sediments, as discussed in the beginning of this section. If necessary, the effects of flocculation can be accounted for by an appropriate flocculation model.

In case the sediment concentration is high, sediment-turbulence interactions start to play a role, affecting the near-bed shear stresses through the formation of fluid mud (e.g. Chapter 6). Moreover, flocculation effects generally become important. However, the sedimentation flux is still described with equ. (5.26), e.g. Le Hir et al. (2001) and Winterwerp (2002).

The occurrence of wash load has been attributed to non-saturation of the sediment transport capacity of a flow. Partheniades (1977) however argued that this concept cannot explain common observations on wash load. He proposed an alternative view in which he accounted for the stochastic behaviour of the bed shear stress. He reasoned that if the entire τ_b-pdf would fall in between the critical shear stresses for deposition and erosion for a particular sediment fraction i, no exchange of that sediment fraction between water and bed would be possible, i.e.:

$$\tau_{d,i} < \tau_b < \tau_{e,i} \qquad (5.27)$$

Hence, (5.27) would yield the condition for the existence of wash load.

However, it is now known that τ_b can be (almost) zero for at least part of the time. Moreover, we have argued that the critical shear stress for deposition does not exist as such. This implies that (5.27) cannot yield the conditions for wash load. Our arguments suggest that the occurrence of wash load is the result of a large potential erosion rate, highly exceeding the depositional flux, or vice versa, a small depositional flux because of small settling velocity and/or small suspended sediment concentration, much smaller than the potential erosion rate. In other words, under wash load conditions, depositing particles are re-entrained immediately by the turbulent flow.

It is noted that very fine material, such as colloidal particles and a major part of organic components, do not settle at all because of Brownian motion, or because the excess density of the particles is too small. Such material will not be found in the bed, hence is classified as wash load as well.

The deposition formula proposed in this chapter allows for a continuous sedimentation flux, except for the finest sediment fractions subject to Brownian motion. Hence, even some of such very fine material may be found in the river or seabed if this fine wash load material is cohesive, or can be entrained into the bed by other mechanisms, such as pore water under-pressure, armouring, etc. However, colloidal material is not likely to be found within the bed, even under very calm conditions. Moreover, suspensions of colloidal particles are largely affected by the chemical composition of the water column.

5.3.2 SEDIMENTATION IN FLOWING WATER

In some cases we can establish the sedimentation rate in flowing water with simple tools. These conditions are:
- low suspended sediment concentration, i.e. no sediment-fluid interaction (e.g. Chapter 6),
- erosion and deposition rates are small in comparison to the horizontal sediment flux per unit area,
- no large gradients in horizontal suspended sediment concentrations, and
- no or little complicated three-dimensional structures in the water movement.

The second condition requires that the source-sink term $\theta_b << Uc$, where $\theta_b = E - W_s c$, E = erosion rate and U is a characteristic horizontal velocity. This condition is generally met in low-concentrated suspensions.

We distinguish between the situation in which the settling velocity is constant, and the situation in which flocculation over the water depth plays a role.

No flocculation: constant W_s

If the horizontal gradients in suspended sediment concentration are neglected, the mass balance of cohesive sediment follows from (2.11):

$$\frac{\partial c}{\partial t} - \frac{\partial}{\partial z}(W_s c) - \frac{\partial}{\partial z}\left(\Gamma_T \frac{\partial c}{\partial z}\right) = 0 \qquad (5.28)$$

We assume that the vertical profile of eddy diffusivity Γ_T is parabolic and the vertical velocity distribution is logarithmic (e.g. Section 2.4), that sediment-fluid interactions do not play a role, and that the flow does not contain complicated three-dimensional structures. In that case the equilibrium solution to (5.28) reads:

$$\frac{c}{c_a} = \left[\frac{a/h(1-z/h)}{z/h(1-a/h)}\right]^\beta \tag{5.29}$$

where c_a is a reference concentration at level a, h is the water depth, and the Rouse number β is defined as:

$$\beta = \frac{\sigma_T W_s}{\kappa u_*} \tag{5.30}$$

where σ_T is the turbulent Prandtl-Schmidt number (e.g. Section 2.3) and u_* the shear velocity. This solution can be found in almost all textbooks on sediment dynamics. Moreover, Van Rijn (1993) also gives solutions to (5.28) for non-parabolic eddy diffusivity profiles. Christensen (1972) proposes a slightly modified formulation of (5.29), based on a slightly modified law of the wall $\left(u = u_*/\kappa \ln\{(z+z_0)/z_0\}\right)$, which is often applied in numerical models to prevent singularities near the bed.

For small β, i.e. $\beta \ll 1$, (5.29) can be integrated over the water depth h to yield a solution of (5.28) as a function of the depth-averaged concentration $\bar{c}$:

$$c = \bar{c}\,\frac{\sin(\pi\beta)}{\pi\beta}\left(\frac{1-z/h}{z/h}\right)^\beta \quad \text{for} \quad \beta \ll 1 \tag{5.31}$$

Note that the Rouse number β is generally small in cohesive sediment suspensions. The solution (5.31) can be used to establish the near-bed suspended sediment concentration c_b to be used in the deposition rate (5.26), provided that $\bar{c}$ is known. The effect of turbulence on the settling rate of cohesive sediment is accounted for in the Rouse parameter β, thus in the near-bed concentration c_b.

When the deposition and erosion rates become relatively large, the vertical concentration profile starts to deviate from (5.29) and (5.31), as was discussed in an elegant way by Teeter (1986).

Flocculation: variable W_s

We can provide no general rules for establishing the settling velocity, hence the deposition rate, in case flocculation plays a role. Yet it is instructive to elaborate on the possible spatial and temporal variability of the settling velocity, for instance over a tidal cycle. This is done for a suspension in an open channel of 8 m depth with 0.5 m/s velocity amplitude, and a depth-mean concentration of 1 g/l. We compute the settling velocity with the flocculation model, as implemented in a 1DV-model (e.g. Appendix E), and the geometrical relation between settling velocity and floc size (5.3), using the model parameters for Ems mud of Table 4.2, and a mean fractal dimension $n_f = 2$. Note that neither deposition nor erosion have been included in these simulations. We compare three situations (see also Winterwerp, 2002):

1. w_s computed with full flocculation model, no hindered settling,
2. w_s and c_{gel} computed both with full flocculation model, with hindered settling, and
3. equilibrium settling velocity $w_{s,e}$ and equilibrium gelling concentration $c_{gel,e}$ computed with (4.32), with hindered settling.

We have not included any sediment-fluid interaction in these simulations (e.g. Chapter 6). The computational results are presented in the Figures 5.12 – 5.14.

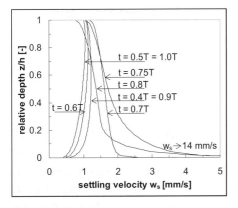

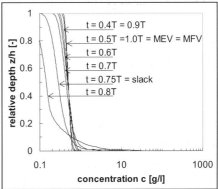

Fig. 5.12a: Computed $w_s(z,t)$, no hindered settling.

Fig. 5.12b: Computed concentration profile $c(z,t)$.

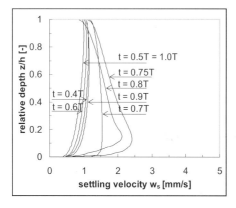 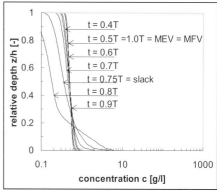

Fig. 5.13a: Computed w_s(z,t) and   *Fig. 5.13b: Computed concentration*
c_{gel}(z,t), with hindered settling.   *profile c(z,t).*

We observe that the computations predict fairly homogeneous profiles for settling velocity and suspended sediment concentration throughout most of the tidal period. Only around slack water larger values and gradients of w_s and c have been computed. In case 1 (no hindered settling) and case 3 (equilibrium conditions), unrealistically large values of w_s, and to a lesser extent of c, are found. This is not so in case 2, where computed settling velocities remain within about 2 mm/s throughout the tidal cycle. Apparently, a full flocculation model is required to obtain realistic values for the settling velocity. This is attributed to relaxation effects which this full model incorporates.

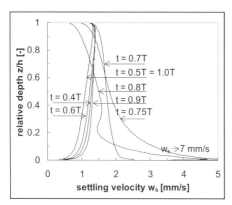 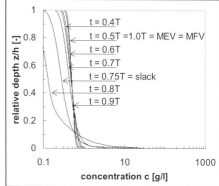

Fig. 5.14a: Computed w_{s,e}(z,t) and   *Fig. 5.14b: Computed concentration*
c_{gel,e}(z,t), with hindered settling.   *profile c(z,t).*

This observation is another manifestation of the time effects of flocculation, discussed in Section 4.9. It is noted that the use of (4.32) yields a very stiff response of the computed settling velocity to varying hydrodynamic conditions. This can be especially troublesome when sediment-fluid interactions are included in such computations.

5.3.3 SEDIMENTATION IN HARBOUR BASINS

Siltation in harbour basins is an important practical problem related to cohesive sediment dynamics. Siltation in these more or less stagnant basins occurs because sediment-poor water within the basin is exchanged with sediment-rich water from the harbour's environment; the sediment in the sediment-rich water can settle under the stagnant conditions in the basin. In this section we present a zero-dimensional method for a first assessment of the siltation rate in harbour basins. It is based to a large extent on the work by Eysink (1989), which was evaluated favourably by Headland (1994). The reader is referred to Van Rijn (2004) for more data and thorough analyses on harbour siltation in general, and for more sophisticated tools to establish the siltation rate.

Our zero-order approach is based on the analytical solution of the mass balance equation for the schematised harbour basin sketched in Fig. 5.15. This basin has a surface area S and depth h; the harbour volume amounts to $V = S{\times}h$. The suspended sediment concentration averaged over the harbour volume is denoted by c_h. The suspended sediment concentration in the ambient water system (e.g. lake, river, estuary or sea) is denoted by c_a, which may vary with time, but at another time scale than c_h, and c_a may therefore be treated as a constant. The exchange flow rate between the ambient water and the harbour basin is given by Q. It is assumed that the harbour basin is perfectly mixed throughout time, and that tidal variations are small, i.e. h is kept constant at zero order. Hence, the mass balance for suspended sediment in the harbour basin becomes:

$$\frac{\mathrm{d}Vc_h}{\mathrm{d}t} = Qc_a - Qc_h - SW_sc_h \tag{5.32}$$

where SW_sc_h is the sedimentation rate F_s within the basin, and W_s the effective sedimentation velocity of the sediment (i.e. reduced by vertical turbulent mixing, etc.). The initial condition at $t = 0$ in the harbour basin amounts to $c_h = c_0$, which yields the following solution to (5.32):

$$c_h = c_e + (c_0 - c_e)\exp\left\{-\left(\frac{Q}{V} + \frac{W_s}{h}\right)t\right\}$$

(5.33)

in which $c_e = Qc_a/(Q+SW_s)$ is the equilibrium concentration in the basin. Hence, the sedimentation rate $F_{s,T}$ within the harbour basin over a time period T (e.g. the tidal period) for (quasi-)equilibrium conditions becomes:

$$F_{s,T} = SW_s \frac{Q}{Q+SW_s} c_a T$$

(5.34)

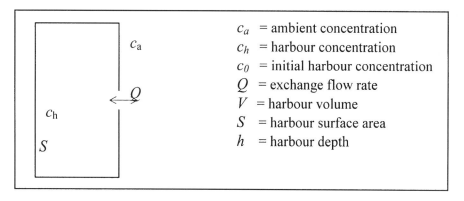

c_a = ambient concentration
c_h = harbour concentration
c_0 = initial harbour concentration
Q = exchange flow rate
V = harbour volume
S = harbour surface area
h = harbour depth

Fig 5.15: Schematic harbour basin.

The siltation or shoaling rate follows from (5.34) and a relation for the dry bed density, such as equ. (3.3.2), used in Section 3.3.2.

The rate of exchange of water Q between a harbour basin and its environment may be governed by a number of processes:

    I.   exchange flow by horizontal entrainment (mixing layer) Q_e,
    II.   exchange flow by tidal filling Q_t,
    III.   exchange flow by fresh-salt driven density currents Q_d,
    IV.   exchange flow by warm-cold driven density currents Q_T, and
    V.   exchange flow by sediment-induced density currents Q_s.

Exchange flow by horizontal entrainment Q_e occurs in any harbour basin, and is most conveniently explained for a basin along a river. The river flow separates at the upstream corner of the basin and forms a turbulent mixing layer, which hits the opposite, downstream corner of the basin. The strength of the mixing layer (level of turbulence) is governed by the geometry of the upstream corner and the stagnation zone at the downstream corner, and determines the entrainment rate, which can be described by:

$$Q_e = f_e A U_r \qquad\qquad (5.35)$$

where A is the cross sectional area of the harbour mouth and U_r is the river flow velocity. The exchange coefficient f_e depends on the harbour mouth configuration and the local flow patterns. Typical values for f_e range from 0.01 to 0.03, depending on the angle of the downstream corner of the harbour mouth (see also Booij, 1986; and Eysink, 1989). However, for unfavourable configurations f_e may be much larger.

The exchange flow rate by tidal filling effects Q_t can be computed straightforward if tidal elevations and basin area are known. It is noted that the effects of tidal filling and entertainment cannot simply be added as the mixing layer will be advected into the harbour mouth during tidal filling (e.g. Eysink, 1989). The analytical solution (5.33) of (5.32) assumes that the sedimentation time scale is large compared to the tidal period. Mehta and Maa (1985) presented an application of siltation in a marina in Florida by tidal filling alone.

The effect of density currents, induced by gradients in salinity or water temperature, on the exchange flow rate can be significant. Such density currents are not only generated by fresh or warm water releases in the harbour basin itself, but also in the case of density gradients in the ambient water (e.g. Eysink, 1989). In particular, the discharge of cooling water within a harbour basin has a dramatic effect on the exchange flow rate. As a rule of the thumb, this exchange flow amounts to three to five times the cooling water discharge.

Sediment, brought into a harbour basin by density-induced exchange processes, may be advected with the density current deep into the basin. This has been elaborated by Lin and Mehta (1997), who present graphs of the sediment penetration as a function of salinity and particle size.

It has been recognised only recently that even at moderate suspended sediment concentrations, as low as 100 mg/l, density currents can be induced by the sediment suspension itself. Winterwerp and Van Kessel (2002) showed that sediment-induced density currents at such low sediment concentrations augment the sediment fluxes Q_s into the harbour area of the Port of Rotterdam by at least factor three. A second effect of these sediment-induced density currents is that they augment the trapping efficiency within the basin largely.

Eysink (1989) derived relations for the exchange flow rates averaged over the tidal period T for harbour basins along estuaries (i.e. including the effects of tide and salinity gradients):

$$\langle Q \rangle = \langle Q_t \rangle + \langle Q_e \rangle + \langle Q_d \rangle + \langle Q_T \rangle \tag{5.36}$$

where triangular brackets imply averaging over the tidal period, $\langle Q \rangle$ is the total exchange flow rate, $\langle Q_t \rangle$ is the tide-induced exchange flow rate, $\langle Q_e \rangle$ is the exchange flow rate by horizontal entrainment, reduced by the effect of tide, $\langle Q_d \rangle$ is the exchange flow rate by density currents, reduced by the effect of tide, and $\langle Q_T \rangle$ is the exchange flow rate by cooling water discharges within the harbour basin, amounting to 3 – 5 times the volume of cooling water released. The value of $\langle Q_t \rangle$ is trivial: $\langle Q_t \rangle = V_t / T$, where V_t = tidal prism (= $2a_0 S$, with a_0 = tidal amplitude and S is surface area of harbour basin). $\langle Q_e \rangle$ should be reduced by the effects of the tide, as during rising water, the mixing layer is advected into the harbour basin:

$$\langle Q_e \rangle = f_e A \frac{\hat{u}_0}{\pi} - f_{t,e} \frac{\langle Q_t \rangle}{2} \tag{5.37}$$

where $\hat{u}_0$ is the amplitude of the tidal velocity and A the cross section of the harbour entrance. Note that under tidal conditions, generally no sediment import into the harbour basin occurs during falling water.

Integration over the tidal period of the salinity-induced exchange flow rate $\langle Q_d \rangle$ gives:

$$\langle Q_d \rangle = f_d A \sqrt{\frac{\Delta \rho_m g h_0}{\rho}} - f_{t,d} \langle Q_t \rangle \tag{5.38}$$

where $\Delta \rho_m$ is the maximum, salinity-induced density difference between harbour and ambient water in front of the harbour entrance and h_0 is the mean water depth in the harbour entrance. The coefficient f_d and the second tide-related term in the right-hand side of (5.38) account for the phase difference between the tide and the salinity distribution, and for the size of the harbour basin in relation to the salinity-induced density current, respectively.

Eysink (1989) proposes the values listed in Table 5.2 for the four coefficients in (5.37) and (5.38), based on a large number of laboratory and field data.

Table 5.2: Coefficients for exchange flow rates (after Eysink, 1989).

f_e	$f_{t,e}$	$f_{t,d}$	f_d
$0.01 - 0.03$	$0.1 - 0.25$	$0.1 - 1.0$	$0.05 - 0.125$

The coefficient f_d varies from 0.125 for vary large harbour basins, when the density gradient over the harbour entrance is maximal throughout the tide, to 0.05 for small harbour basins (tide-integrated density-induced exchange volume typically twenty times the harbour volume). The coefficient $f_{t,d}$ varies from $f_{t,d}$ ≈ 0.1 for $\langle Q_d \rangle / \langle Q_t \rangle = 0.1$ to $f_{t,d} \approx 0.6$ for $\langle Q_d \rangle / \langle Q_t \rangle = 1$ to $f_{t,d} \approx 1.0$ for $\langle Q_d \rangle / \langle Q_t \rangle = 10$. The reader is referred to Eysink (1989) for more details on these coefficients and Headland (1994) for an evaluation of Eysink's analyses.

The effective sedimentation flux $\alpha W_s c_h$ can be obtained with the help of equ. (5.32) when vertical sediment concentration gradients are pronounced.

6.   SEDIMENT-FLUID INTERACTION

The properties of sediment-laden flow in the marine environment are highly affected by the sediment carried by this flow. As a result a pronounced interaction between the (turbulent) flow field and the suspended sediment exists. This is not only the case under rare and exotic conditions: such interactions are fairly common in sediment-laden open channel flow, as discussed in Section 6.1.

Though this book treats the behaviour of cohesive sediment, we include a brief section on the interaction between non-cohesive sediment (i.e. sand) and the flow field in this chapter, as this helps us to understand the behaviour of cohesive sediment.

The major part of this chapter is related to sediment-induced stratification effects. For a general treatise on stratified flows, the reader is referred to for instance Turner (1973). For convenience, we recall two important definitions of the Richardson number, which are used throughout this chapter, and further in this book:

The *gradient Richardson number Ri* is a measure for the stability of stratified flows and defines whether disturbances may grow and thereby break down the stratification, which is the case when Ri exceeds a critical value of 1/4:

$$Ri = -\frac{g\,\partial\rho/\partial z}{\rho(\partial u/\partial z)^2} \qquad (6.1)$$

The *flux Richardson number Ri_f* is a measure for the efficiency of vertical mixing under stratified conditions, and is defined as:

$$Ri_f = -\frac{g\overline{w'\rho'}}{\rho\overline{u'w'}\,\partial u/\partial z} \qquad (6.2)$$

From experiments it has been found that the Ri_f has a maximum value of about 0.15 to 0.2 (see also Section 6.3.1). If we apply Reynolds' analogy, Ri and Ri_f can be related to each other through the turbulent Prandtl-Schmidt number σ_T: $Ri_f = Ri / \sigma_T$. Note that in stratified flows σ_T is not constant but varies with the degree of stratification, i.e. σ_T is a function of Ri (e.g. Turner, 1973; Tennekes and Lumley, 1994; Rodi, 1986).

Various bulk Richardson numbers have been defined in the literature, which correspond to either Ri or Ri_f. These will be defined when introduced.

In the analyses in the next sections, it is assumed that we can treat sediment-water mixtures as one-phase fluids in which all particles follow turbulence movements, except for their settling velocity. Uittenbogaard (1994) argues that this is a correct assumption if $w_s << w'_{rms}$, where w_s is the settling velocity of the sediment and w'_{rms} is a measure for the (rms-value of the) vertical turbulent velocity fluctuations. Because w_s is of the order of 0.1 - 1 mm/s for fine-grained sediment and $w'_{rms} \approx u_*$ in open-channel flow (Nezu and Nakagawa, 1993), where u_* is the shear velocity with typical values of several cm/s, this condition is generally met in estuarine and coastal waters. It is noted that the well-known Rouse number $\beta = w_s / \kappa u_*$, where κ is the von Kármàn constant, appears to be an appropriate parameter to decide whether a sediment suspension may be treated as a single phase fluid.

Uittenbogaard (1994) showed theoretically that even sand particles with a diameter up to 200 µm can properly follow the turbulent movements, typically occurring in tidal flows. This is sustained by Muste and Patel (1995) who measured the fluctuating velocity components of suspended sand particles of 250 µm median diameter in a turbulent flow and found that their rms-value is only about 15 to 20 % smaller than the rms-values of the fluctuating fluid velocity components. Hence, it is concluded that suspensions of fine-grained sediment in the marine environment can indeed be treated as single-phase fluids.

We have included a number of numerical examples in Section 6.3.2 and 6.3.3 showing the practical aspects of sediment-flow interaction, obtained a.o. with the 1DV model described in Appendix E. Here, we also include typical cohesive sediment properties as hindered settling, flocculation, etc., which have been discussed in the preceding chapters.

6.1   INTRODUCTION

Sediment-fluid interaction in the marine environment is a very common phenomenon, and comprises the following effects:

1. Bed-forms (dunes, ripples, banks, etc.) are the result of an interaction between the sediment transport and the water movement. The actual bed forms in a system depend on the flow stage, the presence of short waves, etc. The bed-forms themselves may contribute significantly to the overall hydraulic resistance. The literature contains an overwhelming amount of data and publications on the behaviour of bed-forms for non-cohesive sediment and their effect on the flow, see for instance Van Rijn (1993).

    It is noted that, to the best of our understanding, no systematic studies have been performed on the development and behaviour of bed-forms for cohesive sediments, or mixtures of cohesive and non-cohesive sediment. We can therefore not elaborate in this book on this important feature. We feel that this is a great and important shortcoming in our knowledge on the behaviour of cohesive sediment in the marine environment.

2. Alluvial beds (of sand) are permeable in general, and beds of cohesive sediment can behave as visco-elastic or visco-plastic (e.g. Chapter 8). In either case, the bed structure must have an influence on the properties of the turbulent field: turbulent pressure fluctuation, for instance, may be damped at the soft, c.q. porous bed interface. Almost all our knowledge on boundary layer flow, however, stems from theoretical and experimental studies over rigid, impermeable walls, e.g. Section 2.4 and 2.5. Hence, we cannot elaborate on the effects of an alluvial bed on the development of the turbulent boundary layer.

3. Permeability of the bed also affects its erodibility: inflowing pore water increases bed stability, hence decreases erosion rates, and vice versa, outflowing pore water increases erosion rates (e.g. Chapter 8 and 9).

4. The flow around suspended particles creates a wake behind these particles, consuming energy from the main flow. This effect is not very important at small volumetric concentrations, as cohesive sediment flocs follow the turbulent water movement, hence wake-generating velocity gradients are small. At larger concentrations, the fluid between the flocs is deformed, resulting in an increase in the apparent viscosity, as discussed in Section 5.3.

5. Settling particles generate an upward return flow. This effect becomes important at large volumetric concentrations, and can affect the settling of neighbouring particles, as discussed in Section 5.3. Also this return flow will contribute to deformation of the fluid, affecting the apparent viscosity, as noted above.

6. We have noted that a cohesive sediment-laden flow can be treated as a single-phase fluid. This implies that it is meaningful to define a bulk fluid density, which increases with increasing sediment concentration. The effects mentioned under 3, 4 and 5 are included in our analyses through the concepts of effective viscosity and hindered settling.

7. Sediment particles tend to settle due to gravity. This implies that energy delivered by the flow in the form of mixing is consumed to bring and keep sediment particles in suspension. This effect is explicitly treated in our derivation of an entrainment formula in Section 9.3.

8. The single-phase fluid approach implies that vertical gradients in suspended sediment concentration result in vertical gradients in fluid density. These gradients cause (severe) damping of the vertical transport of momentum and matter (mixing) in open channel flow. This is a key issue in understanding the behaviour of high-concentration mud suspensions, which is treated in the Sections 6.2.3 and 6.3.

9. The stress-strain relations of mixtures of water and cohesive sediment become progressively non-linear (or non-Newtonian as this effect is commonly referred to) with increasing sediment concentration. This non-linearity includes an increase in the bulk viscosity of the mixture, build-up of strength, and often the occurrence of memory effects (thixotropy). Non-Newtonian effects are treated in Chapter 8.

10. A fluid laden with cohesive sediment (clay) often exhibits drag reduction:
    *   At (very) low concentration, clay particles seem to damp the development of streaks and injections – this results in drag reduction and is known as the Toms effect, which is further discussed in Section 6.2.2,
    *   At high suspended sediment concentrations, the aforementioned buoyancy-related effects may also cause a considerable decrease in the effective hydraulic resistance – this is discussed in Section 6.3.2 and 6.4.

11. Because of an increase in bulk viscosity of the suspension, the effective Reynolds number of the flow decreases, possibly down to the transitional regime. In that case, high-Reynolds number turbulence models are no longer valid, and low-Reynolds effects have to be included. We will not elaborate on this subject, as it has not yet been studied in detail – the reader is referred to for instance Toorman (2002) for more information.

12. Gas releases from the seabed can trigger turbulent flow and/or destabilise the seabed. Also dissolved gas, like CO_2, decreases the water density, which may result in local stratification. Because gas is often present in cohesive sediment, it is treated in this book as well (Chapter 11).

We conclude that sediment-fluid interaction is a common feature, and that it has many facets. Some of these interactions occur frequently, such as the formation and influence of bed-forms in alluvial systems, while others are rare. In this chapter we investigate which interactions affect the behaviour of cohesive sediment in the marine environment, and under what conditions. We refer to Section 3.6, where we described cohesive sediment appearances in nature, and introduced definitions and non-dimensional scale numbers.

6.2   SEDIMENT-FLUID INTERACTION – LITERATURE OVERVIEW

6.2.1 NON-COHESIVE SEDIMENT

It is instructive to examine fluid-sediment interaction of non-cohesive sediment first, because the literature describes many studies on the effect of suspended non-cohesive sediment, i.e. sand, on the vertical velocity profile in turbulent open-channel flow. The two classical papers by Vanoni (1946) and by Einstein and Chien (1955) are generally referred to. Vanoni executed experiments for hydraulically rough flow conditions in a flume of 18 m length, 0.85 m width and 0.15 m water depth with sand of 90, 120 and 147 μm median diameter. During the experiments, no sand was allowed to settle on the bed. Vanoni established an effective Von Kármàn constant κ_s for sediment-laden flow from the velocity profile in the lower 50 % of the water column and concluded that κ_s decreases with increasing suspended sediment concentration c. Vanoni hypothesised that this decrease is caused by buoyancy effects, i.e. the damping of turbulence by the suspended sediment. Almost 10 years later, Einstein and Chien reported on their experiments for hydraulically rough flow in a flume of 13.3 m length, 0.3 m width and 0.12 m water depth with sand of 94, 274 and 150 μm median diameter. The suspended sediment concentrations were much larger than Vanoni's. Again no sediment was allowed to settle on the bed. Also Einstein and Chien found decreasing κ_s-values with increasing c.

Gelfenbaum and Smith (1986) re-analysed the experimental data by Vanoni and Einstein-Chien with a semi-analytical approach, based on an empirical description of the eddy viscosity under neutrally buoyant conditions, and a reduction factor, being a function of the local Richardson number, to account for turbulence damping by sediment-induced stratification effects. This approach yields a logarithmic vertical velocity profile with a reduced shear velocity (or reduced κ_s-value) for buoyant conditions ($Ri > 0$). The vertical profile of the suspended sediment concentration simplifies to the well-known Rouse profile for neutrally buoyant conditions. From their analysis, Gelfenbaum and Smith concluded that a proper simulation of the experimental data requires inclusion of sediment-induced damping effects in numerical models.

The decrease in κ_s-values with increasing c was well accepted in sedimentology until it was challenged by Coleman (1981 and 1986). Coleman argued that the logarithmic part of the velocity profile in open channel flow is limited to the lower 10 to 20 % of the water column. Higher in the water column the velocity profile is governed by the law of the wake. The analyses by Vanoni (1946) and Einstein and Chien (1955) would therefore be incorrect: the velocity distribution should be plotted in the defect-form. Coleman carried out a new series of laboratory experiments and re-analysed the data of Vanoni and Einstein-Chien, using the velocity defect law to sustain his arguments. Gust (1984) pointed out that Coleman's analysis for sediment-laden flow is in fact based on one data point only. Later, Valiani (1988) also questioned the correctness of Coleman's conclusion on the basis of an error analysis of the data.

Itakura and Kishi (1980) also re-analysed the experimental data by Vanoni, Einstein-Chien and others. They applied the Monin-Obukhov theory to establish a length scale for sediment-laden flows. This resulted in a log-linear velocity-defect profile, which may be viewed as a defect-formulation with a linear wake profile. Their approach is therefore conceptually similar to Coleman's.

Also McCutchean (1981) applied the Monin-Obukhov theory to analyse the role of suspended sediment on the vertical velocity profile. He presents a concise summary of the role of stratification on the velocity profile, and shows, both theoretically and through experiments, how stratification decreases vertical velocity gradients. Steeper velocity gradients should be attributed to sidewall friction effects.

Lyn (1986, 1988) approached the sediment-flow interaction from another angle. He studied this interaction on the basis of "similarity theory", in which he assigned characteristic length, velocity and concentration scales to the inner and outer part of the flow. According to his experimental data and theoretical analysis, the effect of the suspended sediment on the velocity profile is confined to about the lower 20 % of the water column.

A more theoretical approach of the sediment-flow interaction was presented by Hino (1963) and later by Zhou and Ni (1995). Hino's work was aimed at explaining the reduction in κ_s as a function of the suspended sediment concentration in turbulent flow, including observations by Elata and Ippen (1961) that a suspension of neutrally-buoyant particles also causes a reduction in κ_s together with an increase in turbulent intensity. His analysis started from the turbulent energy equation for clear water flows. To model the effect of sediment would require the inclusion of a buoyancy term to account for the energy necessary to keep the particles in suspension, and a reduction coefficient in the dissipation term. After some tuning of the various model coefficients, Hino was able to reproduce the velocity profiles measured by Vanoni and by Elata and Ippen. However, it is noted that the experiments by Elata and Ippen were not carried out with neutrally-buoyant particles, but with 100 to 150 μm dylene polystylene particles with a specific density of 1.05 and a settling velocity of about 0.1 cm/s. Though the specific density and settling velocity are much smaller than for sand, this does not imply that buoyancy effects are negligible for polystylene particles. In fact, they are as important as for flocs of cohesive sediment (e.g. Section 6.3).

Zhou and Ni (1995) carried out a perturbation analysis on the momentum and continuity equations for the fluid and the sediment. It was assumed that the turbulent flow fluctuations are not affected by the presence of sediment. Sediment-flow interaction was accounted for by an additional force term in the Reynolds stress equations. They showed that the zeroth order approximation of the perturbation equation for the flow yields a logarithmic velocity distribution. The zeroth and first order approximations of the perturbation equation for the concentration distribution yielded a Rousean profile, and the profile found by Itakura and Kishi, respectively. The effect of the sediment on the flow velocity profile to first order appeared to be a parabolic mean flow profile to be superposed on the turbulent flow profile. This would result in a more laminar-flow-like velocity profile with a subsequent decrease in effective κ. They compared a linearised form of their perturbation equations with experimental data reported in the literature (Coleman, 1981 and Einstein-Chien, 1955) and found good agreement after tuning two coefficients in the linearised equation. It is noted that this linearised form can be regarded as a defect law with a quadratic wake function.

It is remarkable that in the discussion on the effective Von Kármàn constant only a few authors refer in their analyses to studies on heat- and/or salinity-induced stratification effects; see for instance Turner (1973). Barenblatt (1953) was possibly the first to elaborate on this analogy. He introduced the Monin-Obukhov length scale ℓ to establish a damping function for the eddy viscosity, which resulted in log-linear velocity profiles. This work was further elaborated by Taylor and Dyer (1977) to establish the effect of various flow and sediment properties on the velocity profile.

Also the approach by Gelfenbaum and Smith (1986), as described above, models the sediment-turbulent flow interaction by the introduction of a damping function in the eddy viscosity profile. This approach was applied to analyse data on high-concentration suspensions as observed on the Columbia River Shelf at water depths of about 50 to 140 m (Kachel and Smith, 1989) and the San Francisco Continental Shelf at water depth between 60 and 70 m during storm periods (Wright et al., 1997 and Wiberg et al., 1994). The sediment bed consisted of a mixture of clay, silt and sand, which generated a suspension with a non-cohesive behaviour, as the clay concentration in the water column remained fairly low. It was shown that during storm periods sediment-induced stratification effects become important and can be properly described with the proposed damping function. The wave-boundary layer appeared to be affected mainly by the coarser sediment fraction, whereas the current boundary layer was affected by the finer sediment fractions. The results also suggest that the upper limit of the amount of sediment that can be contained within the current boundary layer under energetic waves is set by the vertical concentration gradient. This would imply a self-regulating process through a positive feedback between vertical mixing and settling flux. Such a self-regulating mechanism was also proposed by Trowbridge and Kineke (1994).

Further work was presented by Soulsby and Wainwright (1987) in the form of a stability diagram based on the Monin-Obukhov stability parameter $M_\zeta = z/\ell$, where ℓ is the Monin-Obukhov length scale, defined as $\ell \equiv \rho u_*^3 / \kappa g \overline{\rho' w'}$ and z is the vertical co-ordinate: stratification effects would be negligible if $M_\zeta < 0.03$. This diagram was more or less validated with some data from the Thames and the North Sea for suspended sediment concentrations ranging from about 10 to 10,000 mg/l. They concluded that in suspensions of fine sediments stratification effects always commence in the upper part of the water column.

The discussion above is restricted to the effects of suspended sediment on the mean velocity profile with indirect conclusions only of its effects on the turbulent flow properties. Muste and Patel (1997) carried out detailed measurements of the turbulent velocity intensities as a function of suspended sediment concentration. Their experiments were done for hydraulically rough flow in a flume of 30 m length, 0.91 m width and 0.13 m water depth, and with a concrete bottom. Various amounts of sand, sieved to a 210 to 250 μm diameter interval, were injected into the flume; no deposition on the bed was allowed. Though no data on the suspended sediment concentrations are given in their paper, depth-averaged values for the three series of experiments are estimated at about 100, 200 and 400 mg/l. Detailed laser-Doppler velocity measurements revealed no significant effect of the suspended sediment on the rms-values of the horizontal and vertical velocity fluctuations of the fluid. The measured velocity fluctuations of the sediment particles themselves, however, were 15-20 % smaller than those of the flow throughout the depth. Muste and Patel concluded that these experiments do not provide evidence for sediment-induced turbulent damping effects, as this would also affect the turbulent flow properties. Apparently, the only effect in their experiments was the inability of the sand particles to follow the turbulent flow fluctuations completely.

Cellino and Graf (1999) published the results of a detailed experimental study on the influence of suspended sediment on the turbulence properties of open-channel flows. These experiments were carried out with 135 µm sand under hydraulically rough conditions in a 16.8 m long and 0.6 m wide flume at a water depth of 0.12 m. Depth-averaged flow velocities varied between about 0.73 and 0.85 m/s and the suspended sediment concentration was increased from clear water values (non-capacity or starved-bed experiments) in small steps to capacity conditions over a sand bed at a depth-averaged concentration of about 4 g/l. Suspended sediment concentrations were measured by iso-kinetic sampling and mean and turbulent-fluctuating velocities were measured acoustically. These measurements revealed that the Reynolds-stress profile retained a linear form, also for capacity flow, but that the turbulent intensities themselves decreased appreciably relative to clear water values. The vertical profile of suspended sediment concentration became more stratified with increasing sediment load. The vertical eddy diffusivity appeared to decrease by about 50 % for capacity flow with respect to clear water conditions.

An elegant experiment was performed by Lau and Chu (1987) in a tilting flume of 22 m length and 0.67 m width, at a water depth of 0.16 m and a flow velocity of about 1 m/s. They measured the vertical mixing rate of a passive tracer (10 ppt salt, made non-buoyant with methanol) in clear water and in sediment-laden flow, at sediment concentrations of 1 and 5 g/l. From these measurements they deduced a reduction in vertical eddy diffusivity by 57 and 73 % for the tests with sediment, and attributed this reduction to turbulence damping by sediment-induced buoyancy effects.

From this overview we deduce that (non-cohesive) sediment in suspension has an effect on the vertical velocity distribution of the flow and on the vertical diffusivity. Though no conclusive explanation could be deduced, sediment-induced buoyancy effects appear to play a role according to most authors. Therefore, Winterwerp (2001) re-analysed Coleman's experiments with a 1DV model (e.g. Appendix E), incorporating the standard k-ε turbulence closure model described in Section 2.3. He used experimental results published in two papers: Coleman (1981) containing the primary data and analyses, and Coleman (1986) containing additional tables with the raw data and some additional analyses. Some further details were given in Coleman's reply to Gust's (1984) discussion of the results in the 1981-paper.

Coleman's experiments were carried out in a tilting, recirculating flume with a plexiglass channel of 15 m length and 0.36 m width, operated at a water depth of about 0.17 m. The flow rate was measured with a Venturi meter, the flow velocity with a traversible Pitot tube, the bed shear stress with a Preston tube, the surface slope with two point gauges 6 m apart, and the water temperature with a thermometer. Vertical profiles of the suspended sand concentration were also measured with the Pitot tube, deployed to withdraw iso-kinetic samples. The Pitot tube and Preston tube measurements were carried out in the centreline of the flume, 12 m downstream from the inlet. No data on experimental accuracy were given by Coleman.

The sediment used was composed predominantly of quartz and feldspar; the finest sample had a diameter between 88 and 125 μm, with a median diameter of 105 μm, and its settling velocity is established from Stokes' law. The experiments with this finest sediment are analysed in this section.

The bed and wall of the flume were smooth; no roughness elements were installed, and during the experiments, no sand was allowed to settle on the bed. All experiments were carried out at uniform flow conditions.

The results of the 1DV-computations have been described in detail in Winterwerp (1999 and 2001). Here we present only two results that are relevant to our analysis of the sediment-induced effects, as shown in Fig. 6.1 and 6.2.

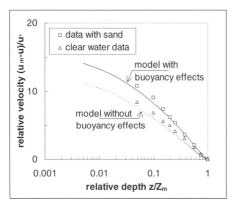
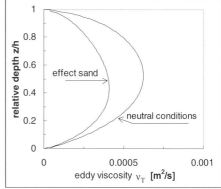

Fig. 6.1: Sediment-induced buoyancy   *Fig. 6.2: Computed eddy viscosity*
effects on velocity profile.   *profiles.*

Fig. 6.1 compares the observed and computed velocity profiles in the defect-law form. This is illuminating, as it is this form of presentation that inspired Coleman to his analysis of the effect of suspended sediment on the wake function of the velocity profile. Figure 6.1 shows that the effect of suspended sand on the measured velocity profile is properly predicted by the 1DV-computations in which a sediment-induced buoyancy term is included in the turbulent energy equation of the model. The effect of sediment-induced buoyancy destruction is further illustrated in Fig. 6.2, showing the computed eddy viscosity profile with and without sand (buoyancy effects). The computed decrease in eddy viscosity is qualitatively similar to the decrease measured by Cellina and Graf (1999).

From this analysis we conclude that the entire velocity profile in sediment-laden flows is affected by sediment-induced buoyancy effects. This is reflected in changes both in the logarithmic part and in the core region of the turbulent flow (defect law). These changes are properly predicted with computations with the standard k-ε turbulence model, including a sediment-induced buoyancy destruction term and the same value of the Von Kármàn parameter κ_s as for buoyant-neural conditions.

We will see in Section 6.3 that these buoyancy effects have major implications for the behaviour of cohesive sediment suspensions, in particular at high concentrations.

6.2.2 LOW-CONCENTRATION MUD SUSPENSIONS

Toms discovered in 1949 that pressure losses in closed pipe circuits decrease by many tens of percent when small amounts of polymers are added to the fluid transported through that pipe circuit. This phenomenon of drag-reduction is known as the Toms-effect. Since its first discovery, the Toms-effect has been applied extensively in industry, and many theoretical and experimental studies have been carried out. Nieuwstadt and Den Toonder (2001) have summarised and analysed these studies, continuing on an extensive survey by Virk (1975). They concluded that, in spite of all this work, no conclusive coherent explanation of the Toms-effect has yet been found.

Gust (1976) noted that drag reduction also occurs in dilute clay-water suspensions. He studied the effect of dilute clay suspensions ($0.5 < c < 4$ g/l) on the velocity profiles of saline water - clay suspensions in a small flume of 8 cm width and 6 cm water depth for hydraulically smooth conditions at various free-stream Reynolds numbers. The clay suspension and the bed in the flume consisted of mud from the German Wadden Sea. Two types of experiments were carried out. In the first series, the bed was non-eroding, and clay concentrations were below 0.5 g/l. In the second series, the bed was eroding, and concentrations went up to about 2 and 4 g/l. Gust observed drag reductions of up to 40 %, which he attributed to a thickening of the viscous sub-layer by a factor of 2 to 5. The turbulence structure further away from the wall seemed not to alter, and the universal logarithmic law appears still applicable beyond the sub-layer with a standard value for the Von Kármàn constant (i.e. $\kappa = 0.4$). His results are presented in Fig. 6.3, showing an increase in sub-layer thickness and an upward shift of the logarithmic part of the velocity profile. Gust compared his results with those of Virk (1975) and concluded that these largely resemble data obtained in pipes with a variety of additives, including polymers, fibers, surfactants and polysaccharides. Therefore, he hypothesised that the observed drag reduction may have a similar physical origin as the Toms effect.

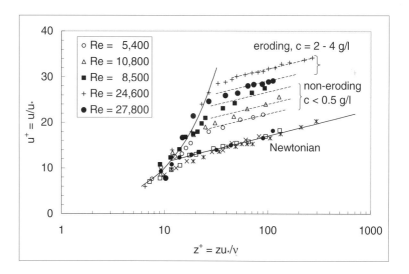

Fig. 6.3: Vertical velocity profiles in open channel flow for various clay concentrations and Reynolds numbers (redrawn from Gust, 1976).

Gust's work inspired Best and Leeder (1993), Wang et al. (1998) and Li and Gust (2000) to carry out similar studies, and they also found considerable drag reduction, as summarised in Table 6.1. However, much larger sediment concentrations were used. Buoyancy effects (e.g. Section 6.3) may therefore have affected the experimental results. In particular Wang et al. observed network formation of water-clay mixtures at concentrations beyond $80 - 100$ kg/m^3. Wang et al. also carried out experiments on hydraulically rough beds, consisting of gravel and rouble mound, at higher concentrations only however.

Table 6.1: Measured drag reduction in clay-water suspensions.

author	Best & Leeder	Wang et al.	Li & Gust
sediment conc. [kg/m^3]	1 – 2	20 – 230	4 & 8
drag reduction [%]	30 – 50	~ 10	40 – 70

In an extensive literature review, Virk (1975) summarised studies on the Toms-effect at that moment, and concluded that the observed drag reduction can be parameterised as:

$$\frac{1}{\sqrt{f}} = \left(4.0 + \delta\right)\log\left\{Re\sqrt{f}\right\} - 0.4 - \delta\log\left\{\left(Re\sqrt{f}\right)_0\right\} \qquad (6.3)$$

where Re is the overall flow Reynolds number, f is the friction coefficient ($\tau_w = 0.5f\rho U^2$, with U = mean flow velocity), δ is the so-called slope increment and $\left(Re\sqrt{f}\right)_0$ depicts the onset of drag reduction. The last two parameters depend on the polymer-water composition, and vary as a function of the polymer concentration: $1 < \delta < 100$ and $300 < (Re\sqrt{f})_0 < 1200$. The validity of (6.3) is bounded by a maximum possible drag reduction found from experiments to be:

$$\frac{1}{\sqrt{f_m}} = 19.0\log\left\{Re\sqrt{f_m}\right\} - 32.4 \qquad (6.4)$$

where f_m is the minimum friction coefficient for maximum drag reduction. According to Nieuwstadt and Den Toonder (2001), equ. (6.3) and (6.4) are still state-of-the-art.

Virk (1975) generalised the effects of polymer additives on the vertical velocity distribution, as shown in Fig. 6.4. This picture suggests that the viscous sub-region is not affected, but that the thickness of the buffer layer increases in drag-reduced flows (see also Nieuwstadt and Den Toonder, 2001). The thickening of the buffer layer was confirmed by experimental and numerical studies by Ptasinski (2001). He carried out physical experiments in a pipe with a solution of partially hydrolysed poly-acrylamide, and numerical experiments through Direct Numerical Simulation. The polymers were modelled as elastic dumbbells. His experimental and numerical results agreed closely, and matched the data by Virk.

Drag-reducing effects in this buffer layer cause the velocity profile to follow the asymptote that corresponds to (6.3). Beyond the buffer layer, in the inertial layer, the logarithmic velocity profile regains its original slope, i.e. at the standard Von Kàrmàn constant ($\kappa = 0.4$). The velocity profile in the inertial layer can be described by (see also (2.19)):

$$\frac{u(z)}{u_*} = \frac{1}{\kappa}\ln\left\{\frac{zu_*}{v}\right\} + 5.6 + \Delta B \qquad (6.5)$$

in which

$$\Delta B = \delta\sqrt{2}\,\log\left\{Re\sqrt{f}\Big/\left(Re\sqrt{f}\right)_0\right\} \qquad (6.6)$$

The process of drag reduction seems to be accompanied by an increase in spacing and decrease in frequency of turbulent bursts in the flow boundary layer (e.g. Best, 1993; and Nieuwstadt and Den Toonder, 2001).

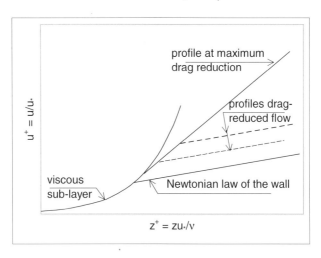

Fig. 6.4: Generalised vertical velocity distribution in drag-reduced flow (redrawn from Virk et al., 1975).

Ptasinski (2002) concluded from his research that two processes cause drag reduction:

1. A major part of the turbulent kinetic energy of the flow is transferred directly
    into elastic energy of the polymers, where it is dissipated by polymer
    relaxation. As a result, turbulent velocity fluctuations and bursting are
    suppressed in the buffer layer.
2. The change in velocity structure close to the wall pushes the largest velocity
    gradients away from the wall, resulting in so-called shear sheltering. Shear
    sheltering was already identified by Lumley (1969, 1973) on the basis of
    physical reasoning. It decouples the motions close to the wall and further in
    the water flow, i.e. at either side of the shear layer.

It is noted that the Toms-effect has been observed and studied for hydraulically
smooth conditions. It is not clear whether drag reduction by additives in general,
and by clay particles in particular, will occur under hydraulically rough
conditions, as in the marine environment. Moreover, two other effects may then
play a role as well:
1. Small amounts of fines in a granular bed of fine to medium sand, often
    encountered in the marine environment, cause a major decrease in
    permeability of the granular skeleton. As a result, the mobility of sand grains
    will reduce considerably. Hence, it is difficult to explain such decreased
    mobility in terms of either drag reduction or decreased permeability.
2. At larger clay concentrations, sediment-induced buoyancy effects begin to
    play a role. It is noted that these effects become important even at fairly
    moderate concentrations (e.g. Section 6.2.3 and 6.3). In that case, the effects
    of drag reduction may be overshadowed entirely.
Yet, it is worthwhile to study the effects of drag reduction further, as the bed may
be hydraulically smooth in muddy environments. In that case, a reduction in bed
shear stress may affect the onset of sediment-induced buoyancy effects (e.g.
saturation – see Section 6.3.1).

A second reason for further studies is the observation that polysaccharides may
also induce drag reduction. Polysaccharides are produced in large quantities on
intertidal mud flats, in particular during the algae bloom season. As a result, a
positive feedback may exist between hydrodynamics, sediment deposition and
algae bloom, resulting in large effects on the transport and fate of cohesive
sediment in the marine environment (see also Chapter 10).

6.2.3 HIGH-CONCENTRATION MUD SUSPENSIONS

We have defined (e.g. Chapter 3) a High-Concentration Mud Suspension
(HCMS) as a sediment suspension for which the flow field is measurably
affected by sediment-induced buoyancy effects, but still exhibits Newtonian

properties. In this section we examine their frequency of occurrence and summarise a number of properties described in the literature.

In The Netherlands, HCMS appear to be fairly common. Van Leussen and Van Velzen (1989) reported on sediment suspensions with (temporary) high concentrations near the river bed in the turbidity maximum of the Rotterdam Waterway near the Botlek harbour basin, in the Borndiep (the tidal channel between Terschelling and Ameland in the Wadden Sea), and in the turbidity maximum in the Ems estuary. Data from the Ems estuary are further elaborated in Section 6.3.2. Measurements carried out in the North Sea around the entrance to Rotterdam Port also revealed HCMS during storm conditions (e.g. Winterwerp et al., 2001).

Van der Ham (1999) and Van der Ham et al. (1998) carried out detailed turbulence measurements in the major channel of the Dollard estuary. They measured mean values and turbulent fluctuations of flow velocity and suspended sediment concentrations at various heights at concentrations of a few 100 to a few 1,000 mg/l. They found that the vertical turbulence structure (i.e. the turbulent stresses, the vertical transport and the flux-Richardson number) is affected significantly by the suspended sediment.

Adams et al. (1990) reported on a series of observations on the vertical sediment structure in a drainage channel in the Namyang Bay tidal flats, on the west coast of South Korea. This area is characterised by strong tidal effects with a tidal range of 4.9 to 7.7 m for neap and spring tide, respectively, with peak spring values up to 9 m. The observations were made during ebb tide at a water depth of the order of 4 m. Continuous measurements of current speed and turbidity revealed a depth-averaged velocity of about 0.6 m/s with a strong gradient around a lutocline of 1 m thickness, and suspended sediment concentrations of the order of 1 g/l in the highly turbulent layer below the lutocline. Only a minor amount of sediment was found in the upper layer. This highly stratified structure is characterised by a gradient Richardson number Ri ≈ 0.33, and by pronounced interfacial waves with high-frequency Kelvin-Helmholz billows filling the entire water depth.

Faas and Wartel (1985) observed concentrations of several 100 mg/l in the turbidity maximum of the Scheldt River near Antwerp, Belgium. Also, off the Belgian coast around the entrance of Zeebrugge Port, high concentrations up to a few g/l have been monitored frequently (Bastin et al., 1982).

Possibly the first publication on HCMS is by Inglis and Allen (1957) describing the sediment dynamics in the Thames, UK (called fluid mud in their paper). The Thames is a meso-tidal river with a distinct turbidity maximum called the Mud-Reaches. During a 13-hour campaign, sediment concentration at 60 % of the water depth was measured to vary between about 50 and 1,000 mg/l. Later, even larger concentrations up to a few g/l have been measured near the bed (Odd, 1988).

West and Oduyemi (1989) measured mean and fluctuating velocity components and sediment concentrations in the Tamar and Conway estuaries at neap, intermediate and spring tide conditions with water depths varying between 1 and 5 m and suspended sediment concentrations ranging from 50 to 4,000 mg/l. From these measurements they deduced a reduction in the vertical turbulent momentum flux by about 80 % at Richardson numbers beyond 0.3 to 0.5 and a reduction in the vertical turbulent sediment flux up to 90 % at Richardson numbers up to unity. The difference in damping between the vertical momentum and sediment flux was attributed to the role of internal waves, which contribute to the velocity fluctuations, but not to vertical mixing (see also Uittenbogaard, 1995).

The UK counts many sedimentologically dynamic estuaries and numerous studies have been published. The estuaries along the east coast in general have meso-tidal conditions, and HCMS-concentrations are reported for Brisbane River (Odd, 1988), and the Tamar and Trent estuaries (Arundale et al., 1997). Along the south and west coast macro-tidal conditions prevail. Many studies have been reported on the Severn estuary, which has a tidal range of about 14.5 m at spring tide and 6.5 m at neap tide (e.g. IOS, 1977; Kirby and Parker, 1980; Hydraulics Research Station, 1980; and Odd and Cooper, 1989). This estuary is notorious for its fluid mud appearances. Data show a strong influence of the spring-neap tidal cycle: during spring tide the vast but loosely consolidated fluid mud layers in the estuary, deposited during neap tide, are eroded slowly. On top of these fluid mud layers the suspension can reach concentrations of a few g/l, with several distinct lutoclines around slack water. The various studies are all dedicated to the behaviour of the fluid mud layers themselves, their formation, mobility and detectability.

The macro-tidal Loire (Teisson and Fritsch, 1988 and Le Hir, 1997) and Gironde estuaries (e.g. du Penhoat and Salomon, 1979) in France are also renowned for their fluid mud occurrences. Allen et al. (1980) studied the influence of the tide on maintaining the turbidity maximum at a concentration of the order of 1 g/l. They concluded that the tidal erosion-deposition cycle is the major mechanism in maintaining this turbidity maximum and that the effects of fresh-saline water induced gravitational circulation is of minor importance. However, the latter process may be responsible for the supply of sediment through long-term residual transport (e.g. Kappenberg, 1995). The processes within the Loire strongly resemble those in the Severn (Le Hir, 1997). The spring-neap tidal cycle drives an erosion-deposition-consolidation cycle generating and re-entraining vast layers of fluid mud yielding a high-concentration suspension of a few g/l in the overlying water column.

Wolanski et al. reported sediment dynamics in the macro-tidal South Alligator River (Wolanski et al., 1988) and the meso-tidal Normanby estuary (Wolanski et al., 1992) in Australia. Both rivers show high concentrations, varying between 1 and 6 g/l. In their analysis, Wolanski et al. focused on the vertical exchange processes and stressed the important role of sediment-induced buoyancy effects on the vertical turbulence structure.

Several studies on HCMS-appearances in East-Asia have been published. Wolanski et al. (1996) reported HCMS in the Mekong river (Vietnam), both in the upstream fresh water part and in the salinity-affected downstream estuary. HCMS in China was reported by Shen et al. (1993) in the Yangtze estuary, by Dong et al. (1997) and Guan et al. (1998) in the macro-tidal Jiaojiang River and by Wright et al. (1990) in the Yellow River. Note that the conditions in the latter river can even be considered as hyper-concentrated with depth-mean concentrations up to 200 g/l (and beyond). Still, in the river mouth HCMS at a few 100 mg/l (locally up to a few g/l) can occur in the water column when fresh deposits are eroded by the tidal current and/or by waves. A thorough study on the sediment dynamics in Ariake Bay, Japan, dedicated to the extension of Kumamoto Port (Kihara et al., 1994) revealed HCMS-occurrences when the soft mud on the sea bed is mobilised by wave activity.

Wells (1983) described HCMS-occurrences (called fluid mud in his paper) at several coastal sites: South Louisiana (Atchafalaya Bay), the north coast of South America (Guyana and Surinam) and the west coast of Korea (Yellow Sea, near the Kum River). Though the tidal and wave conditions at these sites are very different (micro-, meso- and macro-tidal, moderate and high wave energy), the sediment dynamics are very similar. Wells analyses mainly focused on the attenuation of wave energy by the fluid mud and the mobility of the soft and mobilised mud. Unfortunately, no analysis on the sediment processes in the water column was presented.

The results of the two-year AMASEDS-campaign covering low, rising, high and falling river flow on the Amazon Continental Shelf, including a part of the Amazon mouth, have been reported extensively (Kineke, 1993: Kineke and Sternberg, 1995: and Kineke et al., 1996, amongst others). Detailed measurements over the water depth at numerous locations of suspended sediment, flow velocity, salinity and temperature were carried out; unfortunately, no data were collected on settling velocity or wave activity.

This study revealed the importance of fresh-saline water induced gravitational circulation on the horizontal sediment transport processes and the accumulation of the sediments in the turbidity zone, and of the role of the spring-neap tidal cycle in establishing the settling and mixing processes in the water column. Large patches of fluid mud were observed, and almost the entire survey area, covering about 300 to 500 km^2, revealed suspended sediment concentrations of the order of many 100 mg/l to a few g/l, hence in the HCMS-range.

Kineke et al. stress the role of the vertical stratification induced both by the river flow and the sediment suspension, and of hindered settling on sediment dynamics. This is further elaborated in the next section.

Vinzon and Mehta (2003) describe the development of two lutoclines in the sediment suspension on the Amazon Shelf at certain moments of the tide, depending on the relative effects of (hindered) settling and (damping of) vertical mixing. This observation cannot be compared directly to the schematic sketch of Fig. 5.5, as the manifestation of the lower lutocline is affected by mixing and not by bed formation.

Also, data on the effect of HCMS on the effective hydrodynamic drag have been published in the literature. Dong et al. (1997) and Guan et al. (1998) reported a decrease of 30 % in the Manning roughness coefficient $n = 0.015$ sm$^{-1/3}$ in the Jiaojiang estuary in China (mean effective Chézy value of $C \approx 90$ m$^{1/2}$/s). Wang et al. (1998) reported values of the Manning roughness coefficient of $n = 0.035$ sm$^{-1/3}$ in a canal diverting clean water from the Yellow River, and of $n = 0.025$ sm$^{-1/3}$ for hyper-concentration suspensions in the same canal. In the lower reaches of the Yellow River even lower values were found: $n \approx 0.01$ sm$^{-1/3}$.

From an analysis of the data from a large series of tidal stations and numerical simulations of the tidal propagation on the Amazon Shelf, Beardsley et al. (1995) deduced an overall decrease in hydraulic roughness of 50 %. This was explicitly attributed to huge fluid mud appearances in the Amazon mouth. The analyses and simulations revealed a mean effective Chézy value on the outer shelf of $C \approx 60$ m$^{1/2}$/s, a mean effective Chézy value on the inner shelf of about $C \approx 90$ m$^{1/2}$/s, and in the areas of fluid mud appearances of $C \approx 110$ m$^{1/2}$/s.

Gabioux et al. (2003) approached the drag reduction by fluid mud on the Amazon Shelf in a somewhat different way. They assumed that the bed shear stress can be described by laminar oscillating boundary layer theory. From applying a 2Dh model, they concluded that the effective Chézy value increased from about 75 m$^{1/2}$/s for the sandy part of the Shelf to about 160 m$^{1/2}$/s for the fluid mud parts of the Shelf.

A Chézy value of $C = 110$ m$^{1/2}$/s was also found from flow computations of the tidal propagation in the Yangtze River by PDC (1996).

Wolanski et al. (1992) emphasised the influence of high suspended sediment concentrations on the sediment-induced anisotropy of the turbulent flow field. They refer to field observations showing limited vertical mixing of turbidity currents with ambient water and the apparent total collapse of turbulence in laboratory experiments with sediment-laden flow.

We have given a fairly extensive review of literature on the occurrence of High Concentration Mud Suspension and their effect on the flow to establish their importance. From this survey we conclude that suspensions with concentrations of several 100 to a few 1,000 mg/l do occur frequently. Most observers found large interactions of the suspension with the flow field, and significant drag reduction has been reported. These sediment-fluid interactions are further analysed in the next sections.

6.3   SEDIMENT-INDUCED BUOYANCY EFFECTS

6.3.1 THE CONCEPT OF SATURATION

Suspensions of non-cohesive sediment (e.g. sand) under steady state conditions are characterised by an equilibrium concentration, which is a measure of the sediment carrying capacity of the flow. A decrease in flow velocity (or an increase in sediment load) will result in settling of part of the load. The depositing grains form a rigid bed immediately, at which turbulence production remains possible, and the rest of the sediment can be kept in suspension. Hence, a (gradual) decrease in flow velocity will result in a (gradual) decrease in the sediment carrying capacity of this sediment-laden flow. This is elaborated in all textbooks on sediment dynamics.

For cohesive sediment a completely different picture emerges. Consider a sediment-laden flow over a rigid, horizontal bed with an amount of cohesive sediment equal or close to the flow's sediment carrying capacity. When the flow velocity decreases slightly, sediment starts to settle, not to form a rigid bed, but, as a result of the high water content of the settling mud flocs (e.g. Chapter 4 and 5), a layer of fluid mud, thus creating a two-layered fluid system. Initially, i.e. prior to the development of strength, the lower layer remains turbulent because of turbulence production at the rigid bed (Bruens, 2002). At the interface between the two layers turbulent vertical mixing is damped strongly, decreasing the sediment carrying capacity in the upper layer of the flow further. This results in a snowball effect with a catastrophic collapse of the turbulence and the vertical sediment concentration profile. The suspended sediment concentration for cohesive sediment just prior to this collapse is referred to as the "saturation concentration" in this book to distinguish it from the equilibrium concentration for sand.

The first ideas on the existence of such a saturation concentration for cohesive sediment were presented by Teisson et al. (1992); however, at that time no explicit physical meaning was attributed to this phenomenon. The concept of the saturation concentration for cohesive sediment is based on empirical evidence (e.g. Turner, 1973) that a turbulent shear flow field collapses when the flux Richardson number Ri_f exceeds a critical value $Ri_{f,cr}$. Ri_f follows from the turbulent kinetic energy equation and is defined as the ratio of the buoyancy destruction and production term (e.g. Section 2.1.3):

$$Ri_f = -\frac{g\overline{w'\rho'}}{\rho\overline{u'w'}\,\partial u/\partial z} = -\frac{\Delta g\overline{w'c'}}{\rho\overline{u'w'}\,\partial u/\partial z} \qquad (6.7)$$

where a prime denotes the fluctuating part of the horizontal and vertical velocity components u and w, of the bulk density of the water-sediment suspension ρ, and of the suspended sediment concentration by mass c, g is the gravitational acceleration, z is the vertical co-ordinate (positive upward) and Δ ($\Delta = (\rho_s-\rho_w)/\rho_s$) is the relative excess sediment density. The overbar denotes averaging over the turbulent time scale. We have used the (simplified) equation of state (2.13) to relate density and sediment concentration.

Starting with the evaluation of sediment-laden flow at small vertical density gradients, i.e. sub-saturated concentrations, a zero-order approximation is justified, in which the eddy diffusivity is only slightly affected by buoyancy. By assuming a logarithmic velocity profile $(\partial u/\partial z = u_*/\kappa z)$, the corresponding parabolic viscosity profile $(v_T = \kappa u_* z(1-z/h))$, local equilibrium between settling and mixing $(\overline{w'c'} = -w_s c)$, and taking into account hindered settling effects (Richardson-Zaki's formula, (5.6)), equ. (6.7) can be elaborated to:

$$Ri_f \propto \frac{\rho_s - \rho_w}{\rho_w} \frac{ghW_s c_{gel}}{u_*^3} \phi_f(1-\phi_f)^5 \qquad (6.8)$$

where ϕ_f is the volumetric concentration of the suspended sediment, defined as $\phi_f \equiv c/c_{gel}$, with c_{gel} the gelling concentration, i.e. the concentration at which a space filling network develops, and W_s is a reference settling velocity. The gelling concentration c_{gel} depends on the processes of fluid mud formation, and can be established with the flocculation model described in Chapter 4 and 5. From in-situ observations c_{gel} appears to of the order of several tens to about hundred g/l (e.g. Section 3.4.3). For given (but not specified) hydrodynamic conditions the variation of Ri_f as a function of ϕ_f is sketched in Fig. 6.5, showing that Ri_f first increases with increasing ϕ_f and then decreases. The latter trend is caused by hindered settling, reducing sediment-induced buoyancy effects.

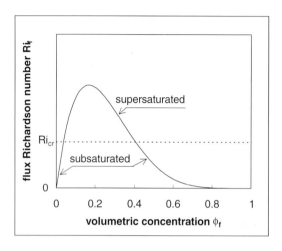

Fig. 6.5: Schematic variation of flux Richardson number with volumetric concentration.

At a specific level in the water column, Ri_f can exceed $Ri_{f,cr}$ and a collapse of the turbulent field above this level occurs. Such a collapse always starts near the water surface, as shown by Soulsby and Wainwright (1987). Here, we refer to these conditions as super-saturation.

The part of the curve for low values of ϕ_f refers to sub-saturated conditions, which are characterised by a measurable, but not collapse-causing interaction between the suspended sediment and the turbulent flow field. It is noted that such sub-saturated conditions are also expected at very high suspended sediment concentrations, which occur for instance in the Yellow River, in turbidity currents generated from fluidisation or liquefaction of consolidated deposits, for instance as a result of geo-mechanical failure at the continental slope, or of wave-induced liquefaction of coastal mud banks.

Substitution into equ. (6.7) of the various formulae for a logarithmic velocity profile and a parabolic eddy viscosity profile, as described above, equ. (6.8), yields a relation for the vertical concentration profile c_s at the critical flux Richardson number:

$$c_s(z) = \frac{Ri_{f,cr}\rho}{\Delta g \kappa} \frac{u_*^3}{hw_s}\left(\frac{h}{z} - 1\right)$$ (6.9)

where u_* is the shear velocity, h the water depth, and w_s the local, effective settling velocity.

Hopfinger (1987) presents an extensive review of the collapse of turbulence in stratified flow, discussing laboratory experiments, oceanographic observations and numerical simulations. His analysis focused on the development of the various length scales relevant for turbulent stratified flow, and he concluded that the onset of collapse occurs when the turbulence integral length scale becomes of the order of the buoyancy length scale.

Turner (1973) analysed the experimental results by Townsend and Ellison, and several others, obtained from experiments in laboratory flumes and observations in the atmosphere. He concluded that a collapse of turbulence occurs when the flux Richardson number Ri_f attains critical values between 0.05 to 0.3, with an average value $Ri_{f,cr} \approx 0.15$. This value is close to the value given by Tennekes and Lumley (1994), who advocate $Ri_{f,cr} \approx 0.2$.

The flux Richardson number Ri_f can be regarded as an efficiency parameter for vertical mixing in stratified flows. Hence, the critical flux Richardson number $Ri_{f,c}$ is a measure for the mixing capacity of flow. Intuitively it follows that saturation occurs when Ri_f becomes critical throughout the water depth (see also Section 6.4). This would imply that if we multiply $c(z)$ in (6.9) with $u(z)$ and integrate over the water depth, we would obtain a transport formula for cohesive sediment at capacity conditions.

For further analysis it is convenient to divide the sediment transport at capacity condition by the water depth h and depth-averaged velocity U, yielding the depth-averaged saturation concentration C_s, which can also be regarded as a scaling parameter for saturated suspensions (Galland et al., 1997):

$$C_s \equiv \frac{1}{hU}\int_0^h uc_s\,dz = K_s\,\frac{\rho}{\Delta g}\frac{u_*^3}{hW_s}$$ (6.10)

where K_s is a proportionality parameter, which may be a function of the Richardson number, as the vertical velocity profile is affected by sediment-induced buoyancy effects. For the time being, we will assume K_s is constant. At depth-mean concentrations beyond C_s (super-saturated), turbulence collapses and the flow is no longer able to carry the sediment in suspension. Hence, C_s is a measure for the sediment load that can be carried by turbulent flow.

It is interesting to note that this relation is very similar to the so-called Knapp-Bagnold criterion (Parker et al., 1986) for the occurrence of submarine turbidity currents:

$$C < \frac{\rho}{\Delta g} \frac{U_t u_*^2}{\delta W_s} \tag{6.11}$$

where U_t is the mean flow velocity of the turbidity current and δ its thickness. This relation is to be interpreted as a necessary condition for a self-sustaining turbidity current; it is also known as the auto-suspension criterion.

The saturation concept is well illustrated by two simulations with a 1DV-model[1] for a hypothetical open-channel flow of 16 m depth, a constant depth-averaged flow velocity of $U = 0.2$ m/s and a constant settling velocity $W_s = 0.5$ mm/s. Initially, the sediment is distributed homogeneously over the water depth. The initial concentration C_0 is increased in small steps until saturation occurs. Other parameter settings and details of the 1DV-model are given in Winterwerp (2001). Fig. 6.6a shows the time-evolution of the suspended sediment concentration in the form of isolutals (i.e. lines of constant sediment concentration) for an initial concentration of $C_0 = 0.023$ g/l and Fig. 6.6b for $C_0 = 0.024$ g/l.

It is clear that the $C_0 = 0.023$ g/l case represents saturation conditions, because a small increase in C_0 results in a catastrophic collapse of the concentration profile, as shown in Fig. 6.6b. The final concentration profile in the $C_0 = 0.023$ g/l case is more or less Rousean, whereas in the $C_0 = 0.024$ g/l case a fluid mud layer is formed. The coefficient in equ. (6.10) appears to have a value of about $K_s \approx 0.7$ for the conditions selected.

Fig. 6.7 shows the time development of the computed vertical distribution of suspended sediment concentration and of the eddy diffusivity for the $C_0 = 0.024$ g/l case (note the logarithmic scales). We observe an almost homogeneous vertical concentration profile at the start of the numerical experiment, together with a parabolic eddy diffusivity profile. Then an interface starts to develop in the upper part of the water column and the vertical concentration gradient

---

[1] The 1DV model uses the equations described in Chapter 2, i.e. it includes the standard k-ε turbulence model with a buoyancy destruction term; a detailed description is presented in Appendix E; all 1DV-computations have been performed with 100 layers.

increases rapidly. At the same time the eddy diffusivity profile starts to deviate significantly from its original shape, developing two more or less parabolic curves at t = 200 min. In the lower eighty percent of the water column turbulence is still generated at the bed and by the velocity shear.

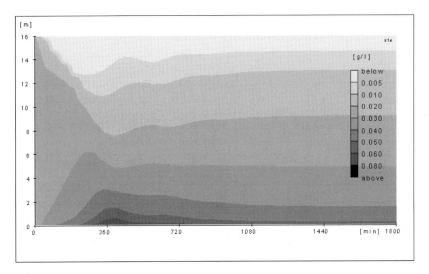

Fig. 6.6a: Computed isolutals for a saturated (C_0 = 0.023 g/l) suspension in steady open channel flow.

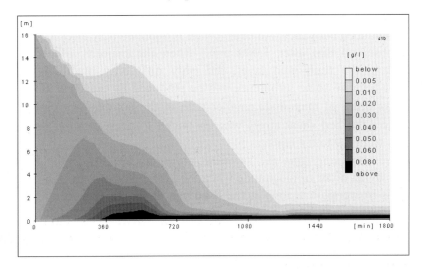

Fig. 6.6b: Computed isolutals for a supersaturated (C_0 = 0.024 g/l) suspension in steady open channel flow.

The maximum value of the eddy diffusivity is about 0.006 m²/s. Around the interface at $z/h = 0.8$, Γ_T almost vanishes, whereas in the upper twenty percent of the water column some mixing can occur again, induced by interfacial stress and local velocity shear; the maximum value of the eddy diffusivity in this part of the water column amounts to about 0.001 m²/s. This behaviour becomes more pronounced during the remaining settling time.

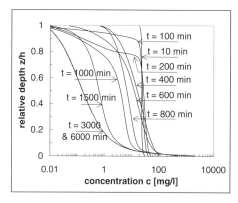 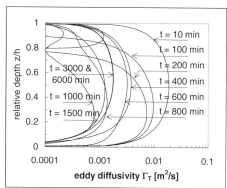

Fig. 6.7: Computed time evolution of vertical profiles of suspended sediment concentration and eddy diffusivity for $C_0 = 0.024$ g/l.

The various profiles show that the eddy diffusivity in the upper part of the water column slowly collapses as a result of turbulence damping due to sediment-induced buoyancy. An equilibrium is obtained only after 3,000 min. It is also observed that after 1,000 min., when the majority of sediment is deposited in the fluid mud layer, the eddy diffusivity profile is restored somewhat because some turbulence can be produced by the shear flow, but it remains an order of magnitude smaller than in the non-buoyant case ($t = 10$ min.). Also, the bed shear stress was computed to decrease considerably: u_* decreases from 0.9 to 0.4 cm/s.

These results imply that the sediment carrying capacity of the flow also decreases by an order of magnitude when the turbulence profile collapses. This collapse is irreversible as long as the fluid mud layer remains soft, i.e. as long as no yield stress builds up, so that no turbulence can be generated at the water-mud interface.

Next, a series of numerical simulations is carried out to verify the scaling law (6.10). This is done for a variety of flow velocities at a constant water depth $h = 16$ m, starting from sub-saturated conditions by increasing the initial sediment concentration in small steps until a collapse of the concentration profile is observed. The results are presented in Fig. 6.8, where the computed vertical flux $W_s \times C_s$ is plotted versus the mean flow velocity U. It is observed that the numerical results follow the functional relationship $W_s C_s \propto U^3$ properly. The influence of numerical parameters such as time-step and number of layers is also examined and appears to be negligible.

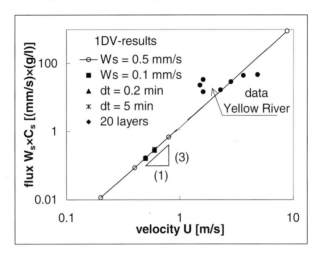

Fig. 6.8: Saturation flux computed with 1DV-model as a function of mean flow velocity, together with some field data of the Yellow River; the fit through the numerical data yields $C_s = 0.023 U^3 / h W_s$.

Though circumstantial evidence is paramount, no data exist at present to validate the saturation concept directly. However, it may be expected that hyper-concentration sediment-laden flow in the Yellow River and tributaries may be near-saturation at times. Therefore, we have included a few Yellow River data points in Fig. 6.8, corrected for hindered settling effects. Though double-logarithmic scales have been used in Fig. 6.8, we may conclude that equ. (6.10) gives a reasonable representation of these field data. Further qualitative evidence is given by Winterwerp et al. (2003).

We may expect that saturation also occurs in tidal flow conditions. In fact, we expect that saturation plays an important role in turbidity maxima in estuaries and in the formation of fluid mud observed in many marine environments.

We have therefore executed a series of 1DV-computations for tidal conditions similar to the procedure described above, i.e. we start at given hydrodynamic conditions with sub-saturated conditions, and increase the sediment load in small steps.

Two typical computational results are shown in Fig. 6.9a and 6.9b for a hypothetical open channel flow of 8 m depth and a tidal velocity amplitude of 0.5 m/s. We observe that a small increase in initial concentration from $C_0 = 0.28$ g/l to $C_0 = 0.29$ g/l leads to a complete collapse of concentration and turbulence field (the latter is not shown).

Under tidal conditions the tidal period T is the governing time scale of the driving force, hence the time scale of all processes should be related to T. In particular the sedimentation depth h_s, defined as the distance a sediment particle may travel during decelerating tide ($h_s = W_s T/2$) with respect to the water depth is important. As a result, we must distinguish between three regimes:

- Regime I: $h \leq h_s$ - All sediment in the water column can settle during decelerating tide to form a fluid mud layer. During the next accelerating tide, the sediment in the fluid mud layer can remix over the entire water depth.

- Regime II: $h > h_s$ - Most of the sediment in the water column can settle during decelerating tide to form a fluid mud layer; the rest remains in the water column. During the next accelerating tide, the sediment in the fluid mud layer and the sediment remaining in the water column can remix over the entire water depth – mixing of the fluid mud layer requires most of the mixing energy.

- Regime III: $h \gg h_s$ – A (small) part of the sediment in the water column can settle during decelerating tide to form a fluid mud layer; the major part remains in the water column. During the next accelerating tide, the sediment in the fluid mud layer and the sediment remaining in the water column can remix over the water depth – mixing of the remaining sediment in the water column requires most of the mixing energy.

We define capacity conditions for tidal flow as those conditions at which all sediment is completely remixed over the entire water column (though vertical concentration gradients may exist). In that case, the saturation conditions can be established with the entrainment model described in Section 9.4. This analysis is elaborated in detail in Winterwerp (1999, 2002), but does not yield simple scaling rules for saturation under tidal conditions. This is illustrated in Table 6.1, showing the dependence of C_S on a number of bulk flow parameters for extreme conditions, where we have defined the Rouse number $\beta \equiv \sigma_T W_s / \kappa u_*$, a bulk Richardson number $Ri_* \equiv \Delta g h / u_*^2$, the relative settling time $T_s' \equiv T_s / T$, and the relative mixing time $T_m' \equiv T_m / T$, where T_m is the actual vertical mixing time.

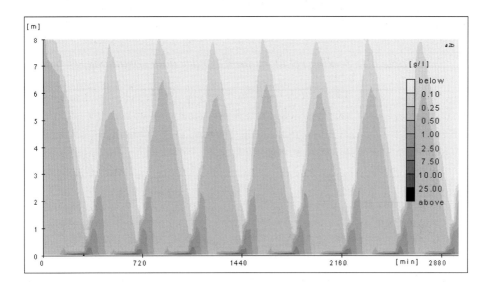

Fig. 6.9a: Computed isolutals for a saturated (C_0 = 0.28 g/l) suspension in tidal open channel flow.

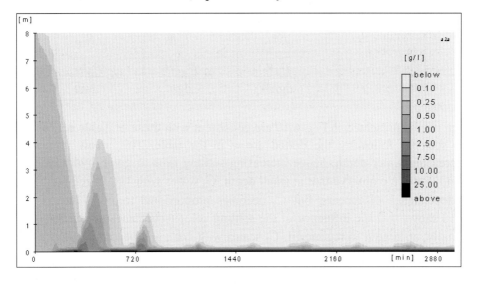

Fig. 6.9b: Computed isolutals for a super-saturated (C_0 = 0.29 g/l) suspension in tidal open channel flow.

Table 6.1 shows that different scaling relations are obtained in the different regimes. This is further illustrated in Fig. 6.10, which shows the computed variation of C_s with h as a function of W_s under tidal flow condition computed with a series of numerical 1DV experiments.

Table 6.1: Scaling relations for C_s under tidal conditions.

	$\beta \ll 1$		
	$T'_m \ll 1$ $T'_m \ll \beta$	$T'_m \ll 1$ $T'_m \gg \beta$	$T'_m \gg 1$ $T'_m \gg \beta$
regime I: $h \leq h_s$	$C_s \propto \dfrac{1}{\beta Ri_*}$ $\propto \dfrac{\rho_w u_*^3}{\Delta g h W_s}$	$C_s \propto \dfrac{1}{T'_m Ri_*}$ $\propto \dfrac{\rho_w T u_*^3}{\Delta g h^2}$	$C_s \propto \dfrac{1}{Ri_*}$ $\propto \dfrac{\rho_w u_*^2}{\Delta g h}$
regime II: $h > h_s$	$C_s \propto \dfrac{T'_s}{\beta Ri_*} \propto \dfrac{\rho_w u_*^3}{\Delta g T W_s^2}$		
regime III: $h \gg h_s$	$C_s \propto \dfrac{1}{\beta Ri_*}$ $\propto \dfrac{\rho_w u_*^3}{\Delta g h W_s}$	$C_s \propto \dfrac{1}{T'_m Ri_*}$ $\propto \dfrac{\rho_w T u_*^3}{\Delta g h^2}$	$C_s \propto \dfrac{1}{Ri_*}$ $\propto \dfrac{\rho_w u_*^2}{\Delta g h}$

The results presented in Fig. 6.10 are consistent with those of Table 6.1, as can be seen by following the dotted arrow in the table. This arrow represents increasing water depth, hence increasing settling time at given hydrodynamic conditions. We observe that at small depth, C_s scales with $1/h$. In regime II, the h-dependence disappears, but C_s becomes proportional to W_s^{-2}. Indeed Fig. 6.10 shows that the influence of the settling velocity becomes more important. Finally, at large h in regime III, C_s scales with $1/h^2$, and becomes independent of W_s, as also shown in Fig. 6.10.

The saturation concept allows a qualitative analysis of the occurrence of mud banks in coastal areas, where wave effects are important. The erosive capacity of waves largely exceeds their mixing capacity. Hence one may expect that under storm conditions, coastal sites may become super-saturated when abundant amounts of mud are available in and mobilised from the seabed.

Mud in the seabed may be mobilised by liquefaction of the bed by stresses induced by waves (e.g. De Wit, 1995; Van Kessel, 1997; and Van Kesteren et al., 1997). If this does not occur, the bed still may be eroded by wave-induced shear stresses at the bed. Such stresses generally exceed the flow-induced shear stresses by an order of magnitude (e.g. Section 9.5.2). Here, only the latter process is accounted for, which however is sufficient for the present analysis.

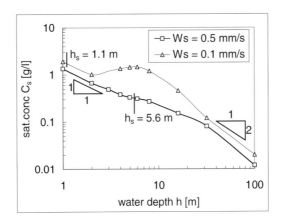

Fig. 6.10: Computed variation of saturation condition with water depth as a function of settling velocity under tidal flow conditions.

Fig. 6.11 presents the results of a 1DV computation for a hypothetical case, showing the effect of 1.8 m waves on the suspended sediment concentration in steady state open-channel flow, initially containing no sediment. Further details on this numerical experiment are given in Winterwerp (2001).

Fig. 6.11 shows that the suspended sediment concentration initially increases with time. However, after about 500 min. the vertical concentration profile starts to collapse and a fluid mud layer is formed. In this period the flow-induced shear velocity decreases from 5.9 to 3.2 cm/s. These results suggest that the flow becomes super-saturated due to the wave action, forming a fluid mud layer. Such a result can only be obtained if sediment-induced buoyancy effects are included in the computations. It is conjectured that this mechanism, which may be referred to as auto-saturation, is responsible for the occurrence of mud banks encountered in many coastal zones (e.g. Kerala coast, India; west coast of Korea; coast of Surinam, etc.).

It is noted that we have used a very simple erosion formula (e.g. equ. 9.2.1), which is probably not valid for wave-induced erosion. This formula implies that flow and waves continue to erode the seabed, as a result of which the amount of sediment in the fluid mud layer continues to grow. In reality, one would expect that waves are damped by the fluid mud layer and that the underling bed becomes protected from further erosion (see also Winterwerp et al., 1998).

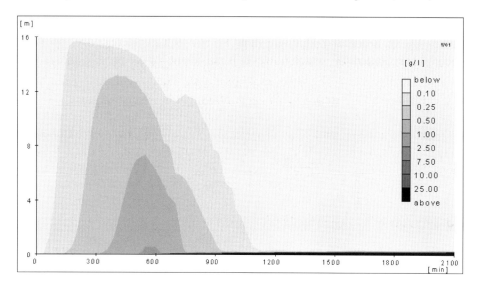

Fig. 6.11: Auto-saturation under waves – results of 1DV computation.

In this section we have elaborated on sediment-induced buoyancy effects on the vertical profiles of sediment concentration and turbulence in flows laden with cohesive sediment. The observed snowball effect, i.e. a complete collapse of the concentration and turbulent flow field, is the result of a fluid mud layer formed upon the deposition of suspended cohesive sediment flocs, which damp the mixing capacity of the turbulent flow field.

These stratification effects are further enhanced by hindered settling, which has been accounted for in the computational examples presented.

The saturation concept is based on theoretical analyses and numerical experiments, but direct empirical evidence is not available at present. The latter should be obtained from laboratory experiments under well-defined conditions. The catastrophic collapse of turbulence and concentration profile, described in this section, may not be observed in the field, because of the following reasons:

1. Sediment-laden flows probably cannot be too close to saturation, as they would be highly unstable,
2. Other generators of turbulence exist in a natural environment, such as river banks, river bends, etc., which are not affected by stratification,
3. The bed is often irregular, for instance due to ripples, small dunes, cavities, etc. Depositing mud will collect in the depressions of these irregularities. Therefore, turbulence production is still possible at the protruding tops of

the ripples, for instance, if the amount of settling mud is limited. A commonly used measure for such irregularities is the Nikuradse roughness height k_s, and we conjecture that an additional scaling parameter yields the ratio δ_m/k_s, where δ_m is the thickness of the fluid mud layer. Only if δ_m/k_s >> 1, saturation can occur.

4. A small amount of consolidation, not accounted for in the current analysis, may suffice to generate turbulence at the water-mud interface.

Yet, even under conditions that full saturation would not occur, sediment-fluid interaction may be important in the marine environment, as it governs the sediment dynamics in turbidity maxima, and triggers sediment-driven density currents, which contribute largely to the horizontal transport of cohesive sediment.

6.3.2 HCMS IN A TURBIDITY MAXIMUM

Let us examine how sediment-fluid interaction affects the sediment dynamics in a turbidity maximum. For this purpose we analyse field measurements carried out in the Ems estuary at the border of The Netherlands and Germany (Van Leussen, 1994) with the use of a 1DV-model, including the effects of sediment-fluid interaction and flocculation. For details, the reader is referred to Winterwerp (2002).

These measurements were carried out in June 1990 along five stretches of the river. At each stretch, two anchor stations at a mutual distance of approximately 1 km were deployed to measure the flow velocity with an Ott meter, the salinity from conductivity and temperature measurements and the suspended sediment concentration with a Partech turbidity meter, or by taking water samples at high suspended sediment concentrations. These parameters were measured during 13 hours, every half-hour at 0.3 m, 0.5 m, 1 m, 2 m, etc. above the bed, till 1.5 m below the water level. Along these stretches, measurements were also taken from a moving vessel, determining the relative flow velocity, salinity and suspended sediment concentration with identical instruments, and the floc size and settling velocity with a video-system (VIS). A summary of the measuring locations is presented in Fig. 6.12, from Van Leussen (1994).

The river discharge during the measurements was about 10 to 25 m³/s, which is very low for the Ems river. As a result the turbidity maximum is located around stretch B, about 15 km upstream of its usual location in the Emder Fahrwasser. Because of this low river discharge, vertical salinity gradients were virtually absent during the entire measuring campaign. Moreover, the salinity at Station 1 and 2 remained almost constant at $S \approx 0.2$ ppt.

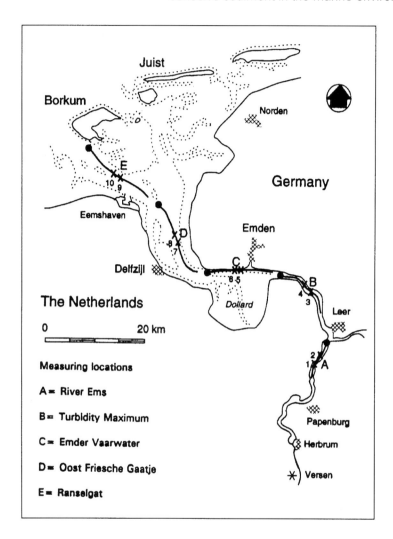

Fig. 6.12: Locations of field measurements in the Ems estuary (after Van Leussen, 1994).

The water level and flow velocity at Station 1 and 2 were almost identical and h and U at both stations were also almost in phase. The suspended sediment concentration c at Station 1 is also almost identical to that at Station 2, though during maximum flow velocity the measured values at Station 1 were somewhat larger than at Station 2. This indicates that, though advection certainly played a role, horizontal gradients in sediment concentration were small. Hence these data are suitable for analysis with the 1DV-model. Maximum values of the depth-mean concentration C varied from about 0.7 to 1.0 g/l during maximum flood and maximum ebb velocity down to about 0.3 g/l during slack water, e.g. Van Leussen (1994) and Winterwerp (2002).

Fig. 6.13, presenting the measurements in the form of isolutals (i.e. lines of equal suspended sediment concentration), shows that during flood the sediment was almost homogeneously mixed over the water depth, whereas during ebb, more stratified conditions occurred. Around HWS, the suspended sediment concentration dropped rapidly. The very large concentrations around $t = 1000$ min. are attributed to instabilities of the mud on the steep riverbanks, supplying large amounts of mud to the river (Van Leussen, 1999), and will be ignored.

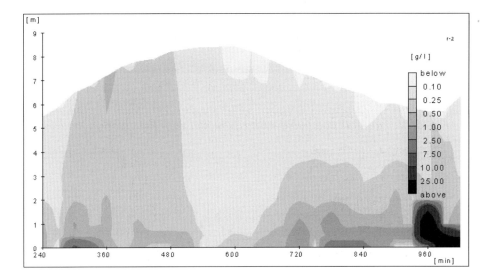

Fig. 6.13: Isolutals from measurements at Station 2.

The measurements in Station 1 and 2 have been simulated with the 1DV-model described in Appendix E, including the effects of sediment-induced buoyancy effects and turbulence-induced flocculation. The measured variation in water level and depth-mean flow velocity is prescribed, and the measured depth-mean suspended sediment concentration is set at $C_0 = 0.61$ g/l, according to the data. All sediment remains in the computational domain, i.e. sedimentation nor erosion is modelled. The results of the computations are shown in Fig. 6.14 in the form of isolutals and as vertical concentration profiles at $t = 1420$ and 1780 min. in Fig. 6.15.

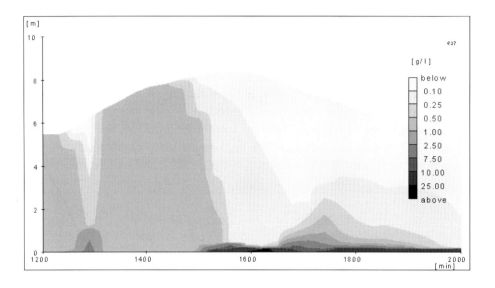

Fig. 6.14: Isolutals computed with 1DV model with flocculation model.

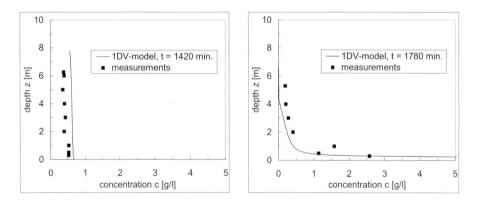

Fig. 6.15: Computed and measured vertical concentration profiles.

The computational results reproduce the vertical structure of the measured concentration distribution properly, including the characteristic rapid settling around HWS ($t = 1500$ min; note that 1000 min. should be subtracted from the values of the 1DV time axis to compare with measured data). It is noted that it was not possible to reproduce the character of the measured data with 1DV-computations with a constant settling velocity (no flocculation) or with a constant fluid density (no sediment-fluid interaction), e.g. Winterwerp (2002).

During flood, at $t = 1420$ min. the sediment is almost homogeneously distributed over the water column, whereas at $t = 1780$ min., around maximum ebb velocity, the concentration profile is highly stratified. This is further elaborated in Fig. 6.16, where the computed eddy viscosity profiles show an almost complete collapse in mixing capacity for the stratified conditions.

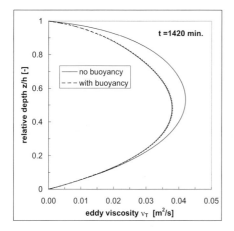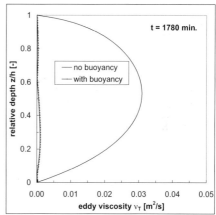

Fig. 6.16: Computed eddy viscosity profiles at t = 1420 and 1780 min.

As the turbulence field varies over the tidal cycle, the floc size, hence settling velocity, is expected to vary over the tidal cycle. This is indeed the case, as depicted in Fig. 6.17, showing the time variation in settling velocity, measured by Van Leussen (1994).

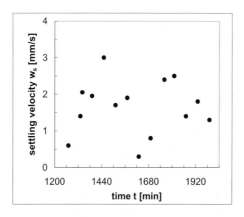

Fig. 6.17: Measured variation in settling velocity.

Such a variation in floc size, hence settling velocity, was also computed with the 1DV-model, as shown in Fig. 6.18, though the time-phase of w_s does not match the measured pattern exactly.

The absolute value of the computed settling velocity, however, agrees very well with the measurements. The large value of w_s around slack water explains the rapid settling of the sediment around slack water ($t = 1500$ min.), as observed during the measuring campaign and predicted with the 1DV-model.

The values of w_s around slack water are limited by the large flocculation time T_f around slack water, as shown in Fig. 6.19 (see also Section 4.4.2): T_f increases to many hours around slack water, as a result of which equilibrium floc sizes cannot be achieved. The large flocculation time around LWS is probably due to the large stratification at that time.

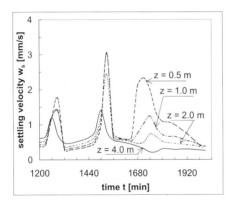 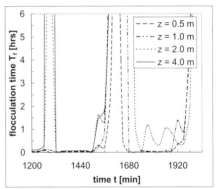

Fig. 6.18: Computed variation in   *Fig. 6.19: Computed variation in*
        settling velocity.   *flocculation time.*

Further, it is important to note that during rising tide (flood), larger settling velocities are found in the upper part of the water column, whereas during falling tide (ebb), the larger settling velocities are found near the bed (e.g. Fig. 6.8). This results in a destabilising effect during flood. The stabilising effect during ebb enhances the sediment-induced buoyancy effects. This explains the stratified conditions during ebb.

The floc size, hence settling velocity, is governed by the turbulence intensity, parameterised by G and the suspended sediment concentration c (see Section 4.3.5). Apparently for the current conditions, the "G-effect" is dominant during flood, whereas the "c-effect" is dominant during ebb. For further details, the reader is referred to Winterwerp (2001).

Next, simulations for Station 3 in the turbidity maximum are discussed (e.g. Fig. 6.12). As the horizontal gradients in concentration are fairly large in this area (Van Leussen, 1994), 1DV-simulations can only give a qualitative indication of the trends. Again, the measured variation in water depth and depth-mean velocity are prescribed, and the depth-mean suspended sediment concentration is set at the measured value of $C_0 = 5$ g/l.

The computed vertical concentration distribution at three times is compared with the measurements in Fig. 6.20, showing that the thickness of the fluid mud layer and the near-bed (fluid mud) concentrations are reasonably predicted. It is noted that the observed near-bed concentrations depicted in Fig. 6.20 can only be reproduced when the full flocculation model is included in the 1DV-computations. If relaxation effects in floc growth and break-up are omitted, the computed fluid mud concentrations would be much smaller (e.g. Section 4.4.1), because the floc size would become unrealistically large.

It is further noted that the computed near-bed concentration values are also in the range of fluid mud concentrations in estuaries and coastal areas, as reported in the literature, and presented in Table 6.3. These values show a considerable scatter, however. The large values on the Amazon Shelf may be attributed to high shear rates in the boundary layer: these large shear rates may reduce the floc size significantly, yielding large fluid mud concentrations.

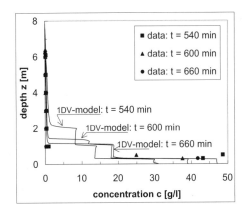

Fig. 6.20: Comparison of computed and measured concentration distributions and formation of fluid mud at Station 3, turbidity maximum.

We have shown that the suspended sediment behaviour in/near the turbidity maximum in the Ems estuary is characterised by a rapid settling around slack water and a highly stratified concentration profile during ebb. Our analysis of

the data with a 1DV-model reveals that this behaviour is governed by two processes:
1. sediment-induced buoyancy effects,
2. flocculation:
   2.a.   the large growth in floc size, hence settling velocity, causes rapid settling around slack water, and
   2.b.   the stable floc size, hence settling velocity profile during ebb enhances the effects of stratification.

The flocculation time, in relation to the time scale of the driving forces, is important in limiting the growth of the flocs, hence the settling velocity around low-turbulent slack water conditions, and the fluid mud concentration (gelling concentration) within a few tens g/l.

Table 6.3: Typical fluid mud concentrations (c_{gel}) in estuaries.

location	reference	fluid mud conc. [g/l]	remarks
Severn	Crickmore ('82) Kirby & Parker ('83)	appr. 10	
Parret	Odd et al. ('93)	40 - 80	above cons. bed
Ems Estuary	Van Leussen ('94)	appr. 40	0.35 from bed
Amazon	Kineke et al. ('95) Kineke et al. ('96)	40 - 250	0 - 2 m from bed
Loire	Le Hir ('97)	appr. 40	1 m thick layers
Jiaojiang River	Guan et al. ('98)	appr. 40	0.35 from bed

6.3.3 HCMS AND RAPID SILTATION

The Directorate-General "Rijkswaterstaat", The Netherlands Ministry of Transport, Public Works and Water Management, installed four semi-permanent anchor stations on the seabed at either side of the Maasgeul, the access channel to the Port of Rotterdam, as part of the SILTMAN-project; see Fig. 6.21. At these anchor stations suspended sediment concentration at 0.15, 0.55, 2 and 7 m and flow velocity $u_{0.35}$ and direction at 0.35 m above the seabed were monitored continuously with an optical turbidity sensor and an EMS-flowmeter, respectively. The instruments were operated continuously during the winter periods 1995/96 and 1996/97. Also data on significant wave height $H_{1/3}$, wind speed and direction and tidal elevation at stations nearby were available.

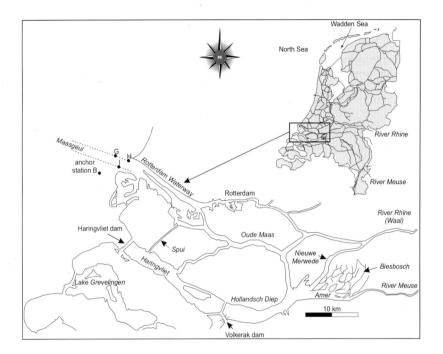

Fig. 6.21: Location of measuring stations around Maasgeul.

The 1995/96 season is characterized by calm weather without long and/or frequent stormy periods. During the entire season the sediment concentration remained small, rarely exceeding the mean summer values of several 10 mg/l. Occasionally, concentrations up to a few 100 mg/l were measured. This season was also marked by low siltation rates in the basins of Rotterdam Port and, consequently, minor dredging needs.

The 1996/97 season was more typical for the meteorological conditions in this part of the North Sea, and dredging operations had to be carried out as usual. During this period very high suspended sediment concentrations were measured frequently, exceeding 10 g/l at the lower measuring stations. Fig 6.22 shows the data at anchor station B for November 13 and 14, 1996, as an example. Wind speed and wave height were high, but not extreme. It is probably the sequence of rough weather conditions that determines the availability of the sediment; once it is available, the local hydro-meteo conditions govern its dynamics.

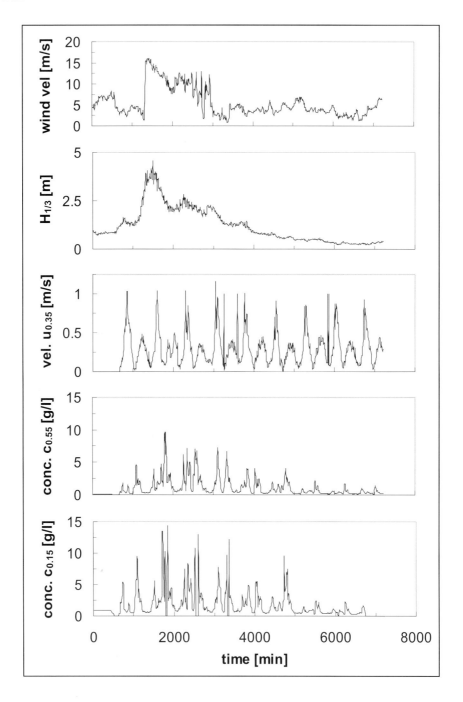

Fig. 6.22: Measurements of wind speed, wave height, flow velocity $u_{0.35}$ and concentrations $c_{0.15}$ and $c_{0.35}$ at Station B, November 13 & 14, 1996.

Inspection of Fig. 6.22 reveals that the peaks in suspended sediment concentration $c_{0.15}$ and $c_{0.55}$ measured at 0.15 and 0.55 m above the bed occur around slack water. We also observe that when $c_{0.55}$ decreases, $c_{0.15}$ increases, and vice versa. This behaviour implies that the suspended sediment is subject to settling and remixing, and that erosion of or deposition on the seabed does not play a role. In other words, all sediment seems to be sustained in the water column. Indeed, the seabed in this part of the North Sea is sandy throughout the year, containing small amounts of fines only.

These measurements are further analysed with the 1DV-model applied in previous sections (and Appendix E). We include sediment-fluid interactions (i.e. sediment-induced buoyancy effects and hindered settling), damping of vertical mixing by vertical salinity gradients induced by the Rhine outflow, and the contribution of waves to the vertical mixing; possible flocculation effects are not accounted for though. We prescribe the measured depth-mean velocity and wave height variation over time, together with the measured vertical salinity profile. The water depth is kept constant, which is acceptable as the tidal amplitude is small compared to the water depth. Again, all sediment remains within the computational domain, i.e. neither erosion nor deposition is included. For further details, the reader is referred to Winterwerp et al. (2001).

The results of the 1DV-computations are presented in Fig. 6.23 and 6.24. Fig. 6.23 shows that the behaviour of the measured suspended sediment concentration is reproduced, i.e. an increase in near-bed concentration when the concentration higher in the water column decreases, and vice versa.

Fig. 6.24 shows that the sediment is not completely mixed over the water column. This is attributed to the reduction in vertical mixing by vertical salinity gradients (e.g. Winterwerp et al., 2001).

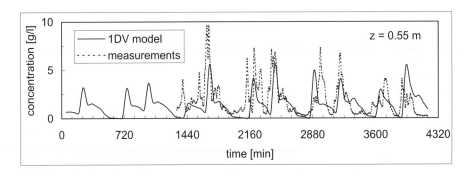

Fig. 6.23: Computed and measured concentrations at 0.55 and 0.15 m above the seabed at Station B, Nov. 13 & 14, 1996.

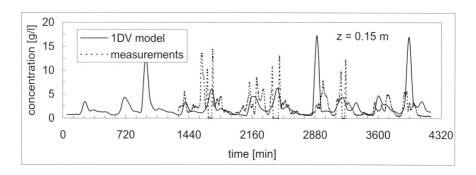

Fig. 6.23: continued

Next, the 1DV-model is used in a Lagrangean sense, i.e. it is used to assess the sediment behaviour in the Maasgeul, the navigational channel to Rotterdam Port just north of anchor station B. This channel is about 24 m deep, whereas the water depth at station B is about 16 m. As a result, the (cross flow) velocity in the channel decreases considerably, and the flow is no longer able to keep the sediment in suspension. This is depicted in Fig. 6.25.

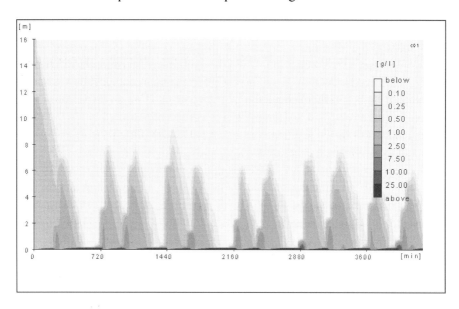

Fig. 6.24: Computed isolutals at Station B, Nov. 13 & 14, 1996; reference conditions: with buoyancy, waves, salinity and hindered settling.

These computational results suggest that, when entering the Maasgeul, the turbulence and concentration profile collapse, forming a sediment-driven density current, which develop into layers of fluid mud within the basins of Rotterdam Port. Indeed, simultaneous measurements in these basins showed concentrations of a few tens mg/l only in the major part of the water column. Only near the bed a very dense suspension was encountered ($c \approx 100$ g/l). On the basis of these 1DV-computations it was hypothesised that such sediment-driven density currents are the cause of the rapid siltation observed in the harbour basins of Rotterdam Port during storm conditions.

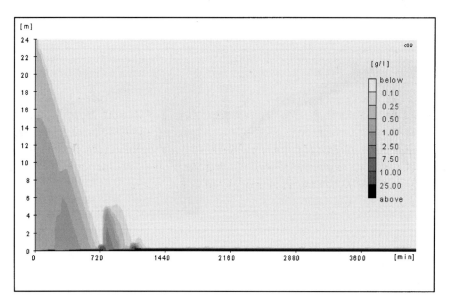

Fig. 6.25: Prognostic simulation of concentration profile in the Maasgeul, Nov. 13 & 14, 1996; reference conditions: with buoyancy, waves, salinity and hindered settling.

This hypothesis was further tested by running a full 3D-model (i.e. the DELFT3D system), simulating the suspended sediment transport in the Maasmond area. These computations were carried out with and without sediment-fluid interaction (i.e. sediment-induced buoyancy effects in the turbulence model, hindered settling and barotropic pressure effects in the momentum equations), e.g. Winterwerp and Van Kessel (2003) for details. Results of tide-averaged suspended sediment concentrations in the lower layer of the computational grid are presented in Fig. 6.26 (see also Fig. 6.21 for harbour lay-out).

The large differences in suspended sediment concentration result in large differences in computed sediment fluxes (hence siltation rates) in the Maasmond area, as depicted in Table 6.4. The cross sections at which the fluxes have been computed are indicated in Fig. 6.26. This table shows that the net flux over a tidal cycle through cross section 1, Maasmond mouth, almost triples. This is the result of an increase in both the gross import and gross export, though the import effect is the larger of the two. The major differences between the two simulations are found in the lower layers: in the case of inclusion of sediment-fluid interaction, the concentration profile becomes highly stratified, and a layer of fluid mud is formed in the harbour basins. It is expected that if non-Newtonian effects and/or consolidation of the fluid mud layers were accounted for, the difference between the two simulations would become even larger.

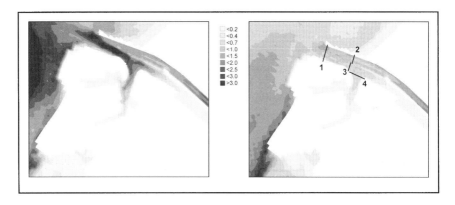

Fig. 6.26: Sediment concentration [g/l] near the bed computed with (left panel) and without (right panel) sediment-fluid interactions.

These computations have been performed with an inflow concentration of 100 mg/l at the model boundaries. This is a characteristic value for mild winter conditions. For typical summer conditions, with a boundary concentration of 10 to 50 mg/l, the effect of sediment-fluid interaction on the sediment fluxes is 10 % only. As the sediment-fluid interactions are highly non-linear, it may be expected that the increase in sediment flux would rapidly become larger when the suspended sediment concentration at sea increases to several 100 mg/l, typical values in the Dutch coast under storm conditions.

In summary, we conclude that sediment-driven density currents resulting from highly non-linear sediment-fluid interaction may significantly affect the net sediment fluxes, hence siltation rates, in navigational channels and harbour basins. These density currents generate and feed the fluid mud layers observed in many basins throughout the world.

Under extreme conditions, siltation rates can be expected to become an _____ order of magnitude larger than predicted with models that do not account for such sediment-fluid interactions.

Table 6.4: Water and sediment fluxes for Rotterdam harbour area;
see Fig. 6.26 for definition cross sections.

s-f inter action	cross section	sediment flux [kg/s]			water flux [m³/s]		
		tidal net	gross import	gross export	tidal net	max. ebb	max. flood
yes	1. Maasmond	884	1503	-620	-1,264	-12,929	10,575
yes	2. R'dam Waterway	76	466	-390	-1,304	-9,067	5,490
yes	3. Calandkanaal	792	828	-35	0	-3,457	5,470
yes	4. Beerkanaal	-404	8.0	-413	0	-2,119	1,481
no	1. Maasmond	325	719	-395	-1,263	-12,886	10,298
no	2. R'dam Waterway	50	365	-315	-1,292	-8,959	5,476
no	3. Calandkanaal	263	307	-44	0	-3,518	5,452
no	4. Beerkanaal	-126	4.8	-131	0	-2,095	1,505

6.4 SEDIMENT-FLUID INTERACTIONS IN THE BENTHIC BOUNDARY LAYER

We have shown in Section 6.2.1 that sediment-induced buoyancy effects influence the entire vertical velocity profile of sediment-laden flow. This implies that both the Von Kármàn constant κ_s and the parameters of the defect-law are affected. This has been elaborated by Toorman (2002, 2003) by integrating $du/dz = u_*/\kappa_s z$ for non-constant κ_s, and modifying the logarithmic velocity profile as follows:

$$\frac{u}{u_*} = \frac{1}{\kappa}\ln\left\{\frac{z}{\alpha z_0}\right\}$$

(6.40)

where u_* is the sediment-affected shear velocity, κ the reference Von Kármàn constant for buoyant-neutral conditions, z_0 the roughness length for buoyant-neutral conditions and α is a drag-reduction factor ($\alpha \leq 1$), which is related to the momentum damping function F (e.g. equ. 6.47)) through an integral relationship.

Equ. (6.40) implies drag reduction through a reduction of z_0. The reduction factor α varies over the water depth and is a function of the Richardson number. On the basis of a series of numerical experiments, Toorman proposes the following explicit relation for α:

$$\alpha = \exp\left\{-\left(1 + a\frac{W_s}{u_*}\right)\left(1 - \exp\{-bRi^n\}\right)\right\} \tag{6.41}$$

where Ri is the gradient Richardson number, and the various coefficients were established as $a = 7.7$, $b = 1.7$ and $n = 0.85$. This relation yields drag reduction up to 70 to 80 % as a result of sediment-induced buoyancy effects.

It is noted that for hydraulically rough conditions a quadratic friction law, based on the log-law of the wall, is generally applied as boundary condition for the integration of the momentum equation (2.2). Toorman argues that sediment-induced changes in the velocity profile, hence in the log-law, should be included in the boundary conditions for determining the vertical velocity profile, i.e. (6.40) should be applied instead of $du/dz = u_*/\kappa z$.

Further to these arguments, Toorman reasons that the boundary conditions for the k-ε turbulence model (and for other turbulence models as well) should also be modified. At equilibrium, a balance exists between turbulence production P, turbulence dissipation ε and buoyancy destruction B: $P = B + \varepsilon$, or $P(1 - Ri_f) = \varepsilon$. As a result, the common boundary conditions for the k-ε turbulence model would become (see also Section 2.3):

$$k\big|_{z=Z_b} = \frac{u_*^2}{\sqrt{c_\mu}}\sqrt{1 - Ri_f}$$

$$\varepsilon\big|_{z=Z_b} = \frac{u_*^3}{\kappa_s z_0}\left(1 - Ri_f\right) \tag{6.42}$$

The flux Richardson number Ri_f can be regarded as an efficiency parameter for vertical mixing in stratified flows. Hence, the critical flux Richardson number

$Ri_{f,c}$ is a measure for the mixing capacity of the flow. Intuitively it follows that capacity conditions (saturation) occur when Ri_f becomes critical throughout the water depth. This assumption has implicitly been used in deriving the saturation conditions in Section 6.3.1.

This reasoning would also imply that Ri_f becomes constant over the water depth at saturation. This is what Toorman (2000, 2003) found from numerical experiments studying the behaviour of sediment-laden flow in an open channel with a $k\text{-}\varepsilon$ turbulence model, and from a re-analysis of the physical experiments by Cellino and Graf (1999). Such a constant $Ri_f (\equiv B/P)$ implies:

$$\frac{dB}{dz} = -Ri_f \frac{dP}{dz} \tag{6.43}$$

Assuming local equilibrium and a constant settling velocity, W_s, this yields:

$$gW_s \frac{\rho_w}{\rho^2} \frac{d\rho}{dz} = -Ri_f u_*^2 \frac{d}{dz}\left(\left(1 - \frac{z}{h}\right)\frac{du}{dz} \right) \tag{6.44}$$

Next, a logarithmic velocity profile is assumed, i.e. $du/dz = u_*/\kappa_s z$. Toorman found that the effective Von Kármàn κ_s is constant over the water depth at saturation conditions, yielding a formulation for the eddy viscosity profile at saturation:

$$v_T = \sigma_T \frac{\rho_w}{\rho} W_s z \left(1 - \frac{z}{h}\right) \approx \sigma_T W_s z \left(1 - \frac{z}{h}\right) \tag{6.45}$$

Note that (6.45) still yields a parabolic profile. By comparison with the "classical" eddy viscosity profile, equ. (6.45) gives a relation for the slope of the logarithmic velocity profile at saturation, or for the effective, sediment-reduced Von Kármàn constant κ_s:

$$\kappa_s = \frac{\sigma_T W_s}{u_*} \tag{6.46}$$

Note that the shear velocity and the turbulent Prandtl-Schmidt number σ_T are both affected by sediment-induced buoyancy effects. The effect on the latter can be described with classical damping functions:

$$\sigma_T = \frac{F(Ri)}{G(Ri)} \sigma_{T,0} \qquad\qquad (6.47)$$

where $\sigma_{T,0}$ is the Prandtl-Schmidt number for buoyant-neutral conditions and F and G are damping functions for eddy viscosity and eddy diffusivity, respectively.

At present, it is not clear how important these modifications are in practical situations, i.e. for predicting the transport and fate of cohesive sediment. It can be argued that cohesive sediment suspensions in nature cannot be too close to saturation, as such sediment-laden flow would be highly unstable. Thus, upon small disturbances in the velocity field, the turbulence field and suspension would collapse, resulting in the formation of fluid mud. Away from saturation, the corrections described above are fairly small and inclusion for engineering purposes may not be necessary. When the flow becomes super-saturated, for instance in a navigation channel, the precise conditions at saturation are not important either.

On the other hand, these alternative formulations would imply some modification of the saturation condition itself (Toorman, 2002). Moreover, it was argued that the behaviour of mud banks in coastal areas is related to a process referred to as auto-saturation. In this case the precise formulations at saturation are important. Note however, that the boundary conditions described above have been derived for rigid walls only. We therefore feel that Toorman's arguments need more attention and should be validated with data and/or direct numerical simulations of sediment-laden flow.

7. SELF-WEIGHT CONSOLIDATION

In this chapter we treat self-weight consolidation, i.e. consolidation of cohesive sediment deposits under the influence of its own over burden. We limit ourselves to the case of a constant or increasing load. Effects of load relaxation, resulting in swell of the sediment are discussed in Chapter 8.

We present a detailed derivation of the equation for self-weight consolidation of soft mud, i.e. accounting for large deformations in the vertical direction. This equation is known as the Gibson equation, and we present its derivation both in a Eulerian reference frame, and in a material reference frame, commonly used in soil mechanics.

We present the classical constitutive relations for permeability and (Terzaghi's) effective stress, and new relations assuming a self-similar structure of the bed. The latter allow formulation of consolidation in the form of an advection-diffusion equation, which can be combined easily with equations for the water movement, including the effects of small amounts of sand. We present a novel technique to determine coefficients of the constitutive equations from simple experiments.

We have included a number of examples on consolidation experiments and their simulation with numerical consolidation models. This enables the reader to get a feeling for the applicability and accuracy of state-of-the-art consolidation modelling, for engineering purposes.

7.1 INTRODUCTION

In Chapter 5 we discussed how and under what conditions mud flocs can settle. When sedimentation continues, more and more mud flocs accumulate on the bed, and the flocs that arrived first are squeezed by the ones on top. Pore water is driven out of the flocs and out of the space between the flocs. This process is known as self-weight consolidation, a process resulting in large (vertical) deformations of the bed. In the (hindered) settling phase discussed in Chapter 5, the flocs are supported wholly by the upward fluid flow. In the consolidation phase, flocs are more and more supported by particle interactions. The upward fluid flow during hindered settling and consolidation is not a priori homogeneously distributed in the horizontal plane, and localisation of flow in channels (piping) may occur. In this chapter we focus on the one-dimensional consolidation process and therefore assume no inhomogeneous pore water flow.

Terzaghi (1943) showed that the stress state in a (consolidating) bed is a function of the difference between the total stress and the pore water pressure; this difference is referred to as the effective stress (see also Section 2.1, equ. (2.5) and Section 8.1). Note that the effective stress is a mathematical concept and not equal to the particle stresses in the soil. It is important to recognise that in soil mechanics the deviatoric stresses in the water phase are neglected. Hence, the superposition of skeleton stresses (not particle stresses) and fluid stresses is reduced to a superposition of isotropic stresses only. Moreover, we are primarily interested in processes in vertical direction, as this is the direction of the largest deformations in the consolidating mud (direction of gravity). As a result, (2.5) becomes (see also (2.4)):

$$\sigma_{zz} = p^{w} + \sigma_{zz}^{sk} \tag{7.1a}$$

which is commonly written in classical soil mechanics as:

$$\sigma_{v} = u + \sigma_{v}' \tag{7.1b}$$

where $\sigma_{zz} = \sigma_{v}$ is the total vertical stress, $\sigma_{zz}^{sk} = \sigma_{v}'$ is the vertical effective stress and $p^{w} = u$ is the pore water pressure.

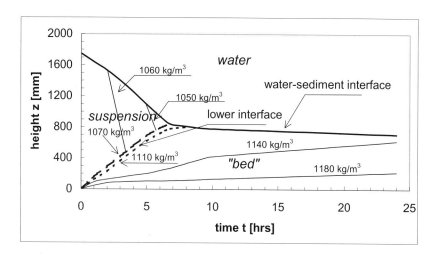

Fig. 7.1: Settling and consolidation measured in a 2 m high settling column (redrawn from Been and Sills, 1981).

Been and Sills (1981) presented an illustrative picture of the settling and consolidation phase of mud flocs on the basis of experiments in a 2 m high consolidation column. Fig. 7.1 shows the lowering of the interface with time, and the development of a second, lower interface between suspension and bed (see Sections 5.3 for the conditions under which this second interface can develop). During settling and consolidation, two phases are recognised: the dense fluid phase and the "bed"-phase – note that the bed can still be very soft. Been and Sills also measured the bulk density of the sediment-water mixture, the results of which are presented in Fig. 7.1 as well. This is further elaborated in Fig. 7.2, showing the vertical distribution of bulk density and pressure measured at 4:45 hrs after the start of the experiment. The left panel clearly shows two interfaces, around z = 1200 mm and around z = 600 mm; near the bed (z = 100 mm), even a third density jump can be distinguished. The pressure distribution in the right panel shows the divergence of total and pore pressure, i.e. the development of effective stresses, supporting the flocs. The actual pore pressure consists of a hydrostatic part and the so-called excess pore pressure p_e. The latter dissipates during the consolidation process, and vanishes when all particles are supported by inter-particle contacts, i.e. by the effective stress.

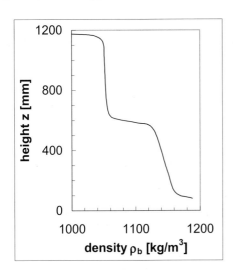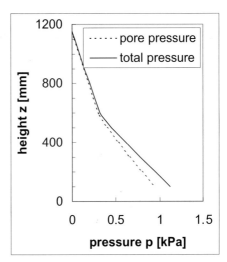

Fig. 7.2: Vertical density and stress distribution measured at 4:45 hrs in a 2 m high settling column (redrawn from Been and Sills, 1981).

In soil mechanics, the state of the consolidating bed is commonly given by the so-called void ratio e, i.e. the ratio of the volume of voids (pores) to the volume of solids. In this book, we prefer the use of mass or volumetric concentration of the solids c or ϕ. Relations between e and c are given by:

$$e = \frac{\rho_s - c}{c} \quad \text{or} \quad \phi_s = \frac{1}{1+e} \quad \text{or} \quad c = \frac{\rho_s}{1+e} \tag{7.2}$$

Chapter 3 and Appendix B contain more relations between a number of parameters often used in soil mechanics and hydrodynamics.

7.2   THE GIBSON CONSOLIDATION EQUATION

Consolidation of soft mud layers is described by the so-called Gibson equation (Gibson et al., 1967), which is generally considered to be the state-of-the-art for engineering applications. We present a detailed derivation of this equation, to be compared with the derivation in Section 7.3. The reader is referred to Mitchel (1976) and Schiffman et al. (1985), amongst others, for further details. Here, we follow Merckelbach (1996), who carried out a thorough analysis of all steps in the derivation.

We start again with the simple wave equation in vertical direction z (e.g. equ. (5.19)), obtained from the mass balance equation (2.11):

$$\frac{\partial \phi_s}{\partial t} + \frac{\partial}{\partial z}(v_s \phi_s) = 0 \tag{7.3}$$

where ϕ_s is the volumetric solids concentration and v_s is the velocity of the mud flocs relative to a fixed reference plane, as shown in Fig. 7.3. The motion of the pore water relative to the same datum is denoted by v_f. Note that v_f is the actual pore water velocity relative to the fixed reference plane. Substitution from (7.2) yields an equation for the void ratio:

$$\frac{\partial e}{\partial t} + (1+e)^2 \frac{\partial}{\partial z}\left(\frac{v_s}{1+e}\right) = 0 \tag{7.4}$$

The actual pore water velocity relative to the flocs is defined by $v_{eff} = v_f - v_s$. In groundwater mechanics it is common to use the specific discharge q defined as the pore water flux per unit cross section, yielding $q = n \times v_{eff}$ [m^3/s/m^2]. The actual pore water flow can be modelled with Darcy's law (also known as Darcy-Gersevanov's law):

$$\frac{e}{1+e}\left(v_f - v_s\right) = -k\frac{1}{g\rho_w}\frac{\partial p_e}{\partial z} \qquad (7.5)$$

where k is the permeability of the soil and p_e is the excess pore water pressure induced by the self-weight overburden, and defined as the difference between the actual pore water pressure p^w and the hydrostatic pore water pressure:

$$p_e = p^w - \int_z^{Z_s} g\rho_w \mathrm{d}z' = p^w + g\rho_w\left(z - Z_s\right) \qquad (7.6)$$

in which Z_s is the level of the water surface. Note that (7.6) implies that interstitial water can flow through all the pores of the mud, i.e. no distinction is made between likely variations in hydraulic resistance within the network-forming flocs and the space in between. This effect is empirically accounted for through the value of the bulk permeability parameter k.

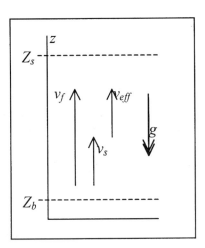

Fig. 7.3: Definition of floc and fluid velocity v_f and v_s, and water surface level Z_s and bed level Z_b.

The vertical gradient in total pressure σ_{zz} follows from the vertical balance at equilibrium of the two-phase mixture:

$$\frac{\partial \sigma_{zz}}{\partial z} = -g\rho = -\frac{e}{1+e}g\rho_w - \frac{1}{1+e}g\rho_s \qquad (7.7)$$

Because of continuity, the net volume flux of the mixture should be zero in the closed system under consideration (i.e. no base flow):

$$\frac{e}{1+e}v_f + \frac{1}{1+e}v_s = 0 \qquad (7.8)$$

Substitution of (7.8) into (7.5), and using (7.6) and the effective stress principle (7.1), the vertical velocity of the flocs v_s becomes:

$$v_s = k + \frac{k}{g\rho_w}\left(\frac{\partial \sigma_{zz}}{\partial z} - \frac{\partial \sigma_{zz}^{sk}}{\partial z}\right) \qquad (7.9)$$

Substitution of (7.7) and (7.9) into (7.4) yields the one-dimensional consolidation equation (the Gibson equation) in Eulerian co-ordinates for the void ratio e:

$$\frac{\partial e}{\partial t} + (1+e)^2\left(\frac{\rho_s - \rho_w}{\rho_w}\right)\frac{\partial}{\partial z}\left(\frac{k}{(1+e)^2}\right) +$$
$$+ \frac{(1+e)^2}{g\rho_w}\frac{\partial}{\partial z}\left(\frac{k}{1+e}\frac{\partial \sigma_{zz}^{sk}}{\partial z}\right) = 0 \qquad (7.10)$$

Numerical solution of this non-linear equation is difficult because of the moving upper interface (boundary conditions). Therefore, it is common practice to rewrite (7.10) in a moving reference frame, the so-called material or Lagrange co-ordinate system. A vertical material co-ordinate ζ is introduced that represents the volume of solids:

$$\zeta(z,t) = \int_0^z \frac{dz'}{1+e(z',t)} \tag{7.11}$$

The time τ in the material co-ordinate system is equal to the time in the original reference frame: $t = \tau$. Transformation of (7.4) and (7.9) into material co-ordinate system (ζ, τ) yields:

$$\frac{\partial e}{\partial \tau} + \frac{\partial v_s}{\partial \zeta} = 0 \tag{7.12}$$

and:

$$v_s = -\frac{\rho_s - \rho_w}{\rho_w} \frac{k}{1+e} - \frac{k}{g\rho_w(1+e)} \frac{\partial \sigma_{zz}^{sk}}{\partial \zeta} \tag{7.13}$$

Substitution of (7.13) into (7.12) yields

$$\frac{\partial e}{\partial \tau} + \left(\frac{\rho_s - \rho_w}{\rho_w}\right)\frac{\partial}{\partial \zeta}\left(\frac{k}{1+e}\right) + \frac{\partial}{\partial \zeta}\left(\frac{k}{g\rho_w(1+e)}\frac{\partial \sigma_{zz}^{sk}}{\partial \zeta}\right) = 0 \tag{7.14}$$

In classical soil mechanics it is assumed that k and σ_{zz}^{sk} are a function of void ratio e only. This implies that history effects, channelling and formation and break-up of flocs are not included. With these assumptions the classical Gibson equation in material co-ordinates is obtained:

$$\frac{\partial e}{\partial \tau} + \left(\frac{\rho_s - \rho_w}{\rho_w}\right)\frac{d}{de}\left[\frac{k}{1+e}\right]\frac{\partial e}{\partial \zeta} + \frac{\partial}{\partial \zeta}\left[\frac{k}{g\rho_w(1+e)}\frac{d\sigma_{zz}^{sk}}{de}\frac{\partial e}{\partial \zeta}\right] = 0 \tag{7.15}$$

The material functions $k(e)$ and $\sigma_{zz}^{sk}(e)$ are discussed in Section 7.4. The coefficient in the third term of (7.15) is also known as the coefficient of consolidation c_v – compare with (7.18), which for small strains reduces to the consolidation coefficient originally defined by Terzaghi:

$$c_v = -\frac{k(e)}{g\rho_w(1+e)}\frac{\mathrm{d}\sigma_{zz}^{sk}(e)}{\mathrm{d}e} = -\frac{k}{g\rho_w(1+e_0)}\frac{\mathrm{d}\sigma_{zz}^{sk}(e)}{\mathrm{d}e} \qquad (7.16)$$

The Gibson equation is a parabolic equation and needs two boundary conditions, one at the water-sediment interface, and one at the rigid bottom. The numerical solution is then straightforward. Moreover, seepage at the base of the consolidating sediment (double drainage) can be included.

It is noted that the advantage of a moving co-ordinate system vanishes when more sediment settles on, or is eroded from the water-sediment interface during the consolidation process: the upper boundary layer becomes complicated again. This is of course exactly what is happening in natural marine environments where sedimentation, erosion and consolidation occur continuously. We therefore elaborate further on the Eulerian form of the consolidation equation in Section 7.6 and further.

7.3   SPECIAL CASES OF THE GIBSON EQUATION

The Gibson equation can be regarded as a generalisation of two special cases developed earlier by Terzaghi and Kynch. This is shown in the next two sections.

7.3.1   TERZAGHI'S CONSOLIDATION THEORY

Terzaghi (1943) derived a theory of consolidation for small deformations, i.e. no self-weight consolidation, which reads in Eulerian co-ordinates:

$$\frac{\partial \sigma_{zz}^{sk}}{\partial t} - c_v\frac{\partial^2 \sigma_{zz}^{sk}}{\partial z^2} = 0 \qquad (7.17)$$

where the consolidation coefficient c_v is defined as:

$$c_v = -\frac{k(1+e)}{g\rho_w}\frac{\partial \sigma_{zz}^{sk}}{\partial e} = \frac{k}{m_v g\rho_w}$$ (7.18)

in which $m_v = -1/\{(1+e)\partial \sigma_{zz}^{sk}/\partial e\}$ is the coefficient of compressibility m_v.

The consolidation coefficient c_v plays an important role in quantifying the response of a cohesive sediment bed to external loading, as discussed in Chapter 8. With a few assumptions, Terzaghi's consolidation equation (7.17) follows directly from (7.10):

1. small deformations, i.e. the void ratio does not deviate much from the initial void ratio, thus the second (advection) term in (7.10) vanishes and higher order, non-linear terms may be neglected, and
2. further to 1., the permeability k is kept constant.

Hence, (7.10) becomes:

$$\frac{\partial e}{\partial t} + \frac{(1+e)k}{g\rho_w}\frac{\partial^2 \sigma_{zz}^{sk}}{\partial z^2} = 0$$ (7.19)

Upon substitution from (7.18) into (7.19), Terzaghi's equation (7.17) is obtained. This equation is applicable in the final phase of consolidation of soft cohesive sediment deposits when deformations become small.

7.3.2   KYNCH'S SEDIMENTATION THEORY

Kynch's equation (1952) was introduced in Chapter 5 (equ. (5.19)), and is repeated here for convenience:

$$\frac{\partial c}{\partial t} + \frac{d(v_s c)}{dc}\frac{\partial c}{\partial z} = 0$$ (7.20)

in which the vertical velocity of the flocs is a function of concentration, as discussed in Section 5.3. Kynch used the following relation:

$$v_s = W_s = W_{s,r}\left(1 - \phi_s\right) \tag{7.21}$$

where ϕ_s is the volumetric concentration of the solids. Hence, Kynch accounted for the (linear) effects of return flow only (e.g. Section 5.3).

Substitution of (7.2) into (7.20) yields the continuity equation (7.4). Darcy's law (7.9) can be rewritten as:

$$v_s = -k\left[\frac{1}{1+e}\frac{\rho_s - \rho_w}{\rho_w} - \frac{1}{g\rho_w}\frac{\partial \sigma_{zz}^{sk}}{\partial z}\right] \tag{7.22}$$

Next, it is assumed that v_s is a function of void ratio only (Been, 1980). Hence the second term on the right-hand side of (7.22) is omitted. Substitution into (7.20), c.q. (7.4) then yields:

$$\frac{\partial e}{\partial t} + \left(1+e\right)^2\left(\frac{\rho_s - \rho_w}{\rho_w}\right)\frac{\partial}{\partial z}\left(\frac{k}{\left(1+e\right)^2}\right) = 0 \tag{7.23}$$

which is exactly Gibson's equation in Eulerian co-ordinates (7.10) after omitting its last term. Hence, in Kynch's model it is assumed that no effective stresses develop. Kynch's equation is therefore applicable in the earlier phase of the consolidation of soft mud deposits, when effective stresses are small.

After some algebra we find that the settling function $F(\phi)$, introduced in Section 5.3, for the Kynch equation with Been's settling term becomes:

$$F(\phi) \propto \frac{dk\phi^2}{d\phi} \tag{7.24}$$

It can be shown that the commonly used material functions imply that $dF/d\phi$ is always negative, i.e. that two interfaces always develop in the early consolidation regime, using those material functions (e.g. Section 5.2.2).

7.4   MATERIAL FUNCTIONS FOR THE GIBSON EQUATION

The consolidation model (either (7.10), (7.15) or (7.19)) can only be solved when relations for k and σ_{zz}^{sk}, the so-called material functions or constitutive relations, are known. Alexis et al. (1992) collected a large amount of data on permeability and effective stress, available at that time. Though the data showed a general trend of decreasing permeability and increasing effective stress with decreasing void ratio, Alexis could not obtain a general relation.

At present, it is common to correlate experimental data with the following material functions (e.g. Schiffman et al., 1985; Hu, 1990; Townsend and McVay, 1990; and Liu and Znidarcic, 1991):

$$e = A_p \left(\sigma_{zz}^{sk} / \sigma_{zz,ref}^{sk} - \sigma^* \right)^{-B_p} \qquad \text{where} \quad \sigma_{zz,ref}^{sk} = 1 \text{ Pa}$$

$$\left(k / k_{ref} \right) = A_k e^{B_k} \qquad \text{where} \quad k_{ref} = 1 \text{ m/s} \tag{7.25}$$

The reference effective stress $\sigma_{zz,ref}^{sk}$ and permeability k_{ref} provide the dimensionless coefficients A_p and A_k. The parameter σ^* is introduced by Znidarcic (1991) to reflect a zero effective stress condition at finite void ratio. It can be used to include over-consolidated conditions in the material functions. Typical values for the coefficients in (7.25) show a large scatter as shown in Table 7.1.

Table 7.1: Typical values of coefficients in material functions (7.25).

A_p	B_p	A_k	B_k	σ^*
1 - 100	0.1 - 1	$10^{-8} - 10^{-14}$	1 - 10	0-10^5

The actual value of these coefficients depends on grain size distribution (clay content), organic content, activity and pore size distribution. A classification of soils based on these parameter is treated in Chapter 3. This classification can be used to quantify the coefficients A and B in equ. (7.25). The plasticity limits (PL and LL) represent a certain undrained shear strength, e.g. effective stress and permeability, as shown in Fig. 3.18 and 3.19. The coefficients in (7.25) can be expressed as a function of the clay content ξ^{cl} and activity A for a specific regime in the plasticity chart (Fig. 3.16). Fig. 7.4 shows the ranges for the coefficients as a function of ξ^{cl} for inorganic cohesive sediment, which are bounded by the "A" and "B"-line[1] in Fig. 3.16. Some experimental values are also included in Fig. 7.4.

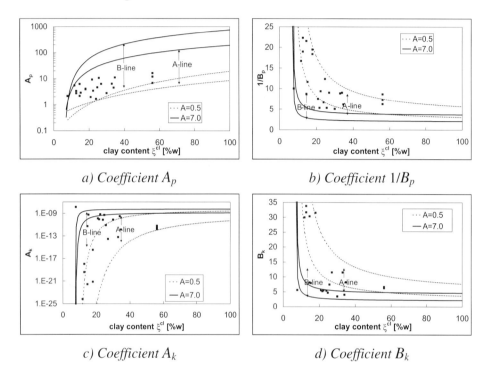

a) Coefficient A_p

b) Coefficient $1/B_p$

c) Coefficient A_k

d) Coefficient B_k

Fig. 7.4: Range of coefficients as a function of clay content ξ^{cl} and activity A based on geotechnical soil classification (e.g. Chapter 3).

The reciprocal value of the coefficient B_p has been plotted (Fig. 7.4b) to compare with B_k (Fig. 7.4d). The ranges in $1/B_p$ and B_k appear to be very similar. In Fig. 7.5 experimental data for both exponents are correlated, showing that they are of the same order of magnitude for the majority of cohesive sediments that range from natural salt and fresh water deposits to industrial sediments like mine tailings, waste water and drinking water plants (see also Section 7.3). There is a lower bound for sand dominated sediments, with a slope less than unity. Above this line a wide range of data points is found for all kinds of cohesive sediments. However when only the data points from settling columns are considered (Merckelbach, 2000; Townsend and McVay, 1990; and Liu and Znidarcic, 1991), it appears that the data points are located

---

[1] The A-line distinguishes between inorganic and organic/silt-rich sediment, and the B-line envelops natural sediments (e.g. Section 3.3).

on the line $1/B_p = B_k$, which is the relation predicted by fractal theory (see Section 7.6). The other data points are obtained from oedometer tests and Seepage Induced Consolidation tests on in-situ samples (e.g. Appendix C.5), which results in initial stress states differing from those in a deposited bed in a settling column. Only some of these data points are found on the "fractal theory" line. An exception is the data point for the Sidere experiment in a settling column (Bartholomeeusen et al., 2002), but these column tests were performed on sandy silt with only 3% clay content.

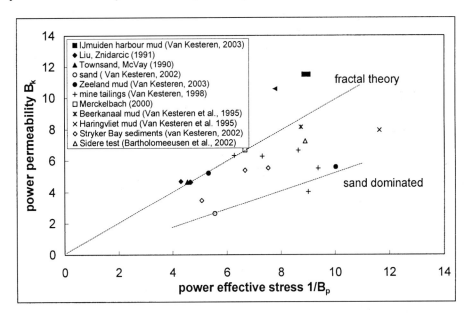

Fig. 7.5: Comparison power for permeability and effective stress.

Instead of the power functions (7.25), also exponential functions are widely used (Wichman, 1999 and Bartholomeeusen et al., 2002):

$$\sigma_{zz}^{sk} = \exp\{m_1 + m_2 e\} + m_3 \exp\{m_4 + m_5 e\}$$
$$k = \exp\{m_6 + m_7 e\}$$
(7.26)

in which m_i ($i = 1,7$) are coefficients to be determined experimentally. Two exponential functions for the effective stress in (7.26) are necessary to describe a wide range of void ratio's from very soft mud to a stiff clay. They represent the bimodal relation between undrained shear strength and water content, as found for cohesive sediments in general (see Chapter 3). For water contents above the liquid limit or below the plastic limit only one exponential function suffices.

The validity range for the power functions (7.25) is much larger than that for the exponential functions (7.26). However, as indicated in Chapter 3, two different power functions may also be necessary to reflect the bimodal relationship over the full range of water content from soft mud to stiff clay (Fig. 3.18). Moreover, we advocate the application of the power functions because there is a physical explanation based on the nature of skeleton fabric of cohesive sediments during consolidation as explained in Section 7.6.

Toorman (1996, 1997, 1999) proposes to integrate the hindered settling and consolidation regime. To that purpose, he rewrote the Gibson equation in Eulerian co-ordinates as:

$$\frac{\partial \phi_s}{\partial t} - \frac{\partial}{\partial z}\left[w_0(1-\varepsilon)\phi_s + (D_e + D_B)\frac{\partial \phi_s}{\partial z} \right] = 0 \qquad (7.27)$$

where:

$$w_0 = W_{Stokes}\exp\left\{ -\frac{\phi_s}{\phi_{max}}(1-\phi_s) \right\},$$

$$\varepsilon = \left(\exp\left\{ \frac{\phi_s}{\phi_{max}} - 1 \right\} \right)^b$$

$$D_e = w_0(1-\varepsilon)\frac{\phi_s}{\phi_{max}}\lambda H_\infty \exp\left\{ \lambda\left(1 - \frac{\phi_s}{\phi_{max}}\right) \right\}$$

in which W_{Stokes} is the Stokes settling velocity, ϕ_{max} is the maximal volumetric particle concentration, λ is a dimensionless model parameter, H_∞ is the final bed thickness and D_B is the Brownian diffusion coefficient. The only constitutive material parameters are λ and ϕ_{max}.

The formulation of the diffusion coefficient D_e corresponds to a Peclet number Pe for pore water dissipation, that is only a function of ϕ_s/ϕ_{max}:

$$Pe = \frac{w_0 H_\infty}{D_e} = \left[(1-\varepsilon) \frac{\phi_s}{\phi_{max}} \lambda \exp\left\{ \lambda\left(1 - \frac{\phi_s}{\phi_{max}} \right) \right\} \right]^{-1} \tag{7.28}$$

The Peclet number Pe is a measure for the degree (or rate) of consolidation: for $Pe < 1$, the process is drained (final phase of consolidation; pore water pressures dissipated for more than 85%), and for $Pe > 10$, the process is undrained (initial phase of consolidation; pore water pressures dissipated for less than 15%); see also Section 8.1.2.

7.5   APPLICATION OF GIBSON'S EQUATION

Townsend and McVay (1990) presented a series of numerical consolidation experiments that have been used for an intercomparison of various numerical consolidation models in Florida's mining industries. It is regarded as a benchmark in the literature, and is widely used to verify the proper implementation of the consolidation equation in a numerical code.

All computations have been carried out with numerical models based on the classical Gibson equation, which is generally accepted to yield a proper description of the consolidation process. The test case referred to as "Scenario A" is presented here. This case consists of a numerical single-drained, self-weight consolidation experiment with a slurry of initial height $\delta_0 = 9.55$ m and initial uniform void ratio $e_0 = 14.8$, which is beyond the structural density. This implies that hindered settling effects do not play a role. The material functions in non-dimensionless form are given by:

$$k = 0.2532 \cdot 10^{-6} e^{4.65}, \text{ and} \tag{7.29a}$$

$$e = 7.72 \sigma_{zz}^{sk -0.22}, \text{ yielding } \sigma_{zz}^{sk} \approx 10.8 \cdot 10^3 e^{-4.55} \tag{7.29b}$$

where k has the dimension [m/day] and σ_{zz}^{sk} has the dimension [kPa]. The results of these numerical experiments are presented in Fig. 7.6, showing that most computational results collapse to one solution, indicating a proper numerical discretisation of the equations.

Toorman and co-workers (1991, 1993, 1997 and 1999) have published on a number of experiments in consolidation columns, their analyses and numerical simulation. Here we show results from column tests and their numerical simulation with mud from the Scheldt River, Belgium, as presented in Toorman and Berlamont (1991). The initial suspension height amounted to 1.96 m and the initial density to 1053 kg/m^3. The sand fraction was estimated at 3 %. The material functions were derived from direct measurements in the column.

A comparison of measurements and numerical computations is given in Fig. 7.7, showing excellent agreement between observations and computations. It can be concluded that in this particular case (i.e. 1D consolidation in confined space) the Gibson equation describes the consolidation process adequately, when the material functions are known in detail.

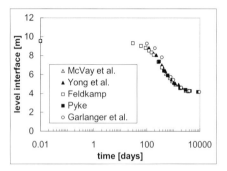 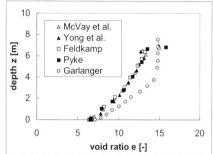

Fig. 7.6: Interface as a function of time and vertical void ratio distribution from numerical experiments (after Townsend and McVay, 1990).

In 2002, the University of Oxford took the initiative to set up a series of consolidation experiments with mud from the River Scheldt to serve as a data-bank for testing consolidation models. A number of researchers were invited to model these experiments. The results are published in Bartholomeeusen et al. (2002).

The series of experiments consisted of two sub-sets. The measured interface level and vertical density profiles of four experiments were provided to be analysed and to establish the material functions. Next, the initial conditions for a fifth experiment with the same mud were provided, and the researchers were invited to predict the consolidation process, using the material functions derived from the calibration tests. The measured interface level and density profiles were revealed only after the modelling exercise. The results for this fifth experiment are presented in Fig. 7.8 and 7.9, showing large differences between predictions and observations.

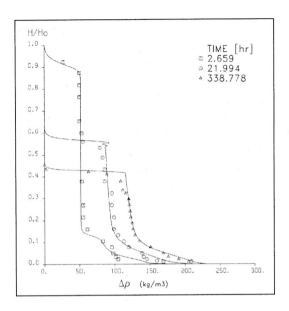

Fig. 7.7: Observed and predicted vertical density profiles in a consolidation column (after Toorman and Berlamont, 1991).

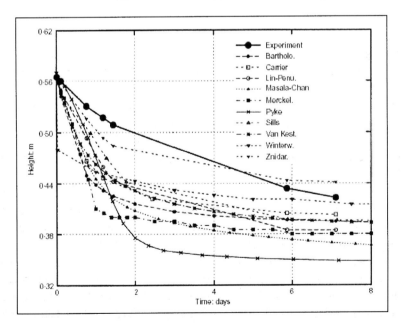

Fig. 7.8: Comparison of observed and predicted settling curves (after Bartholomeeusen et al., 2002).

This initiative can be regarded as the most objective way to test the state-of-the-art of consolidation modelling, and we must conclude this state is far from satisfactory. One of the reasons behind the unfavourable performance is the large sensitivity of the permeability and effective stress to the void ratio: small variations in material functions yield large differences in consolidation behaviour. Moreover, the effects of creep are not being accounted for. This is inherent to the current approach of describing the consolidation process with Gibson's equation and corresponding material functions.

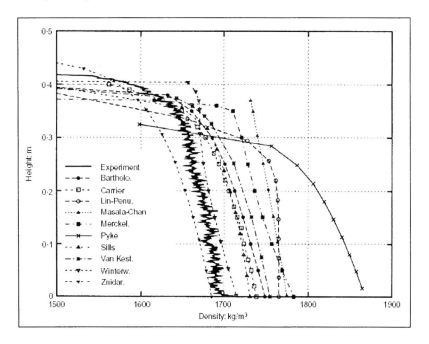

Fig. 7.9: Comparison of observed and predicted vertical density profiles after seven days (after Bartholomeeusen et al., 2002).

7.6   FRACTAL DESCRIPTION OF BED STRUCTURE

7.6.1   EFFECTIVE STRESS

Kranenburg (1994) proposed to treat flocs and beds of cohesive sediment as self-similar fractal structures. He noticed that it would be naive to expect that mud, with its large variability in composition and properties would form aggregates that are exactly self-similar. Nevertheless, he showed that many properties of mud can be described with power law relations, which is characteristic of fractal structures.

Further to this approach, Merckelbach (2000) and Merckelbach and Kranenburg (2003a) used this fractal concept to derive material functions for the consolidation equation of sediment-water mixtures, which may contain small amounts of sand. This approach is outlined in this section, and applied in the derivation of a consolidation equation in the next section.

Merckelbach (2000) and Merckelbach and Kranenburg (2003a) assume that the bed structure is built of a network of flocs consisting of clay and water (e.g. Chapter 3). Silt particles are enclosed within the flocs – this assumption is consistent with observations that the clay-silt ratio in most natural systems is constant. If the amount of sand in the mixture is below its critical value (Chapter 3), sand particles fill space in the clay-water network, without affecting the network structure or properties. This picture is schematically sketched in Fig. 7.10.

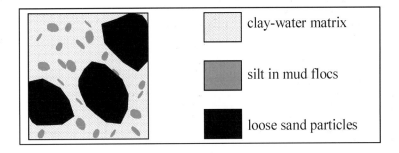

Fig. 7.10: Schematic structural picture of clay-silt-sand-water mixture (after Merckelbach, 2000 and Merckelbach and Kranenburg, 2003a).

Further to our analysis in Section 4.2, the volumetric concentration ϕ_s^m of clay particles, including silt particles in flocs of size D_f is given by:

$$\frac{\phi_s^m}{1-\phi_s^{sa}} = \lambda \left(\frac{D_p}{D_f} \right)^{3-n_f} \qquad (7.30)$$

where ϕ_s^{sa} is the volumetric concentration of the sand particles, D_p is the size of the primary particles (i.e. the coagulates, not of the clay particles, e.g. Chapter 3 and Section 4.2), n_f is the fractal dimension and λ is a shape parameter.

It is argued that the isotropic effective stress p^{sk} in a (consolidating) bed is linearly related to the number of bonds per unit area (see also Mitchell, 1976):

$$p^{sk} = a(S-b)$$
(7.31)

where S is the number of bonds per unit area, a is the strength of a single bond and b is the number of inactive bonds; b may vary with time as a result of (un)loading of the bed. The strength of the network is determined by the strength of the flocs, thus the number of effective bonds in critical yield planes, referred to as critical bonds. In self-similar structures, the number of critical bonds s is independent of the size of that structure, hence:

$$S = sD_f^{-2}$$
(7.32)

This implies also that the yield strength σ_y of the structure scales as:

$$\sigma_y \sim D_f^{-2} \quad \text{or} \quad \sigma_y \sim \left(\Delta\rho_f\right)^{\frac{2}{3-n_f}}$$
(7.33)

in which we used equ. (4.6), and $\Delta\rho_f$ is the relative floc density. This relation was derived by Kranenburg (1994), and he showed that (7.33) compares favourably with data from literature. Substitution of (7.32) and (7.31) into (7.30) yields a constitutive equation for the isotropic effective stress:

$$p^{sk} = \frac{as}{D_p^2} \lambda^{-\frac{2}{3-n_f}} \left(\frac{\phi_s^m}{1-\phi_s^{sa}}\right)^{\frac{2}{3-n_f}} - ab$$
(7.34)

The vertical isotropic stress follows from the assumption that effective stresses in horizontal direction $\sigma_h^{sk} = \sigma_{xx}^{sk} = \sigma_{yy}^{sk}$ are equal (see also Section 2.1):

$$\sigma_{zz}^{sk} = \frac{3}{1+2K_0} p^{sk} \tag{7.35}$$

where $K_0 = \sigma_h^{sk} / \sigma_{zz}^{sk}$. This description is appropriate when lateral strains and/or displacements are small or absent. Substitution into (7.34) yields:

$$\sigma_{zz}^{sk} = \frac{3}{1+2K_0} \lambda^{-\frac{2}{3-n_f}} \frac{as}{D_p^2} \left(\frac{\phi_s^m}{1-\phi_s^{sa}} \right)^{\frac{2}{3-n_f}} - \frac{3ab}{1+2K_0} \tag{7.36}$$

which is parameterised to yield a constitutive equation for the effective stress for water-mud mixtures with small amounts of sand in case the bed structure can be regarded as self-similar:

$$\sigma_{zz}^{sk} = K_p \left(\frac{\phi_s^m}{1-\phi_s^{sa}} \right)^{\frac{2}{3-n_f}} - K_{p,0} \tag{7.37}$$

The coefficient $K_{p,0}$ (hence the time-variation of b) can be used to account for the effects of creep.

Note that also Butterfield (1979) proposed the use of a power function relation between effective stress and void ratio (in fact $(1 + e)$) on the basis of an analysis of consolidation data. His analysis also yields a constant consolidation coefficient.

7.6.2 PERMEABILITY

The fractal approach used in Section 7.6.1 can also be applied to develop a constitutive relation for the permeability of a (consolidating) bed, e.g. Merckelbach (2000) and Merckelbach and Kranenburg (2003a).

The basic idea of modelling permeability of a bed of cohesive sediment is assuming that (the) pores in the bed are connected by a network of tubes – this assumption is of course necessary to expel pore water from a consolidating bed.

The flow through these tubes will be laminar and its velocity v can be described as Poiseuille flow:

$$\frac{dp^w}{ds} = -\frac{\mu}{\ell^2} v \qquad (7.38)$$

where p^w is pore water pressure, s is the co-ordinate along a tube with diameter ℓ, and μ is the pore water viscosity. In a fixed frame of reference (7.38) becomes:

$$\xi_s \frac{dp_e}{dz} = -\frac{\mu}{\ell^2} (v_f - v_s)(1 - \phi_s) \qquad (7.39)$$

where ξ_s is a coefficient accounting for the curvature of the tubes ($0 < \xi_s < 1$), also known as tortuosity, and v_f and v_s have been defined in Fig. 7.3. Because of continuity, the relation between v_f and v_s reads (see also (7.8)):

$$\phi_s v_s + (1 - \phi_s) v_f = (\phi_s^m + \phi_s^{sa}) v_s + (1 - \phi_s^m - \phi_s^{sa}) v_f = 0 \qquad (7.40)$$

From substitution of (7.40) into (7.39) and combination with Darcy's law, equ. (7.4) gives an expression for the permeability of the (consolidating) bed:

$$k = g\rho_w \frac{\xi_s \ell^2}{\mu} \qquad (7.41)$$

The assumption of scale invariance implies that the diameter of the larger pores, hence ℓ, scales with D_f (Kranenburg, 1994). Thus, after substitution from (7.29) we obtain:

$$k = \lambda^{\frac{2}{3-n_f}} \frac{\xi g \rho_w D_p^2}{\mu} \left(\frac{\phi_s^m}{1 - \phi_s^{sa}} \right)^{-\frac{2}{3-n_f}} \qquad (7.42)$$

which is parameterised to yield a constitutive equation for the permeability in case the bed structure can be regarded as self-similar:

$$k = K_k \left(\frac{\phi_s^{'m}}{1 - \phi_s^{sa}} \right)^{-\frac{2}{3-n_f}}$$

(7.43)

Note that in deriving (7.42) it is sufficient to account for the larger pores only, as the Poiseuille flow rate q scales as: $q \sim v_s \ell^2 \sim \ell^4$.

Typical values for soft mud deposits, as obtained by Merckelbach (2000), amount to: $K_k \approx (1 - 400) \cdot 10^{-18}$ m/s and $K_p \approx (3 - 100) \cdot 10^{12}$ Pa at $n_f = 2.70$ to 2.75. These values were found from experiments with mud from the Caland-Beerkanaal in The Netherlands.

In Section 3.3.1 we showed that the Atterberg limits of soil are related to the type of clay mineral and its clay content, and that the permeability k and undrained shear strength c_u are a function of these Atterberg limits. Furthermore, we argued that c_u and σ_{zz}^{sk} are related, and that k and σ_{zz}^{sk} are a function of the soil structure, which can be described with a fractal dimension. The latter is sustained by the results presented in Section 7.4 (equ. (7.25) and Fig. 7.5), showing that the power laws describing permeability and effective stress are inversely related. From these arguments we infer that the structure of the bed is a function of the type of clay mineral and its clay content only. This hypothesis is to be elaborated further, as it would provide us with a powerful tool to predict the consolidating behaviour of cohesive sediment on the basis of its mineral composition only.

7.7   CONSOLIDATION AS AN ADVECTION-DIFFUSION PROCESS USING FRACTAL THEORY

In this section we derive a set of equations for the consolidation of sand-mud mixtures at low sand concentration, and include the effects of hindered settling in a heuristic way, as proposed by Toorman (1996, 1999). Part of this derivation is a repetition of that in Section 7.2, but it is illustrative to show all steps. We use the reference frame of Fig. 7.3.

We start from the three-dimensional mass balance equation (2.11) in Eulerian co-ordinates, but with the volumetric sediment concentration, defined as:

$$\phi_s = \sum_j \phi_s^j = \phi_s^m + \phi_s^{sa} = \sum_i c_i^m / \rho^m + \sum_i c_i^{sa} / \rho^{sa}$$

where superscript $\cdot^m$ relates to mud (i.e. clay and silt) and superscript $\cdot^{sa}$ relates to sand, as the dependent variable, and subscript $\cdot_i$ to sediment fraction i. For convenience we assume $\rho^m = \rho^{sa} = \rho_s$. The mass balance for the a sand-mud mixture then reads:

$$\frac{\partial(\phi_s^m + \phi_s^{sa})}{\partial t} + \frac{\partial}{\partial x_i}\left((u_i - \delta_{i3}\Xi_s^m)\phi_s^m + (u_i - \delta_{i3}\Xi_s^{sa})\phi_s^{sa}\right) +$$

$$-\frac{\partial}{\partial x_i}\left(D_s \frac{\partial(\phi_s^m + \phi_s^{sa})}{\partial x_i}\right) = 0 \tag{7.44}$$

In (7.44) u_i and ϕ_p represent instantaneous quantities (fluctuating because of turbulence), and D_s is the molecular diffusion coefficient. The vertical velocity of the sediment particles with respect to the fixed reference frame is given by Ξ_s, where Ξ_s represents either the (hindered) settling velocity of the particles in the water column, denoted by w_s, or the particle velocity v_s within the fluid mud layer (i.e. during consolidation). In the consolidation regime, $\Xi_s^m = \Xi_s^{sa}$, whereas in the hindered settling regime segregation can occur: $\Xi_s^m \neq \Xi_s^{sa}$.

The mass balance for the mud fraction in the consolidation regime reads:

$$\frac{\partial \phi_s^m}{\partial t} + \frac{\partial}{\partial z}\left(v_s^m \phi_s^m\right) = \frac{\partial \phi_s^m}{\partial t} + \frac{\partial}{\partial z}\left(v_s \phi_s^m\right) = 0 \tag{7.45a}$$

and the mass balance for the sand fraction in the consolidation regime reads:

$$\frac{\partial \phi_s^{sa}}{\partial t} + \frac{\partial}{\partial z}\left(v_s^{sa} \phi_s^{sa}\right) = \frac{\partial \phi_s^{sa}}{\partial t} + \frac{\partial}{\partial z}\left(v_s \phi_s^{sa}\right) = 0 \tag{7.45b}$$

where v^m and v^{sa} are the vertical velocities of the mud and sand particles relative to a fixed reference plane (e.g. Fig. 7.3). The pressure gradient in the vertical direction becomes:

$$\frac{\partial \sigma_{zz}}{\partial z} = -\rho g = -\left(\phi_s^m + \phi_s^{sa}\right) g \rho_s - \left(1 - \phi_s^m - \phi_s^{sa}\right) g \rho_w \qquad (7.46)$$

where σ_{zz} is the total vertical stress and ρ, ρ_s and ρ_w are the bulk density of the suspension, the density of the sediment and the density of the water, respectively. In the consolidation regime, the relative vertical position of the various sediment particles remains unaltered. Therefore the velocities of the mud and sand particles are identical: $v_s^m = v_s^{sa} = v_s$. The relative velocity of the fluid v_f in vertical direction with respect to the sediment particles in our fixed reference frame is given by Darcy's law and reads:

$$\left(v_f - v_s\right)\left(1 - \phi_s^m - \phi_s^{sa}\right) = -\frac{k}{g\rho_w}\frac{\partial p_e}{\partial z} \qquad (7.47)$$

where k is the permeability of the fluid mud layer and p_e is the excess pore water pressure defined as the difference between the actual pore water pressure p^w and the hydrostatic pore water pressure:

$$p_e = p^w - \int_z^{Z_s} g\rho_w \mathrm{d}z' = p^f + g\rho_w\left(z - Z_s\right) \qquad (7.48)$$

in which Z_s is the level of the water surface. Note that (7.47) implies that interstitial water can flow through all the pores of the mud, i.e. no distinction is made between likely variations in hydraulic resistance within the network-forming flocs and the space in between. This effect is empirically accounted for through the value of the bulk permeability parameter k. Furthermore, because of continuity, the relation between v_f and v_s reads:

$$\left(\phi_s^m + \phi_s^{sa}\right)v_s + \left(1 - \phi_s^m - \phi_s^{sa}\right)v_f = 0 \text{ , hence}$$
$$\left(v_f - v_s\right)\left(1 - \phi_s^m - \phi_s^{sa}\right) + v_s = 0 \qquad (7.49)$$

Combining (7.47), (7.48) and (7.49) and introducing the effective stress concept (7.1) yields:

$$v_s = \frac{k}{g\rho_w} \frac{\partial \left(p^w + g\rho_w \left(z - Z_s \right) \right)}{\partial z} = k + \frac{k}{g\rho_w} \left(\frac{\partial \sigma_{zz}}{\partial z} - \frac{\partial \sigma_{zz}^{sk}}{\partial z} \right) \qquad (7.50)$$

Substituting from (7.45) gives:

$$v_s = -k \frac{\rho_s - \rho_w}{\rho_w} \left(\phi_s^m + \phi_s^{sa} \right) - \frac{k}{g\rho_w} \frac{\partial \sigma_{zz}^{sk}}{\partial z} \qquad (7.51)$$

In the consolidation phase, no diffusion of particles occurs and the diffusion coefficient should be set to zero. Substituting v_s from (7.51) into Ξ_s in (7.44) gives an equation for self-weight consolidation processes in Eulerian co-ordinates for the volumetric concentration of a sand-mud mixture:

$$\frac{\partial \phi_s^m}{\partial t} - \frac{\partial}{\partial z} \left(k \frac{\rho_s - \rho_w}{\rho_w} \left(\phi_s^m + \phi_s^{sa} \right) \phi_s^m \right) - \frac{\partial}{\partial z} \left(\frac{k \phi_s^m}{g\rho_w} \frac{\partial \sigma_{zz}^{sk}}{\partial z} \right) = 0 \qquad (7.52a)$$

$$\frac{\partial \phi_s^{sa}}{\partial t} - \frac{\partial}{\partial z} \left(k \frac{\rho_s - \rho_w}{\rho_w} \left(\phi_s^m + \phi_s^{sa} \right) \phi_s^{sa} \right) - \frac{\partial}{\partial z} \left(\frac{k \phi_p^{sa}}{g\rho_w} \frac{\partial \sigma_{zz}^{sk}}{\partial z} \right) = 0 \qquad (7.52b)$$

If we use the material function (7.36) for the effective stress, the effective stress gradient term in (7.52) becomes:

$$\frac{k \phi_s^m}{g\rho_w} \frac{\partial \sigma_{zz}^{sk}}{\partial z} = \Gamma_c \left[\frac{\partial \phi_s^m}{\partial z} + \frac{\phi_s^m}{1 - \phi_s^{sa}} \frac{\partial \phi_s^{sa}}{\partial z} \right] \qquad (7.53a)$$

$$\frac{k \phi_s^{sa}}{g\rho_w} \frac{\partial \sigma_{zz}^{sk}}{\partial z} = \Gamma_c \left[\frac{\phi_s^{sa}}{\phi_s^m} \frac{\partial \phi_s^m}{\partial z} + \frac{\phi_s^{sa}}{1 - \phi_s^{sa}} \frac{\partial \phi_s^{sa}}{\partial z} \right] \qquad (7.53b)$$

with

$$\Gamma_c = \frac{2}{3 - n_f} \frac{K_k K_p}{g\rho_w} \qquad (7.54)$$

We note that (7.54) is identical to the consolidation coefficient c_v used in classical soil mechanics. This implies that $\Gamma_c = c_v = $ constant if the bed structure is truly self-similar.

Our approach implies that we assume that the consolidating fluid mud layer is permanently at equilibrium. Substituting (7.41) and (7.53) into (7.52) yields an equation for the mass balance of cohesive sediment in the consolidating fluid mud layer.

Next, we incorporate the effects of hindered settling (Section 5.3) in Ξ_s of the advection term, and rewrite the equations in an advection-diffusion form with a source-term in the right-hand side. We introduce the heuristic parameter η to obtain a smooth transition between the descriptions for hindered settling and permeability. For the mud fraction we find:

$$\frac{\partial \phi_s^m}{\partial t} - \frac{\partial}{\partial z}\left(\Xi_s^m \phi_s^m\right) - \frac{\partial}{\partial z}\left((D_s + \Gamma_T + \Gamma_c)\frac{\partial \phi_s^m}{\partial z}\right) =$$

$$= \frac{\partial}{\partial z}\left(\Gamma_c \frac{\phi_s^m}{1-\phi_s^{sa}}\frac{\partial \phi_s^{sa}}{\partial z}\right)$$

(7.55a)

in which we define:

$$\Xi_s^m = f_{hs}^m + \frac{f_c}{1+\eta f_c}, \qquad \text{with}$$

$$f_{hs}^m = \frac{\left(1-\phi^m - \phi_s^{sa}\right)\left(1-\phi_s^m - \phi_s^{sa}\right)}{1+2.5\phi^m}W_{s,r}^m, \quad \text{and}$$

(7.55b)

$$f_c = \frac{\rho_s - \rho_w}{\rho_w}k\left(\phi_s^m + \phi_s^{sa}\right)$$

and for the sand fraction we find:

$$\frac{\partial \phi_s^{sa}}{\partial t} - \frac{\partial}{\partial z}\left(\Xi_s^s \phi_s^{sa}\right) - \frac{\partial}{\partial z}\left(\left(D_s + \Gamma_T + \Gamma_c \frac{\phi_s^{sa}}{1-\phi_s^{sa}}\right)\frac{\partial \phi_s^{sa}}{\partial z}\right) =$$

$$= \frac{\partial}{\partial z}\left(\Gamma_c \frac{\phi_s^{sa}}{\phi_s^m}\frac{\partial \phi_s^m}{\partial z}\right)$$

(7.56a)

in which we define:

$$\Xi_s^{sa} = f_{hs}^{sa} + \frac{f_c}{1+\eta f_c}, \qquad \text{with}$$

$$f_{hs}^{sa} = \frac{\left(1-\phi^m-\phi_s^{sa}\right)}{\left(1-\phi^m\right)}\frac{\left(1-\phi_s^m-\phi_s^{sa}\right)}{1+2.5\phi^m}\left(W_{s,r}^{sa}-\phi^m W_{s,r}^m\right), \qquad \text{and}$$

(7.56b)

$$f_c = \frac{\rho_s-\rho_w}{\rho_w}k\left(\phi_s^m+\phi_s^{sa}\right)$$

where $\eta = 10^5$ s/m, which value was found by trial and error. In our approach we have prescribed a zero sediment flux at the base and water surface Z_b and Z_s, but double drainage is easily implemented through an additional velocity in (7.45). The diffusion term in (7.55) and (7.56) consists of a contribution of molecular diffusion D_s, eddy diffusivity Γ_T and the coefficient of consolidation Γ_c. Of course, D_s and Γ_T are zero in the consolidating bed, whereas Γ_c is zero in the turbulent sediment suspension.

Equ. (7.55) reduces to a simple advection-diffusion equation with constant diffusion term (and right-hand side equal to zero) in case of settling in still water of mud only. Beyond the hindered settling regime, we obtain the following consolidation equation:

$$\frac{\partial \phi_s^m}{\partial t} - \frac{\partial}{\partial z}\left(\frac{\rho_s-\rho_w}{\rho_w}k\phi_s^m\phi^m\right) - \Gamma_c\frac{\partial^2 \phi_s^m}{\partial z^2} = 0$$

(7.57)

It is illustrative to elaborate on a consolidation equation for the total amount of sediment. Let $\phi_s^m = \alpha\phi_s$ and $\phi_s^{sa} = (1-\alpha)\phi_s$, where ϕ_s is the total volumetric solids concentration. Substitution into (7.55a) and (7.56a) and summation gives:

$$
\frac{\partial \phi_s}{\partial t} - \frac{\partial}{\partial z}\left(\frac{\rho_s - \rho_w}{\rho_w} k\phi_s\phi_s \right) - \frac{\partial}{\partial z}\Gamma_c\left(\frac{1}{1-(1-\alpha)\phi_s}\frac{\partial \phi_s}{\partial z} \right) =
$$

$$
= \frac{\partial}{\partial z}\Gamma_c\left(\frac{(1-\phi_s)\phi_s}{\alpha(1-(1-\alpha)\phi_s)}\frac{\partial \alpha}{\partial z} \right)
$$

(7.58)

To solve (7.58), we need an equation for α as well. As $\alpha = \alpha(z,t)$ there is no real advantage to solve the consolidation equation for total solids (7.58), and we prefer to apply the two-equation model (7.55) – (7.56).

7.8   MATERIAL FUNCTIONS FOR FRACTAL APPROACH

The material functions (7.36) and (7.42) can of course be determined from direct measurements of the permeability and effective stress within a consolidating mud layer. Merckelbach (2000) and Merckelbach and Kranenburg (2003b) propose another method in which the various parameters are determined from observations of the evolution of the interface only, i.e. no permeability or stress measurements have to be carried out. In the following we present this method briefly; for further details the reader is referred to Merckelbach and Kranenburg (2003b).

   This method is based on the observation that in the initial phase (Phase I) of the consolidation, effective stresses are small, so that the consolidation process is governed by the permeability of the soil mainly. In other words, the diffusion term in the consolidation equation vanishes (Kynch-regime). It is convenient to set Phase I beyond the time of contraction t_c, as the solution of the consolidation equation is then independent of the initial conditions.

   In the final phase (Phase II) of consolidation, the consolidation rate becomes small and the interface does not lower much any more. In this phase, effective stresses govern consolidation, and the advection term in the consolidation equation vanishes. This is schematically shown in Fig. 7.11.

   It is convenient to superpose (7.52a) and (7.52b), which is valid, as Phase I and II are beyond the hindered settling phase, to yield the consolidation equation for the total amount of solids ϕ_s:

$$\frac{\partial \phi_s}{\partial t} - \frac{\partial}{\partial z}\left(\frac{\rho_s - \rho_w}{\rho_w} k \phi_s^2 \right) - \frac{\partial}{\partial z}\left(\frac{k \phi_s}{g \rho_w} \frac{\partial \sigma_{zz}^{sk}}{\partial z} \right) = 0 \qquad (7.59)$$

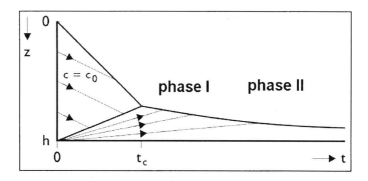

Fig. 7.11: Schematic consolidation diagram, showing the two phases to establish material functions in relation to time of contraction t_c.

Phase I:

We start with a consolidating mixture without sand and substitute the material function (7.43) in the consolidation equation (7.59), which, as we have seen, reduces to the simple wave equation in the initial phase of consolidation:

$$\frac{\partial \phi_s}{\partial t} - K_k \frac{\rho_s - \rho_w}{\rho_w}(2-n)(\phi_s)^{1-n} \frac{\partial \phi_s}{\partial z} = 0 \qquad (7.60)$$

where $n = 2/(3-n_f)$. The simple wave equation can be solved with the method of characteristics (e.g. Section 5.3) to yield:

$$\phi_s = \left((n-2) K_k \frac{\rho_s - \rho_w}{\rho_w} \frac{t}{z} \right)^{\frac{1}{n-1}} \qquad (7.61)$$

This is valid beyond $t > t_c$, when the solution to (7.60) is no longer dependent on the initial conditions. As before, we assume that the sand particles only fill space in the mud-water mixture, and do not affect the network structure. In that case, we can define a reduced Gibson height ζ_m, accounting for the total clay and silt solids and ζ_s for sand only:

$$\zeta_m = \int_0^h \frac{\phi_s^m}{1 - \phi_s^{sa}} dz \quad \text{and} \quad \zeta_s = \int_0^h \phi_s^{sa} dz \qquad (7.62)$$

From integration of (7.61), we obtain:

$$h(t) - \zeta_s = \left(\frac{2-n}{1-n}\zeta_m\right)^{\frac{1-n}{2-n}}\left((n-2)K_k \frac{\rho_s - \rho_w}{\rho_w}\right)^{\frac{1}{2-n}} t^{\frac{1}{2-n}} \qquad (7.63)$$

This allows determination of n (thus n_f) and K_k, hence the permeability k from measuring the height $h(t)$ of the interface as a function of time.

Phase II:

In this phase of the consolidation process, deformations are small, and the advection term in (7.60) can be neglected. Then, for equilibrium conditions ($\partial \phi_p / \partial t = 0$), the consolidation equation reduces to (e.g. (7.51)):

$$\frac{\rho_s - \rho_w}{\rho_w}\phi_s = -\frac{1}{g\rho_w}\frac{\partial \sigma_{zz}^{sk}}{\partial z} \qquad (7.64)$$

Substitution from material function (7.37) and introducing the reduced Gibson height (7.62), (7.64) can be integrated to obtain the vertical density distribution in the final phase of consolidation:

$$h - \zeta_s - z = \frac{n}{n-1}\frac{K_p}{g(\rho_s - \rho_w)}(\phi_s)^{n-1} \qquad (7.65)$$

Integration of (7.65) over the bed thickness yields an equation for the final thickness of the bed h_∞:

$$h_\infty = \zeta_s + \frac{n}{n-1} \frac{K_p}{g(\rho_s - \rho_w)} \left(\frac{g(\rho_s - \rho_w)}{K_p} \zeta_m \right)^{\frac{n-1}{n}} \qquad (7.66)$$

where the Gibson height depends on the initial conditions only.

If the vertical density distribution is known, K_p can be determined from (7.65) (n was determined in Phase I). It is even possible to establish K_p from the final bed level height (7.66) alone. In that case, the material functions for k and σ_{zz}^{sk}, including the consolidation coefficient c_v can be established from a simple consolidation experiment by only following the level of the interface.

Merckelbach and Kranenburg (2003b) provide more details and the two examples presented in Fig. 7.12 and 7.13, in which the method is demonstrated against a series of column data obtained with mud from the Rhine-Meuse estuary in The Netherlands.

The reduced Gibson height was $\zeta_s = 8.7 \cdot 10^{-3}$ m. The values of the other parameters obtained with this method are given in Table 7.2.

Table 7.2: Values of material parameters obtained from Fig. 7.12 & 13.

n_f	K_k [m/s]	K_p [Pa]
2.76	$3.68 \cdot 10^{-14}$	$7.55 \cdot 10^{9}$

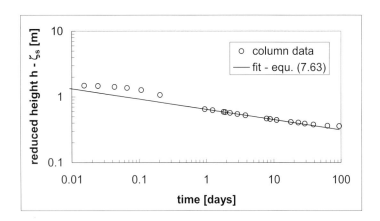

Fig. 7.12: Example of assessment of n and K_k from initial consolidation phase I (after Merckelbach and Kranenburg, 2003b).

Merckelbach and Kraneburg also present a comparison of the parameters of Table 7.2 with values obtained from direct measurements.

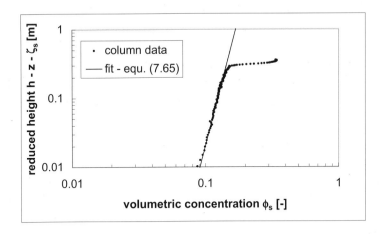

Fig. 7.13: Example of assessment of K_p from initial consolidation phase II (after Merckelbach and Kranenburg, 2003b).

7.9   APPLICATION OF FRACTAL APPROACH

In this section we present a few applications of the consolidation equations (7.55) and (7.56), using the fractal material functions of Section 7.6, implemented in the 1DV-model described in Appendix E. First we show that the numerical implementation of these equations matches the Gibson equation in material co-ordinates properly. For this purpose we use Townsend and McVay's numerical benchmark experiments described in Section 7.5. The material functions of (7.29) are rewritten as power law functions, yielding the parameters of Table 7.3. The numerical results are compared with Townsend and McVay's numerical data in Fig. 7.14, showing excellent agreement between the Gibson approach and our numerical implementation of the model described in Section 7.7.

Next, our approach is compared with data obtained by Merckelbach (2000) in a consolidation column of 1.54 m high with mud from the approach channel to the Port of Rotterdam. The initial conditions were in the hindered settling regime, and the values for the hindered settling formulae were obtained by trial and error – Merckelbach applied Richardson-Zaki's formula in his hindcast. The parameters in the consolidation regime were obtained with the method described in Section 7.8 – e.g. Table 7.3. The model results are compared with the experimental data in Fig. 7.15, showing a fair agreement. The computations show a slight overshoot near the interface – this is a numerical anomaly resulting from the numerical discretisation of the equations.

Table 7.3: Parameters to simulate consolidation experiments.

	Townsend & McVay (1990)	Merckelbach (2000)	Winterwerp (2004)
sand fraction	-	0.23	0.12
$C_{0,m}$ [kg/m^3]	167.7	81	53
$C_{0,s}$ [kg/m^3]	-	24	7
$W_{s,m}$ [mm/s]	-	0.1	0.5
c_{gel} [kg/m^3]	$< C_{0,m}$	200	100
$D_{50,s}$ [μm]	-	30	40
n_f	2.62	2.72	2.70
K_k [m/s]	$4.12 \cdot 10^{-13}$	$1.8 \cdot 10^{-14}$	$1.0 \cdot 10^{-14}$
K_p [Pa]	$1.42 \cdot 10^8$	$3.2 \cdot 10^8$	$1.0 \cdot 10^9$

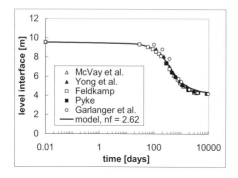 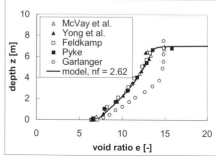

Fig. 7.14: Comparison of "fractal" consolidation model with benchmark experiment of Townsend and McVay (1990).

The pronounced tail near the base of the column, with concentrations of 700 to 900 kg/m^3 consists mainly of sand and is fully attributed to segregation during the hindered settling phase. It is therefore interesting to rerun the model for the hypothetical case that the initial concentration is beyond the gelling point. The results are presented in Fig. 7.16, showing that the sand-mud ratio now remains constant with time, hence no segregation occurs. Note that the advection terms in (7.56a and 57) decrease monotonically with ϕ, as a result of which a near-bed interface is always formed in the initial phase of consolidation when the effective stresses (diffusion term in equ. (7.56a and 57)) are small.

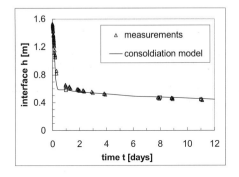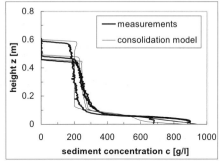

Fig. 7.15: Comparison of "fractal" consolidation model with experimental data by Merckelbach (2000).

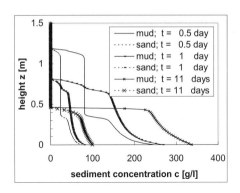

Fig. 7.16: As Fig. 7.15, but with the assumptions $C_{m,0} > c_{gel}$.

Winterwerp (2004) incorporated the consolidation equation (7.55) and (7.56) in an integrated model to simulate (hindered) settling, consolidation and re-entrainment of cohesive sediment around slack water in the marine environment. A simple relation between the sediment strength and effective stresses is used. Here, we only present the results of the settling and consolidation phase. The material parameters, as given in Table 7.3, were derived by trial and error, as the available data did not allow the analysis presented in Section 7.8. The results are presented in Fig. 7.17, showing a fair agreement between observations and computations. Segregation between sand and mud is clearly visible.

It is noted that the fractal dimension, describing the structure of the bed varies between about 2.61 and 2.75, i.e. n_f is much larger than the values typical for flocs in the water column (e.g. Chapter 4). This must be due to the squeezing of the flocs in the bed under self-weight consolidation. However, physical-mathematical descriptions describing this transition in floc structure are not available at present. This implies that integrated settling-consolidation-erosion models, as mentioned above, require input of two structural parameters, i.e. one value of the fractal dimension n_f for the mud flocs in the water column and one value of n_f for the mud flocs in the consolidating bed.

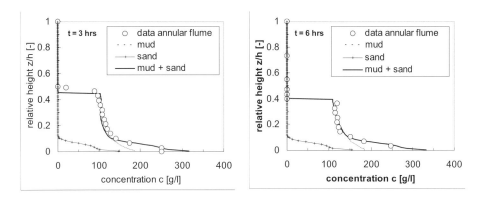

Fig. 7.17: Vertical concentration profiles measured in a rotating annular flume compared with computations.

7.10 APPROXIMATED SOLUTION OF CONSOLIDATION EQUATION

Numerical solution of the consolidation equation, as discussed in the previous sections, requires a detailed vertical discretisation and small time step to obtain accurate results. Yet, the settling-consolidation-erosion sequence described above is representative only for a short period in the long-term transport and fate of cohesive sediment in the natural environment. However, a complete integration of the settling and consolidation models in full three-dimensional sediment transport models is not feasible at present from a computational time point of view, in particular when the bed level is increasing and decreasing in a continuous sequence of deposition and erosion. Therefore, De Boer et al. (2003) have derived an approximate solution to the consolidation equation.

For this purpose, the Gibson equation in material co-ordinates (7.14), is rewritten for the volumetric concentration ϕ_s, using (7.2):

$$\frac{\partial \phi_s}{\partial \tau} - \left(\frac{\rho_s - \rho_w}{\rho_w}\right)\phi_s^2 \frac{\partial}{\partial \zeta}(k\phi_s) - \frac{\phi_s^2}{g\rho_w}\frac{\partial}{\partial \zeta}\left(k\phi_s \frac{\partial \sigma_{zz}^{sk}}{\partial \zeta}\right) = 0 \qquad (7.67)$$

With the use of the material functions (7.37) and (7.43) the consolidation equation (7.67) can be rewritten with the effective stress σ_{zz}^{sk} as dependent variable:

$$\frac{\partial \sigma_{zz}^{sk}}{\partial \tau} + A_1\left(\sigma_{zz}^{sk}\right)^{2-n_f}\left(\frac{\partial \sigma_{zz}^{sk}}{\partial \zeta}\right)^2 + A_2\left(\sigma_{zz}^{sk}\right)^{2-n_f}\frac{\partial \sigma_{zz}^{sk}}{\partial \zeta} +$$

$$- A_3\left(\sigma_{zz}^{sk}\right)^{3-n_f}\frac{\partial^2 \sigma_{zz}^{sk}}{\partial \zeta^2} = 0 \qquad (7.68)$$

where the coefficients A_1, A_2 and A_3 are defined as follows:

$$A_1 = \frac{1}{g\rho_w}\frac{n_f - 1}{3 - n_f}K_k K_p^{n_f - 2}$$

$$A_2 = \frac{\rho_s - \rho_w}{\rho_w}\frac{n_f - 1}{3 - n_f}K_k K_p^{n_f - 2} \qquad (7.69)$$

$$A_3 = \frac{1}{g\rho_w}\frac{2}{3 - n_f}K_k K_p^{n_f - 2}$$

Using the effective stress concept, we expand the effective stress σ_{zz}^{sk} around its equilibrium $\sigma_{zz,equ}^{sk}$. If we consider only its first order deviation, i.e. the excess pore water pressure p_e $\left(\sigma_{zz}^{sk} = \sigma_{zz,equ}^{sk} + p_e\right)$, we obtain:

$$\frac{\partial p_e}{\partial \tau_*} + \frac{n_f - 1}{2} \frac{1}{\sigma_{zz}^s} \left(\frac{\partial p_e}{\partial \zeta_*} \right)^2 + \frac{n_f - 1}{2} g(\rho_s - \rho_w) \frac{\zeta_*}{\sigma_{zz}^s} \frac{\partial p_e}{\partial \zeta_*} - \frac{\partial^2 p_e}{\partial \zeta_*^2} = 0 \quad (7.70)$$

where we have included the following dimensionless co-ordinates, using the material height (Gibson height) ζ_m of the consolidating layer:

$$\zeta_* = \frac{\zeta}{\zeta_m} \quad \text{with} \quad \zeta_m = \int_0^{h_i} \phi_s dz$$

$$\tau_* = \frac{\tau}{T} \quad \text{with} \quad T = \frac{1}{\Gamma_c} \left(\frac{K_p}{\sigma_{zz}^{sk}} \right)^{3-n_f} \zeta_m^2 \quad (7.71)$$

If we consider thin consolidating layers (small ζ_m), consolidation times are small, and we may neglect the first phase of consolidation (Phase I, Section 7.3.4). This implies that the non-linear second and third advection terms of (7.70) may be neglected, which yields a diffusion equation for the excess pore water pressure, similar to Terzhagi's equation (7.17):

$$\frac{\partial p_e}{\partial \tau_*} - \frac{\partial^2 p_e}{\partial \zeta_*^2} = 0 \quad (7.72)$$

Note that Lee and Sills (1981) derived equ. (7.72) by assuming $\sigma_{zz}^{sk} \propto e$. This equation was solved with the same Fourier series as described below. However, application of the simple linear constitutive relation implies that their solution always yields a linear vertical density profile. Moreover, because of this assumption it is not easy to regain the physical quantities in a Eulerian reference framework. Therefore, we prefer our own approach in which we omit the advection terms only in a later phase of the derivation of the simplified consolidation equation.

At the top of the consolidating layer we apply a Neumann boundary condition. If we apply a Dirichlet boundary condition at the base of the consolidating layer, double drainage conditions are included, i.e. we can treat a multiple layer consolidation process:

$$p_e(1, \tau_*) = 0 \qquad \tau_* > 0$$
$$\frac{\partial p_e(0, \tau_*)}{\partial \zeta_*} = 0 \qquad \tau_* > 0 \tag{7.73}$$

The initial condition reads:

$$p_e(\zeta_*, 0) = p_0(\zeta_*) \qquad 0 < \zeta_* < \zeta_m \tag{7.74}$$

The solution to the diffusion equation (7.72) with these initial and boundary conditions is well known from literature, and is written in the form of a Fourier series (e.g. Carslaw and Jeager, 1959):

$$p_e(\zeta_*, \tau_*) = \sum_{k=1}^{\infty} f_k \exp\left\{\left(-\pi^2\left(k - \tfrac{1}{2}\right)^2 \tau_*\right) \cos\left(\pi\left(k - \tfrac{1}{2}\right)\zeta_*\right)\right\} \tag{7.75}$$

where the coefficients f_k are given as:

$$f_k = 2\int_0^1 p_0(\zeta_*) \cos\left(\pi\left(k - \tfrac{1}{2}\right)\zeta_*\right) d\zeta_* \tag{7.76}$$

and where subscript $*_0$ refers to initial conditions. If we consider a monotonic consolidation process (i.e. neither deposition nor erosion), f_k can be obtained from analytical integration of (7.76):

$$f_k = \frac{8g(\rho_s - \rho_s)\zeta_m}{\pi^2(2k - 1)^2} \tag{7.77}$$

In other cases, (7.76) has to be assessed numerically. From some numerical experiments it was concluded that accurate results (from a physical point of view) are obtained when a few (De Boer et al. used four components) Fourier components are used.

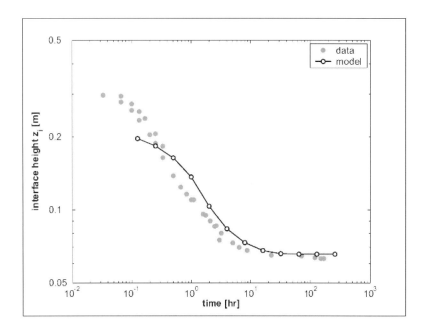

Fig. 7.18: Comparison computed and measures interfaces.

The approximate solution (7.75) is tested against data from a consolidation experiment performed in a settling column by Kuijper et al. (1990). In this experiment both the settling of the mud interface and the density profiles have been measured as a function of time for a 1-day and a 7-day consolidation experiment. Using the method described in Section 7.3.4, the parameters of the material functions have been determined from the settling rate of the interface and from the equilibrium density profiles. The computed and measured interface level and density profiles are compared in Fig. 7.18 and 7.19.

The method presented in this section allows long-term computations of the bed development at acceptable computational effort, also when the total amount of sediment varies over time as a result of sedimentation and/or erosion. Note that this approach is valid for homogeneous sediment mixtures only, i.e. segregation of sand and mud should not occur. However, a small sand fraction may be included in the consolidating sediment layer.

It is observed that the measured interface height and density profiles are reasonably predicted. Deviations are due to the neglect of the advection terms, which are important in the initial phase of consolidation and near the water-sediment interface, and, more important, to the inaccuracy of the material functions – these were not determined explicitly during the execution of the experiments.

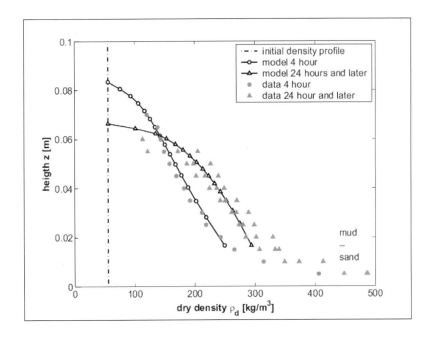

Fig. 7.19: Comparison computed and measures density profiles.

8. MECHANICAL BEHAVIOUR

Chapter 3 characterises cohesive sediment as granular material, in which sediment particles of different size and shape form a skeleton with a complementary pore system. In the marine environment the pores are generally filled with water, though gas can be present as well. Both the skeleton and pore system characteristics determine the mechanical behaviour of granular materials. Especially the characteristics of the pore-filling medium play an important role, as the cohesion of the sediment **and** the pore medium are responsible for our perception of cohesive sediment being cohesive. The flow of the pore medium through the skeleton during deformation consumes energy and retards deformations of the skeleton. This explains why cohesive sediments in the marine environment can behave sometimes as solids and sometimes as liquids.

The mechanical behaviour of sediment is governed by a combination of skeleton and pore medium characteristics. This is not only the case for static or slow loading conditions and episodic events, but also under dynamic loading (cyclic and flow) conditions. As a result the physics responsible for the "soil mechanical" behaviour under static or slow loading conditions are the same as under "flow mechanical" behaviour, better known as "rheology", under dynamic loading and flowing conditions (e.g. Keedwell, 1984). Therefore, non-dynamic and dynamic conditions are treated together in this chapter.

Some of the theory presented in this chapter has been developed for cemented material, e.g. concrete, but has never been applied to cohesive sediment. We feel that by combining the developments in theories on cemented material with those on granular material, a complete picture of the mechanical behaviour of cohesive sediment deposits in the marine environment can be obtained.

Soils may behave either as homogeneous or as heterogeneous material. In general, three basic failure modes of granular material can be distinguished:
- ductile failure: large-scale failure without localisation (see Fig. 8.1a);
- shear failure: local failure along shear planes (see Fig. 8.1b);
- tensile failure: local failure by cracks (see Fig. 8.1c).

In Section 8.1 we discuss the behaviour of pore water by treating the seabed as a multi-phase system. The stress-strain relations at macro-scale, obtained by treating the sediment-water-gas mixture as a continuous medium, are presented in Section 8.2. These relations are most appropriately described with tensors, and we present a brief summary of tensor analysis in Appendix D. In Section 8.3 we summarise the possible modes of failure. Ductile behaviour is referred to when the seabed fails at large scale without localisation. Often however, the bed contains discontinuities and inhomogenities, as a result of which shear and tensile failure are initiated by cracks in the material. We apply the theories discussed in Section 8.2 and 8.3 to describe the behaviour and failure of the seabed under cyclic loading in Section 8.4. Finally, in Section 8.5 we describe the strain-rate dependent behaviour of the bed, relating the theory in the earlier sections to classical rheological descriptions.

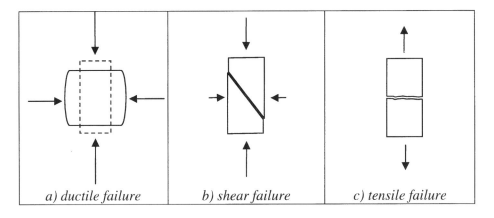

Fig. 8.1: Modes of failure.

8.1   THE SEAFLOOR AS A MULTI-PHASE SYSTEM

As described in Chapter 3, cohesive sediment is a mixture of a solid phase (minerals, organic matter), a liquid phase (in general water) and sometimes a gas phase (methane, carbon dioxide). The solid phase consists of granular particles sizing in the range from mm's to particles less than $1\mu m$ that can form flocs, which is typical for cohesive sediments. However, the cohesive behaviour of sediment is not only determined by bond strengths in the skeleton of the solid phase, but also by the flow of pore water and the pore pressures generated. In this section, we discus the behaviour of a two-phase system of granular solids with water as the pore fluid. The role of gas will only be accounted for as far as pore water and solid characteristics are concerned. The role of gas with respect to bubble formation in the sediment skeleton is treated in Chapter 11.

8.1.1 EFFECTIVE STRESS CONCEPT

The fluid phase consists of inter-particle fluid (fluid within the flocs, c.q. primary particles or flocculi, e.g. Chapter 4) and free pore fluid. The inter-particle fluid transfers forces within the skeleton and is immobile except at very high stresses (MPa). The two-phase system of pore fluid and solid's structure, skeleton or fabric can be modelled as parallel bodies or Kelvin-Voigt bodies, i.e. superposition of stresses is assumed (e.g. Section 2.1.1):

$$\sigma_{ij} = \sigma_{ij}^{w} + \sigma_{ij}^{sk} \tag{8.1}$$

in which, σ_{ij} is the total stress tensor, and the superscripts $\bullet^{w}$ and $\bullet^{sk}$ refer to pore water and skeleton, respectively (e.g. Appendix A). The stress tensor can be decomposed into a deviatoric and isotropic part (see Appendix D):

$$\sigma_{ij} = \tau_{ij} + \delta_{ij} p$$
$$\sigma_{ij}^{w} = \tau_{ij}^{w} + \delta_{ij} p^{w} \tag{8.2}$$
$$\sigma_{ij}^{sk} = \tau_{ij}^{sk} + \delta_{ij} p^{sk}$$

in which, τ_{ij} is the deviatoric stress tensor, p is the isotropic stress and δ_{ij} is the unity tensor or Kronecker delta. Note that the isotropic stress and the normal stress components of the stress tensor are defined as **positive for compression**. We have decomposed the stress tensor into an isotropic and deviatoric part:

$$p = p^{w} + p^{sk} \tag{8.3}$$

and

$$\tau_{ij} = \tau_{ij}^{w} + \tau_{ij}^{sk} \tag{8.4}$$

In soil mechanics equ. (8.3) is known as the effective stress concept of Terzaghi (1943):

$$p = p' + u \tag{8.5}$$

in which p^{sk} is replaced by p' (and the vertical component by c'_v) called the effective isotropic stress, and p^w is replaced by the pore water pressure u.

If the stresses in the pore fluid are small, τ_{ij}^w is small compared to τ_{ij}^{sk} and can be neglected:

$$\tau_{ij} \approx \tau_{ij}^{sk} \tag{8.6}$$

When the volume concentration and the shear strain rate are sufficiently small, no structural effects occur, hence τ_{ij}^{sk}, thus τ_{ij} is almost zero, and the viscosity of the liquid phase governs the behaviour of the material. This is the case in fluid mud for instance.

The effective stress concept has been developed theoretically by Biot (1956) (see e.g. Barends, 1980), strictly only holding for the case when the solid phase is fully surrounded by a Newtonian pore fluid. It is also implicitly assumed that the solids compressibility can be neglected with respect to the compressibility of the soil skeleton. Therefore, several corrections of the effective stress concept have been proposed for cemented materials. Most of these account for a reduced effective pore water area (see e.g. Skempton, 1961). Nur and Byerlee (1971) experimentally verified the following correction on (8.3), based on the compressibility ratio of solids and skeleton:

$$p = p^{sk} + \left(1 - \frac{C_s}{C_{sk}}\right)p^w \tag{8.7}$$

in which C_s [Pa^{-1}] is the compressibility of the solids and C_{sk} [Pa^{-1}] is the compressibility of the skeleton. The compressibility is reciprocal to the bulk modulus and is defined as the ratio of the increments in isotropic stress Δp and volume strain $\Delta \varepsilon_v$:

$$C^{-1} = \frac{\Delta p}{\Delta \varepsilon_v} \tag{8.8}$$

However, equ. (8.7) does not account for the interrelation between the volumetric changes in skeleton, pore water and solids under increasing stress levels. Therefore, it is more straightforward to express the changes in total isotropic strain in terms of total stress and pore water pressure changes. For cohesive sediments the generation and dissipation of pore water pressures is a key factor in the mechanical behaviour of the material, as described next.

8.1.2 DRAINED AND UNDRAINED BEHAVIOUR

When pore water pressures are generated, two conditions can be distinguished:

drained: generated pore water pressures are dissipated by pore water flow.

undrained: virtually no pore water flow through the pores can occur and generated pore water pressures cannot dissipate.

When we consider loading by an isotropic stress increment Δp on a volume of sediment V, the volume balance in terms of volume strain increment $\Delta \varepsilon_v$ reads (note that compression is positive):

skeleton	pore water	solids	transport

$$\Delta \varepsilon_v = \quad \Delta \varepsilon_v^{sk} \quad = n \Delta \varepsilon_v^{w} + \quad (1-n) \Delta \varepsilon_v^{s} \quad + \quad \Phi / V =$$

$$= \Delta p^{sk} C_{sk} + \Delta p^{w} C_s = n \Delta p^{w} C_w + (1-n) \Delta p^{w} C_s + \alpha \Delta p^{sk} C_s + \left(\int_A \underline{Q} . \underline{n} \, dA \right) \Big/ V \tag{8.9}$$

in which ε_v is the total volume strain, C_w, C_s, C_{sk} is the compressibility of pore water, solids and skeleton, respectively, n is the porosity, α is a coefficient for compression of the solids due to skeleton stresses, and Q is the specific discharge of pore water through the surface of V. Note that under drained conditions $\Delta p^{w} \to 0$.

The total volume strain increment $\Delta\varepsilon_v$ equals the volume strain increment of the skeleton $\Delta\varepsilon_v^{sk}$, which comprises:

1. An increase in effective isotropic stress p^{sk}, and
2. An increase in pore water pressure p^w, as a result of which solid particles may decease in volume. This would yield the same decrease in volume of the skeleton

In the undrained case, by definition, the transport of pore water Φ in equ. (8.9) is zero, and the incremental volume change of pore water and solids must balance the total volume change. Two contributions to the volume strain of the solids may be relevant:

1. Due to an increase in pore water pressure the solid particles may decease in volume, and
2. Due to an increase in effective stress the volume of the solid particles may decrease non-isotropically (the strain in vertical direction is generally larger than in horizontal directions, which is characterised by a Poisson ratio ν ratio smaller than unity).

The isotropic effective stress increment as function of the pore water pressure is found from the balance equation (8.9). Together with the effective stress concept (8.3) the ratio of the isotropic stress increment and the generated pore water pressure increment is obtained:

$$\frac{\Delta p}{\Delta p^w} = 1 + n \frac{C_w - C_s}{C_{sk} - \alpha C_s} \tag{8.10}$$

The coefficient α reflects possible compression of the solids in the skeleton due to skeleton stresses. When the solids consist of spherical particles α is almost zero, while in the complementary case of spherical voids, α equals unity (Schatz, 1976). The latter case occurs in porous rock, and can become relevant in cemented cohesive sediment. In general, $\alpha = 0$ for un-cemented cohesive sediments.

Typical values for the compressibility of various sediment components are listed in Table 8.1. In general, the compressibility of the solid phase is so small, that only the compressibility of the pore water and skeleton are relevant. When the pore water does not contain free gas, the compressibility of the pore fluid is also small with respect to the skeleton. In that case, an external isotropic stress increment will result in an identical increment in pore water stress. In the undrained case, an external isotropic stress increment has no effect on the effective (skeleton) stresses.

However when gas is present, the compressibility of pore water increases tremendously and an external stress increment generates both a pore water pressure increment and an effective stress increment. The effect of gas on pore water compressibility is discussed in more detail in the next section. In case that also the solids contain gas or compressible organic matter, the solid's compressibility increases and may even become larger than the compressibility of the pore water. In that case the second term in the right hand side of equ. (8.10) becomes negative and an increase in external isotropic stress results in a decrease of the isotropic effective stress. As a result, the sediment becomes very sensitive to liquefaction.

Table 8.1: Compressibility data (Van Kesteren, 1995).

material	phase	compressibility $[10^{-9}$ Pa$^{-1}]$
water	pore phase	0.48
water + 5% gas	pore phase	100-500
quartz	solid phase	0.027
organic	solid phase	0.5-100
dense sand	skeleton	20
mud	skeleton	1000
soft clay	skeleton	100
stiff clay	skeleton	10
limestone	skeleton	0.55
sandstone	skeleton	0.093

The increase in pore water pressure Δp^w, divided by the pore water pressure increment in undrained loading as a result of a stationary external loading Δp, is shown in Fig. 8.2 for drained and undrained conditions as a function of the Péclet number for pore water pressure dissipation Pe_w. This number is defined as:

$$Pe_w = \frac{V \ell}{c_i} \qquad (8.11)$$

in which V is a velocity scale, ℓ is a length scale and c_i is the isotropic consolidation coefficient (subscript $\bullet_i$ is added to distinguish from the vertical consolidation coefficient c_v). The consolidation coefficient c_i is a material parameter for the dissipation of pore water pressure, defined as:

$$c_i = \frac{k_i}{\rho_w g} \frac{1}{C_{sk} - \alpha C_s + n(C_w - C_s)}$$ (8.12)

in which k_i is the isotropic permeability [m/s] and ρ_w is the water density.

Because permeability and compressibility depend on the isotropic effective stress and stress history, c_i varies with the stress state. However, this variation is much smaller than variations in permeability and compressibility, as c_i is a function of the ratio of permeability and compressibility (e.g. equ. (8.12)).

Fully drained conditions are found near the origin of Fig. 8.2 at $Pe_w < 1$, represented by a straight line in the diagram, tangent to the full curve.

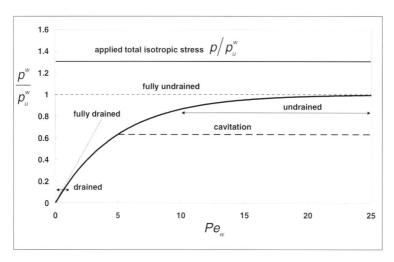

Fig. 8.2: Response of pore water pressure to external loading for drained and undrained conditions.

At high Pe_w ($Pe_w > 10$) the curve asymptotically approaches fully undrained conditions. The pore water pressure for fully undrained conditions can be computed with equ. (8.10). It is lower than the applied total isotropic stress increment, depending on the ratios of the compressibilities.

It is important to realize that drained conditions do not imply that no pore water pressures are generated: these are always generated, but their magnitude depends on the velocity and length scales within the soil. In sand, drained conditions occur in general. For instance in densely packed sand, shear failure generates dilation of the skeleton, resulting in negative pore water pressures. An increase in shear failure rate and/or size of the deformation zone will generate larger pore water underpressures. These pressures may become so negative that cavitation of the pore water occurs (e.g. Van Os et al., 1985).

Because of the low permeability and high compressibility of cohesive sediment, loading generally generates an undrained response. Such undrained behaviour is responsible for the behaviour typical of cohesive sediment. When the time of loading is not constrained, the behaviour of cohesive sediment becomes drained. An example is the process of surface erosion, which can be regarded as a drained failure process (see Chapter 9).

8.1.3   COMPRESSIBILITY OF GASEOUS PORE WATER

The compressibility of gas-water mixtures has been studied extensively, especially with respect to bubble dynamics and cavitation (e.g. Young, 1984 or Brennen, 1995). Barends (1980) gives a formulation for the compressibility of air-water mixtures C_{aw} as a function of water pressure, gas content and surface tension:

$$C_{aw} = C_w + \frac{\left[\left(1-\phi_g'\right)\left(1-\Omega\right)^{-1}\right]-1}{p - p_{vap} + \dfrac{2\sigma_{st}}{3r}\left(2 - \dfrac{\Omega}{\left(1-\Omega\right)\phi_g'}\right)} \tag{8.13}$$

in which C_w is the compressibility of water [Pa^{-1}], p and p_{vap} are the water, c.q. vapour pressure, Ω [-] is the solubility coefficient of gas in water, ϕ_g' is the gas volume fraction $\left(\phi_g' = \phi_g/\left(1-\phi_s\right)\right)$ relative to the total volume of pore water, σ_{st} is surface tension [N/m] and r is bubble radius. The solubility coefficient Ω can be expressed in Henry's coefficient for insolubility H by (see also Chapter 11):

$$\Omega = \frac{RT}{H V_{mol}^w} \tag{8.14}$$

in which H is Henry's coefficient [Pa], R is the gas constant (= 8.2 J/mol/^{0}K), T is the absolute temperature [^{0}K] and V_{mol}^w is the molar volume of water (= $18 \cdot 10^{-6}$ m^3/mol).

When the solubility of gas in water, combined with the associated surface tension, is neglected ($\Omega = 0$ and $\sigma_{st} = 0$), equ. (8.13) can be approximated by:

$$C_{aw} = C_w + \phi_g' / p \qquad (8.15)$$

This relation is similar to those suggested by Verruijt (1969) and Van Wijngaarden (1972).

Equ. (8.13) can be used to determine changes in compressibility as a function of the pore water pressure. When the pore water pressure increases, the bubble radius and therefore the gas fraction reduces. If we assume that the pore water pressure increases so slowly that gas bubbles can dissolve and diffuse in the pore water to a new equilibrium, the relation between water pressure and bubble radius is governed by the preservation of mass of gas. This is shown in Fig. 8.3.

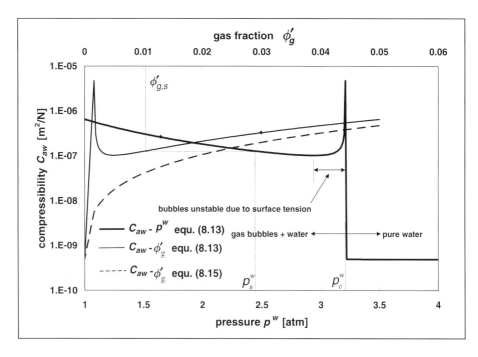

Fig. 8.3: Compressibility as function of pressure and gas fraction.

Consider an initial gas fraction $\phi'_{g,0} = 0.05$, an initial pressure of $p_0 = 1$ bar and an initial pore radius of $r_0 = 5$ μm. If the pore pressure is slowly increased, the compressibility reduces to a minimum at around 3 bar, then increases and subsequently collapses to the pure water compressibility of $0.48 \cdot 10^{-9}$ Pa^{-1}. The increase in compressibility occurs when the term with the surface tension in the denominator of equ. (8.13) becomes negative. This is the case when the gas fraction becomes smaller than $\phi'_{g,s}$, which is given by:

$$\phi'_{g,s} = \frac{\Omega}{2(1-\Omega)} \tag{8.16}$$

where $\phi'_{g,s}$ can be regarded as the maximal amount of gas that can be dissolved in the pore water at given conditions. The water pressure at this condition is:

$$p^w_{min} = p_{vap} - \frac{2\sigma_{st}}{r_0} \sqrt[3]{\frac{2(1-\Omega)}{\Omega}} \phi'_{g,s} +$$
$$+ \frac{2(\Omega(1-\phi'_{g,s}) + \phi'_{g,s})}{3\Omega}\left(p_0 - p_{vap} + \frac{2\sigma_{st}}{r_0}\right) \tag{8.17}$$

When the water pressure amounts to about 1.3 times this value, known as the critical water pressure p^w_c (e.g. Fig. 8.3), the gas phase vanishes, as gas dissolves in the pore water, i.e. only bubbles occur for $p^w < p^w_c$.

The relation between the gas fraction and compressibility is also shown in Fig. 8.3. The exact formulation equ. (8.13) and its approximation equ. (8.15) are both given. It appears that the approximation predicts the compressibility to the same order of magnitude, except in the region where the gas bubbles become unstable at small p^w.

The gas fraction $\phi'_{g,s}$ can also be read from the C_{aw} - ϕ'_g curve in Fig. 8.3 as the ϕ'_g-value just prior to the minimum in the pore pressure p^w.

In case that also gas bubbles are adhered to the solid surfaces, Barends (1980), amongst others, proposes to include their effect in the compressibility of the pore water. However, this gas can also be considered as part of the solids building the skeleton. We therefore favour to include the effects of gas adhered to the solids in the compressibility of these solids.

8.2 STRESS-STRAIN RELATIONS

We need a constitutive equation relating internal forces and deformations to solve the equations for preservation of mass and momentum. In general, constitutive equations are given as a relation between stress and strain increments, in which it is implicitly assumed that the relevant length scales are much larger than the particle size of the sediment. Hence, we assume that the soil behaves as a homogeneous material. These constitutive equations are complex because most solids "remember" their history of plastic deformations, except in the fully elastic regime. By definition, plastic deformations (e.g. strains) are irreversible and accumulate during loading. Therefore, plastic strains are the material's memory of load history. Typical examples are over-consolidation of cohesive sediment deposits and liquefaction due to cyclic loading. For sediments with a gas phase, the constitutive relations are even more complex. Some details are discussed in Chapter 11. Here we assume that gas is not affecting the constitutive behaviour of the sediment skeleton, but only affects the compressibility of the pore fluid.

In this section several constitutive models applicable to cohesive sediment are discussed, including memory effects as a result of over-consolidation and cyclical behaviour. Given the predominant role of pore water, a coupled two-phase system is considered, in which elasto-plastic behaviour of the skeleton is accounted for.

8.2.1 ELASTIC AND PLASTIC STRAINS

As discussed in Appendix D, the actual stress-induced deformations of a material are given by the finite deformation tensor of Green $\underline{\underline{C}}$, or, when relative deformations are considered, by the strain tensor $\underline{\underline{E}}$. These tensors represent the material averaged deformations over length scales much larger than the grain size of the sediment. Therefore, they do not represent the actual deformations of the granular skeleton at particle scale. In general, the relative displacements of particles differ from the large scale deformations. The deformations at particle scale are important when localised failure of the skeleton occurs, as discussed in Section 8.3.

When an incremental change in the skeleton or effective stresses occurs, an incremental change in strains is generated. In the so-called elasto-plastic approach, the skeleton is modelled as a Maxwell-body type material to establish the skeleton's response. In this approach, the total strain is decomposed in an elastic part (which is the recoverable or reversible strain) and a plastic part (which is the irreversible strain). The elastic part is obtained when the stress retrieves its initial state upon relaxation (see Fig. 8.4).

Elastic strains in soft soils are an order of magnitude larger than those in solids like steel and rock (ca. 1 – 5 %), and are non-linear in general. Still, these elastic strains are small compared to the plastic strains, if these occur, and therefore total deformations have to be considered.

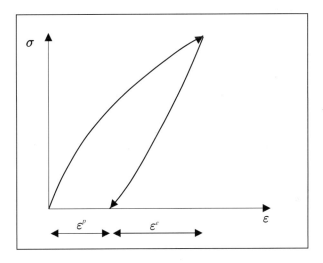

Fig. 8.4: Elastic and plastic strain ε^e and ε^p.

The total deformation of a line element is given by the deformation gradient tensor $\underline{\underline{F}}$ (D.18). Because the elastic part of a loading increment is obtained by unloading, the deformation gradient is decomposed into a plastic deformation first, followed by an elastic deformation:

$$\mathrm{d}\underline{x} = \underline{\underline{F}}^e \, \underline{\underline{F}}^p \, \mathrm{d}X \qquad\qquad (8.18)$$

where superscripts $\bullet^{e,p}$ refer to elastic and plastic deformations, respectively. For incremental loading we need the deformation rates. The spatial velocity gradient $\underline{\underline{L}}$, given by equ. (D.27), becomes:

$$\underline{\underline{L}} = \dot{\underline{\underline{F}}} \underline{\underline{F}}^{-1} = \dot{\underline{\underline{F}}}^{e} \underline{\underline{F}}^{e-1} + \underline{\underline{F}}^{e} \dot{\underline{\underline{F}}}^{p} \underline{\underline{F}}^{p-1} \underline{\underline{F}}^{e-1} \equiv \underline{\underline{L}}^{e} + \underline{\underline{L}}^{p} \qquad (8.19)$$

The first term of the very right-hand side is the elastic part of the velocity gradient tensor $\underline{\underline{L}}^{e}$ and the last term is its plastic part $\underline{\underline{L}}^{p}$. These can be decomposed into the deformation rate tensor $\underline{\underline{d}}$ (D.29), the symmetric part of $\underline{\underline{L}}$, and the spin tensor $\underline{\underline{w}}$ (D.31), which is the skew part of $\underline{\underline{L}}$:

$$\underline{\underline{d}} = \underline{\underline{d}}^{e} + \underline{\underline{d}}^{p} \; ; \; \underline{\underline{w}} = \underline{\underline{w}}^{e} + \underline{\underline{w}}^{p}$$

$$\underline{\underline{d}}^{e} = \frac{1}{2} \Big[\underline{\underline{L}}^{e} + \underline{\underline{L}}^{eT} \Big]; \quad \underline{\underline{d}}^{p} = \frac{1}{2} \Big[\underline{\underline{L}}^{p} + \underline{\underline{L}}^{pT} \Big] \qquad (8.20)$$

$$\underline{\underline{w}}^{e} = \frac{1}{2} \Big[\underline{\underline{L}}^{e} - \underline{\underline{L}}^{eT} \Big]; \quad \underline{\underline{w}}^{p} = \frac{1}{2} \Big[\underline{\underline{L}}^{p} - \underline{\underline{L}}^{pT} \Big]$$

Similar to the stress tensor (see equ. (D.9b)), the deformation rate tensor can be decomposed in an isotopic part $\dot{\varepsilon}_{v}$ for the volumetric strain rate and a deviatoric part $\dot{\gamma}_{ij}$ for the shear strain rate:

$$\frac{1}{2} \dot{\gamma}_{ij} = d_{ij} - \frac{1}{3} \delta_{ij} d_{kk} \; ; \quad \dot{\varepsilon}_{v} = d_{kk} \qquad (8.21)$$

which can be further decomposed into an elastic and plastic part. The factor ½ is added to maintain consistency with the classical definition for shear deformation. The deviatoric strain rate $\dot{\gamma}$ in the π-plane (Appendix D.1, Fig. D.4) is given by:

$$\dot{\gamma} = \sqrt{\dot{\gamma}_{ij} \dot{\gamma}_{ij}} \qquad (8.22)$$

To describe plastic deformations in granular material, the ratio of plastic volume strain rate $\dot{\varepsilon}_{v}^{p}$ and the plastic deviatoric strain rate $\dot{\gamma}^{p}$ is important, given by the dilatancy ratio β:

$$\beta = \dot{\varepsilon}_v^p / \dot{\gamma}^p \qquad\qquad (8.23)$$

8.2.2 ELASTIC STRESS-STRAIN RELATIONS

The elastic behaviour of solids (such as construction steel) is of great importance in mechanical engineering, where in general the stress states should be far below failure levels. The accompanying deformations, such as deflection of supporting beams, should be small with respect to the dimensions (i.e. length scales) of the construction. Therefore most of the constitutive relations for elastic behaviour of such solids concern very small strains only (of the order of less than 0.1 %), and a linear strain tensor can be used.

Also in cohesive sediments elastic strains are relatively small (< 10 %), but in general the stress relation is non-linear. Therefore constitutive relations between stress and strain for cohesive sediment are given in increments for strain rate and stress rate. The constitutive relation between the stress and strain rate tensors $\dot{\sigma}_{ij}$ and d_{kl}^e, being second order tensors, is given by the fourth order stiffness tensor D_{ijkl}^e:

$$\dot{\sigma}_{ij} = D_{ijkl}^e d_{kl}^e \qquad\qquad (8.24)$$

The coefficients of D_{ijkl}^e are described by Hooke's law for isotropic material (e.g. Malvern, 1969 and Timoshenko and Goodier, 1970):

$$D_{iijj}^e = \lambda_L + 2\delta_{ij}\mu_L = \frac{\nu E}{(1+\nu)(1-2\nu)} + 2\delta_{ij}G \qquad\qquad (8.25)$$

$$D_{ijij}^e = G = \mu_L = \frac{E}{2(1+\nu)} \quad ; \quad D_{ijkl}^e = 0$$

where λ_L and μ_L are the Lamé constants, implicitly defined in (8.25), E is Young's modulus, G is the shear modulus and ν is the Poisson ratio, defined as $\nu = -\dot{\varepsilon}_j / \dot{\varepsilon}_i$. From equ. (8.25) and the definitions for the deviatoric stress t (equ. (D.11)) and deviatoric strain γ (equ. (8.22)) it follows that:

$$G = \frac{i}{\dot{\gamma}^e} \qquad (8.26)$$

The elastic moduli can be measured with triaxial tests (see Appendix C) on cylindrical samples. The horizontal stress is kept constant in standard triaxial tests, while the vertical stress is increased. The Young's modulus E is found from the ratio of the vertical stress increment and the vertical strain increment (see Fig. 8.5). The Poisson ratio v is found from the ratio of horizontal and vertical strain increments. Other elastic moduli can be obtained from these two elastic parameters, such as the compression modulus K, the cylindrical constrained modulus M_{cyl} and the plane strain constrained modulus M_{ps} (see Fig. 8.5):

$$K = \frac{E}{3(1-2v)} \; ; M_{cyl} = \frac{(1-v)E}{(1+v)(1-2v)} \; ; M_{ps} = \frac{v\,E}{(1+v)(1-v)} \qquad (8.27)$$

The reciprocal of the compression modulus K is known as the compressibility of the skeleton C. The reciprocal of the constrained modulus M_{cyl} equals the compressibility coefficient in oedometer tests m_v.

Hooke's law (equ. (8.25)) can still be applied for large elastic deformations, i.e. finite strains. But as the stress rate in equ. (8.24) should only be related to the deformations in the material, it must be replaced by (equ. (D.36) and (D.37)):

$$\frac{D}{Dt}\underline{\underline{\sigma}} \equiv \overset{o}{\underline{\underline{\sigma}}} = \dot{\underline{\underline{\sigma}}} - \underline{\underline{\Omega}}\,\underline{\underline{\sigma}} + \underline{\underline{\sigma}}\,\underline{\underline{\Omega}} \qquad (8.28)$$

where the open dot refers to the co-rotational Jaumann stress rate. This operation neutralises the rotation of the material or reference frame with respect to the observer. The constitutive relation in spatial coordinates then reads:

$$\overset{o}{\sigma}_{ij} = D^e_{ijkl} d^e_{kl} \qquad (8.29)$$

When the stress and strain rate tensors are considered with respect to the material coordinates, the second Piola-Kirchoff stress tensor $\hat{\underline{T}}$ (equ. (D.46)) and the strain tensor of Green $\underline{\underline{E}}$ are used (equ. (8.17)), representing the work done within the sediment $\left(= \int_V \int_t \hat{T}_{ij} \dot{E}^e_{ij} \mathrm{d}t\mathrm{d}V \right)$:

$$\dot{\hat{T}}_{ij} = \hat{D}^e_{ijkl} \dot{E}^e_{kl} \tag{8.30}$$

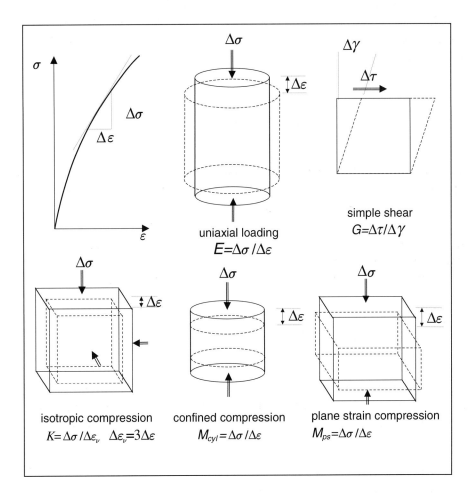

Fig. 8.5: Elastic moduli.

The Green strain rate is related to the spatial deformation rate $\underline{\underline{d}}^e$ through equ. (D.29):

$$\dot{\underline{\underline{E}}}^e = \underline{\underline{F}}^T \underline{\underline{d}}^e \underline{\underline{F}} \tag{8.31}$$

The material stiffness tensor $\hat{D}^e_{ijkl}$ then reads:

$$\hat{D}^e_{ijkl} = \lambda_L C^{-1}_{ij} C^{-1}_{kl} + \left(\mu_L - \lambda_L \ln J\right)\left(C^{-1}_{ik} C^{-1}_{jl} + C^{-1}_{il} C^{-1}_{jk}\right) \tag{8.32}$$

where J is Jacobi's determinant (equ. (D.14)) and C_{ij} is the component of Green's deformation tensor ($C_{ij} = 2E_{ij} + \delta_{ij}$, see equ. (8.17)).

8.2.3 ELASTO-PLASTIC STRESS-STRAIN RELATIONS

In elasto-plastic stress-strain models superposition of the elastic and plastic deformations rate d^e_{kl} and d^p_{kl} is assumed (equ. (8.20)). The stress state at a certain deformation d_{kl} can then be expressed with equ. (8.24) as:

$$\dot{\sigma}_{ij} = D^e_{ijkl}(d_{kl} - d^p_{kl}) \tag{8.33}$$

For finite strains, all material derivatives (denoted by a dot) must be replaced by the co-rotational Jaumann stress rate (denoted by an open dot) in order to neutralise rotation of the material with respect to the spatial coordinates, as discussed in the previous section.

Plastic deformation (i.e. local failure of the soil) is controlled by a yield criterion and a flow rule. Yield criteria are a function of the stress rate, generally described by a yield surface f in the principal stress space (see Fig. 8.6a), given by:

$$f\left(\sigma_{ij}, h_{ij}\right) = 0 \text{ and } \dot{f} \equiv \frac{\partial f}{\partial \sigma_{ij}} \dot{\sigma}_{ij} + \frac{\partial f}{\partial h_{ij}} \dot{h}_{ij} = 0 \tag{8.34}$$

Here h_{ij} is the so-called hardening tensor, which reflects the growth of the yield surface due to plastic strain history. Within the yield surface, $f < 0$ and non-linear elastic behaviour occurs, while for $f = 0$ plastic deformations (flow) occurs. The second part of equ. (8.34) is the so-called consistency condition.

The evolution of the hardening tensor as a function of the evolution of the stress state, the plastic strain rate d_{ij}^p and the accumulated plastic strain is called the hardening rule (e.g. equ. (8.46) and Section 8.4). In case of zero strain rate, the hardening tensor is a function of the stress state and the accumulated plastic strain only. When the yield surface grows isotropically, i.e. when the yield surfaces in the π-plane remain concentric (e.g. Appendix D), the hardening process is called isotropic. Examples of isotropic hardening models are those by Von Mises and by Drucker-Prager, developed for metals (see Fig. 8.6b and c). In general however, evolution of the yield surface in stress space is not isotropic, but moves through the principal stress space depending on the load history. This so-called kinematic hardening results in anisotropic properties of the sediment, which is important for cyclic loading and anisotropic pre-stressed (over-consolidated) conditions.

In soil mechanics, plastic deformations are referred to as flow, described with so-called flow rules, yielding relations between the plastic strain rate and the local stress state. In general, it is assumed that flow can be described by coaxial tensors, which implies that the principal directions of the stress tensor coincide with the principal directions of the plastic strain rate tensor. When a plastic potential g is defined as:

$$g\left(\sigma_{ij}\right)=0 \;\Rightarrow\; \dot{g}=\frac{\partial g}{\partial \sigma_{ij}}\dot{\sigma}_{ij}=0, \tag{8.35}$$

the plastic deformations in the bed are governed by the following flow rule:

$$d_{ij}^p = \lambda\frac{\partial g}{\partial \sigma_{ij}} \;;\; \lambda > 0 \tag{8.36}$$

where $\dot{\lambda}$ is a plastic multiplier, proportional to the rate of the total accumulated plastic strain. When $f < 0$ no plastic flow occurs, and $\dot{\lambda} = 0$.

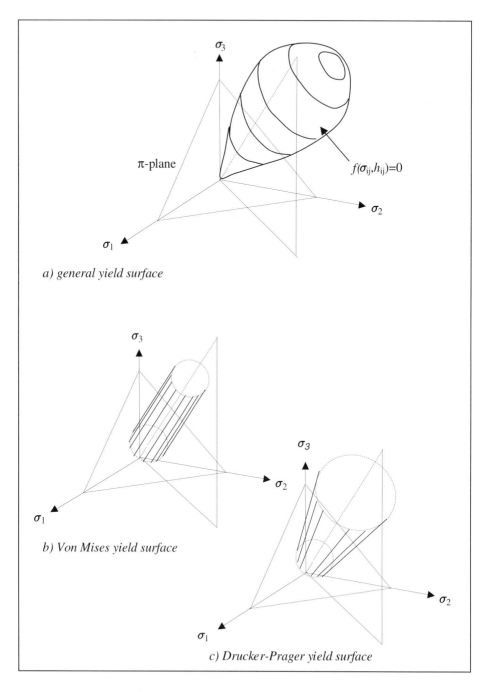

a) general yield surface

b) Von Mises yield surface

c) Drucker-Prager yield surface

Fig. 8.6: Yield surface in principal stress space.

In the principal stress space, the flow rule describes a plastic strain rate perpendicular to the plastic potential surface. When the plastic potential surface g (equ. 8.35)) coincides with the yield surface, the deformations are referred to as associated flow, and the stress-strain relation is given by a so-called associated flow rule. In that case the plastic strain rate is perpendicular to the yield surface f (equ. 8.34)) as well. Stress increments along the yield surface do not dissipate energy in this case. Hence, at these associated flow conditions, deformations take place at minimum energy dissipation as the flow direction is perpendicular to the yield surface.

Combining equ. (8.33), (8.34) and (8.36) yields a general constitutive relation between the stress increment $\dot{\sigma}_{mn}$ and the total strain increment d_{ij} (Zienkiewicz et al. 1989, Hölsher, 1995):

$$\dot{\sigma}_{mn} = \left[D_{mnij} - \frac{1}{H} D_{mnpq} \frac{\partial g}{\partial \sigma_{pq}} \frac{\partial f}{\partial \sigma_{kl}} D_{klij} \right] d_{ij} \tag{8.37}$$

in which the hardening parameter H is given by:

$$H = \frac{\partial f}{\partial \sigma_{mn}} D_{mnij} \frac{\partial g}{\partial \sigma_{ij}} - \frac{\partial f}{\partial h_{ij}} \frac{\dot{h}_{ij}}{\dot{\lambda}} \tag{8.38}$$

The term between brackets in equ. (8.37) is the so-called tangent stiffness tensor. The last term in equ. (8.38) is called the plastic hardening modulus.

8.2.4 COHESIVE SEDIMENT CONSTITUTIVE RELATIONS

Two failure conditions formed the basis for the development of the so-called Double Hardening or Double Cap models describing the yield surface for sand, clay and rock:

- the Mohr-Coulomb failure condition, for which the shear-induced yield stresses depend on the isotropic stresses, and
- plastic deformation conditions during compression.

One of the first Double Cap models was the Cam-Clay model developed in Cambridge (Roscoe and Schofield, 1963). This isotropic Double Cap model consists of a Drucker-Prager like yield surface for shear failure and an elliptical cap for failure by compression. This model is depicted in the principal stress space in Fig. 8.7. Instead of plotting the yield surface as a three-dimensional graph in principal stress space, it is more convenient to represent the stress-strain conditions in two diagrams:

- the p-q diagram in a cross section of a plane containing the principal space diagonal and one of the principal axes (Fig. 8.8a), with the isotropic stress p (the diagonal in the principal stress space) and deviatoric stresses (a stress vector in the π-plane - see also the definitions in equ. (D.9a) and (D.11))
- the cross section of the yield surface in the π-plane (Fig. 8.8b); associated flow (page 276) is given by flow vectors perpendicular to the cross sections in Fig. 8.8b.

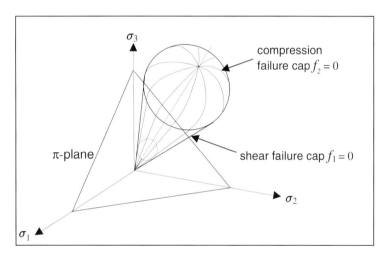

Fig. 8.7: Double cap yield surfaces in principal stress space.

The failure surfaces in Fig. 8.7 are given by (see for instance Groen, 1995):

$$f_1 = \frac{q}{R_1(\theta)} - \frac{6\sin\varphi}{3-\sin\varphi}\left(p^{sk} + p_{coh}^{sk}\right) = 0 \qquad \text{Mohr Coulomb shear failure}$$

$$(8.39)$$

$$f_2 = \alpha\left[\frac{q}{R_2(\theta)}\right]^2 + \left(p^{sk} + p_{coh}^{sk}\right)^2 - \left(p_c^{sk} + p_{coh}^{sk}\right)^2 = 0 \qquad \text{compression failure}$$

in which p_{coh}^{sk} is the isotropic stress representing cohesion at zero isotropic skeleton stress, p_{c}^{sk} is the isotropic consolidation stress (e.g. Fig. 8.8a), φ is the internal friction angle (i.e. the envelope of the stress-strain circles in the Mohr-diagram), and $R(\theta)$ is a function representing the non-circular shape of the yield surface in the π-plane as a function of the Lode angle θ (e.g. Appendix D.1, Fig. D.4).

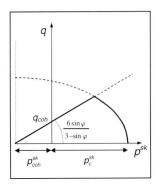
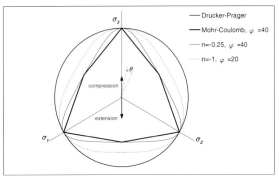

a) The p-q-diagram.   *b) Cross sections of yield surfaces in π-plane.*

Fig. 8.8: Yield surfaces of double cap model.

Experiments have shown that failure surfaces in soil are not axi-symmetric around the principal stress space diagonal: there is a difference between the deviatoric failure stress for axial compression and radial compression (extension). This results in deviations from a circular failure surface, as depicted in Fig. 8.8b, which can be modelled by (Van Eekelen, 1980):

$$R(\theta) = \left[\frac{1 - b\cos 3\theta}{1 - b}\right]^{n} \quad ; b = \frac{\left[\dfrac{3 + \sin\varphi_e \, \sin\varphi_c}{3 - \sin\varphi_c \, \sin\varphi_e}\right]^{-1/n} - 1}{\left[\dfrac{3 + \sin\varphi_e \, \sin\varphi_c}{3 - \sin\varphi_c \, \sin\varphi_e}\right]^{-1/n} + 1} \tag{8.40}$$

where φ_c and φ_e are friction angles measured in respectively triaxial compression and extension tests. Fig. 8.8b shows the shape of the failure surface for several values of the power n and friction angle φ. For $n = -0.25$ (Van Eekelen, 1980) the surface is very close to the Mohr-Coulomb failure surface for sand (straight solid lines in Fig. 8.8b).

The Double Cap approach, discussed above, requires a flow rule for each cap. Therefore the flow rule (8.36) must be extended with the two plastic potentials g_1 and g_2 to describe plastic deformations in the soil as a function of the stresses σ_{ij}:

$$d_{ij}^p = \dot{\lambda}_1 \frac{\partial g_1}{\partial \sigma_{ij}} + \dot{\lambda}_2 \frac{\partial g_2}{\partial \sigma_{ij}} \qquad (8.41a)$$

where g_1 and g_2 are given by (Groen et al., 1995):

$$g_1 = q - \frac{6\sin\psi}{3-\sin\psi}\left(p^{sk} + p_{coh}^{sk}\right) = 0 \qquad \text{Mohr Coulomb failure by shear}$$

$$\qquad (8.41b)$$

$$g_2 = \alpha q^2 + \left(p^{sk} + p_{coh}^{sk}\right)^2 - \left(p_c^{sk} + p_{coh}^{sk}\right)^2 = 0 \qquad \text{failure by compression}$$

in which ψ is the dilatancy angle. This angle is a measure for the plastic volume increase induced by deviatoric shear strain increments. The dilatancy angle is obtained by substitution of the plastic potential in the flow rule:

$$\sin\psi = -\sqrt{\frac{8}{3}}\frac{\dot{\varepsilon}_v^p}{\dot{\gamma}^p} \qquad (8.42a)$$

The dilatancy angle ψ is the angle at which particles are displaced when dilated (see also Section 8.3); ψ can be related to the internal friction angle φ by Rowe's stress-dilatancy theory (Rowe, 1962):

$$\sin\psi = \frac{\sin\varphi - \sin\varphi_{cv}}{1 - \sin\varphi\sin\varphi_{cv}} \qquad (8.42b)$$

where φ_{cv} is the internal friction angle at which shear deformation is possible without volume changes ($\psi = 0$): the so called **critical state** (Schofield and

Wroth, 1968). This critical state forms the basis for the erosion formula described in Chapter 9.

According to equ. (8.42b) plastic potentials do generally not coincide with failure surfaces, i.e. the flow rule is non-associative with the failure surface. Associated flow is only possible when the plastic potentials coincide with the failure surfaces ($\psi = \varphi$), which is the case only for the trivial situation $\varphi_{cv} = 0$, or the unrealistic case $\varphi = \pi/2$.

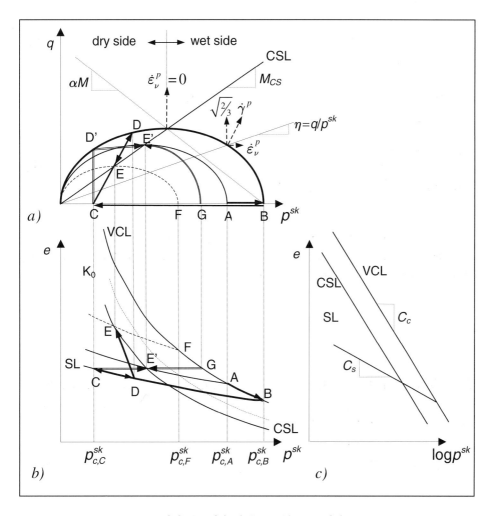

Fig. 8.9: Modified Cam-Clay model.

The evolution of the shear yield surface (equ. (8.39a)) is determined by φ and is therefore a function of the accumulated plastic shear strain. The consolidation stress p_c^{sk}, which is a function of the plastic volume strain, determines the evolution of the compression yield surface (equ. (8.39b)).

This constitutive model can properly simulate triaxial test on sand and normal consolidated clays, but it cannot cope with the softening behaviour of over-consolidated clay. Therefore, the Cam-Clay model was extended using critical state theory (Schofield and Wroth, 1968), commonly known as the Modified Cam-Clay model (Roscoe and Burland, 1968), in which one elliptical yield surface is defined for both shear and compressive failure (see p-q diagram in Fig. 8.9a).

The size of the elliptical yield surface in the p-q diagram is determined by the isotropic consolidation stress p_c^{sk}. Changes in skeleton stresses appear as lines in the p-q diagram; these lines are called stress paths. For instance, consider the stress path AB on the horizontal axis of Fig. 8.9a, i.e. along the diagonal in the principal stress space. This isotropic stress path starts on the yield surface through point A and depicts a plastic volume strain, during which the skeleton is compressed to a lower void ratio, squeezing out pore water (consolidation). This results in growth of the elliptic yield surface to the one through point B. The relation between void ratio and isotropic consolidation stress p_c^{sk} is depicted by the so-called Virgin Compression or Consolidation Line (VCL) in Fig. 8.9b.

We have reasoned in Chapter 3 that for cohesive sediment the deviatoric stress at failure (i.e. the remoulded shear strength c_u) is uniquely related to the void ratio or water content of the sediment. Therefore, a unique failure line parallel to the VCL can be constructed (see Fig. 8.9b), along which shear failure is possible without volume changes, known as the Critical State Line (CSL). The CSL in the p-q diagram intersects the top of the set of elliptical yield surfaces.

Unloading after isotropic consolidation along stress path BC in the p-q-diagram (Fig. 8.9a) yields a non-linear elastic response, which follows the Swelling Line (SL) in Fig. 8.9b. The ratio of the isotropic consolidation stress in B and C is defined as the over-consolidation ratio OCR:

$$\text{OCR} = p_{c,B}^{sk} \Big/ p_{c,C}^{sk} \qquad\qquad\qquad (8.43)$$

The VCL, CSL and SL are generally modelled as exponential functions of the void ratio e and therefore appear as straight lines when the isotropic stress is plotted on logarithmic scale (see Fig. 8.9c). The slope of the VCL and CSL on log-scale is defined as the compression index C_c, and the slope of the SL is known as the swell index C_s.

However, power functions seem to cover a larger range of void ratio's in experimental data and should therefore be preferred. Moreover, such power functions agree favourably with our fractal description of mud flocs and cohesive sediment beds (e.g. Chapter 4 and 7; see also Buterfield, 1979). For power functions Fig. 8.9c still holds, when the void ratio is also plotted logarithmically.

The CSL divides the yield surface (Fig. 8.9a) in a "wet side" right of the top of the yield surface and a "dry side" at its left. Stresses exceeding the yield surface on its wet side result in plastic compaction (e.g. consolidation), during which water is expelled from the pores (stress path AB in Fig. 8.9a,b). When the yield surface is exceeded on the dry side, plastic dilation occurs and water is sucked into the pores until the critical state is reached.

Fig. 8.9a,b also shows the non-isotropic stress path CDE, following the loading path AB and the unloading path BC. The path CDE is typical for triaxial tests, where the vertical load σ_3 is increased, while the horizontal stresses σ_1 and σ_2 remain constant. From C to D the stress path is non-linear elastic until the yield surface is reached in point D. The plastic dilation along path DE yields an increase in void ratio e reaching the critical state in point E, corresponding to a smaller elliptical yield surface through point F in Fig. 8.9a.

In summary, when the yield surface is followed along its wet side the yield surface is growing (hardening), whilst a path along the dry side yields shrinking (softening) of the yield surface.

The stress paths ABC and CDE represent drained conditions. In case of undrained conditions the void ratio e remains constant, and any stress path would be depicted by a horizontal line in Fig. 8.9b. If the triaxial test described above would be executed under such undrained conditions, the stress path will be quite different from CDE as plotted in Fig. 8.9a in the form of a double line. In the elastic part, the isotropic skeleton stress remains constant and the stress path follows C to D'. As D' is on the dry side of the yield surface, there is a tendency for plastic dilation. This results in an increase of the isotropic skeleton stress until the critical state is reached in point E', where shear failure is possible at constant volume. The deviatoric stress at failure is twice the undrained shear strength c_u (e.g. Appendix D.1). The same undrained shear strength is obtained when the undrained stress path starts from the virgin compression line VCL at the same void ratio in point G in Fig. 8.9b. Because this point is on the wet side of the yield surface through G, there is a tendency for plastic compaction, which results in a decrease of isotropic skeleton stresses, and an increase of the yield surface until the critical state is reached in point E' (Fig. 8.9a). The fact that the same critical state point can be reached for normal (starting in G) and over-consolidated conditions (starting in C) enables one to relate the undrained shear strength to the over-consolidation ratio OCR:

$$\frac{c_u}{p^{sk}} = \frac{1}{2} M_{cs} \frac{\alpha}{1+\alpha} \text{OCR}^{\Lambda} \quad \text{where} \quad \Lambda = \frac{C_c - C_s}{C_c} \tag{8.44}$$

where α is a coefficient for the location of the VCL with respect to the CSL ($\alpha = 1$ for an elliptical yield surface), M_{CS} is given in Fig. 8.9 and Λ is called the critical state parameter.

Hence, we conclude that under drained conditions one yield surface is followed until failure at the critical state, whereas under undrained conditions, many yield surfaces may be followed until failure.

The elliptical yield surfaces (solid and dotted curves in Fig. 8.9a) are described by:

$$f = \left[\frac{q}{R(\theta)} \right]^2 - M \, p^{sk} \left(p_c^{sk} - p^{sk} \right) \; ; \; M = \frac{6 \sin \varphi}{3 - \sin \varphi} \tag{8.45}$$

The function $R(\theta)$ is set to unity in the modified Cam-Clay model (i.e. a circular yield surface in the π-plane).

The hardening tensor h_{ij} (equ. (8.34)) yields a relation between the isotropic plastic volume strain $\dot{\varepsilon}_v^p$ and the isotropic consolidation stress p_c, based on the definition of VCL and SL, i.e. $h_{ij} = \delta_{ij} p_c$. The hardening rule is given by:

$$\dot{\varepsilon}_v^p = \frac{C_c - C_s}{(1+e) \ln 10} \frac{\dot{p}_c^{sk}}{p_c^{sk}} \quad \Leftrightarrow \quad \dot{h}_{ii} = h_{ii} \frac{(1+e) \ln 10}{C_c - C_s} \dot{\varepsilon}_v^p \tag{8.46}$$

The plastic potential g (equ. (8.35) and (8.42)) in the Cam-Clay model coincides with the yield surface, which implies associated flow. This is consistent with the definition that the CSL intersects the top of the yield surface (Fig. 8.9a), where plastic volume strain increments do not occur at associated flow.

The dilatancy ratio β (equ. (8.23)) is found from the associated flow rule (8.36) and the deviatoric plastic strain increment:

$$\beta = \frac{\dot{\varepsilon}_v^p}{\dot{\gamma}^p} = \frac{1}{\sqrt{6}} \frac{M^2 - \eta^2}{2\eta} \tag{8.47}$$

where we use (8.21) and (8.22) and η is the stress ratio q/p^{sk} (see Fig. 8.9a).

Within the elastic region (i.e. the semi-ellipses in Fig. 8.9a) the bulk modulus K is obtained from the swelling line SL:

$$K \equiv \frac{\dot{p}^{sk}}{\dot{\varepsilon}_v^e} = \frac{\ln 10}{C_s} p^{sk} (1 + e) \tag{8.48}$$

The shear modulus G is obtained from the elastic relations in (8.25) and (8.27):

$$G \equiv \frac{\dot{i}}{\dot{\gamma}^e} = \sqrt{\frac{2}{3}} \frac{\dot{q}}{\dot{\gamma}^e} = \frac{3}{2} \frac{1 - 2v}{1 + v} K \tag{8.49}$$

The stress path in the seafloor due to self-weight consolidation is important for cohesive sediment. The horizontal strains in a horizontal seafloor are zero and therefore a confined compression path (see Fig. 8.9b), called the K_0-stress path, is followed when the load of sediment increases by sedimentation. The accompanying stresses and strains are axial-symmetric around the vertical axis (Fig. 8.5b). The horizontal and vertical (elastic and plastic) strains and stresses are the principal strains and stresses in the bed. The horizontal stresses in both directions are equal ($\sigma_1 = \sigma_2$), and the ratio between horizontal stress and vertical stress is referred to as K_0 ($\sigma_{1,2} = K_0 \sigma_3$). K_0 is also known as the coefficient of earth pressure at rest. Hence we find for the stress ratio η:

$$\eta = \frac{q}{p^{sk}} = \frac{\sigma_3 - \sigma_1}{\frac{1}{3}(2\sigma_1 + \sigma_3)} = 3\frac{1 - K_0}{1 + 2K_0} \tag{8.50}$$

The confined compression path for constant K_0 is a line parallel and in between the CSL and VCL (see Fig. 8.9b). During confined compression the deviatoric and volume plastic strain are:

$$\dot{\gamma}^p = \sqrt{\tfrac{8}{3}}\left(d^p_{33} - d^p_{11}\right) \;\; ; \;\; \dot{\varepsilon}^p_v = 2d^p_{11} + d^p_{33} \tag{8.51}$$

From equ. (8.49), (8.46) and the confined compression condition (which implies that the sum of the horizontal plastic and elastic strains must be zero), K_0 can be computed as a function of the internal friction angle φ, the Poisson ratio ν and the critical state parameter Λ (see equ. (8.44)).

Computed values of K_0 with the modified Cam-Clay model ($\nu = 0.3$, $\Lambda = 0.4$) are compared with experimental data (Kumbhojkar et al., 1993) in Fig. 8.10 (Van Eekelen et al., 1994), showing the dependency of K_0 on the internal friction angle φ. By increasing the curvature of the elliptical failure surface in compression, like f.i. the Delft-Egg-model (Van Eekelen et al., 1994), a better result is obtained. Experimental data are reasonably described by $K_0 = 1 - \sin\varphi$ (Jâki , 1944) as shown in Fig. 8.10.

In general, K_0 is not constant but increases with increasing isotropic stress. This is partly because the internal friction angle φ decreases with increasing isotropic stress.

However at very low isotropic stresses, as in fresh deposits of very soft cohesive sediments, K_0 becomes zero (Van Kesteren, 2004) and the behaviour of the sediment becomes very similar to that of fresh snow. The flocculated structure causes a collapse of the structure in load direction only, whilst the structure in lateral direction remains unaffected. Hence, the stress history is "recorded" in the sediment by the orientation and amount of damage of the internal structure. This results in material properties that vary in different directions, i.e. the material becomes anisotropic. In that case, isotropic hardening models are not applicable anymore. Instead, kinematic hardening models should be used depending on the plastic strain history reflecting the direction-dependent damage that occurred in the material. These models are more complex, but are necessary to describe complicated processes such as cyclic loading by waves (e.g. Section 8.4).

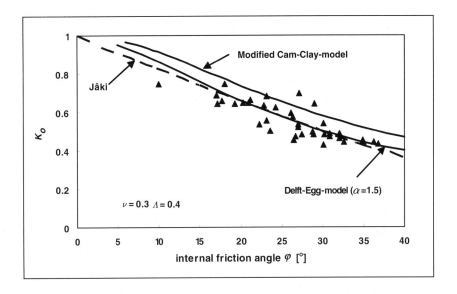

Fig. 8.10: K_0 for various clays (redrawn from Van Eekelen, 1994; data from Kumbhojkar et al., 1993).

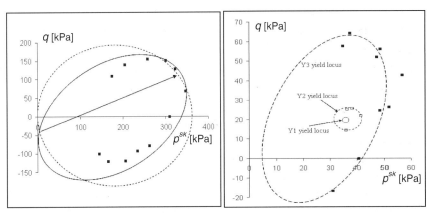

a) K_0 oriented yield surface (after Al-Tabbaa, 1984).

b) Local, midfield and total kinematic hardening.

Fig. 8.11: Kinematic hardening (see Rouainia, M. et al., 2002)

The anisotropy in K_0-loading can be illustrated (Fig. 8.11a) by comparing the modified Cam-Clay model with experimental data of yield surfaces in clays. These yield surfaces are merely coupled to the K_0-loading path than to the diagonal in the principle stress space (Al Tabbaa, 1984). Furthermore, when stress increments are applied in all directions, starting at a certain stress point, plastic strains result within the local yield surfaces (Fig. 8.11b, Smith et al., 1992). As plastic deformations in the material (damages of the material) accumulate, the yield surface expands when the load is further increased.

Today, the Modified Cam-Clay model is accepted as a basis for further developments. Examples are the Delft-Egg-model (Van Eekelen, 1994), and Desai's model (Desai, 1991). Desai's model is applied in our analysis of cyclic loading in Section 8.4.
  A second important loading condition for marine sediment is due to waves and currents. Under such conditions, the principle directions of the stresses are not stationary, but rotating. This is discussed in Section 8.4, which deals with cyclic behaviour and liquefaction. However, failure modes are discussed first in Section 8.3.

8.3   FAILURE MECHANISMS

8.3.1 TYPE OF FAILURE

In Section 8.1 we have described three modes of failure (e.g. Fig. 8.1): tensile, shear and ductile failure. These modes may occur in un-cemented granular material such as sand, mud and clay, but also in cemented granular material such as rock. The actual mode of failure occurring is determined by the isotropic skeleton stresses as depicted in the p-q-diagram in Fig. 8.12 (Van Kesteren, 1995).
  This p-q-diagram is drawn for axial-symmetric conditions, where the radial principal skeleton stresses σ_1^{sk} and σ_2^{sk} are equal. The principal axes are also given in Fig. 8.12. The horizontal axis (isotropic skeleton stress) divides the p-q-diagram in an upper area for compression where the vertical principal stress σ_3^{sk} is larger than σ_1^{sk} and σ_2^{sk}, and a lower area for extension, where the vertical principal stress σ_3^{sk} is smaller than σ_1^{sk} and σ_2^{sk}.
  In granular material, two yield surfaces can be distinguished, as discussed in the previous section: a shear failure cap and a compressive cap. In the tensile region, where all three principle skeleton stresses are negative or zero, the limit

stress state for both compression and extension is often schematized by a tension cut-off with a certain tensile strength (see shaded area in Fig. 8.12).

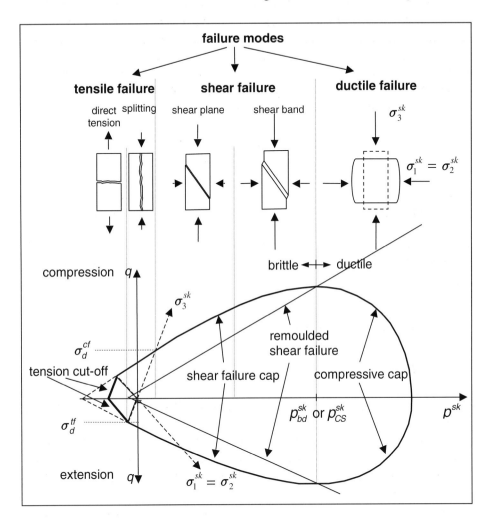

Fig. 8.12: Failure mechanisms in p-q-diagram (Van Kesteren, 1995).

There exists a strong analogy between the failure mechanisms in un-cemented and in cemented granular materials, such as rock. It is therefore illustrative to describe some failure processes in rock, as published in literature.

The isotropic stress at the transition from shear failure to compression (ductile failure) is called the brittle-ductile transition stress p_{bd}^{sk} in rock mechanics. For $p^{sk} < p_{bd}^{sk}$ shear failure occurs, characterised by localization of shear deformation due to the growth and coalescence of micro-cracks, i.e. cracks at particle scale. At low isotropic stresses, with positive $\left(\sigma_1^{sk} = \sigma_2^{sk} > 0,\ \sigma_3^{sk} > 0\right)$ principal skeleton stresses, coalescence of micro-cracks results in a dilating shear plane of *en echelon* wing-shaped cracks (Fig. 8.13a) (Brace et al., 1966 and Gramberg, 1989). At stresses close to p_{bd}^{sk} there is no distinct shear plane, but a wide shear zone or shear band, where the *en echelon* wing-shaped cracks are less dilatant and friction in the planar crack (inset in Fig. 8.13a) dominates failure. After failure, a residual stress state is reached, which is dominated by friction in the shear plane or shear band.

When one of the principal skeleton stresses becomes zero or negative, tensile failure occurs, characterised by coalescence of wing-shaped cracks into a planar crack perpendicular to the lowest principal stress. Splitting tensile failure occurs when $\sigma_1^{sk}, \sigma_2^{sk} < \sigma_3^{sk}$, while direct tensile failure occurs in case $\sigma_1^{sk}, \sigma_2^{sk} > \sigma_3^{sk}$.

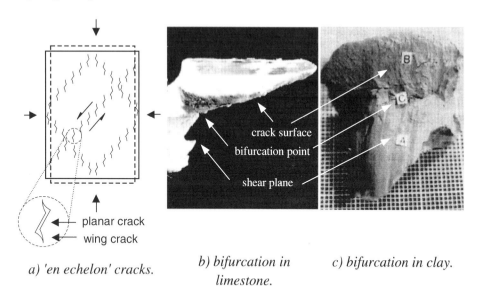

a) 'en echelon' cracks.   *b) bifurcation in limestone.*   *c) bifurcation in clay.*

Fig. 8.13: Bifurcation shear plane to tensile crack (Van Kesteren, 1995).

For $p^{sk} > p_{bd}^{sk}$ (Fig. 8.12) failure during compression occurs at the compressive cap. As a result of this mode of failure, the internal structure of the material is damaged or crushed in such a way that plastic volume strains (compaction) occur. This results in a process called strain hardening. Crushing of the skeleton in rock enables large plastic shear deformations without localization. Because of this ductile behaviour the isotropic stress p_{bd}^{sk} is called the brittle-ductile transition stress in rock mechanics. Due to the crushing of the rock skeleton the mechanical behaviour changes irreversible into that of un-cemented soils, characterised by a growing yield surface during further compression of the crushed material with an increasing isotropic stress at critical state $\left(p_{CS}^{sk} \right)$.

Fabric analysis of clay samples shows that, similar to rock, a shear plane or shear band is formed after coalescence of opening *en echelon* micro-cracks (e.g. Morgenstern et al., 1968 and Yong et al., 1971). Experimental evidence for this failure mechanism follows from the bifurcation of shear planes into tensile failure planes, as is shown in Fig. 8.13b and c, respectively, for pilot tests on limestone and Westerwald clay (Van Kesteren, 1995). A shear plane is observed in the crushed zone in the limestone, bifurcating into a tensile crack when the shear plane propagates towards the stress free surface. A similar bifurcation pattern develops in the clay sample.

The generation and development of micro-cracks is an important mechanism, as dilatancy and localization of shear deformation can start before the yield surface is reached. As a result, failure can occur at stresses well below the strength of the material. The orientation and density of the cracks are stress-path dependent and influence the position of yield surfaces and stress-strain behaviour (kinematic hardening).

The shear failure cap and tension cut-off are determined by crack initiation and unstable crack growth. This crack initiation is determined by the size, orientation and density distribution of micro-cracks, which are generated during small strain loading and unloading, and which are stress-path dependent. Macro-cracks become unstable, when the incremental elastic energy, released during unloading of the material, exceeds the fracture energy necessary for macro-crack growth. As the amount of energy stored in the material increases with its size, the growth rate of macro-cracks is dependent on the volume of sediment considered. Therefore, next to the limit stress states and yield surfaces for tensile failure, also the fracture energy necessary to reach the residual stress state, must be determined. Hence, tensile strength is not a pure material parameter, but is also dependent on the required fracture energy and the amount of sediment under consideration.

The strength of the bonds between particles and the reversibility of these bonds in un-cemented materials, such as sand and clays, differ from those of cemented granular materials, such as rock. Bonds between particles in both cemented and un-cemented granular materials reduce the mutual mobility of these particles. Inhomogeneous deformations on micro-scale or particle-scale are necessary for homogeneous plastic deformation at macro-scale, though the latter are treated as a continuum. This apparent paradox is visualised in Fig. 8.14 through the homogeneous deformation of a sample to half its original height. Contrary to continuum deformation theory, particles have to be rearranged upon deformation of granular material, and the size and shape of the particles determine the number of particles in undeformed and deformed line elements.

Such a rearrangement is only possible by inhomogeneous shear deformation by particle sliding and/or rolling (Vermeer, 1983), or by particle crushing. When such rearrangement is impossible because of particle bonds or particle interlocking, homogeneous plastic deformation at macro-scale is not possible, and localisation of deformations will occur in the form of tensile cracks or shear bands depending on the isotropic stress level.

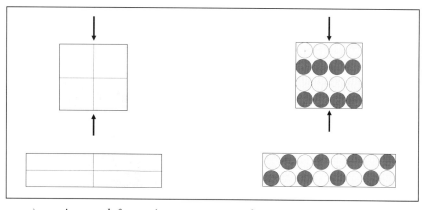

a) continuum deformation.   *b) particle rearrangement.*

Fig. 8.14: Homogeneous deformation at continuum and particle scale.

Therefore, a necessary condition for ductile behaviour is bond or particle failure. Furthermore, additional plastic shear strains at particle scale is required so that particles can be rearranged. Such rearrangements are only possible when the stresses for bond or particle failure are at or beyond the compressive cap of the skeleton. Note that particle failure is always accompanied by changes in the skeleton (e.g. cemented calcareous soils, Kao and Bell, 1987).

In soft cohesive sediment particle failure is not to be expected given the low stress levels with respect to particle strength. In such soils, bond failure can occur however at both sides of the compression cap for plastic deformations in the skeleton. When the bond strength is large with respect to the strength of the granular skeleton, bond failure is always accompanied by skeleton failure. This is the case in most soft porous rocks, normally consolidated clays and in soft cohesive sediments. When the bond strength is small with respect to the strength of the granular skeleton, bond failure may occur without skeleton failure (which may be the case for cemented densely packed sand and over-consolidated clays).

8.3.2 TENSILE FAILURE

To the knowledge of the authors, tensile failure in saturated soft cohesive sediments has never been described in the literature. Most literature on tensile failure in mud is dedicated to desiccation (drying) and tensile failure in soft and stiff clays. Nevertheless tensile failure has been observed under several testing conditions during erosion tests (e.g. Chapter 9) and expansion of mud beds due to gas bubble growth (e.g. Chapter 11). Here, we apply tensile failure theory for clays (Van Kesteren, 1995, 2004) to elaborate on the conditions at which tensile failure may occur in the seabed.

As discussed in the previous section, tensile failure starts with the opening of micro-cracks. In granular material, such micro-cracks are often formed by discontinuities in the skeleton, for instance at interfaces of large particles, where no effective stresses occur. When these micro-cracks grow and coalescence with neighbouring cracks, tensile failure occurs. This mechanism is depicted in Fig. 8.15 for a notched bar under tensile loading, showing stresses concentrating in front of the notch (dashed line in Fig. 8.15). The stresses around the micro-cracks in the zone in front of the notch cause the crack front of two neighbouring, growing micro-cracks to propagate away from each other (Ingraffea, 1987). Coalescence of micro-cracks occurs under shear failure in the undisturbed material between the cracks. When coalescence in front of the notch is complete, a macro-crack forms and the failure process proceeds in a zone in front of the macro-crack.

This process also occurs during tensile tests on clay. Fig. 8.16a shows the results of a tensile test on clay at low deformation rates (10^{-4} s^{-1}); these are however large enough to prevent drainage (undrained conditions). Failure occurs at stresses much smaller than the undrained shear strength c_u, and a planar macro-crack forms, denoted as a brittle crack. Due to shear failure between adjacent micro-cracks, the surface of the macro-crack is characterised by a ray pattern starting at the location where the macro-crack is initiated (marked by a dot in Fig. 8.16b).

When several macro-cracks are initiated and coalesce, individual three-dimensional shell-shaped fractures (Fig. 8.17a) propagate in a two dimensional ray pattern (Fig. 8.17b).

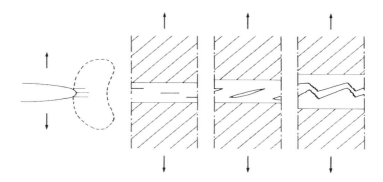

Fig. 8.15: Crack propagation in cohesive sediment.

      a) brittle tensile crack.   *b) radial pattern brittle crack surface.*

Fig. 8.16: Tensile test on clay.

Brittle cracks are no longer formed when the tensile deformation rate exceeds a certain level. In that case, ductile behaviour occurs up to very large strains (Fig. 8.18). Due to the large deformations, the growth of micro-cracks is visible on the surface of the sample as opened flaws, though without coalescence (Fig. 8.18b). Beyond a certain strain, fracturing occurs in the neck (Fig. 8.18c). The surface of the fracture now differs largely from the brittle failure patterns: there is no radial pattern anymore, but the rupture (Fig. 8.18 d) forms a crater-like surface instead. The failure stress is equal to the undrained failure stress under compression. Therefore, this type of crack is denoted as a ductile crack.

An increase in backpressure causes an increase in the strain at failure. The crater-like surface in Fig. 8.18d is evidence for the generation of cavitation, and subsequently the formation of enclosed cavities (vacuoles). The final failure surface forms when the material between the bubbles is torn off.

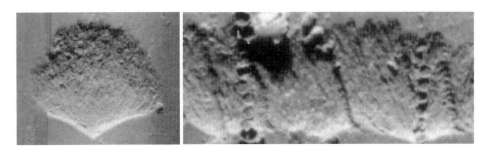

a) initial 3D crack.   *b) 2D crack propagation.*

Fig. 8.17: Coalescence of macro-cracks in 2-dimensional crack propagation.

To analyse this tensile behaviour of cohesive sediment, we consider an initially penny-shaped crack in a porous medium (see Fig. 8.19a). Loading a single phase material with a vertical tensile stress results in stress concentrations at the edge of the crack. According to linear elastic fracture mechanics (LEFM) stress concentrations at different loadings can be superimposed as shown in Fig. 8.19b (Broek, 1982). According to this linear theory, water over-pressure inside the crack would lead to the same stresses around the edges of the crack as an equally large tensile stress loading.

If the pore system of the soil and the penny shaped crack is saturated with water, loading with a tensile stress σ^t initially generates a pressure drop in the crack equal to σ^t, as a result of which no stress concentrations around the edges of the crack are generated (see Fig. 8.19b). The isotropic total stress in the surrounding sediment matrix drops with $\sigma^t/3$ (axial symmetric through-flow), together with the pore water pressure. As the pore water pressure in the penny shaped crack is now lower than in the surrounding pores, pore water flows towards the crack, and the water pressure inside the crack increases and, subsequently, stress concentrations around the edge of the crack will appear and increase. The crack volume then increases with the inflowing pore water.

This drainage of cracks is governed by the size of the cracks and the consolidation coefficient of the surrounding sediment.

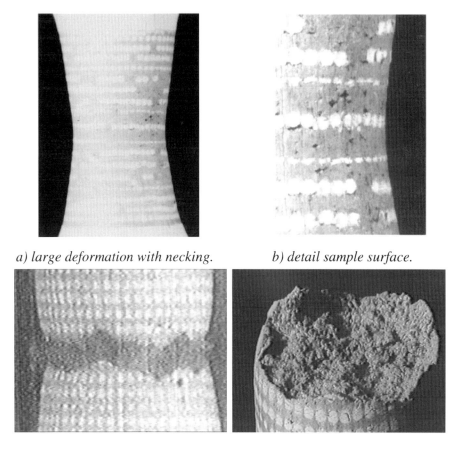

a) large deformation with necking.   b) detail sample surface.

c) detail necking and rupture.   d) ductile crack surface.

Fig. 8.18: Tensile failure in clay at high deformation rate.

For a constant tensile loading rate $\dot{\sigma}^t$ the water pressure in the crack p_{cr}^w is given by (Van Kesteren, 2004):

$$\frac{p_{cr}^w}{\dot{\sigma}^t t} = \frac{2(1-K_0)}{3Fo_w}\left[2\sqrt{\frac{Fo_w}{\pi}} - 1 + \exp\{Fo_w\}\mathrm{erfc}\{\sqrt{Fo_w}\}\right] + \frac{1+2K_0}{3} \tag{8.52a}$$

in which Fo_w is a dimensionless time scale, known as the Fourier number for pore water pressure dissipation (in soil mechanics also known as dimensionless consolidation time T), defined as:

$$Fo_w = \frac{c_v t}{a^2}$$ (8.52b)

with a the radius of the penny-shaped crack, which is generally measured visually, for instance from cross sections of the deformed material and/or from the size of the surface flaws.

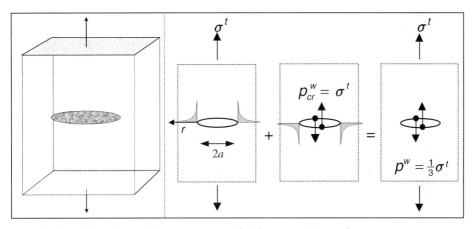

a: Penny shaped crack.   b: Superposition of stresses.

Fig. 8.19 Crack initiation in cohesive sediment.

Fig. 8.20 shows the time evolution of the pore water pressure in a crack, according to equ. (8.52a) for $K_0 = 0$ (i.e. no horizontal stresses). At small times, i.e. $Fo_w \rightarrow 0$, the pressure within the crack remains equal to the tensile stress and therefore no stress concentrations occur at the edges of the micro-crack. This implies that the cracks are not opening and ductile behaviour results by definition. At large times, i.e. for $Fo_w > 100$, the water underpressure in the crack has dissipated for more than 90 % and stress concentrations are almost fully developed around the edges of the crack. Equ. (8.52b) gives a characteristic time scale for the development of stress concentrations of about 100 s for $a = 0.1$ mm and $c_v = 10^{-8}$ m^2/s. At this time scale, the Fourier number for pore water pressure dissipation at the length scale of the tensile test specimen (in the order of 0.1 m) amounts to about 10^{-4}. The response of the soil to tensile loading is then still undrained, as for drained tensile test conditions, $Fo_w > 1$. This corresponds to a time scale of 10^6 s or a strain rate of the order of 10^{-8} s^{-1}. Crack initiation and crack growth can therefore be regarded as locally drained processes, while the overall loading conditions for the soil are undrained.

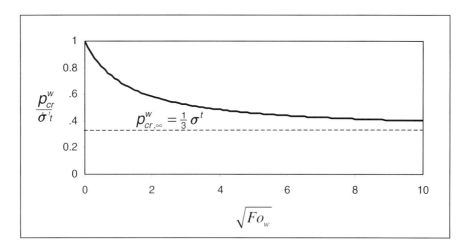

Fig. 8.20: Drainage of micro-crack at constant loading rate ($K_0 = 0$).

We can determine the evolution of stress concentrations at the edge of the crack with equ. (8.52a) as function of loading rate. In LEFM the stress distribution at the edge of the crack is given by (Broek, 1982):

$$\sigma_3 = -\frac{K_I}{\sqrt{2\pi r}} \qquad (8.53a)$$

where r is the distance from the crack tip (see Fig. 8.19b) and K_I the stress intensity factor defined by (note that tensile stress is defined negative):

$$K_I = -\alpha \sigma_t \sqrt{\pi a} \qquad (8.53b)$$

The coefficient α includes crack geometry: for a circular penny shaped crack $\alpha = 1$ and for a half penny shaped crack (surface flaw) $\alpha = 1.12$. The subscript $\bullet_I$ refers to the mode of crack tip loading (see Fig. 8.21):
- mode I: tensile loading,
- mode II: shear loading, and
- mode III: lateral shear or tear loading.

In general it is assumed that when K_I exceeds a critical value K_{Ic}, the so-called fracture toughness, the crack tip fails and cracks grow in an unstable way as a result of the release of elastic energy.

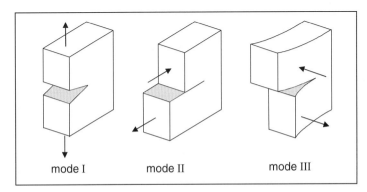

mode I   mode II   mode III

Fig. 8.21: Crack tip loading modes.

In the case of loading at a constant rate $\dot{\sigma}^t$ and pore water draining towards the crack, the stress intensity factor is given by the superposition of the external tensile stress and the pore water pressure inside the crack:

$$K_I = \left(\frac{p_{cr}^w}{\dot{\sigma}^t t} - 1 \right) \sigma^t \sqrt{\pi a} \tag{8.54}$$

For $K_I = K_{Ic}$, the undrained tensile strength σ_u^{tf} can be computed with equ. (8.54) as a function of the loading rate by substitution of equ. (8.52) in (8.54); the constant fracture toughness K_{Ic} is a material parameter defined in (8.57). Because the theory is based on LEFM the loading rate can be described by a strain rate, assuming a certain strain at failure. In Fig. 8.22 the computed failure stress is shown and compared with tensile tests on Boom clay (Van Kesteren, 2004).

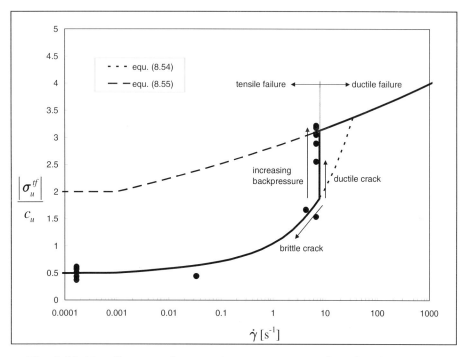

Fig. 8.22: Tensile strength vs. strain rate: transition from brittle to ductile failure (data Boom clay).

The data were obtained with submerged triaxial tests and variable backpressure. The theory on draining micro-cracks predicts the observed strain rate effects quite well up to a strain rate of $\dot{\gamma} = 7 \text{ s}^{-1}$. This strain rate marks the transition from the development of brittle cracks to ductile cracks. Ductile cracks are determined by growth of spherical voids, which requires $\sigma^t > 1.2c_u$ at the given strain rate (Van Kesteren, 2004). Because cavitation of pore water plays a role in the growth of spherical voids, an increase in backpressure will result in an increase of the tensile strength up to the dynamic failure strength during compression. The step at $\dot{\gamma} = 7 \text{ s}^{-1}$ in Fig. 8.22 corresponds to the point where the tensile stress, computed with equ. (8.54), exceeds 1.2 times the undrained shear strength. The measured data in the step have been obtained by increasing the backpressure from just above vapour pressure (lowest point) to 1 MPa (10 atm) (upper point).

The strain rate of undrained ductile failure in compression σ_u^{cf} can be expressed as a power relation (e.g. Van Kesteren, 1992):

$$\frac{\sigma_u^{cf}}{c_u} = 2\left[\frac{\dot{\gamma}}{\dot{\gamma}_s}\right]^n \qquad\qquad (8.55)$$

where c_u is the undrained shear strength, $\dot{\gamma}_s$ is the deviatoric shear strain rate at the transition from static to dynamic loading and n is an exponent with typical values of $n = 0.01 - 0.2$. Shear failure localised in a shear plane (Fig. 8.12) occurs somewhere in between the solid and dashed lines in Fig. 8.22, depending on the isotropic stress.

From a physical point of view, soft cohesive sediments behave very similar to more consolidated clays. Therefore, generation of tensile cracks can be expected in mud as well. Examples are presented in Fig. 8.23, showing cracks in a mud layer flowing over a horizontal bottom, and shear deformation of a placed mud bed in an erosion flume. In both cases, cracks occur in the direction of shear, induced by gravity in Fig. 8.23a and by hydraulic friction in Fig. 8.23b. The occurrence of these cracks can be explained from large deformations during shear, and are treated in Chapter 8.3.4.

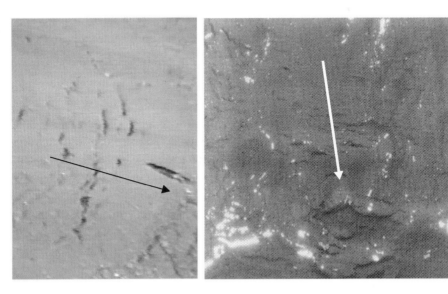

a) flowing mud layer.   b) flowing water (erosion test).

Fig. 8.23: Cracks in shear deformation (arrows give direction of shear).

8.3.3 TENSILE FAILURE PARAMETERS

From the analysis in Section 8.3.2 on stress concentrations and crack formation, we must conclude that the undrained tensile strength σ_u^{tf} is not a material parameter, as it depends on the length scale in the material, its geometry and type of loading. Therefore, experiments to measure the various stress parameters must be carried out with great care. Only under drained test conditions, the drained unconfined tensile strength σ_d^{tf} for clays can be related to the drained unconfined shear strength in compression σ_d^{cf} (see Fig. 8.12 and Fig. 8.24), as predicted by the modified Griffith theory (Irwin, 1948 and Bishop et al., 1969):

$$\sigma_d^{tf} = -\frac{1}{4}\sigma_d^{cf}\left(\sqrt{1+\tan^2\varphi_{cr}} - \tan\varphi_{cr}\right) \tag{8.56}$$

in which φ_{cr} is the internal friction angle in the micro-cracks. Typical values for the ratio $\sigma_d^{tf}/\sigma_d^{cf}$ for clays vary between 0.13 and 0.2 (e.g. for London Clay and Boom Clay).

For soft cohesive sediment it is difficult to perform drained tensile tests and therefore drained extension tests are used. In a triaxial extension test, the vertical stress decreases after isotropic consolidation of the sample. Examples of drained and undrained triaxial tests on mud from Lake Ketel (The Netherlands) are shown in the p-q-diagram in Fig. 8.24 (Van Kessel et al., 2002). The drained tests appear as straight lines with a 3:1 slope, while for the undrained tests the skeleton or effective stress path is curved due to the effects of pore water pressures. These curves show that above a certain axial strain ($\varepsilon_3 \approx 15$ % both in extension and compression) or stress ratio ($q/p^{sk} \approx 1.2$) the isotropic skeleton stress p^{sk} increases after the decrease at $\varepsilon_3 < 15$ %. This change in behaviour is caused by the tendency to dilate at $\varepsilon_3 > 15$ % and is known as the Phase Transformation Point (PTP) (Tatsuoka and Ishihara, 1974). According to Been and Jefferies (1985) the stress ratio at PTP coincides with the critical state stress ratio M_{CS} (see Fig. 8.9). The failure envelope around all tests (solid lines in Fig. 8.24) yields $\sigma_d^{tf}/\sigma_d^{cf} \approx 0.2$.

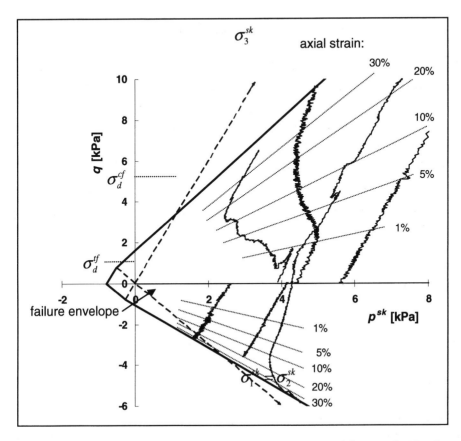

Fig. 8.24: Drained and undrained triaxial tests on mud from Lake Ketel
(Van Kessel et al., 2002).

Under undrained tensile loading, the tensile failure strength σ_u^{tf} is determined by crack initiation and propagation. For unstable growth the amount of elastic energy released during crack growth must exceed the specific fracture energy. This so-called fracture energy, denoted by G_{Ic} [N/m], is the amount of energy [Nm] necessary to create 1 m^2 of new crack surface area. It can be considered as a material parameter for crack propagation during tensile failure as crack length, test sample geometry and boundary conditions do not affect G_{Ic}. An example is presented in Fig. 8.25, where the fracture energy is plotted as a function of crack length (Lee et. al, 1982) for a tensile beam test on clay (inset in Fig. 8.25), showing that G_{Ic} does not vary with crack length.

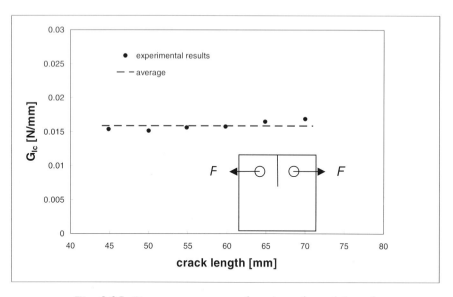

Fig. 8.25: Fracture energy as function of crack length
(redrawn from Lee, 1982).

For large plastic deformations the fracture energy is obtained from the integral of the stress-displacement curve (so-called J-integral, see Fig. 8.26). Under large plastic deformations, a main part of the fracture energy is consumed by the irreversible plastic strains in the crack tip zone. When the size of the plastic zone in front of the crack tip has similar dimensions as the crack size or as the size of the sample, the size of the plastic zone is not constant during crack growth, and therefore also the fracture energy is not constant (see Fig. 8.26). In that case, crack growth is stable (i.e. the growth rate becomes constant) for $J >$ J_{Ic} only, as more energy is required for further growth. For small plastic strains, the critical fracture energy equals G_{Ic} which can be expressed with the linear elastic fracture toughness K_{Ic} (Broek, 1982):

$$\text{plane stress: } G_{Ic} = \frac{K_{Ic}^2}{E} \text{ ; plane strain: } G_{Ic} = \left(1 - v^2\right)\frac{K_{Ic}^2}{E} \tag{8.57}$$

where E is Young's modulus and v is the Poisson's ratio. From measurement on clays (Van Kesteren, 2004), it appeared that the fracture energy for undrained tensile loading scales with the undrained shear strength:

$$G_{Ic} = \alpha c_u \quad ; \; 0.1 \cdot 10^{-3} \text{ m} < \alpha < 0.3 \cdot 10^{-3} \text{ m} \tag{8.58}$$

in which c_u the undrained shear strength.

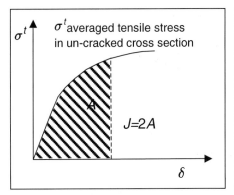
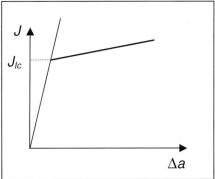

a) stress displacement diagram.   *b) J-integral versus crack growth.*

Fig. 8.26: Fracture energy for large plastic deformation (J-integral).

This observation is supported by data from Lee (1982), who reported values of $\alpha = 0.17 \cdot 10^{-3}$ m. Also in the extensile tests on soft mud from Lake Ketel (Van Kessel et al., 2002) $\alpha = 0.2 \cdot 10^{-3}$ m was found.

According to Rice (1968) α can be interpreted as the crack tip opening displacement prior to crack growth (see Fig. 8.27a). Typical crack opening displacements in cohesive sediments would then amount to 100 to 300 μm. This is supported by transmission images in Fig. 8.27 of thin slices of cracked clay samples after arresting of crack growth during cutting tests.

An alternative way of determining the fracture energy of soft cohesive sediment, and in particular useful for in-situ conditions, is by means of injecting water with a "micro-probe" (Van Kesteren, 2004). By controlling the discharge during injection and measuring the water pressure, a relation is found between pressure and injected volume. At low discharges this relation depicts the permeability of the sediment. Increasing the discharge results in hydraulic fracture conditions: the local water pressure necessary to squeeze the water through the pore system becomes too high for the skeleton surrounding the injection point.

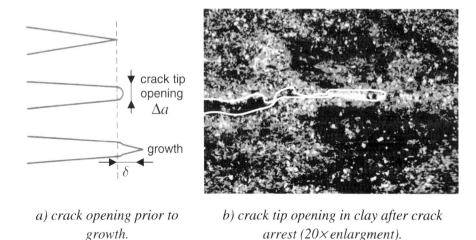

a) crack opening prior to   b) crack tip opening in clay after crack
growth.   arrest (20×enlargment).

Fig. 8.27: Crack opening displacement in clays.

Contrary to the stable growth of gas bubbles (Chapter 11), there is no water-gas
meniscus stabilising the skeleton, and cracks are initiated when water-filled
cavities are created. This process is called hydraulic fracture and is visualised
in Fig. 8.28a by means of injecting water with blue dye into a transparent mud
made of the artificial clay mineral laponite (Chapter 3). Fig. 8.28b shows an
example of the pressure measured with a micro-probe as function of the water
displacement. At the peak of the curve, a crack is initiated and its growth is
determined by the energy release stored in the sediment and injection system.

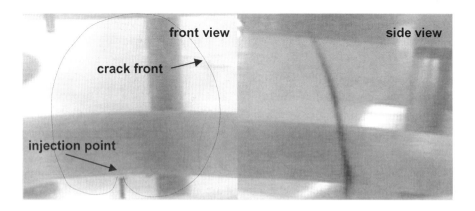

a) Hydraulic fracture in laponite.

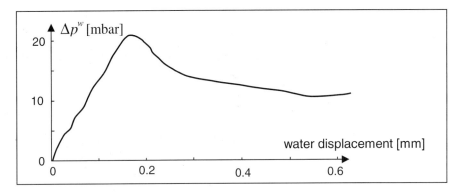

b) Pressure-discplacement curve during hydraulic fracture test.

*Fig. 8.28: Determining fracture energy of mud in hydraulic fracture from
area under stress-curve (Van Kesteren, 2004).*

Before the pressure-displacement curve can be integrated for determining the
fracture energy, a correction is necessary for the viscous energy dissipation
involved in the Darcian flow through the pore system. The fracture energies
obtained with this method are in the same range as predicted by equ. (8.58).

8.3.4 SHEAR FAILURE

Localised shear deformation in a shear plane or shear band (see Fig. 8.12) is
similar to tensile failure, determined by drainage of the shear band and energy
release from the material surrounding the shear band supplying the required
"fracture" energy for increasing shear in the band. Dilation effects play an
important role in shear failure. Fig. 8.29a shows a shear band under
compression. The plastic shear strain rate tensor is given by the flow rule equ.
(8.36) and a plastic potential function such as equ. (8.41b):

$$d_{11}^{p} \propto 1 + \sin \psi \; ; \; d_{22}^{p} = 0 \; ; \; d_{33}^{p} \propto -1 + \sin \psi \; ; \; d_{ij}^{p} = 0 \qquad (8.59)$$

where ψ the dilatancy angle.

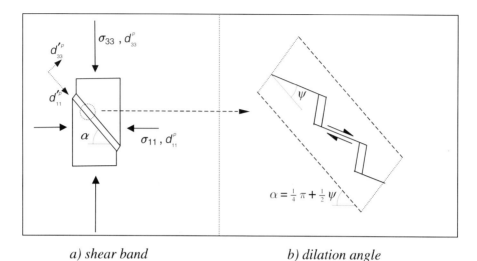

a) shear band   *b) dilation angle*

Fig. 8.29: Failure in a shear band.

When this strain rate tensor is rotated over α in the direction of the shear band the strain components read:

$$d_{11}'^{p} \propto \cos 2\alpha + \sin 2\psi$$
$$d_{33}'^{p} \propto -\cos 2\alpha + \sin \psi \qquad (8.60)$$
$$d_{13}'^{p} \propto 2 \sin 2\alpha$$

Compatibility between shear band and the ambient material requires that line elements in the direction of the shear band remain constant in length during shear band formation. Therefore, the incremental plastic strain rate $d_{11}'^{p}$ equals zero, yielding a relation for the shear band angle α:

$$\alpha = \frac{1}{4}\pi + \frac{1}{2}\psi \qquad (8.61)$$

Failure lines along this angle are called zero-extension lines. Substitution of equ. (8.61) in (8.60) yields for the strain rate components:

$$d_{11}'^p = 0 \; ; \; d_{33}'^p \propto 2\sin\psi \; ; \; d_{13}'^p \propto 2\cos\psi \tag{8.62}$$

The strain rate $d_{33}'^p$ represents the velocity gradient normal to the shear plane, while $d_{13}'^p$ represents the shear velocity or velocity gradient parallel to the shear plane. Hence, equ. (8.62) describes the relative velocity of the material in the shear band in the direction of the dilation angle with respect to the material surrounding the shear band, as depicted in Fig. 8.29b. This dilatant behaviour occurs at many locations in the shear band, resulting in deformations, which can be considered as micro shear cracks. When these shear cracks coalesce a shear band is formed.

Large strains occur in the shear band and its failure can be described with the actual deformation tensor of Green C_{ij} (see Appendix D, equ. (D.13)). In Fig. 8.30a shear deformation in a shear band is schematised with the C_{11} axis in the direction of the shear band. The components of the deformation tensor are given by:

$$C_{11} = C_{22} = 1 \; , \quad C_{33} = \frac{1}{\cos^2 \Omega_{13} \left(1 - \tan \Omega_{13} \tan \psi\right)^2} \tag{8.63}$$

$$C_{12} = C_{23} = 0 \; , \quad C_{13} = \sqrt{C_{11}C_{33}} \sin \Omega_{13} = \frac{\tan \Omega_{13}}{\left|1 - \tan \Omega_{13} \tan \psi\right|}$$

where ψ is the dilation angle, Ω_{13} is the shear angle between the undeformed line element OA and the deformed line element OA' (see Fig. 8.30a). The diagonal components of C_{ij} are equal to the ratios of the lengths squared of the orthogonal line elements after and before deformation. Furthermore, $C_{11} = C_{22} = 1$ because of compatibility with the material surrounding the shear band. The component C_{33} follows from the squared ratio of the lengths of OA' and OA. The principal components are given by:

$$C_1 = \tfrac{1}{2}\left(C_{11} + C_{33}\right) - \sqrt{\tfrac{1}{4}\left(C_{11} - C_{33}\right)^2 + C_{13}^2}$$

$$C_2 = 1 \tag{8.64}$$

$$C_3 = \tfrac{1}{2}\left(C_{11} + C_{33}\right) + \sqrt{\tfrac{1}{4}\left(C_{11} - C_{33}\right)^2 + C_{13}^2}$$

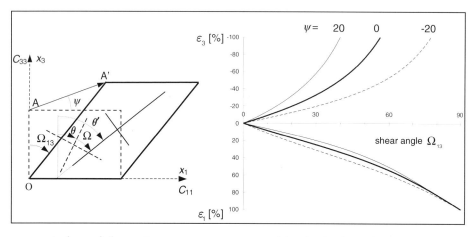

a) shear deformation   *b) principal strains*

Fig. 8.30: Large strain analysis in shear band (Van Kesteren, 2004).

The principal direction in the reference (undeformed) system is found from:

$$\tan 2\theta = \frac{2C_{13}}{C_{33} - C_{11}}$$

(8.65)

where θ is the angle with the normal to the shear band (Fig. 8.30a). Orthogonal line elements remain orthogonal in the principal direction after deformation and should contain the same material points. The orthogonal line elements for the deformation are plotted in Fig. 8.30a in the principal direction. The rigid rotation Ω of the orthogonal line elements is given by:

$$\tan \theta' = \tan \Omega_{13} + \tan \theta \left(1 - \tan \Omega_{13} \tan \psi \right) \quad , \quad \Omega = \theta' - \theta$$

(8.66)

where θ' is the principal direction in the deformed state (see Fig. 8.30a). For comparison with small strain analysis, principal strains are plotted as a function of the shear angle Ω_{13} in Fig. 8.30b. These principal strains are defined as:

$$\varepsilon_i = -\sqrt{C_i} + 1$$

(8.67)

where ε_i is positive in compression and negative in tension.

Fig. 8.30b shows that up to shear angles of 20^0, the tensile and compressive strains are almost linear, but at higher shear angles the tensile strain increases rapidly. The shear deformation becomes more elongated (stretching). The effect of the dilation angle ψ on this elongation is shown in Fig. 8.30b for $\psi = 20^0$ (extension) and -20^0 (compaction). When the shear angle decreases in case of dilation ($\psi = 20^0$), tensile strains increase rapidly, which results in tensile failure in the shear band at smaller shear angles, i.e. more brittle behaviour results. At negative dilation angles (compaction), the tensile strain decreases and larger shear angles are possible before tensile failure occurs, i.e. more ductile behaviour is the result.

As the principal strain direction rotates during shearing, micro-cracks with an orientation between 45^0 and 90^0 (i.e. perpendicular to the shear plane at large shear deformations) may open. Whether this will happen is determined by the stress state. Therefore, the stability of a micro-crack in different orientations with respect to the "far field" principal stresses is considered (see Fig. 8.31a). Loading of shear cracks involves tensile and shear stresses, i.e. a combination of mode I and II loading (see Fig. 8.21), with residual shear stresses within the micro-shear crack.

The stress intensity factor K (e.g. equ. (8.53)) for the mixed mode I and II loading can be assessed with LEFM (e.g. Broek, 1986):

$$K_{I,II} = K_I \cos^3\left(\theta_w/2\right) - 3K_{II} \cos^2\left(\theta_w/2\right)\sin\left(\theta_w/2\right) \tag{8.68}$$

where θ_w is the angle between the planar shear crack and the wing crack initiated at the edge of the shear crack (see Fig. 8.31a). When the crack propagates perpendicular to the local maximum principal tensile stress, θ_w is given by (Sih, 1991):

$$\tan\left(\theta_w/2\right) = \frac{1}{4}\frac{K_I}{K_{II}} - \frac{1}{4}\mathrm{sign}\left(K_{II}\right)\sqrt{\frac{K_I^2}{K_{II}^2} + 8} \tag{8.69}$$

K_I and K_{II} are the stress intensity factors for mode I and II loading defined by:

$$K_{\mathrm{I}} \equiv \sigma'_{33} \sqrt{\pi a} \qquad K_{\mathrm{II}} \equiv \sigma'_{13} \sqrt{\pi a} \tag{8.70}$$

where a is half the length of the initial shear crack, and σ'_{33} and σ'_{13} are the far field normal and shear stress with respect to the orientation of the crack (see Fig. 8.31a). These stresses are found from the principal stress with (see Appendix D, equ. (D.8)):

$$\sigma'_{33} = \sigma_1 \sin^2 \alpha + \sigma_3 \cos^2 \alpha$$
$$\sigma'_{13} = \tfrac{1}{2}(\sigma_1 - \sigma_3) \sin 2\alpha \tag{8.71}$$

When $\sigma'_{33} > 0$, normal forces occur inside the crack and the stress intensity factor K_{I} becomes zero. In addition, the mode II stress intensity factor K_{II} reduces depending on the friction angle inside the shear crack. Application of the superposition principle for stresses inside the crack and the far field stresses (see Fig. 8.19) yields:

$$K_{\mathrm{II}} = \left(\sigma'_{13} - \sigma'_{33} \tan \varphi_{cr} \right) \sqrt{\pi a} \tag{8.72}$$

where φ_{cr} is the friction angle inside the crack. In case $\sigma'_{33} \tan \varphi_{cr} = \sigma'_{13}$, K_{II} becomes zero and no stress concentrations at the edges of micro-cracks occur.

An example of the stress intensity factor $K_{\mathrm{I,II}}$ as a function of the crack angle α is shown in Fig. 8.31b for two loading conditions: uni-axial tension and uni-axial compression, both with zero lateral stress ($\sigma_1 = 0$). The friction angle inside the crack φ_{cr} is set at 20^0. The stress intensity factor $K_{\mathrm{I,II}}$ is made dimensionless with the stress intensity factor $K_{\mathrm{I}}(0)$ for a horizontal crack ($\alpha = 0$) in mode I loading. Fig. 8.31b shows that a maximum in $K_{\mathrm{I,II}}$ is reached at $\theta_w \approx 22^0$ in uni-axial tension, and at $\theta_w = 55^0$ in uni-axial compression. For uni-axial compression the direction of the wing crack θ_w is constant and at its maximum value (70.5^0), because σ'_{33} is positive for all crack angles, hence only mode II occurs.

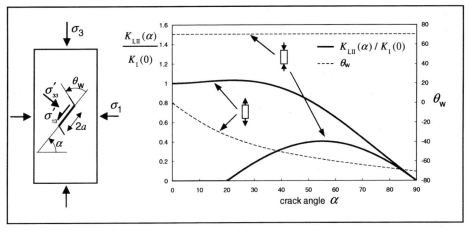

a) shear crack.   *b) stress intensity factor.*

Fig. 8.31: Stress intensity factor as function of crack angle.

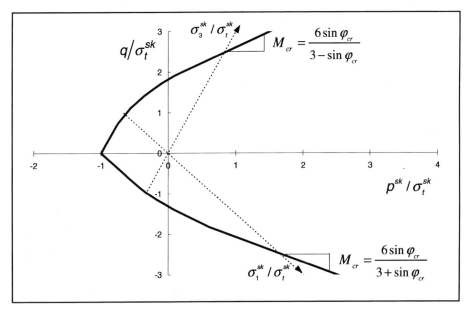

Fig. 8.32: Locus for growth micro-cracks in p-q-diagram.

The conditions for growth of micro-cracks can be read from the *p-q*-diagram in
Fig. 8.32 by varying the principal skeleton stresses σ_1, σ_3 in the sediment.
Wing cracks are initiated when the maximum in $K_{I,II}$ equals a constant fracture

toughness K_{Ic}, which is the strength of the material against fracture. The stresses are normalised with the tensile stress (absolute value) at which a horizontal micro-crack ($\alpha = 0$) in tension is initiated (note that this stress is smaller than the tensile strength determined by coalescence and growth of macro-cracks). Micro-cracks remain closed inside the envelope plotted in Fig 8.32, and plastic strains show ductile behaviour. Outside the envelope, the stress state results in growth and coalescence of micro-cracks forming macro-cracks. Depending on the amount of elastic energy released, macro-crack growth can become unstable, yielding failure in tension or shear.

The locus found by applying LEFM shows a remarkable resemblance with the yield surfaces shown in Fig. 8.12. The slope of the locus M_{cr} at higher isotropic compressive stresses is constant and equal to $6\sin\varphi_{cr}/(3-\sin\varphi_{cr})$. Under extension $M_{cr} = 6\sin\varphi_{cr}/(3+\sin\varphi_{cr})$. These lines should intersect the yield surfaces at their critical state, as isotropic stresses larger than the critical state ductile stress occur.

8.4   CYCLICAL BEHAVIOUR

Stress-strain relations describing the behaviour of cohesive sediment have to be related to the type of sediment and to the loading conditions. For instance, in case of one-dimensional, self-weight consolidation the load increases monotonically. This can be described with a simple one-dimensional consolidation equation, such as Gibson's finite strain equation (7.10). The loading path becomes more complex under load reversal and rotation of principle stresses, as may occur under cyclic loading by waves. In that case it is necessary to account for the loading history (or better, the failure history) in the stress-strain relations with respect to the time scales of the relevant loadings. The period of loading limits pore water drainage during each cycle determining the accumulation of excess pore pressure. Moreover, as the strain rate peaks are much larger than their cycle-averaged values, viscous effects may become important.

In this section the constitutive behaviour for soft cohesive sediment in shallow marine environments is presented, where waves are often important. We also present an elasto-plastic approach. Constitutive models of cyclical behaviour are discussed in Section 8.4.1 with some examples for wave loading conditions in Section 8.4.2. The rate dependent visco-plastic behaviour is treated in Section 8.5, where also the elasto-plastic approach is discussed.

We are concerned with large deformations only. This implies that we will only briefly treat visco-elastic models, which are applicable for small strain rates only (Chou et al., 1989, 1993).

8.4.1 CYCLIC STRESS-STRAIN RELATIONS

To model the stress-strain relations for soft cohesive sediment we follow the basic ideas of the Critical State theory and the Modified Cam-Clay elasto-plastic model treated in the previous sections. However, we will take into account damaging and weakening effects during tensile failure, and the anisotropic behaviour under compression failure.

Drained and undrained triaxial tests (see Fig. 8.24) and simple undrained cyclical shear tests on mud (see Fig. 8.33, Cornelisse and Van Kesteren, 1995) reveal that the failure envelope under shear approximates the Mohr-Coulomb failure envelope. Therefore, the elliptical yield surfaces of the Modified Cam-Clay model (equ. (8.45)) are replaced by the continuous yield surface proposed by Desai (e.g. Desai et al., 1991):

$$ f = \frac{I_2}{\left(p_c^{sk}\right)^2} - \left[\lambda \frac{I_1^2}{\left(p_c^{sk}\right)^2} - \alpha \frac{I_1^n}{\left(p_c^{sk}\right)^n} \right] R^2(\theta) = 0 \qquad (8.73) $$

where α and λ are coefficients for the yield surfaces for compression and shear respectively, p_c^{sk} is the isotropic consolidation stress in the skeleton, $R(\theta)$ is a function representing the non-circular shape of the yield surface in the π-plane as a function of the Lode angle θ (see Section 8.2.4, equ. (8.40) and Appendix D), and the exponent n describes the curvature of the yield surface. For $n > 2$ the yield surface approaches a Mohr-Coulomb like failure surface at low isotropic skeleton stress.

The Lode angle θ can be expressed as a function of the stress invariants of the (Couchy) skeleton stress tensor σ_{ij}^{sk} (see Appendix D):

$$\cos 3\theta = -\frac{3\sqrt{3}}{2} \frac{-I_3 + \frac{1}{3}I_1 I_2 - \frac{2}{27}I_1^3}{\left(\frac{1}{3}I_1^2 - I_2\right)^{3/2}} \qquad (8.74)$$

The stress invariants I_1, I_2 and I_3 are defined as (see Appendix D):

$$I_1 = \sigma_{ii}^{sk} \quad I_2 = \frac{1}{2}\left[I_1^2 - \sigma_{ij}^{sk}\sigma_{ji}^{sk}\right] \quad I_3 = \det \underline{\underline{\sigma}} \qquad (8.75)$$

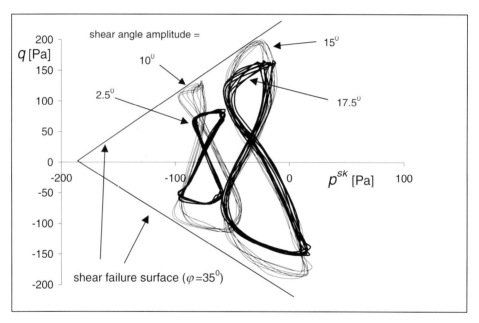

Fig. 8.33: Variation of shear stress and water pressure during cyclical shear tests on Haringvliet mud at various shear angle amplitudes (Van Kesteren and Cornelisse, 1995).

The skeleton stresses σ_{ij}^{sk} are also known as effective stresses in soil mechanics (see Chapter 7 and Section 8.1). The stress invariants I_1 and I_2 can be replaced by the isotropic skeleton stress p^{sk} and the deviatoric stress q, defined by (see Fig. 8.9 and Appendix D):

$$p^{sk} = I_1/3 \quad q = \sqrt{\tfrac{3}{2}}\, t = \sqrt{I_1^2 - 3I_2} \qquad\qquad (8.76)$$

Substitution into equ. (8.73) yields a Desai-type yield surface for cohesive sediment (Van Kesteren, 2004) as shown in Fig. 8.34:

$$f = \left(\frac{q}{p_c^{sk} R(\theta)} \right)^2 - M_{\varphi m} \left(\left(\frac{p^{sk} + p_{coh}^{sk}}{p_c^{sk}} \right)^2 - \left(\frac{p^{sk} + p_{coh}^{sk}}{p_c^{sk}} \right)^n \right) = 0 \qquad (8.77)$$

where p_c^{sk} is the isotropic consolidation stress, p_{coh}^{sk} is a shift of the yield surface and $M_{\varphi m}$ is the tangent at zero isotropic skeleton stress. The subscript $\bullet_{\varphi m}$ of the tangent M refers to the maximal friction angle under shear failure corresponding to the tangent of the yield surface in p-q space for triaxial compression conditions (see Fig. 8.8), and is defined in a similar way as the critical state tangent M_{CS} by:

$$M_{CS} = 6\sin\varphi_{cv}\big/\left(3 - \sin\varphi_{cv}\right) \;\; ; \;\; M_{\varphi m} = 6\sin\varphi_m\big/\left(3 - \sin\varphi_m\right) \qquad (8.78)$$

(see also equ. (8.45), which is repeated here for convenience), where φ_m is the maximal friction angle and φ_{cv} is the friction angle at constant volume (critical state).

The tangent at zero isotropic skeleton stress under extension (lower part of the yield surface in Fig. 8.34) follows from the function $R(\theta)$. For a Mohr-Coulomb like shear failure surface, $R(\theta)$ is given by equ. (8.40) and for $\theta = \pi$ we find $M_{\varphi m} = 6\sin\varphi_m\big/\left(3 + \sin\varphi_m\right)$.

The shift of the yield surface by p_{coh}^{sk} accounts for the drained strength (cohesion) at zero isotropic skeleton stress, which is, by definition, characteristic for cohesive sediment. Furthermore a tension cut-off should be included accounting for drained tensile failure, as treated in Section 8.3.

The exponent n in (8.77) is defined by the ratio of the isotropic effective stress p_{NC}^{sk} of normally consolidated mud and at its critical state this stress p_{CS}^{sk} (see Fig. 8.35) is given in an implicit form:

$$n = 2\left(p_{NC}^{sk} / p_{CS}^{sk} \right)^{n-2} \tag{8.79}$$

The kinematic hardening by self weight consolidation along the K_0–line (Fig. 8.34) can be taken into account by introducing an asymmetric shift with the gradient M_{Ko} :

$$f = q^2 + \left(M_{Ko} p^{sk} \right)^2 - 2q M_{Ko} p^{sk} \cos\theta + \\ -\left(R(\theta) M_{\varphi m} - M_{Ko} \cos\theta \right)^2 \left(\left(p^{sk} \right)^2 - \left(p^{sk} \right)^{2-n} \left(p^{sk} \right) \right) = 0 \tag{8.80}$$

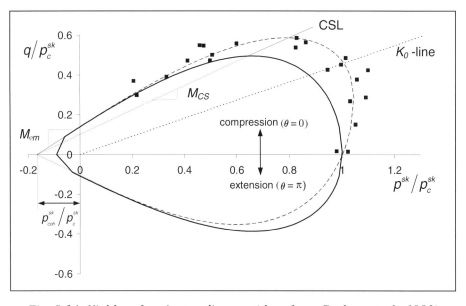

Fig. 8.34: Yield surface in p-q diagram (data from Graham et al., 1983).

The yield surface measurements in clay by Graham et al. (1983) are also plotted in Fig. 8.34, showing that the asymmetric Desai yield surface (equ. (8.80)) represents the data on the actual yield surface quite well ($M_{Ko} = 0.25$, and $M_{\varphi m} = 0.85$).

The parameters M_{Ko}, M_{φ} and p_c^{sk} are to be regarded as parameters of state for hardening and/or softening during plastic strain deformations. For cyclical shear loading, the Lode angle $\theta = 90^0$ and $R \approx 1$, and the tangent μ to the yield surface is then given by:

$$\mu = -\frac{\partial f / \partial p^{sk}}{\partial f / \partial q} = M_{K_0} + \left(M_\varphi - M_{K_0}\right) \frac{2p^{sk} - n\left(p_c^{sk}\right)^{2-n}\left(p^{sk}\right)^{n-1}}{\sqrt{\left(p^{sk}\right)^2 - \left(p_c^{sk}\right)^{2-n}\left(p^{sk}\right)^n}} \qquad (8.81)$$

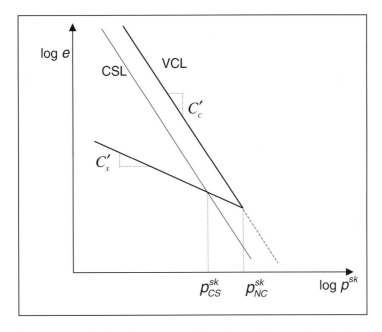

Fig. 8.35: Power relation between void ratio and isotropic effective stress.

The flow rule, which determines the direction of plastic strain increments, is defined by a plastic potential function g, which is similar to the yield surface function f, except that non-associative flow is allowed during shear failure:

$$g = \left(q - M_{K_0} p^{sk}\right)^2 - \left(M_\psi - M_{K_0}\right)^2 \left(\left(p^{sk}\right)^2 - \left(p_c^{sk}\right)^{2-n}\left(p^{sk}\right)^n\right) = 0 \qquad (8.82)$$

The tangent M_φ (e.g. Fig. 8.34) of the local yield surface function f in the apex at zero deviatoric stress is replaced by the tangent M_ψ for the plastic potential function g. The subscript $\cdot_\psi$ refers to the dilation angle in shear failure:

$$M_\psi = 6\sin\psi / (3 - \sin\psi) \tag{8.83}$$

The dilatancy angle ψ can be related to the internal friction angle with Rowe's dilatancy theory (Rowe, 1962; see equ. (8.43b)):

$$\sin\psi = \frac{\sin\varphi - \sin\varphi_{cs}}{1 - \sin\varphi \sin\varphi_{cs}} \tag{8.84}$$

where subscript $\cdot_{cs}$ refers to the critical state. M_ψ (which is the tangent of the critical state line, e.g. Fig. 8.34) can be expressed as a function of M_φ and M_{cs} with equ. (8.64) and (8.69). The plastic volume and deviatoric strain increments are obtained from the flow rule and are expressed by:

$$\dot{\varepsilon}_v^p = C \frac{\partial g}{\partial p^{sk}} \qquad \dot{\gamma}^p = C \frac{\partial g}{\partial q} \tag{8.85}$$

in which the reciprocal of the coefficient C resembles viscosity. The direction of flow is represented by the ratio $\beta = \dot{\varepsilon}_v^p / \dot{\gamma}^p$ of the plastic volume strain increment and the deviatoric strain increment:

$$\beta = \frac{\partial g / \partial p^{sk}}{\partial g / \partial q} = -M_{K_0} - \left(M_\psi - M_{K_0}\right) \frac{2 p^{sk} - n \left(p_c^{sk}\right)^{2-n} \left(p^{sk}\right)^{n-1}}{\sqrt{\left(p^{sk}\right)^2 - \left(p_c^{sk}\right)^{2-n} \left(p^{sk}\right)^n}} \tag{8.86}$$

By substitution of equ. (8.66), β can be expressed as a function of the tangent μ:

$$\beta = -\mu + \left(\mu - M_{K_0}\right) \frac{M_\varphi - M_\psi}{M_\varphi - M_{K_0}} \tag{8.87}$$

When $M_\psi = M_\varphi$, the associative flow rule (8.36) is retrieved. In case $M_{K_0} = 0$, equ. (8.87) can be simplified to:

$$\beta = -\mu M_\psi / M_\varphi \tag{8.88}$$

The kinematic hardening function determines the amount of deviatoric shear flow during failure (e.g. Nishi, 1990):

$$\dot{\gamma}^p = \frac{\zeta}{H}\left(\dot{q} - \mu \dot{p}^{sk}\right) \quad \text{and} \quad \zeta = n_H \left(\frac{M_r - M_\varphi}{M_r + M_{\varphi m}}\right)^{n_H - 1} \tag{8.89}$$

in which H (e.g. equ. 8.38)) is the hardening modulus [Pa], n_H is the power of the hardening function, $M_{\varphi m}$ is the maximum tangent of the yield surface at apex and M_r is the tangent of the yield surface at apex during unloading (reversal), and reflects the stress history of the material.

The hardening rule for a plastic volume strain increment is found from the elasto-plastic compaction behaviour of the material, which can be determined from triaxial (e.g. Appendix C.5.5) or oedometer tests (e.g. Appendix C.5.4). In case of very soft cohesive sediment ($c_u < 1$ kPa) a Seepage Induced Consolidation test (Appendix C.5.4) can be applied. These tests generally depict a power law decay of void ratio with isotropic effective stresses, yielding a hardening rule, similar to that for the Cam-Clay model given by equ. (8.46):

$$\dot{\varepsilon}_v^p = \left(C_c' - C_s'\right) \frac{e}{1+e} \frac{\dot{p}_c^{sk}}{p_c^{sk}} \tag{8.90}$$

where C_c' and C_s' are the compression and swell index for the power law relation between void ratio and isotropic stress (see Fig. 8.35). Similarly, the elastic strains are computed (equ. (8.48) and (8.49)):

$$K_{sk} \equiv \frac{\dot{p}^{sk}}{\dot{\varepsilon}_v^e} = \frac{p^{sk}}{C_s'}\frac{1+e}{e} \tag{8.91}$$

$$G_{sk} \equiv \frac{\dot{i}}{\dot{\gamma}^e} = \sqrt{\frac{2}{3}}\frac{\dot{q}}{\dot{\gamma}^e} = \frac{3}{2}\frac{1-2v}{1+v}K_{sk} \tag{8.92}$$

where v is the Poisson ratio. The volume balance in equ. (8.9) is required to solve this set of constitutive equations. We assume a water-saturated system. Thus the bulk stiffness of the mud skeleton is much smaller than that for water and solids, and equ. (8.9) reduces to:

$$\Delta\varepsilon_v = \Delta\varepsilon_v^e + \Delta\varepsilon_v^p = \Phi/V \tag{8.93}$$

Under undrained conditions, no pore water fow occurs and the flux $\Phi = 0$. The elastic volume strain increment $\dot{\varepsilon}_v^e$ can be expressed as a function of the isotropic skeleton stress increment $\dot{p}^{sk}$ with equ. (8.91), whereas the plastic volume strain increment $\dot{\varepsilon}_v^p$ can be expressed as a function of the isotropic and deviatoric skeleton stress increments $\dot{p}^{sk}$ and $\dot{q}$ with equ. (8.86) and (8.89):

$$\dot{\varepsilon}_v = \dot{\varepsilon}_v^e + \dot{\varepsilon}_v^p = \frac{\dot{p}^{sk}}{K_{sk}} + \beta\dot{\gamma}^p = \frac{\dot{p}^{sk}}{K_{sk}} + \frac{\beta\zeta}{H}\left(\dot{q} - \mu\,\dot{p}^{sk}\right) = 0 \tag{8.94}$$

Hence, for undrained conditions we find a relation between the isotropic and deviatoric skeleton stress increments:

$$\dot{p}^{sk} = -\frac{\beta\zeta}{H/K_{sk} - \beta\mu\zeta}\,\dot{q} \tag{8.95}$$

The deviatoric stress increment can be related to the total deviatoric strain increment with equ. (8.75) and (8.78):

$$\dot{\gamma} = \dot{\gamma}^e + \dot{\gamma}^p = \sqrt{\frac{2}{3}}\frac{\dot{q}}{G_{sk}} + \frac{\zeta}{H}\left(\dot{q} - \mu\,\dot{p}^{sk}\right) \quad \text{or} \tag{8.96a}$$

$$\dot{\gamma} = \dot{q}\left[\sqrt{\frac{2}{3}}\frac{1}{G_{sk}} + \frac{\zeta}{H}\left(1 + \frac{\mu\beta\zeta}{H/K_{sk} - \mu\beta\zeta}\right)\right]$$ (8.96b)

Equ. (8.96) can be used to determine the deviatoric stress increment when the deviatoric shear strain is given, or, vice versa, to determine the deviatoric strain increment when the deviatoric shear stress is given.

8.4.2 EXAMPLES OF CYCLICAL LOADING

Fig. 8.36 shows some examples of the cyclical constitutive model (8.96) for undrained simple shear loading of an over-consolidated mud sample in response to a cyclical shear strain given by:

$$\gamma = \hat{\gamma}\sin\omega t \;\; ; \;\; \dot{\gamma} = \omega\hat{\gamma}\cos\omega t$$ (8.97)

The input parameters for the model computations are given in Table 8.2.

The isotropic total stress is kept constant at $p = 30$ Pa corresponding to an over-consolidation ratio OCR = 6.7 (see Table 8.2), which was defined in Section 8.3 as OCR $= p_{NC}^{sk}/p^{sk}$ (e.g. Fig. 8.9 and equ. (8.43)). Fig. 8.36 shows stable oscillations in p-q-space for shear strain amplitudes $\hat{\gamma}$ = 0.01, 0.02 and 0.03 rad, which is approximately equivalent to 1, 2 and 3 %.

A dynamic equilibrium is reached after 5 to 10 cycles. The arrows indicate the direction of the stress path. The characteristic 'butterfly' curves are also obtained with the simple shear tests shown in Fig 8.33. These curves are the result of the tendency of skeleton compaction in the loading phase following a non-linear elastic unloading phase. Plastic shear deformation in the early reloading phase results in an increase in pore water pressures and a reduction of isotropic skeleton stresses in case the total isotropic stress remains constant. During the reloading phase in opposite direction, plastic shear failure occurs with a tendency to dilate, depending on the level of isotropic skeleton stresses with respect to the isotropic consolidation stresses. During failure, pore water pressures decrease and the isotropic skeleton stresses increase again.

Fig. 8.37 presents the computed evolution of the stress path in p-q space for two initial isotropic stresses. Case 1 comprises almost normally consolidated mud at $p = 150$ Pa (OCR = 1.3) and Case 2 comprises over-consolidated mud at $p = 30$ Pa (OCR = 6.7). In both cases the deviatoric shear strain amplitude was set to $\hat{\gamma} = 0.03$ rad. It is shown that excess pore water pressures build up rapidly for normally consolidated conditions. The stresses become stable close to the isotropic skeleton stress of the over-consolidated test.

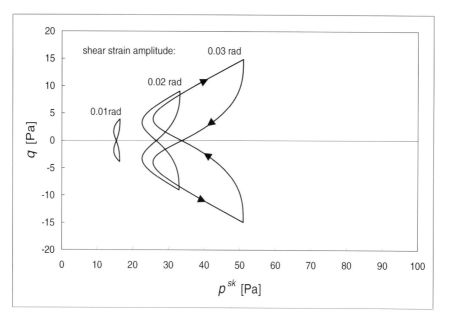

Fig. 8.36: Cyclic behaviour for over-consolidated mud (OCR = 6.7, computed with equ. (8.96)).

Fig. 8.38 shows the computed corresponding excess pore water pressures as a function of time. It is obvious that the normally consolidated mud sample is much more sensitive to liquefaction than the over-consolidated sample.

This behaviour is also observed in wave flume tests. Fig. 8.39 presents the pore water pressures under the influence of waves measured in a 25 cm thick bed of silty sediment placed in a flume (Foda and Tzang, 1994). The uniformity of the sediment D_{60}/D_{10} is estimated at 6.5, hence the maximal porosity is at most 45 %, e.g. Fig. 3.24. As the in-flume porosity measured between 49.4 % and 52.5 % (values given by Foda and Tzang), the sediment placed in the flume is expected to be sensitive to liquefaction. Indeed, Fig. 8.39 shows that the mean water pressure, measured 20 cm below the sediment-water interface, increases to about 1.3 kPa. For $n = 49.4 - 52.5$ %, full liquefaction of the sediment would yield a water over-pressure of about 1.6 kPa. Hence the sediment is almost fully liquefied.

The time delay between the onset of wave action and increase of pore pressure of about 20 s is remarkable – we hypothesise that this is caused by an initial elastic response of the bed to the wave stresses.

Table 8.2: Input parameters model computations undrained cyclical response.

parameters	symbol	value	unit
yield surface			
isotropic stress normally consolidation (NC)	p_{NC}^{sk}	200	[Pa]
isotropic stress at critical state (CS)	p_{CS}^{sk}	133	[Pa]
isotropic stress at apex	p_0^{sk}	0	[Pa]
friction angle at apex	φ	35	[°]
tangent at apex (equ. (8.64))	M_φ	1.42	[-]
stress ratio at self weight consolidation	M_{Ko}	0	[-]
exponent in Desai-model (equ. (8.67))	n	3	[-]
elastic moduli			
swell index power law	C'_s	0.04	[-]
Poisson ratio	v	0.3	[-]
plastic moduli			
compression index power law	C'_c	0.21	[-]
hardenings modulus	H	200	[Pa]
power hardenings function	n_H	3	[-]
initial conditions			
void ratio	e	3	[-]
initial isotropic stress (OCR = 6.7)	p_{in}^{sk}	30	[Pa]
initial isotropic stress (OCR = 1.3)	p_{in}^{sk}	150	[Pa]
permeability	k	$1.e^{-10}$	[m/s]
time period	T	6	[s]
amplitude shear strain	$\hat{\gamma}$	1, 2, 3	[%]

It is noted that Foda and Tzang also present results where the amplitude in pore pressure exceeds the amplitude of the applied wave forcing. This is explained by resonance as a result of "channeling" of the seepage flow. If this effect would explain the increase in pressure amplitude in silty sediment, it is probably not important in cohesive sediment, and we do not elaborate on this.

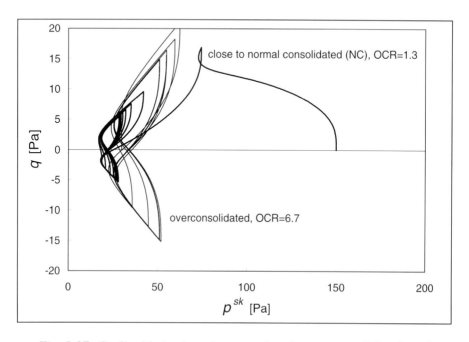

Fig. 8.37: Cyclical behaviour for normal and over-consolidated mud.

De Wit and Kranenburg (1997) carried out similar experiments on a Westerwald clay suspension, deposited in a wave flume. Also these authors measured a strong increase in pore pressure at 55 mm below the sediment water interface as a result of liquefaction of the bed. However, contrary to Foda and Tzang (1994), De Wit and Kranenburg measured a decay in pore pressure with time after an initial strong pressure rise, as expected. Most probably, Foda and Tzang's experiment was not continued long enough to measure a similar decay.

It can be reasoned that recent deposits in the marine environment have no or little over-consolidation (OCR ≈ 1), and are therefore close to their critical state, hence sensitive to liquefaction under the influence of cyclical loading. This liquefaction may take place rapidly, typically within several tens cycles of loading only. Upon liquefaction, the pore water pressure increases and the effective (skeleton) stresses decrease: the sediment loosens its strength. If the material remains immobile, the original normal-consolidation state will be regained when the cyclical loading stops. This consolidation process and the relevant time scales have been elaborated in Chapter 7.

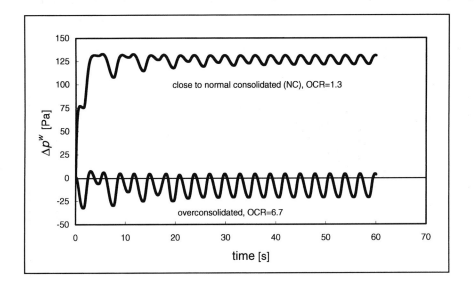

Fig. 8.38: Computed excess pore water pressures in normally and over consolidated mud beds (equ. (8.95)).

When the sediment lies on a slope, it may start to flow upon liquefaction in the form of a turbidity current. In that case the skeleton is entirely broken up and skeleton (effective) stresses become zero. However, when the sediment comes to rest, consolidation starts again, and the resulting stresses differ not much form those in the original sediment deposits.

Only on geological time scales, significant OCR's are encountered as a result of creep, ice cover, erosion of sediment from the upper layers, etc. Sediments at such high OCR are generally not sensitive to liquefaction.

The OCR of a specific sample can be measured with triaxial tests, as explained in Appendix C.5.5.

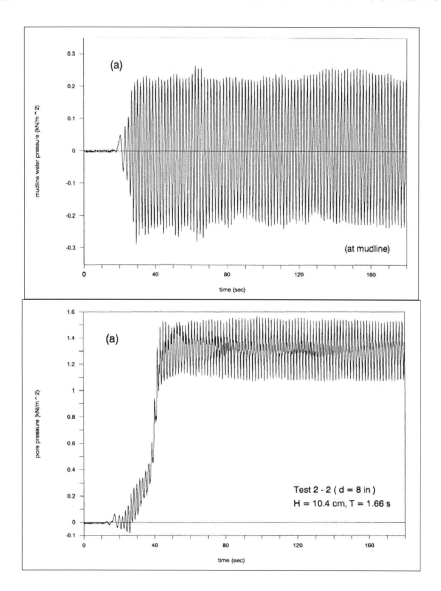

Fig. 8.39: Wave-induced water pressures measured just above sediment bed and at 8 inch below interface (after Foda and Tzang, 1994).

8.5   STRAIN-RATE DEPENDENT BEHAVIOUR

8.5.1 RANGE OF STRAIN RATES

In general, the stress state in sediment deposits is a function of the state and the rate of deformation. The deformation rate is largely governed by the transition from drained to undrained behaviour (see Section 8.2). Under undrained conditions the shear strength increases with increasing deformation rate. This is illustrated in Fig. 8.40 for three different measurements: a rotoviscometer test on IJmuiden Port mud, and cone penetration and undrained triaxial tests on Boom clay and Haaften clay. The undrained shear strength is normalised with the undrained shear strength at rest, amounting to 50 Pa for IJmuiden Port mud, 50 kPa for the Haaften clay and 150 kPa for the Boom clay. These results show that the sensitivity of the shear strength to strain rate decreases with undrained shear strength.

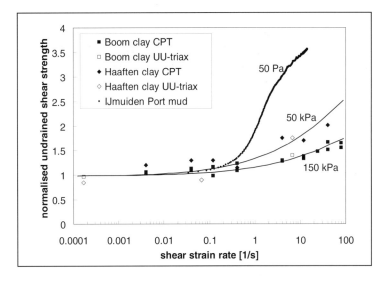

Fig. 8.40: Strain rate dependency of undrained shear strength
(Van Kesteren, 2004).

Fig. 8.41 shows a schematic diagram of the shear strength of cohesive sediment as a function of shear strain rate over a wide range of strains (Van Kesteren, 2004). The transition from drained to undrained behaviour is located in a zone in the middle part of this diagram. Note that the transition between drained and undrained behaviour is determined by the Péclet number Pe_w for pore pressure dissipation (see Section 8.2), thus by the shear strain rate times a length scale squared. At larger length scales the transition zone shifts towards the left of the diagram, i.e. to lower shear strain rates. At these low shear strain rates creep becomes important under drained conditions.

As said, the shear strength under undrained conditions increases with increasing shear strain rate. At high shear strain rates inertia may become important with respect to viscosity, in which case transition to turbulent flow occurs. This transition is determined by Reynolds number. Similar to the Péclet number for pore pressure dissipation, the Reynolds number is proportional to the shear strain rate times the length scale squared. The shear strain rates at transition are given by:

$$\dot{\gamma}_d = Pe_{w,d}\frac{c_i}{\ell^2} \; ; \; \dot{\gamma}_u = Pe_{w,u}\frac{c_i}{\ell^2} \; ; \; \dot{\gamma}_{tur} = Re_{tur}\frac{\eta}{\rho\ell^2} \qquad (8.98)$$

in which $Pe_{w,d}$ and $Pe_{w,u}$ are the Péclet numbers for drained and undrained condition respectively, Re_{tur} is the Reynolds number for transition to turbulent flow, c_i is the isotropic consolidation coefficient [m^2/s], ℓ is a length scale [m], η is the dynamic viscosity [Pa·s] and ρ is the bulk density [kg/m^3].

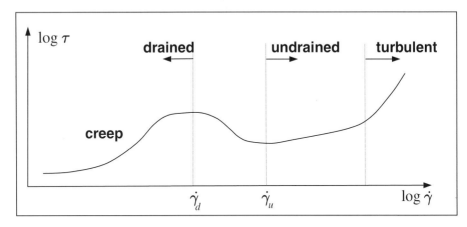

Fig. 8.41: Shear strength as a function of shear strain rate.

From Fig. 8.2 we find $Pe_{w,d} \approx 1$ and $Pe_{w,u} \approx 10$. The transition to turbulent flow for a Newtonian fluid amounts to $Re_{tur} \approx 3{,}000$. For non-Newtonian materials the rheological behaviour must be taken into account, as explained in the next section.

8.5.2 RHEOLOGICAL BEHAVIOUR

In general, rheological behaviour is described through a relation between the deviatoric stress tensor τ_{ij} and deviatoric shear strain rate tensor $\dot{\gamma}_{ij}$. Two main flow regimes can be distinguished: shear flow and elongated flow. In shear flow the principle stresses and strains rotate, while in elongation flow the orientation of principle stresses and strains does not alter (see Section 8.3). As most flows in the marine environment, where water and cohesive sediment interact, are boundary layer flows, we focus on shear flow, for which Newtonian, pseudo-plastic (shear thinning), dilatant (shear thickening), Bingham and visco-plastic flow can be distinguished (see Fig. 8.42a; see also Chapter 3 and Coussot, 1997). The corresponding viscosity η and apparent viscosity η_a $\left(= \tau/\dot{\gamma}\right)$ are plotted in Fig. 8.42b.

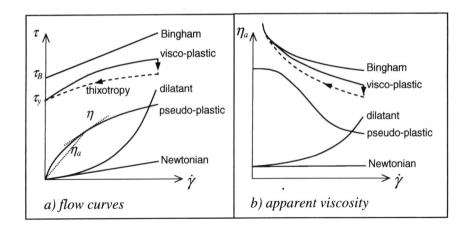

Fig. 8.42: Schematised rheological flow curves.

The flow curve for IJmuiden Port mud in Fig. 8.40 reveals visco-plastic behaviour when the yield stress is exceeded, characterised by a decreasing viscosity with increasing strain rate (shear-thinning). In general, cohesive sediment depicts such shear-thinning behaviour, while sand-dominated sediment exhibits dilatant behaviour. Visco-plastic behaviour of soft cohesive sediment is often accompanied by time effects, known as thixotropy, yielding a decrease in viscosity with time at a constant shear strain rate, and recovery after a period of rest. Thixotropic behaviour in cohesive sediment is caused by a breakdown of skeleton structure; at constant shear strain rate, breakdown and recovery balance, which results in hysteresis of the visco-plastic flow curves (see Fig. 8.42a,b).

A variety of rheological models has been developed describing the time and rate dependent behaviour of soils, based on different arrangements of three basic elements: a spring (elastic effects), a dashpot (viscous effects) and a slider (plastic effects; see also Keedwell, 1984), as shown in Fig. 8.43. Serial coupling of the elements (the so-called Maxwell model) yields equal stresses in the elements and superposition of strains, while parallel coupling of the elements (the so-called Kelvin-Voight model) corresponds to equal strains in the elements and superposition of stresses. Many other combinations have been reported in the literature.

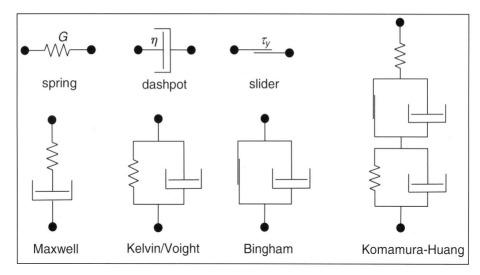

Fig. 8.43: Rheological models for time dependent stress-strain relations.

The elasto-plastic models discussed in the previous sections are based on a superposition of elastic and plastic strains, and can therefore be classified as Maxwell-type models.

The effective stress concept of superposition of pore water pressure and skeleton stresses is an example of a Kelvin-Voight-type model. For visco-plastic behaviour, as shown in Fig. 8.42, the Bingham model is often used, which is composed of a parallel combination of a slider and a dashpot, i.e. superposition of stresses involved in plastic failure (yield stress) and viscous flow. An example of combinations of the three basic types is the model by Komamura and Huang (1974) used for drained and undrained creep of soils.

A theoretical approach for rate dependency is based on the rate process theory developed in thermodynamics (e.g. Mitchell, 1976). This theory describes thermodynamic equilibrium in chemical reactions, which can be applied if soil dynamic failure is regarded as a thermally activated process, describing local displacements in the soil skeleton. This theory expresses the rate of shear strain as a function of the shear stress τ and Bingham strength τ_B (e.g. Houlsby et al., 2002):

$$\dot{\gamma}^p = A \sinh\left\{ B\left(\tau - \tau_B \right) \right\}$$ (8.99)

in which A and B are functions of temperature. This function corresponds well with observed relations between shear stress and the logarithm of shear strain rate in creep phenomena.

In general power law relations are found between shear stress and strain rate, such as the Hershel-Bulkley model, which is a generalisation of the Bingham model (see also Section 3.4.3):

$$\tau = \tau_y + K\left(\dot{\gamma}^p \right)^n \quad \Leftrightarrow \quad \left(\dot{\gamma}^p \right)^n = \frac{\tau - \tau_y}{K}$$ (8.100)

in which τ_y is the yield stress, K is the consistency index and n is the flow index. For $n = 1$ the original Bingham model is retrieved and K represents the dynamic viscosity.

To model thixotropic effects, a time dependent structural model can be used in combination with rheological models, such as the Bingham model. Moore (1959) proposed a simple model to describe such thixotropic effects through a structure parameter λ_s (see also Coussot, 1997). This model can be extended (Winterwerp, 1993, see also Toorman, 1997) by incorporating this structure parameter also in the yield strength:

$$\tau = \lambda_s \tau_y + \left(\eta_\infty + c\lambda_s \right) \dot{\gamma}$$ (8.101)

where τ_y is the yield strength for fully flocculated sediment, η_∞ is the viscosity for fully deflocculated sediment, and c is a viscosity parameter. The structure parameter $\lambda_s = 1$ for fully flocculated sediments (all bonds active) and $\lambda_s = 0$

for fully deflocculated sediment (all bonds destroyed). The rate of flocculation is given by:

$$\frac{d\lambda_s}{dt} = a(1-\lambda_s) - b\lambda_s \dot{\gamma} \tag{8.102}$$

where a and b are flocculation and floc break-up parameters, to be determined empirically. The equilibrium flow curve for (8.101) and (8.102) reads:

$$\tau_{equ} = \frac{a}{a+b\dot{\gamma}}\tau_y + \left(\eta_\infty + \frac{a}{a+b\dot{\gamma}}c\right)\dot{\gamma} \tag{8.103}$$

and its slope (i.e. effective viscosity):

$$\frac{d\tau_{equ}}{d\dot{\gamma}} = \frac{ab}{(a+b\dot{\gamma})^2}\tau_y + \eta_\infty + \frac{a^2c}{(a+b\dot{\gamma})^2} \tag{8.104}$$

The Moore model was extended differently by Cross (1965):

$$\frac{d\lambda}{dt} = a(1-\lambda_s) - b\lambda_s \dot{\gamma}^m \tag{8.105}$$

The reader is referred to for instance Coussot (1997) for the effect of various stress-strain relations on the vertical velocity profiles of mudflows.

The effects of plastic strain rate and stress state in various rheological models, such as the Moore model, the Hershel-Bulkley model (8.99) or the rate process theory equ. (8.103), can be accounted for in elasto-plastic models by means of an additional term in the plastic potential or flow potential g (Section 8.2.3 & 8.2.4, see also Houlsby et al. 2002). This additional term reads for the Hershel-Bulkley model:

$$g_{HB} = \frac{nK}{1+n}\left[\frac{\tau-\tau_y}{K}\right]^{\frac{n+1}{n}} \tag{8.106}$$

For the rate process theory model this term reads:

$$g_{RPT} = \frac{A}{B}\left[\cosh\left\{B\left(\tau-\tau_y\right)-1\right\}\right]$$

(8.107)

and for the Moore model we find:

$$g_M = \frac{\left(\tau-\tau_y\right)^2}{2\left(\eta_\infty+a_t\lambda_s\right)}$$

(8.108)

At high shear strain rates, the flow can become turbulent. Similar to the transition between viscous and turbulent flow for Newtonian fluids, the transition for non-Newtonian fluids can is measured with an effective Reynolds number, which is shown in the Moody diagram $\left(\text{Fig. 8.44; } \tau_w=\left(f/8\right)\rho\bar{u}^2\right)$.

The effective Reynolds number (see also equ. (3.18), Section 3.4.1), for instance for a Herschel-Bulkley model, is given by the generalised Reynolds and Hedstrøm number defined by (Metzner and Reed 1955):

$$\mathrm{Re}' = \frac{\rho V^{2-n}\ell^n}{8^{n-1}K\left[1+3n/4n\right]^n} \qquad He' = \frac{\tau_y}{\rho V^2}\left(Re'\right)^2$$

(8.109)

in which V is the flow velocity and ℓ is a length scale, f.i. the mud layer thickness. If the shear strain rate is approximated as $\dot\gamma=V/\ell$, the shear strain rate for turbulent flow of a visco-plastic fluid can be determined:

$$\dot\gamma_{tur}^{2-n} = \frac{8^{n-1}K\left[1+3n/4n\right]^n}{\rho\ell^2}Re'_{tur}$$

(8.110)

in which Re'_{tur} is the critical generalised Reynolds number for transition to turbulent flow of a visco-plastic fluid. Re'_{tur} increase with the Bingham yield stress τ_B. This is depicted in Fig. 8.44, where the Fanning friction coefficient f is plotted against Re'_{tur} for various Hedstrøm numbers. Instead of the Hedstrøm number, the coefficient $\tau_y/\rho V^2$ in equ. (8.82) can be used. As $\tau_y \sim c_u$, this ratio reflects the level of undrained shear strength with respect to inertia stresses. In

Fig. 8.44 lines of constant values of $\tau_y/\rho V^2$ are plotted, showing that for $\tau_y/\rho V^2$ < 0.002 turbulent flow occurs, when Re'_{tur} > 3,000 (e.g. Liu and Mei, 1989 and Van Kessel, 1997). After transition, the turbulent stresses exceed the viscous stresses largely and the stress-strain relations are governed by the eddy viscosity instead of the molecular effects.

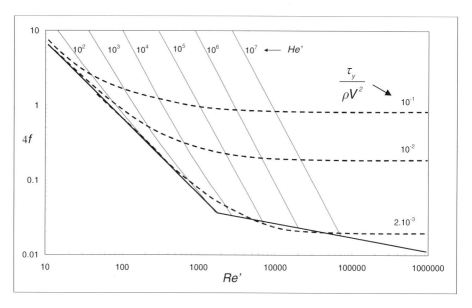

Fig. 8.44: Moody diagram for the Fanning friction coefficient f for visco-plastic fluids (Van Kesteren, 2004).

8.5.3 CREEP

The small rates of deformation shown in Fig. 8.40 and 8.41 at low values of the strain rate $\dot{\gamma}$ of about 10^{-4} to 10^{-8} s^{-1} are referred to as creep. Creep can occur under drained or undrained conditions, depending on the length scales in the deforming soil.

Secondary consolidation during self weight consolidation is one of the most common creep phenomena, during which excess pore water pressures of the primary consolidation phase are dissipated. Fig. 8.45 shows a schematic picture of the void ratio as function of the isotropic stress for normal consolidation upon sedimentation – the most right line is the virgin compression line VCL (see also Bjerrum, 1972). After primary consolidation, plastic volume deformation can occur with strain rates decreasing in time. This deformation is sketched in

Fig. 8.45 as a decrease in void ratio with time at constant isotropic stress p_0, which is the said creep effect.

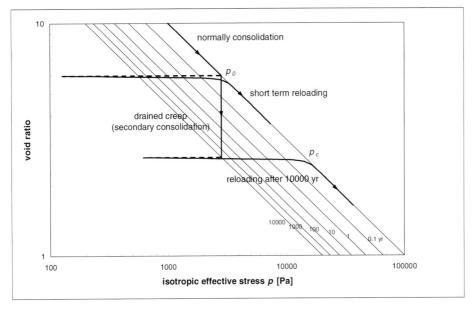

Fig. 8.45: Variation in void ratio with isotropic effective stress during drained creep during secondary consolidation (after Bjerrum, 1972).

When after a long time, say 10,000 years, the isotropic stress increases again, the "aged" sediment will behave as an over-consolidated soil. If the volume deformation upon reloading is elastic, the normal (virgin) consolidation line VCL is followed, when the isotropic stress exceeds the apparent consolidation stress p_c. This means that creep does not change the constitutive behaviour of the sediment, but only changes the strain history at a given stress state.

When the void ratio instead of the isotropic stress is kept constant (i.e. undrained conditions), creep results in a reduction in isotropic stress, which is known as stress relaxation (see Fig. 8.45). Stress relaxation under undrained conditions after primary consolidation at p_0 results in an increase of pore water pressure. If undrained conditions occur after partly secondary consolidation, the relaxation and generated pore water pressures are smaller than in the case without secondary consolidation. An example of such undrained isotropic creep is shown in Fig. 8.46a for San Francisco Bay mud (Holzer et al., 1973).

Creep does not only occur under constant isotropic stress conditions, but may also occur under shear, i.e. at constant deviatoric stress. This results in plastic shear strains increasing in time, as depicted in Fig. 8.47a. After a phase of primary creep a more or less "steady state" creep phase occurs during which the strain rate drops inversely with time. When the plastic shear strain exceeds a certain critical level, the strain rate starts to increase again until failure occurs.

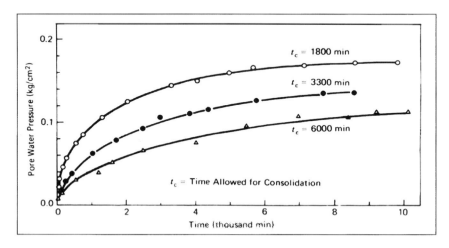

a) undrained isotropic creep.

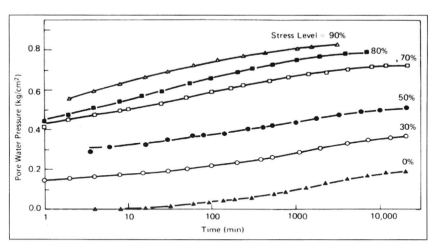

b) undrained deviatoric creep.

Fig. 8.46: Excess pore water pressure due to undrained creep in San Francisco Bay mud (Holzer et al,. 1973); the t_c = 1800 min line in the upper panel is identical to the 0 % stress level line in the lower panel.

The magnitude of the shear strain rate during creep depends on the deviatoric stress levels with respect to the deviatoric stress at failure (Fig. 8.47b). Strain rates are small at low deviatoric stress levels, while close to the deviatoric stress at failure, plastic shear strain rates increase until failure. The plastic shear strains generated during creep induce plastic volume strains as well, which may be dilatant at low effective stresses.

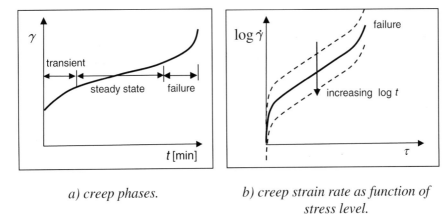

a) creep phases.   b) creep strain rate as function of
                                                     stress level.

Fig. 8.47: Schematised creep behaviour.

The main part of the plastic volume strain during creep at constant deviatoric stress is caused by isotropic creep induced by effective stresses. This is depicted in Fig. 8.46b, showing undrained creep for San Francisco Bay mud at various deviatoric stress levels (Holzer et al., 1973). The creep-induced pore water over-pressures at the various deviatoric stress levels are more or less constant and equal to the pore water over-pressures generated by undrained isotropic creep (0 % line). The physical processes governing the strain rate dependency are quite similar at the various strain rate levels: at deviatoric stress levels below the failure level, the stresses are continuously redistributed because of local failure resulting in additional plastic shear and volumetric strains. Under undrained conditions and high shear strain rates, the plastic volumetric strain generates excess pore water pressures, which can be negative when dilation occurs.

8.5.4 DYNAMIC RESPONSE

When cohesive sediment is subjected to cyclical loading, e.g. waves, the dynamic response is not only determined by the elasto-plastic behaviour of the sediment as treated in Section 8.4, but also strain rate or viscous effects can be important. An example is wave damping over a muddy seabed, where the elasto-plastic behaviour of the mud may result in liquefaction by pore water pressure generation, while its viscous behaviour is responsible for wave energy dissipation (e.g. Dalrymple et al., 1978).

The behaviour of a visco-elasto-plastic material is analysed under undrained conditions under cyclical simple shear deformation to assess the contribution of strain rate effects (see Fig. 8.48). Consider an oscillating shear strain with amplitude $\hat{\gamma}$ and angular frequency ω given as a complex variable by:

$$\gamma = \hat{\gamma}e^{i\omega t} \tag{8.111}$$

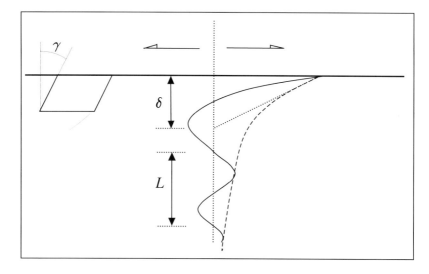

Fig. 8.48: Dynamic visco-elastic response.

The shear stress in this visco-elastic material will oscillate at the same frequency, but with a phase shift:

$$\tau = \hat{\tau}e^{i(\omega t + \theta)} \tag{8.112}$$

where $\hat{\tau}$ is the shear stress amplitude and θ is the phase shift between stress and strain.

The ratio between τ and γ is defined as the complex shear modulus, with a real component G' known as storage modulus and G'' as the loss modulus:

$$G^* = \frac{\tau}{\gamma} = \left|G^*\right| e^{i\theta} = G' + iG'' \Rightarrow \frac{G''}{G'} = \tan\theta \qquad (8.113)$$

When the visco-elastic behaviour is modelled as a Kelvin-Voigt model (see Fig. 8.43) the viscous and elastic stresses are superimposed:

$$\tau = G\gamma + \eta\dot{\gamma} \qquad (8.114)$$

where G is the elastic shear modulus [Pa] and η is the dynamic viscosity [Pa·s]. Substitution of equ. (8.111) and (8.112) yields for the storage and loss modulus and the phase shift:

$$G' = G \; ; \quad G'' = \eta\omega \; ; \quad \tan\theta = \frac{G''}{G'} = \frac{\eta\omega}{G} \qquad (8.115)$$

In the case of a Maxwell model (see Fig. 8.43) the viscous and elastic strains are superimposed:

$$\dot{\gamma} = \frac{\dot{\tau}}{G} + \frac{\tau}{\eta} \Leftrightarrow \tau = -\eta\frac{\dot{\tau}}{G} + \eta\dot{\gamma} \qquad (8.116)$$

For the Maxwell model the storage and loss modulus and phase shift are given by:

$$G' = \frac{G(\eta\omega)^2}{G^2 + (\eta\omega)^2} \; ; \quad G'' = \frac{G^2\eta\omega}{G^2 + (\eta\omega)^2} \; ; \quad \tan\theta = \frac{G''}{G'} = \frac{G}{\eta\omega} \qquad (8.117)$$

The relations (8.115) and (8.117) have been made dimensionless by G and are depicted in Fig. 8.49.

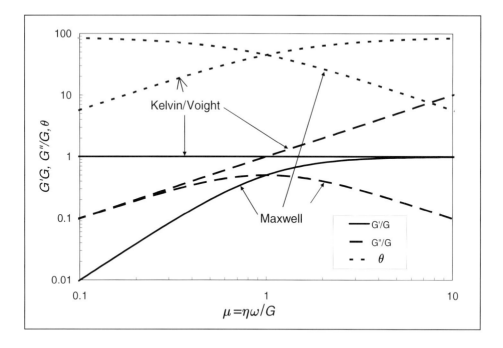

Fig. 8.49: Visco-elastic moduli and phase shift.

Hence, superposition of viscous and elastic stresses (Kelvin-Voigt) shows a quite different response of the storage and loss modulus and phase shift than superposition of viscous and elastic strains (Maxwell). The storage and loss modulus and the phase shift can be determined with standard oscillating rheometer tests (see Appendix C.4). In Fig. 8.50 an example is shown of measurements by Chou et al. (1993) on kaolinite suspensions. The storage and loss modulus G' and G'' are plotted as a function of the strain amplitude in Fig. 8.50 for two different solid concentrations. It shows that with increasing strain amplitude, G' remains constant for small strains and decreases above a certain strain amplitude. Also the loss modulus G'', being a measure for the viscosity, decreases. This behaviour can be related to an elastic response at small strains and failure at larger strains, when plastic strains increase with higher strain amplitudes, thus at lower stiffness of the skeleton.

A similar result is shown in Fig. 8.51 for tests on kaolinite suspensions by De Wit (1995), showing that at decreasing strain amplitude (i.e. angular frequency), the reduced stiffness remains more or less constant until the strain amplitudes become so small, that elastic strains become dominant again, resulting in an increase of G'.

It can be concluded that plastic failure behaviour plays an important role in the dynamic response of cohesive sediment beds. Furthermore, the measured storage and loss modulus as a function of angular frequency, as depicted in Fig. 8.51, show little resemblance with the Maxwell or Kelvin-Voigt visco-elastic models in Fig. 8.49. This supports our conclusion that plastic behaviour is more important than elastic behaviour in the dynamic response of cohesive sediments in the marine environment.

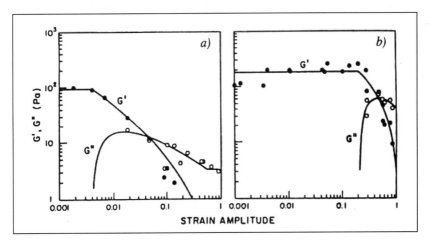

Fig. 8.50: Loss and storage modulus G′ and G″ versus strain amplitude for kaolinite with concentrations 29.3% (a) and 51.5% (b) (Chou et al., 1993).

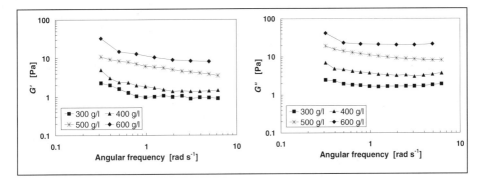

Fig. 8.51: Loss and storage modulus G′ and G″ versus angular frequency for kaolinite suspensions (De Wit, 1995).

The dynamic response of cohesive sediment subjected to waves may cause liquefaction of the bed, which enables both bed transport of the sediment and its resuspension by currents (tidal currents, density driven currents). Furthermore, this dynamic response is important for adsorbing wave energy. In literature, visco-elastic modelling is often used to assess wave damping on mud banks (e.g. Maa and Metha 1987, 1988, 1990; Chou, 1989, 1993; Feng, 1993).

As shown in Section 8.4, the dynamic response under undrained cyclical loading is primarily determined by the plastic behaviour both with respect to the generation of pore water pressures as to the reduction in strength. Among others, De Wit (1995) showed that wave damping can be assessed quite well by modelling the mud as a Newtonian fluid. In Fig. 8.52, experiments are compared with models based on the work of Gade (1958) and by Dalrymple and Liu (1978), for which Gade's model was extended for arbitrary wave lengths. It appeared the viscous approach gives reasonable results, though we omitted plastic effects, which certainly play a role. This is probably caused by the fact that the energy consumed in the mud layer is determined by the depth-averaged horizontal velocity amplitude of the mud layer and the bottom shear stress at the interface with the un-deformed mud. In these models, is does not matter how the energy is dissipated within the layer: by viscous or by yield stresses. When elastic behaviour is important, energy can be stored and recovered resulting in less wave damping.

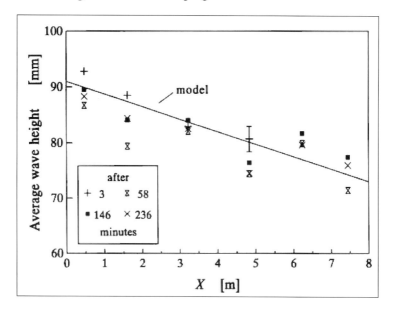

Fig. 8.52: Wave damping in China clay suspension (De Wit., 1995).

In summary, we conclude that dynamic, c.q. cyclical loading, as may occur under wave action in coastal areas, will affect a seabed consisting of cohesive sediment by:
- liquefaction of the bed, and
- (viscous) damping of the waves.

The liquefied mud behaves as a dense fluid, damping vertical mixing, thus affecting the turbulent flow field as well (see Chapter 6). Hence, we expect a complex interaction between the (turbulent) flow field, the wave field and the sediment dynamics.

9. EROSION AND ENTRAINMENT

The transport and fate of cohesive sediment in the marine environment is governed to a large extent by water-bed exchange processes, i.e. deposition and erosion. In this chapter we treat erosion processes of different beds, ranging from old, well-consolidated deposits to fresh, soft deposits. The erosion rates of such deposits as a function of the local hydrodynamic conditions (flow and waves) may vary by orders of magnitude, depending on the sediment–pore water composition and properties, stress history, etc. Moreover, often only thin layers of a few mm to a few cm of the bed are eroded in one tidal cycle. Yet, when mixed over the water column, the eroded sediment may increase the local suspended sediment concentration by tens to hundreds of mg/l, depending on the local water depth. We have shown in Section 3.1.3, Fig. 3.9 that within such thin layers, large gradients in physico-chemical properties may occur, affecting the erodibility of sediment deposits. This explains, amongst other things, why the literature contains so many different erosion formulae with large ranges in parametric values.

We present a conceptual picture of the erosion process in Section 9.1. Section 9.2 contains a summary of relevant literature. In Section 9.3 we propose a classification scheme for bed erosion, and descriptions of the various modes of erosion are presented in Sections 9.4 and 9.5.

9.1 PHENOMENOLOGICAL DESCRIPTION OF EROSION

Various modes of erosion can be distinguished, depending on the bed properties, such as strength and permeability, and the stresses by water movement. In this book, we distinguish four modes: entrainment, floc erosion, surface erosion and mass erosion[1]. These are summarised in Fig. 9.1.

Entrainment occurs when the mud is so soft that it behaves as a viscous fluid. This may be the case when deposits are formed by rapid siltation, for instance forming fluid mud in a navigational channel (e.g. Section 3.4 and 6.3.3), or by liquefaction of mud deposits by wave action (e.g. Section 8.4). When the upper layer is turbulent and the mud layer not, as in case of fluid mud, the mud is entrained by the water layer. If the mud layer is turbulent, as in case of turbidity currents, water from the upper water layer is entrained by the mud layer. If both layers are turbulent, turbulent mixing between the two layers may

---

[1] Note that also other definitions are used in literature.

occur. Currently, entrainment is properly understood, as summarised in Section 9.4.

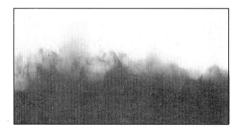

Fig. 9.1a: Entrainment of a dense mud layer by turbulent water flow (entrainment = "one-way" mixing).

Fig. 9.1b: Floc erosion = disruption and break-up of individual flocs or part of flocs from the bed surface.

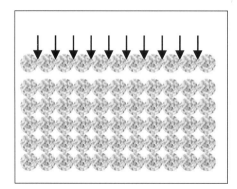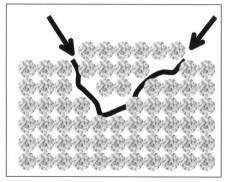

Fig. 9.1c: Surface erosion = drained process: eroding sediment particles are replaced by water.

Fig. 9.1d: Mass erosion = undrained process: local failure within the bed (crack formation).

Floc erosion is referred to when (part of the) flocs at the bed surface (is) are individually disrupted by the water movement. This may occur when the flow-induced (peak) bed shear stresses exceed the strength of the flocs or the adhesion of flocs to the bed. Floc strength and adhesion may decrease in time by cyclical stresses in turbulent flow and/or by wave action. Floc erosion rates are usually not large, but this mode of erosion is a continuous process that may therefore contribute to the overall erosion process considerably. At present, this erosion process is poorly understood, e.g. Section 9.5.1

Continuity requires that sediment particles removed from the bed surface be replaced by water. This implies that flow of ambient water into, or of pore water within the bed, is an important process. Surface erosion can be regarded as a drained failure process. The top of the bed liquefies as a result of swelling when the bed is over-consolidated, and/or as a result of hydrodynamic pressure fluctuations induced by the turbulent flow and/or waves. This mode of erosion is reasonably well understood at present, e.g. Section 9.5.1.

Mass erosion can be regarded as an undrained failure process. In this case, external stresses exceed the undrained shear strength of the bed, and lumps of material are removed. This mode of erosion occurs in particular in case of turbulent flow or waves over irregular beds. Cliff erosion is a well-known form of mass erosion. This mode of erosion is poorly understood at present, e.g. Section 9.5.1.

Waves can have a large effect on erosion. Firstly, wave-induced stresses can liquefy the bed, as discussed in Section 8.4. The liquefied mud then may become subject to the entrainment processes described above. Secondly, waves induce large bed shear stresses, often well-exceeding flow-induced stresses. This is further elaborated in Section 9.5.2.

If the eroding agent is not clear water or a low concentrated mud suspension, but for instance a dense turbidity current, replacement by water of sediment particles at the bed surface is limited by the low permeability of the turbidity current. We may refer to this process as hindered erosion, e.g. Section 9.5.3.

9.2   LITERATURE ON EROSION

9.2.1 EROSION FORMULAE

Partheniades (1962, 1965, 1986) was the first to carry out erosion experiments on marine cohesive sediments in a systematic way. These experiments were carried out with mud from San Francisco Bay in a straight, recirculating flume with Bay water. Later, Partheniades also used artificial clays. The first series of erosion experiments were carried out on a placed, remoulded bed at in-situ density, and the second series was done on so-called deposited beds, formed through deposition and subsequent consolidation from a cohesive sediment suspension. The results are presented in Fig. 9.2.

Both experiments showed a non-zero erosion rate at the smallest bed shear stresses applied. This indicates that a threshold shear stress (critical shear stress for erosion) did not exist, or was very small. In his analysis of the data, Partheniades assumed a Gaussian bed shear stress distribution, and obtained the following formula for the erosion rate E:

$$E = \frac{AD_{50}\rho_s}{t^{(\tau_b)}}\left[1 - \frac{1}{\sqrt{2\pi}}\int_{-\frac{c}{k\eta_b\bar{\tau}_b}-\frac{1}{\eta_b}}^{\frac{c}{k\eta_b\bar{\tau}_b}-\frac{1}{\eta_b}}\exp\left\{-\frac{\omega^2}{2}\right\}\mathrm{d}\omega\right] \tag{9.1}$$

in which A and k are coefficients, D_{50} is the median diameter of the bed forming flocs, $t^{(\tau_b)}$ is the time that the time-varying bed shear stress exceeds the cohesive forces within the bed, c is the cohesion due to interparticle forces, $\bar{\tau}_b$ is the mean bed shear stress and $\eta_b\bar{\tau}_b$ is its variance.

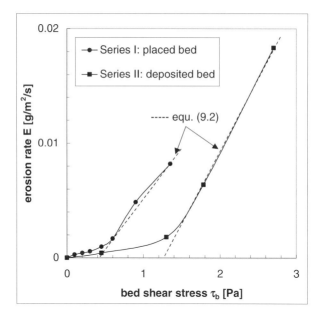

Fig. 9.2: Erosion rates measured by Partheniades (redrawn from Partheniades, 1986).

Similar results were found by Christensen and Das (1965) and Croad (1981), with different coefficients however.

Ariathurai (1974) parameterised Partheniades' results:

$$E = M\left(\frac{\tau_b - \tau_e}{\tau_e}\right) \quad \text{for} \quad \tau_b > \tau_e \tag{9.2}$$

where M is an erosion rate parameter, τ_b the turbulent-mean bed shear stress, and τ_e a critical (threshold) shear stress for erosion. This formula was combined with Krone's deposition formula (Section 5.3.1, equ. (5.21)) to compute the water-bed exchange rate in a numerical model for the transport of cohesive sediment. This description of the water-bed exchange processes is known as the Krone-Partheniades bed-boundary condition. We have argued in Section 5.3.1 that Krone's deposition formula in fact already describes the full erosion-deposition bed boundary condition. As a result, we advocated the use of the near-bed sediment flux only to establish the gross sedimentation rate.

Note that the inclusion of τ_e in the denominator of (9.2) is correct from a dimensional point of view (see also equ. (9.3)). However, applying equ. (9.2) involves inaccuracies in establishing M from erosion experiments, as M becomes sensitive to small errors in τ_e.

Following the work by Partheniades, several other erosion experiments were carried out, in particular in the 1970's and 80's. Many of these experiments were done in (rotating) annular flumes. This is a circular flume with a rotating lid to drive the flow. Often, the flume itself can rotate in opposite direction to minimise secondary currents. This flume was applied by Mehta and Partheniades (1975) on deposition experiments, and later for erosion experiments on kaolinite suspensions (Mehta and Partheniades, 1979). Rotating annular flumes are in use throughout the world, and we have included a brief description in Appendix C.

The test procedure generally applied to study erosion in the (rotating) annular flume is as follows. A placed or deposited bed (e.g. Appendix C) is prepared in the flume and the flow velocity is increased in small steps. Each step is maintained for a fixed time period, generally one hour, and the increase in suspended sediment concentration in the flume (i.e. the amount of eroded material) is measured. A typical example of the results of such experiments is given in Fig. 9.3, showing the increase in suspended sediment concentration in an annular rotating flume as a function of the bed shear stress, as measured by Kuijper et al. (1990a) for mud from the Western Scheldt, The Netherlands.

The erosion formula (9.2) has been generalised (Harrison and Owen, 1971; Kandiah, 1974; Mehta, 1981; Lick, 1982; Sheng, 1984), and is used (in mathematical models) throughout the world, no doubt because of its simplicity:

$$E = M\left(\frac{\tau_b - \tau_e(z,t)}{\tau_e(z,t)}\right)^n \quad \text{for} \quad \tau_b > \tau_e \quad\quad\quad (9.3)$$

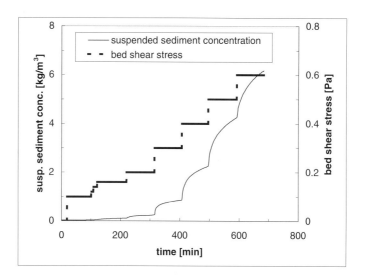

Fig. 9.3: Typical example of applied bed shear stress and suspended sediment concentration in rotating annular flume (after Kuijper et al., 1990).

The exponent n is generally unity, though other values have been found (Harrison and Owen, 1971; Kusuda et al., 1985). The critical bed shear stress for erosion τ_e is generally assumed to be a constant material parameter, like M, but often varies with depth and time because of consolidation and physico-chemical effects. Typical values given in literature are: 0.1 Pa $< \tau_e < 5$ Pa. Also, the erosion parameter M should vary with time and depth, but is generally taken constant; typical values are: $0.01 \cdot 10^{-3}$ kg/m^2/s $< M < 0.5 \cdot 10^{-3}$ kg/m^2/s. A summary of parameter values for mud from English estuaries is given by Whitehouse et al. (2000). It should be noted that τ_e and M can be much larger, cq. smaller, in particular for consolidated clay deposits (e.g. Section 9.5).

Equ. (9.3) is commonly applied to well-consolidated, homogeneous beds, in which case τ_e and M are more or less constant throughout the bed. This type of erosion is sometimes referred to as Type I (unlimited) erosion in the literature.

For soft beds with (strong) gradients in strength, an alternative formula was advocated by Mehta and Partheniades (1979):

$$E = E_f \exp\left\{\alpha\left(\frac{\tau_b - \tau_e(z)}{\tau_e(z)}\right)^{\beta}\right\} \tag{9.4}$$

where E_f is referred to as a floc erosion rate, α and β are material dependent parameters and $\tau_e(z)$ is a depth-varying critical shear stress for erosion. This type of erosion is sometimes referred to as Type II (depth-limited) erosion in the literature. Contrary to (9.3), equ. (9.4) predicts a decreasing erosion rate with time (depth), as often observed during erosion experiments. This formula was applied by Parchure and Mehta (1985); Amos et al. (1992);and Chapalain et al. (1994). Typical values for the various parameters are: $0.003 \cdot 10^{-3}$ kg/m^2/s $< E_f < 5 \cdot 10^{-3}$ kg/m^2/s; $0.5 < \beta < 1$; $5.0 < \alpha < 15.0$ and 0.01 Pa $< \tau_e < 0.1$ Pa.

Sanford and Maa (2001) showed that (9.4) and (9.2) may yield a similar erosion rate under the assumptions that in equ. (9.2) $\tau_e = \tau(z)$, $M = M(z) = \eta \times \rho_b(z)$, η = constant, and $d\tau_e/dz$ = constant. They argue further that the time-scale of the eroding forces in relation to the time-scale of erosion determines whether the erosion process is unlimited or depth-limited. This observation is qualitatively similar to our analysis of the role of the permeability of the soil, as discussed in Section 9.3 and 9.5.1.

9.2.2 EFFECT OF PHYSICO-CHEMICAL PARAMETERS

We have analysed bed strength and flocculation as a function of the physico-chemical properties of the water-sediment mixture in Chapter 3 (Section 3.2). The response of the diffusive double layer around the sediment particles was the key to this analysis, and forms the basis of our understanding of the effects of physico-chemical parameters on the erosion of cohesive sediment beds.

A number of experiments were carried out, in particular in the 1960's and 70's, to establish the effects of physico-chemical parameters on the erodibility of cohesive sediments. In particular, the work of Kandiah (1974) has to be mentioned because of his systematic approach. It is noted that most studies and information available are related to erosion formula (9.3).

An increase in (dry) bed density generally increases the number of bonds between the particles (flocs), hence increases bed strength. A number of researchers have established highly empirical correlations between the critical (threshold) shear stress for erosion τ_e and the (bulk) density of the bed:

$$\tau_e = \alpha_1 \rho_b^{\beta_1} \quad \text{or} \quad \tau_e = \alpha_2 \rho_{dry}^{\beta_2} = \alpha_2 c_b^{\beta_2} \tag{9.5}$$

Values for the parameters in equ. (9.5) are given in Table 9.1:

Table 9.1: Values for the coefficients in equ. (9.5). Bulk and dry bed density ρ_b and ρ_{dry} are in SI-units (except Krone: g/cm³). Further, [1] Migniot assumed that $\tau_e \propto \sqrt{\tau_y}$ and [2] Krone assumed that $\tau_e \propto \tau_B$ (Bingham strength).

	α_1	β_1	α_2	β_2
Migniot[1] (1968)			-	2.5
Owen (1970)	5 to 7	2.3 to 2.4		
Krone[2] (1984)			-	2.5

The relationship by Owen was extended by Mitchener et al. (e.g. Whitehoue et al., 2000) to cover a wider range of cohesive bed types (parameters are given in SI-units):

$$\tau_e = 0.015(\rho_b - 1000)^{0.73} \tag{9.6}$$

Smerdon and Beasley (1959) related the threshold for erosion τ_e to the plasticity index PI (e.g. Section 3.3.1):

$$\tau_e = 0.163 \text{PI}^{0.84} \quad \text{PI in \%} \tag{9.7}$$

This relation was found from a series of laboratory experiments on soft clay with strength varying from $c_u = 0.1$ kPa to 10 kPa. Because of these high strengths, τ_e in (9.7) can be regarded as the true critical shear stress for erosion τ_{cr}, e.g. Section 9.3.

Natural sediments often consist of a mixture of sand and mud (e.g. silt and clay). In Chapter 3 we have presented a conceptual framework to analyse and classify these sediment mixtures.

A few studies on the behaviour of sand-mud mixtures have been reported in literature. Kandiah (1974) showed empirically that the effect of sand content on the erodibility of cohesive soils is a function of the Sodium Adsorption Ratio (SAR) of the pore water. Kandiah reasoned that the strength of a bed is determined by clay-clay bonds and clay-sand bonds. At high SAR, the clay-clay bonds would be weaker than the clay-sand bonds, and an increase in clay content would result in a decrease in overall bed strength, i.e. an increase in erodibility (see next pages). At small SAR the opposite would happen, and an increase in clay content yields an increase in overall bed strength.

In the 1990's a number of laboratory experiments on sand-mud mixtures were carried out, e.g. Torfs (1995;) and Williamson (1993 – see Chesner and Ockenden, 1997 and Whitehouse et al., 2000). Chesner and Ockenden (1997) schematised the results in a simple diagram, that was used in a depth-averaged sediment transport model of the Mersey estuary. In this diagram, τ_e first increases and then decreases with increasing mud content, starting from a pure sand bed.

The various experimental data were further analysed by Torfs et al. (2001), who found that τ_e increased with increasing mud content, except at mud contents of a few percent, when a small decrease in τ_e was observed, as shown in Fig. 9.4. Note that Fig. 9.4 suggests an off-set in erosional behaviour, similar to the definition of ξ_0 in Section 3.3.1 (Fig. 3.17).

The empirical relation (9.7) is also shown in Fig. 9.4 in case the mud fraction equals the clay fraction, as for the experiments of Torfs. Using the definition of the activity A of the clay minerals (equ. (3.7)), relation (9.7) is plotted in Fig. 9.4 for various values of A. The measured data of Torfs appear to correspond well with $A = 1.5$.

Van Ledden (2003) presented a thorough analysis of the various studies on sand-mud mixtures (see also Chapter 3). On the basis of this analysis, and on the classification of sand-mud beds, he proposed two heuristic erosion formulae. In the non-cohesive regime (e.g. Section 3.3.2), sand and mud particles behave independently and the individual sediment components do not affect the erodibility of the other constituent. Hence:

$$E^{sa} - D^{sa} = \gamma W_s \left(c_e^{sa} - c^{sa} \right)$$

$$E^m = \xi^m M \left(\frac{\tau_b - \tau_{e,n}}{\tau_{e,n}} \right) \quad \text{for} \quad \tau_b > \tau_{e,n}$$

$$(9.8)$$

where E and D are the erosion and deposition rate, W_s is the settling velocity, c is the suspended sediment concentration, M an erosion parameter, τ_b the bed shear stress, c_e^{sa} is the equilibrium concentration for sand and γ is a form coefficient; superscripts $\bullet^m$ and $\bullet^{sa}$ refer to the mud and sand fractions, respectively.

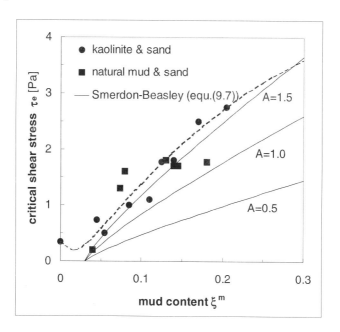

Fig. 9.4: Critical shear strength for erosion as a function of mud content ξ^m (after Torfs et al., 2001) and compared with Smerdon and Beasley (9.7).

In the cohesive regime, erosion is governed by the cohesive nature of the bed, and sand particles are passive: they are eroded with the mud particles at a rate proportional to its fraction:

$$E^{sa} = \xi^{sa} M \left(\frac{\tau_b - \tau_{e,c}}{\tau_{e,c}} \right) \quad \text{for} \quad \tau_b > \tau_{e,c}$$

$$E^m = \xi^m M \left(\frac{\tau_b - \tau_{e,c}}{\tau_{e,c}} \right) \quad \text{for} \quad \tau_b > \tau_{e,c}$$

$$(9.9)$$

Note that the thresholds for erosion of a cohesive and non-cohesive bed, $\tau_{e,c}$ and $\tau_{e,n}$ respectively, may be different. Also the erosion rates for cohesive and non-cohesive beds will differ. This is depicted in Fig. 9.5 showing the erodibility of different sand-mud mixtures. These mixtures consist of 80 and 120 μm sand, mixed with various amounts of kaolinite and bentonite, varying the permeability k of the sample, whereas the porosity of the samples is the same. All samples have a non-cohesive behaviour (e.g. Chapter 3) and were subject to an eroding velocity of 1 m/s. Fig. 9.5 shows that the erosion velocity V_e scales as $V_e \propto k^n$, where $n \approx 0.3$ (see also Section 9.5.1).

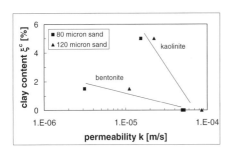 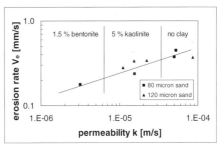

Fig. 9.5: Variation of permeability with clay content and variation of erosion rate with permeability (Van Kesteren, 2004).

The Sodium Adsorption Ratio (SAR) is defined as the ratio of the concentration of one-valence Sodium cations to two-valence Calcium and Magnesium cations (e.g. Section 3.2.1). This implies that at small SAR, two-valence cations are abundant, and the diffusive double layer is thin, hence attractive forces are relatively large. An increase in SAR, hence an increase in single-valence cations causes an increase in diffusive double layer thickness, and attractive forces become smaller. This was shown through experiments by Kandiah (1974), Arulandan (1975), Ariathurai (1978) and Hayter (1984). As an example, the results by Kandiah (1974) are shown in Fig. 9.6.

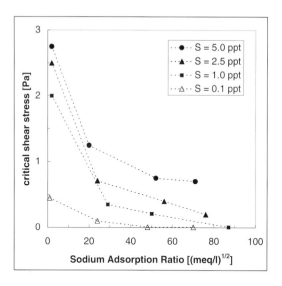

Fig. 9.6: Critical shear stress for erosion for illite clays as function of SAR at various salinities (redrawn from Kandiah, 1974).

It should be noted that an increase in salinity means an increase in cation concentration and therefore a decrease in diffusive double layer thickness, hence an increase in attraction. As a result, floc strength increases and erodibility decreases with increasing salinity (e.g. Fig. 9.6). However, two- and higher-valence cations are more effective, which explains the SAR dependency of τ_e.

An increase in pH results in a decrease in H^+-ions, hence an increase in diffusive double layer thickness and therefore higher repulsive forces. Thus, a decrease in critical shear stress, hence larger erodibility is to be expected. This is depicted in Fig. 9.7, showing an increase in erodibility and decrease in critical shear strength for erosion with increasing pH.

Experiments at Delft Hydraulics in a rotating annular flume revealed a decrease in τ_e when S was increased, which is in agreement with Kandiah's finding at high SAR (Wijdeveld, 1997).

Note that the concentration of Fe^{2+} and Mn^{2+} cations in the bed pore water may have a strong effect on the properties of the diffusive double layer, hence on the strength of the bed (e.g. Wiedemyer and Schwamborn, 1996; and Sigg et al, 1991). These cations however are not included in the definition of SAR, and therefore SAR is not a sufficient parameter to describe the influence of pore water on bed stability.

30 % kaolinite, cation conc = 10 meq/l,   30 % montmorilonite, cation conc = 10
        SAR = 15.5 (meq/l)$^{1/2}$ meq/l, SAR = 7.5 (meq/l)$^{1/2}$

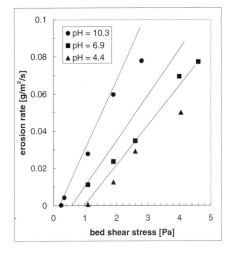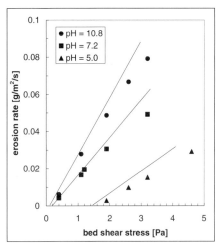

Fig. 9.7: Erosion rate for kaolinite (left panel) and montmorilonite (right panel) clays as function of pH (redrawn from Kandiah, 1974).

It is further noted that the time scale over which pore water properties (SAR) change as a result of changing water properties (e.g. salinity) is of the order of a few weeks. Hence, it is expected that erodibility will not change much over a spring-neap cycle with subsequent salinity variations, but may change over seasons, with variations in river run-off, hence salinity in the water column.

The relation obtained by Kandiah (1974) between the critical (threshold) shear stress for erosion and the sediment-pore water properties was discussed in Section 3.2.1, Fig. 3.12. For convenience this relation is given again in Fig. 9.8. It shows that τ_e increases with increasing CEC at small SAR, but decreases at high SAR. At small SAR the diffusive double layer is thin, and the particles can approach each other closely. Hence, at small SAR a stronger bed is to be expected than at large SAR. An increase in CEC at low SAR yields a further reduction in the thickness of the diffusive double layer, hence a further increase in bed strength. In contrast, at increasing CEC at large SAR, repulsive forces prevail, resulting in a decrease in bed strength.

Some authors have studied the effect of temperature on the erodibility of cohesive sediment beds, amongst whom Kandiah (1974); Ariathurai et al. (1978); and Croad (1981). Kelly and Gularte (1981) carried out experiments in a closed flume on a remoulded illite clay at various temperatures. They found an increase in erosion rate by about a factor three when the temperature increased from 10 to 30 °C. Though the results of the various experiments are not very conclusive (and inconsistent), we would expect an increase in erodibility with increasing temperature, as the (pore) water viscosity decreases with increasing temperature, resulting in larger permeability, hence stronger pore water flow, and higher free-stream Reynolds numbers. Note that Kelly and Gularte explain their findings from a decrease in particle bonds with increasing temperature, as a result of a change in energy barrier (e.g. Section 3.2.1).

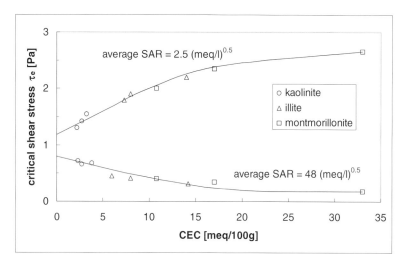

Fig. 9.8: Critical shear strength for erosion as function of SAR and CEC
(redrawn from Kandiah, 1974).

For convenience, we also present Fig. 3.9 again in Fig. 9.9, showing the large gradients in chemical parameters in the upper part of the bed (see also Jørgensen and Boudreau, 2001). In particular the gradients in Fe are of importance. Particulate Fe^{3+} is stable under oxygenated conditions, i.e. in the upper part of the bed, and is mainly bound to the clay minerals or present in the form of (hydr)oxides, which appear as amorphous particles and negatively charged coatings (e.g. Tipping et al., 1981; and Tipping and Cooke, 1982). These particles and coatings stabilise the bed by decreasing the bed's permeability and by cementing effects between the sediment particles (e.g. Schwertmann and Taylor, 1989; and Mitchell, 1993).

The Fe-(hydr)oxides can be reduced rapidly at the redox-interface (e.g. Fig. 9.9), increasing the erodibility of the bed. This has been confirmed by Kuijper et al. (1990b) who observed that the erosion resistance of mud from the

Hollandsch Diep, an estuary branch in The Netherlands, was initially quite high ($\tau_e > 0.3$ Pa), with virtually no erosion at bed shear stresses less than τ_e. A small increase in bed shear stress beyond that threshold resulted in rapid erosion of the upper part of the bed at an erosion rate of about 0.4 g/m²/s, which was an order of magnitude larger than the erosion rate of sediments deeper in the bed. This upper part of the bed was characterised by a light ochre colour due to oxidised Fe, contrary to the deep brown colour of the rest of the bed.

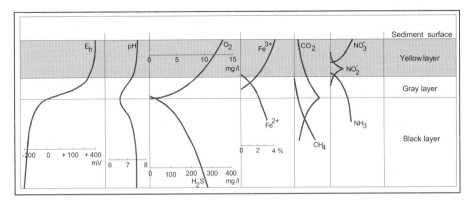

Fig. 9.9: Characteristic profiles of various chemical parameters in the upper part of the bed (after Fenchel and Riedl, 1970).

The effects of biology on sediment erodibility are discussed in some detail in Chapter 10.

9.3   CLASSIFICATION SCHEME FOR EROSION

In this section we present a scheme to distinguish between the four modes of erosion discussed in Section 9.1. Formulae for these modes are presented in the Sections 9.4 and 9.5. The mode of erosion of a bed of cohesive sediment is determined by three factors:
1. the stress state of the sediment in the bed and its history,
2. the type and magnitude of the eroding forces,
3. the time scale of the response of the bed in relation to the time scale of the driving forces.

the stress state of the bed sediment

A sediment "bed" behaves as a viscous fluid when the effective stresses σ_{ij}^{sk} (e.g. Section 7.1 and 8.1) within the bed are small or zero. When σ_{ij}^{sk} is not small, the fluid exhibits a yield strength and behaves as a visco-plastic material (e.g. Section 8.2). Such cohesive sediment appearances may occur during settling and consolidation from cohesive sediment suspensions, or as a result of liquefaction by waves for instance (e.g. Section 8.4), or due to fluidisation, for instance as a result of groundwater flow. These appearances are commonly known as fluid mud, consolidating bed or turbidity currents (Section 3.4).

When the stress state of a bed is characterised by the Virgin Compression Line, as under normal (virgin) consolidation, its stress state is given by the K_0-line in Fig. 8.9. Such a bed is on the wet (right) side of its critical state (Section 8.2), which implies that shearing results in water overpressure without volume changes. The strength of the bed is characterised by the remoulded shear strength c_u and the bed behaves as an elasto-plastic material. The adhesion of flocs to the bed surface is given by the strength τ_{cr}.

When part of the load on a bed at its virgin state is removed, as under the influence of erosion or excavation, the bed becomes over-consolidated. In that state, the bed tends to swell as water flows into the bed pores to adapt to the lower stress state (Section 8.2). This swelling process takes time, as a result of which the bed remembers its previous stress state. The swelling of the bed is accompanied by a decrease in strength; however, the bed strength will never fall below τ_{cr} defined in the previous paragraph.

the type and magnitude of the eroding forces

Bed erosion and failure are the result of flow- and wave-induced shear and normal stresses (stagnation pressure = $\frac{1}{2}\rho u^2$). We have shown that because of the stochastic character of turbulence (e.g. Section 2.2.2 and 5.4.1) erosion can take place when the mean value of the stresses is below the strength of the bed, i.e. c_u or τ_{cr} (see Fig. 5.9).

Cyclical loading by for instance waves can result in liquefaction of the bed (Section 8.4), increasing p^w and decreasing σ_{ij}^{sk}. In turn, the soft liquid mud may damp the waves through viscous dissipation, as a result of which a complex interaction between waves and bed occurs.

time scale effects

When the deformation rate of the bed is small compared to the time scale for pore water flow, gradients in pore water pressure generate pore water flow. This state is referred to as drained behaviour. When on the other hand the deformation rate is high, pressure gradients will not be compensated by pore water flow, and the behaviour of the bed is referred to as undrained (see Section 8.1.2). The transition from drained to undrained conditions is characterised by the Peclet number Pe_w for pore pressure dissipation (e.g. Section 8.1.2).

If we combine these three aspects we can elaborate on the physical processes responsible for the four modes of erosion described in Section 9.1:

entrainment
Entrainment of soft mud layers (e.g. fluid mud) occurs when
- the bed behaves as a viscous fluid, i.e. effective stresses are small, and
- the stress at the interface between the fluid mud layer and the overlying water column exceeds the yield strength τ_y of the mud, if any.

floc erosion
Floc erosion occurs when individual flocs are disrupted from the bed, i.e. the flow-induced stresses should exceed local strength of the bed τ_{cr}, which can therefore be regarded as the true threshold for erosion τ_{cr}.

The often applied threshold stress for erosion τ_e (e.g. critical shear stress for erosion) is normally measured from experiments during which the onset of floc movement is determined from visual observations. In the erosion formulae (9.3) and (9.4), the erosion rate is given as a function of the mean bed shear stress $\overline{\tau}_b$ and τ_e. However, the onset of floc movement is governed by the peaks in the bed shear stress. The threshold value τ_e should therefore be regarded as an apparent critical shear stress for erosion, which is (much) smaller than the true critical shear stress for erosion τ_{cr}.

It is further noted that the magnitude of the apparent critical shear stress for erosion τ_e also depends on the methodology deployed to measure τ_e, as demonstrated by Gust and Müller (1997) in studies of the shear stress distribution in a variety of erosion devices.

The magnitude of the true critical shear stress for erosion τ_{cr} is determined by the strength of the bed-forming flocs and the adhesion between the bed and these flocs. Floc erosion occurs when the mean bed shear stress is of same order of magnitude or a little larger than the apparent threshold for erosion τ_e.

surface erosion
When the mean flow- and/or wave-induced stresses are considerably larger than the true critical shear stress for erosion τ_{cr}, large layers of sediment are eroded and mobilised, a process referred to as surface erosion. Because of continuity,

eroding layers of flocs have to be replaced by water. This is possible as surface erosion is a slow process and this mode of erosion can therefore be considered as a drained process.

During this erosion process, the bed becomes (locally) over-consolidated and the inflow of water decreases the strength of the bed, but not below τ_{cr}. The weakened bed is then easily eroded.

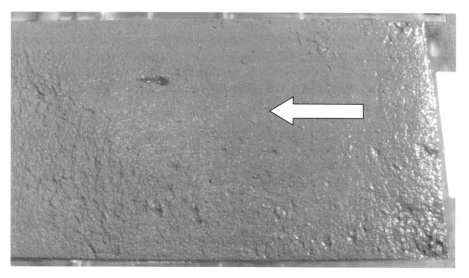

a) Surface erosion on IJmuiden mud; c_u = 60 kPa.

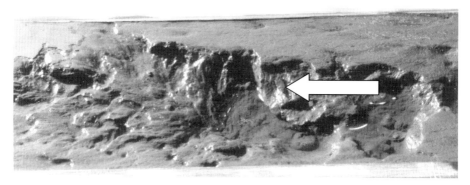

b) Mass erosion on Lake Ketel mud; c_u = 154 Pa.

Fig. 9.10: Photographs of sediment surface after surface and mass erosion; white arrows give flow direction (U = 0.94 m/s) (Van Kesteren, 2004).

mass erosion

Mass failure occurs when the mean flow- and/or wave-induced stagnation pressures are much larger than the true critical shear stress for erosion τ_{cr} and become of the order of the remoulded shear strength c_u or larger. The deformation rate is so large that pore water cannot compensate for the induced water pressure gradients. Mass erosion is therefore an undrained process.

Fig. 10a and 10b show photographs of the surface of mud samples after surface and mass erosion, respectively. Fig. 9.10a shows surface erosion and swelling of the bed, whereas Fig. 9.10b shows large scale rupture features resembling crack patterns observed with undrained tensile tests (Section 8.3).

With this analysis yields a qualitative classification scheme can be constructed for the various modes of erosion. The elaboration of the various modes of erosion in Section 9.4 and 9.5 allows for a further quantification of this scheme.

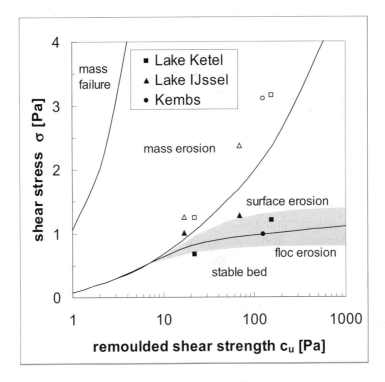

Fig. 9.11: Classification of erosion modes for cohesive sediment; solid symbols refer to surface erosion and open symbols refer to mass erosion.

This erosion scheme is presented in Fig. 9.11, based on a series of measurements by the authors on fresh water mud from Lake Ketel and Lake IJssel (The Netherlands) and Kembs reservoir (France). The material properties are given on the x-axis in the form of the remoulded shear strength of the sediment c_u. The eroding stresses σ are given on the y-axis. Floc and surface erosion are governed by the (mean) bed shear stress $\overline{\tau}_b$, whereas mass erosion and mass failure are controlled by normal and stagnation pressures $\overline{\sigma}_{ii}$.

Mehta (1991) presented a similar analysis where he distinguished between a stable bed, surface erosion and mass erosion in a τ_b - ρ diagram, constructed for organic-rich fresh water mud.

9.4   ENTRAINMENT OF FLUID MUD LAYERS

Entrainment occurs in systems with two layers when a turbulent layer erodes a second non-turbulent (or less turbulent) layer, e.g. Turner (1973). Scarlatos and Mehta (1990); Mehta and Srinivas (1993);and Winterwerp and Kranenburg (1997a) showed that soft mud layers may behave as a viscous fluid, in which case they may be subject to entrainment.

Kranenburg (1994) and Kranenburg and Winterwerp (1997) derived equations for entrainment for two conditions (see also Bruens, 2003):

Case I: Entrainment of a fluid mud layer by the turbulent water layer above,

Case II: Entrainment by a turbulent fluid mud layer (i.e. HCMS or turbidity current) of the water layer above.

This is further depicted in Fig. 9.12, which also contains the various definitions used in the derivation of the entrainment equation. For Case I, the upper turbulent layer erodes the lower mud layer, and the sediment-water interface is lowered: the sediment concentration in the lower mud layer remains constant, and the concentration in the upper layer slowly increases. Case I conditions occur when wind-induced surface shear stresses generate entrainment, in the case of the entrainment of patches of fluid mud in local depressions (navigation channels) by turbulent ambient water, and in many laboratory experiments.

For Case II, the lower turbulent mud layer entrains water (or LCMS) from the upper layer, and the sediment-water interface rises: the sediment concentration in the lower layer decreases, and remains constant in the upper layer. Case II conditions occur in the case of flowing mud layers, either in a river channel

(Loire, e.g. Le Hir, 1997) or on slopes, and in the case of sub-marine turbidity currents.

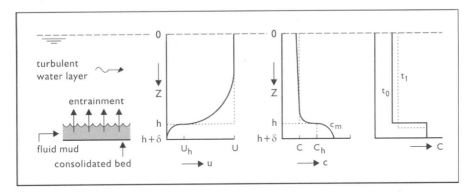

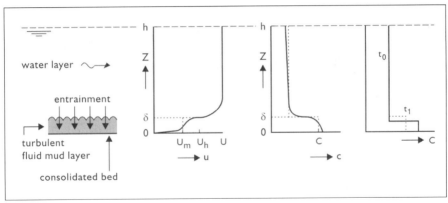

Fig. 9.12: Case I (upper panel: upper layer turbulent) and Case II (lower panel: lower layer turbulent) entrainment.

The entrainment equation is obtained by integrating the 1DV momentum equation, sediment balance equation and turbulent kinetic energy equation (TKE) over the turbulent layer with thickness h (upper layer water depth, Case I) or δ (lower layer mud thickness, Case II), using the equation of state (see also Section 2.1):

$$\frac{\partial u}{\partial t} + \frac{\partial}{\partial z}\overline{u'w'} + \frac{1}{\rho_0}\frac{\partial p}{\partial x} + \frac{2}{W}\frac{\tau_w}{\rho_0} = 0 \qquad (9.10)$$

$$\frac{\partial c}{\partial t} + \frac{\partial}{\partial z} w_s c + \frac{\partial}{\partial z} \overline{w'c'} = 0 \tag{9.11}$$

$$\frac{\partial k}{\partial t} + \overline{u'w'} \frac{\partial u}{\partial z} + \frac{2}{W} \frac{\tau_w}{\rho_0} u - \alpha \frac{g}{\rho_0} \overline{w'c'} - \frac{\partial D}{\partial z} + \varepsilon = 0 \tag{9.12}$$

$$\rho = \rho_0 + \frac{\rho_s - \rho_0}{\rho_s} c = \rho_0 + \alpha c \tag{9.13}$$

In these equations, and the ones to follow in this section, a prime denotes turbulent quantities, turbulence-averaged quantities are denoted by non-capital symbols, and depth- and turbulence-averaged quantities are denoted by capital symbols. All symbols are defined in Chapter 2 and in Appendix E. The fifth term in (9.12) is a redistribution (diffusion) term for TKE. The settling velocity w_s may be affected by hindered settling effects (Section 5.2.2).

We have added the effects of side-wall friction to enable the analysis of laboratory flume experiments; W is the width of the flume and the side-wall friction τ_w is modelled as $\tau_w = \lambda \rho_0 u^2$, where λ is a friction coefficient.

Turbulence production takes place mainly at the water surface (Case I) or at the rigid, consolidated bed (Case II). The boundary conditions at the water surface or rigid bed read:

$$w_s c + \overline{w'c'} = 0 ; \quad \overline{u'w'} = u_*^2 ; \quad D = 0 \tag{9.14}$$

The boundary conditions at the water-sediment interface at δ^+, i.e. just beyond the interface, read:

$$w_s c = \overline{w'c'} = 0 ; \quad \overline{u'w'} = u_\delta^2 ; \quad D = 0 ; \quad k = 0 ; \quad c = c_\delta ; \quad u = U_\delta \tag{9.15}$$

where subscript $\cdot \delta$ refers to interfacial conditions and U_δ is the flow velocity at $z = \delta$ resulting from viscous shear flow in the lower (Case I) or upper (Case II) layer, and u_δ is the friction velocity at $z = \delta$ (interfacial shear).

We elaborate Case I, and only present the results for Case II, as the derivation of the entrainment equation for Case II is similar to that of Case I. Integration of the momentum equation (9.10) over the turbulent layer with thickness h, using the boundary conditions (9.14) and (9.15) yields:

$$\frac{d}{dt}hU \approx u_*^2 - u_\delta^2 + U_\delta \frac{dh}{dt} - \frac{h}{\rho_0}\frac{dp}{dx} \qquad (9.16)$$

where U is the mean velocity in the mixed layer (Fig. 9.11). The third term on the right-hand side of (9.16) represents transfer of momentum from the mud layer by entrainment. Integration of the mass balance equation (9.11) over time t and the turbulent layer h, using the boundary conditions (9.14) and (9.15) yields:

$$\int_0^h cdz \equiv hC = h_0 C_0 + \int_0^h c_b(z)dz \qquad (9.17)$$

where c_b is the concentration in the mud layer. Integration of the TKE-equation (9.12) yields:

$$\frac{d}{dt}\int_0^h kdz + \int_0^h\left(\overline{u'w'}\frac{\partial u}{\partial z}\right)dz - \frac{2\lambda}{W}hU^3 - \alpha\frac{g}{\rho_0}\int_0^h\overline{w'c'}dz + \int_0^h\varepsilon\,dz = 0 \quad (9.18)$$

$$\quad\;\; \text{I} \qquad\qquad \text{II} \qquad\qquad \text{III} \qquad\qquad \text{IV} \qquad\quad \text{V}$$

Kranenburg and Winterwerp (1997) show that k cannot be modelled as u_*^2 in the case of fluid mud flow, but that term I in (9.18) should be modelled as:

$$\text{term I} \sim \frac{d}{dt}\left(hU\frac{dh}{dt}\right) \qquad (9.19)$$

The shear production term II is obtained by integration in parts and becomes, using the momentum equation (9.10) and (9.16) to account for the pressure term:

$$\text{term II} \approx -\left[U_\delta^2(U - U_\delta) + u_*^2(U - U_\delta) + \frac{1}{2}(U - U_\delta)^2 \frac{dh}{dt} + \beta u_s^3 \right] \qquad (9.20)$$

where β is a coefficient of about unity. The last term in (9.20) is the so-called stirring term, which becomes important at large Richardson numbers: $u_s^3 = u_*^3 + u_\delta^3 + 2hu_w^3/W$, with $u_w^2 = \lambda U^2$.

Equ. (9.20) is multiplied with z, integrated (in parts) over the turbulent layer with thickness h and substituted into term IV:

$$\text{term IV} \approx -\alpha \frac{g}{\rho_0} \left[\frac{1}{2} h(c_\delta - C) \frac{dh}{dt} + hw_sC \right] \qquad (9.21)$$

The first term in the right-hand side of (9.21) can be regarded as work to bring sediment in suspension and the second term is the work to keep the sediment in suspension.

The dissipation term V is neglected, as it is difficult to model; it is implicitly accounted for in the coefficients of the final entrainment equation.

Next, the effects of a yield strength τ_y of the mud layer is introduced by relating the various shear velocities to a "yield velocity" $u_y^2 = \tau_y/\rho$. The entrainment equation for Case I then becomes:

$$c_q \frac{d}{dt}\left(hU \frac{dh}{dt} \right) +$$

$$-c_s\left[2\langle u_\delta^2 - u_y^2 \rangle(U - U_\delta) + \left((U - U_\delta)^2 - c_y u_y^2\right)\frac{dh}{dt} \right] + \qquad (9.22)$$

$$+c_s'\langle u_*^2 - u_y^2 \rangle(U - U_\delta) - c_\sigma\langle u_s^2 - u_y^2 \rangle u_s +$$

$$-2c_w \frac{h}{W}\langle u_w^2 - u_y^2 \rangle U + B\frac{dh}{dt} + 2\frac{\alpha C}{\rho_0}gw_sh = 0 \qquad \text{for } \frac{dh}{dt} > 0$$

in which B is the total buoyancy, i.e. $B = \alpha gh(c_\delta - C)/\rho_0$, and the terms with angular brackets are set to zero when the expression between these brackets is negative. The various coefficients were established by Kranenburg and Winterwerp (1997) as: $c_q = 5.6$; $c_s = c_s' = 0.25$; $c_\sigma = 0.42$; $c_w = 0.07$.

In the case of no viscous effects ($u_\delta = U_\delta = 0$), thus no stirring ($u_s = 0$), as in the case of small Richardson numbers, the equilibrium solution (i.e. at large time) to 9.23 yields an explicit expression for the initial entrainment velocity w_e, hence initial entrainment rate E_*:

$$E_* = \frac{1}{u_*}\frac{dh}{dt} = \frac{w_e}{u_*} = \left(\frac{2c_s}{c_q + Ri_*}\right)^{1/2} \tag{9.23}$$

where we have defined $Ri_* = B/u_*^2 = (\rho_b - \rho_0)gh_0/\rho_0 u_*^2$, which is the bulk Richardson number. Note that shear is generated at the water surface (Case I).

At large Ri_* the stirring term u_s becomes important, and (9.22) becomes:

$$E_* = \frac{1}{u_*}\frac{dh}{dt} = \frac{w_e}{u_*} = c_\sigma \frac{u_s^3/u_*}{B} \approx \frac{c_\sigma}{Ri_*} \tag{9.24}$$

where u_* is measured at the water-mud interface.

In flumes with side-wall friction, the equilibrium solution to (9.22) at large Richardson numbers reads (e.g. Winterwerp and Kranenburg, 1997b):

$$E_* = \frac{1}{u_*}\frac{dh}{dt} = \frac{w_e}{u_*} = \frac{c_w}{(2\lambda h/W)(c_q + Ri_*)} \tag{9.25}$$

The effects of large Ri_* and side-wall friction explain the apparent inconsistencies in literature between $w_e \propto Ri_*^{-1}$ or $w_e \propto Ri_*^{-1/2}$.

For Case II we give the results without derivation – the reader is referred to Kranenburg (1994) for further details:

$$
c_q \frac{\mathrm{d}}{\mathrm{d}t}\left(\delta U_m \frac{\mathrm{d}\delta}{\mathrm{d}t}\right) +
$$

$$
-c_s\left[2\langle u_\delta^2 - u_y^2\rangle(U_\delta - U_m) + \left((U_\delta - U_m)^2 - c_y u_y^2\right)\frac{\mathrm{d}\delta}{\mathrm{d}t}\right] +
$$

$$
+c_s'\langle u_*^2 - u_y^2\rangle(U_\delta - U_m) - c_\sigma\langle u_s^2 - u_y^2\rangle u_s +
$$

$$
-2c_w \frac{\delta}{W}\langle u_w^2 - u_y^2\rangle U_m + B\frac{\mathrm{d}\delta}{\mathrm{d}t} + 2\frac{\alpha C}{\rho_0} g w_s \delta = 0 \qquad \text{for} \quad \frac{\mathrm{d}\delta}{\mathrm{d}t} > 0
$$

$$(9.26)$$

Equation (9.26) yields the same asymptotic solutions as (9.22), i.e. equ.'s (9.23), (9.25) and (9.25), but h is replaced by δ, and u_* is now related to the bed shear stress.

The entrainment model for Case I (equ. 9.23) was used to re-analyse the entrainment experiments by Kantha et al. (1977) on stable fresh-saline water two-layer systems, as presented in Fig. 9.13. It is shown that equ. (9.22) predicts the observations properly, including the change in E_* vs Ri_* slope. This change in character with increasing Ri_* is fully explained by the effects of viscosity and side-wall friction. Details of this analysis are given in Winterwerp and Kranenburg (1997b).

We have also plotted the initial entrainment rates for Case I entrainment as measured by Winterwerp and Kranenburg (1997b) and for Case II entrainment as measured by Bruens (2003). These measurements were carried out in a rotating annular flume (Appendix C) with kaolinite and mud from the Caland Canal (entrance to the Port of Rotterdam). It is shown that these entrainment rates agree quite well, in spite of their difference in nature, and follow the $w_e \propto Ri_*^{-1}$ law, indicating that viscous effects and side-wall friction were important in these experiments.

Winterwerp and Kranenburg (1997a) and Bruens (2003) showed that the entrainment rates of fluid mud-water systems and of salt-fresh water systems are very similar. This implies that fluid mud layers behave as viscous fluids, at least during part of their existence. This was also observed by Scarlatos and Mehta (1990) and Mehta and Srinivas (1993).

The effects of yield stress on the entrainment are illustrated in Fig. 9.14, showing decay in entrainment rate with time as a result of increasing strength with depth. The experiment against which the entrainment model is compared was carried out in a rotating annular flume with mud from the Caland Canal.

Results of other, similar experiments are given in Winterwerp and Kranenburg (1997b), including the results of a "tidal" experiment, during which the applied shear stress was varied sinusoidally with time.

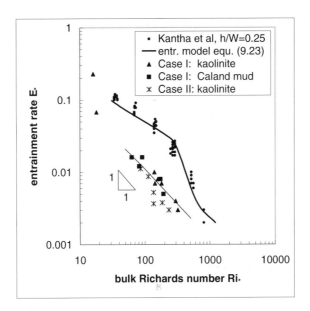

Fig. 9.13: Measured and predicted (equ. 9.23) initial entrainment rates
(after Winterwerp and Kranenburg, 1997b and Bruens, 2003).

Winterwerp (1999) showed that Case I entrainment experiments can also be simulated with a numerical model with a standard k-ε turbulence model, taking into account the effects of buoyancy destruction and hindered settling. The success of those simulations wes attributed to the fact that in Case I entrainment, turbulence is mainly externally produced, i.e. not in the mud layer, but at the water-mud interface.

For Case II entrainment, turbulence production takes place mainly in the lower mud layer and at the water-mud interface. The standard k-ε turbulence model is not appropriate for these conditions because of, for instance, low-Reynolds number effects, and the effects of internal waves, which transport momentum, but not sediment.

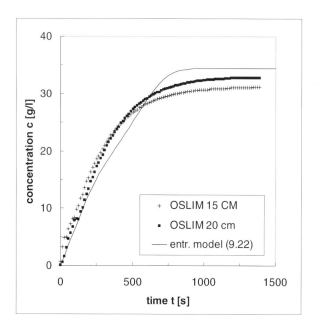

*Fig. 9.14: Time-variation of suspended sediment concentration above a fluid
mud layer as a result of entrainment of that layer
(after Winterwerp and Kranenburg, 1997b).*

9.5   EROSION AS A DRAINED/UNDRAINED PROCESS

9.5.1 FLOW-INDUCED EROSION

In this section we discuss floc erosion, surface erosion and mass erosion
induced by flow.

Floc erosion is a process, in which flocs are disrupted individually from the
bed. The floc erosion rate is a function of:

- the number of flocs exposed – at the scale of the flocs the sediment bed is
  quite irregular, as sketched in Fig. 9.1b and shown in Fig. 9.15,
- the strength of the flocs adhering to the bed, which is measured by τ_{cr}, as
  discussed in Section 9.3, and
- the probability density function of the disruptive forces, i.e. peaks in the
  turbulent stresses (e.g. Section 2.2.2 and 5.4.1).

Theoretically it should be possible to derive a formula for floc erosion on the
basis of the last two issues, though the irregular character of the bed surface is a

complicating factor. Moreover, pdf's for disruptive stresses are only known for flat, impermeable walls. Therefore, such an erosion formula has not yet been developed, and at present we advocate using the parameterised erosion formula by Partheniades (9.2):

$$E = M\left(\frac{\tau_b - \tau_e}{\tau_e}\right) \quad \text{for} \quad \tau_b > \tau_e \tag{9.27}$$

where M is an erosion rate parameter, τ_b the mean bed shear stress and τ_e the apparent critical shear stress for erosion. Both M and τ_e should account for the stochastic character of both the sediment surface and the bed shear stress. As mentioned, τ_e is (much) smaller than τ_{cr} and is in the range of 0.2 Pa $< \tau_e <$ 0.8 Pa, whereas 0.01·10^{-3} kg/m^2/s $< M <$ 0.5·10^{-3} kg/m^2/s.

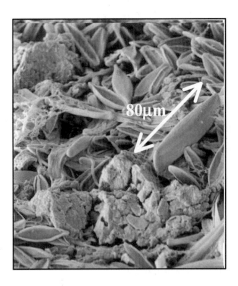

Fig. 9.15: Electron-microscopy photograph of sediment sample from intertidal mud flat, showing mineral particles, diatoms and EPS-filaments – see also Chapter 10 (courtesy by D.M. Paterson, 2003).

Surface erosion can occur when the bed is at or below its critical state, i.e. the bed water content has to increase before failure (erosion) can occur. To increase the bed water content, water has to enter the bed, in case of surface erosion by inflow through the bed surface. This process is known as swelling, and the location of the swelling front δ_s, increasing with time t, is a function of the swelling properties of the sediment, which can be described as a diffusion process:

$$\delta_s = \sqrt{\pi c_v t} \tag{9.28}$$

where c_v is the consolidation coefficient in vertical direction (Section 7.3 and 8.1). All sediment above the swelling front is below its critical state and can be eroded if the hydrodynamic stresses are large enough. This implies that the maximum erosion velocity V_E is given by the propagation velocity of the swelling front:

$$V_E = \frac{d\delta_s}{dt} = \frac{\pi c_v}{2\delta_s} \approx \frac{c_v}{\delta_s} \tag{9.29}$$

At the swelling front, the bed strength is at its critical state, equal to the sediment's remoulded shear strength c_u (e.g. Fig. 3.16). At the bed surface, the bed strength decreases as a result of swelling and the residual strength is determined by the strength of the individual flocs and their adherence to the bed. This strength is denoted by τ_{cr}, which reflects the stress necessary to disrupt flocs from the bed. This threshold value is referred to as the true critical shear stress for erosion (e.g. Section 9.3), which is a function of the sediment properties, for instance through the empirical relation by Smerdon and Beasley (1959), e.g. equ. (9.7).

We assume a linear distribution of the strength within the bed as an approximation of the actual bed strength distribution, as depicted in Fig. 9.16. The erosion depth δ_e then becomes:

$$\delta_e = \delta_s \frac{\tau_b - \tau_{cr}}{c_u - \tau_{cr}} \approx \delta_s \frac{\tau_b - \tau_{cr}}{c_u} \quad \text{as} \quad c_u \gg \tau_{cr} \tag{9.30}$$

where τ_b is the bed shear stress.

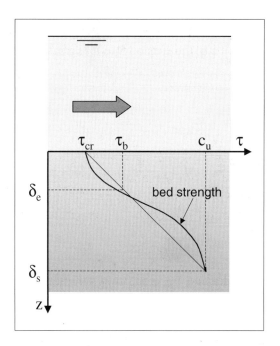

Fig. 9.16: Sketch of strength distribution in a swelling bed under erosion.

The swelling depth δ_s scales with the size of the bed-forming flocs. Moreover, in a continuum approach, the swelling depth should be at least an order of magnitude larger than the size D_f of the primary particles (flocculi), including their interstitial water. So δ_s is approximated as $\delta_s = 10D_f$. We therefore hypothesise that the swelling depth scales as:

$$\delta_e = D_f \left(\frac{10D_{50}(1+e_0)}{D_f} \right)^\alpha = D_f \left(\frac{10D_{50}(1+W_0\,\rho_s/\rho_w)}{D_f} \right)^\alpha =$$

$$= D_f \left(\frac{10D_{50}}{\phi_{s,0} D_f} \right)^\alpha \tag{9.31}$$

in which e_0, W_0 and $\phi_{s,0}\left(=1/(1+W_0\rho_s/\rho_w)\right)$ are the initial void ratio, water content and volume concentration. From a series of experiments we conclude that $\alpha \approx 1$ (see below).

Substituting (9.31) and (9.30) into equ. (9.29) yields an expression for the erosion velocity V_e [m/s], c.q. the erosion rate E [kg/m^2/s] in the case of surface erosion:

$$V_e = \frac{d\delta_e}{dt} = \frac{c_v \phi_{s,0}}{10 D_{50}} \frac{\tau_b - \tau_{cr}}{c_u} \qquad \text{or}$$

$$E = \frac{c_v \phi_{s,0} \rho_{dry}}{10 D_{50}} \frac{\tau_b - \tau_{cr}}{c_u} = M_E (\tau_b - \tau_{cr}) \qquad \text{for} \quad \tau_b > \tau_{cr} \qquad (9.32)$$

$$\text{where} \qquad M_E = \frac{c_v \phi_{s,0} \rho_{dry}}{10 D_{50} c_u}$$

in which we have also defined the erodibility parameter M_E. Note that c_u, $\phi_{s,0}$ and ρ_d vary with depth, but that all other parameters are constant when the sediment composition does not change. The erosion formula (9.32) is very similar to the so-called Partheniades' formula for erosion (9.2). The erosion rate V_e appears to scale with the consolidation coefficient c_v, hence with k (e.g. Section 7.3), in contrast to the erosion rate of non-cohesive sediment (e.g. Section 9.2).

Table 9.2: Parameters erosion experiments in erosion flume.

	LI [-]	D_{50} [μm]	c_v [m^2/s]	c_u [kPa]	W_0 [%]	τ_{cr} [Pa]	σ_m [Pa]	M_E [m/Pa·s]
Lake Ketel	0.54	7.3	$2 \cdot 10^{-8}$	0.154	214.9	1.20	3.15	$0.23 \cdot 10^{-6}$
Lake Ketel	3.28	7.3	$2 \cdot 10^{-8}$	0.022	313.9	0.76	1.24	$1.43 \cdot 10^{-6}$
IJmuiden	0.27	2.5	$5 \cdot 10^{-8}$	0.069	209.3	1.28	2.37	$2.26 \cdot 10^{-6}$
IJmuiden	3.24	2.5	$5 \cdot 10^{-8}$	0.017	302.6	1.01	1.26	$6.61 \cdot 10^{-6}$
Kembs	1.6	21	$5 \cdot 10^{-8}$	0.126	83.7	2.44	3.11	$1.60 \cdot 10^{-6}$

Erosion formula (9.32) has been validated through experiments in an erosion flume at Delft Hydraulics, the results of which have not been published before. The experiments were executed with sediments from Lake Ketel and the Port of IJmuiden in The Netherlands, and from the Kembs Reservoir in France. The erodibility parameter M_E has been established from measuring the erosion rate as a function of bed shear stress. Table 9.2 summarises some of the experimental results. The measured results are compared with equ. (9.32) in Fig. 9.17, showing a favourable agreement between data and predictions.

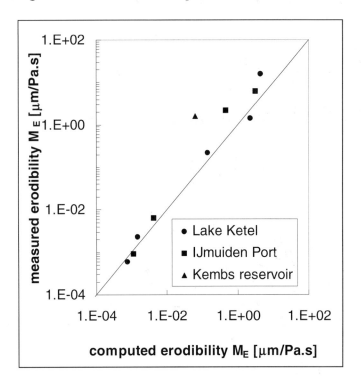

Fig. 9.17: Measured and computed erodibility M_E.

We have defined surface erosion as a drained process in which water gets time to enter the bed, and failure occurs at the critical state of the bed. This is not always the case: failure can also occur under undrained conditions. In that case, two modes can occur:

- Crack formation, i.e. local failure of the bed (Fig. 9.1), resulting in mass erosion. Cliff erosion, as often observed in intertidal areas due to wave attack, is a form of mass erosion. A typical result of cliff erosion is shown in Fig. 9.18.
- Liquefaction, i.e. a very large part of the bed becomes fluid, and may flow as a turbidity current. Such turbidity currents are often observed on the continental slope (e.g. Section 8.4).

Fig. 9.18: Cliff (mass) erosion on mud flat (photo by E.A. Toorman).

Mass erosion is related to flow-induced pressures. Experiments (Van Kesteren, 2004) suggest that mass erosion can occur when:

$$\sigma_m \equiv \frac{1}{2}\rho U^2 > \left(2 \text{ to } 5\right)c_u \tag{9.33}$$

where we have defined the stagnation stress for the onset of mass erosion σ_m. The flow velocity U is the flow velocity of the eroding fluid. Because of the stochastic behaviour of mass erosion, no mass erosion rate can be given. Empirical formulae for the mass erosion rate for engineering application can of course be established, but are highly site-specific and applicable for the experimental test conditions only.

We have included the various data of Table 9.2 in Fig. 9.11, distinguishing between four modes of failure, c.q. erosion:
1. Full liquefaction at large σ and small c_u,
2. Mass erosion at moderate to large σ and small to moderate c_u,
3. Surface erosion at moderate to large $\sigma(\tau_b)$ and moderate to large c_u,
4. Floc erosion at the lower bounds of the surface erosion regime, and
5. Stable bed at small $\sigma(\tau_b)$ and moderate to large c_u.

Such erosion diagrams are site–specific and specific for the local sediment composition as well (including pore water properties).

9.5.2 WAVE-INDUCED EROSION

It is noted that our understanding of wave-induce erosion is still very limited. Further fundamental research is urgent to substantiate the fairly qualitative picture described in this section.

Waves may cause three forms of failure:
1. Surface erosion,
2. Mass erosion,
3. Massive liquefaction.

The third failure mechanism is the result of cyclic loading of the bed by wave-induced stresses; it has been described in detail in Section 8.4. The liquefied layer may be entrained by the upper water layer (e.g. Section 9.4), or may develop into a turbidity current (Section 9.5.3) on, for instance, the continental slope.

The liquefied mud may damp waves (e.g. Section 8.5) and/or interact with the turbulent flow field. In the latter case, the flow may become saturated (auto-saturation, e.g. Section 6.3.1), resulting in a very stable situation.

In the case of surface erosion, it should be appreciated that wave-induced bed shear stresses exceed flow-induced bed shear stresses easily by an order of magnitude. Moreover, as a result of non-linear effects, the net bed shear stress induced by waves and currents is larger than their arithmetic sum (e.g. Section 2.3). This was nicely depicted by Soulsby et al. (1993), as shown in Fig. 9.19.

The friction factors for hydraulically rough and hydraulically smooth conditions were given in Section 2.3.

Fig. 9.20 gives an example of the magnitude of wave-induced bed shear stresses for locally generated waves on shallow (intertidal) areas. The method to compute wave height and period as a function of wind speed, water depth and fetch is due to Groen and Dorrestein (1976). We observe that, in particular at water depths of a few dm, wave-induced bed shear stresses exceed a few 0.1 Pa easily at already very moderate wind conditions.

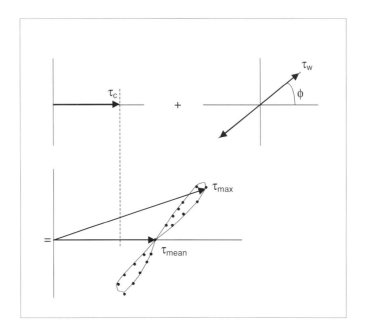

Fig. 9.19: Mean and maximum bed shear stress as a result of waves and currents (redrawn from Soulsby et al., 1992).

In case of very large wave-induced disruptive stresses, pore water cannot flow into the bed and undrained failure takes place. This process is referred to as mass erosion by waves, and similar to (8.35) this occurs when:

$$\sigma_{w,m} \equiv \frac{1}{2}\rho V^2 > (2 \text{ to } 5)c_u \tag{9.34}$$

where $\sigma_{w,m}$ are the wave-induced stagnation stresses and V is the velocity of the approaching water mass in the crest of the waves (Van Kesteren, 2004). In general, this velocity equals the wave velocity c:

$$V = c = \sqrt{\frac{g}{k}\tanh\{kh\}} \tag{9.35}$$

with the water depth h, the wave number $k = 2\pi/\lambda$, and λ is the wave length. For the impact of breaking waves, an empirical expression is commonly used to model the eroding stresses:

$$V = \alpha\sqrt{g(h + H_{rms}/2)} \qquad\qquad (9.36)$$

where α is an empirical parameter with values $1 < \alpha < 3$ (mean value $\alpha = 1.5$) and H_{rms} is the rms-value of the wave height, e.g. Wood (1997).

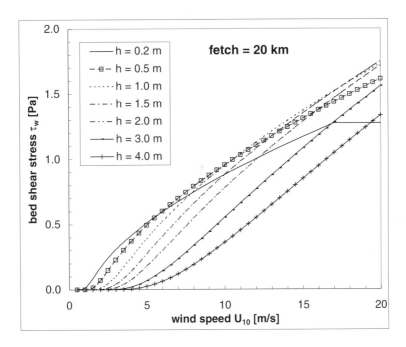

Fig. 9.20: Example of wave-induced bed shear stresses in shallow water computed with Groen and Dorrestein (1976).

9.5.3 EROSION BY TURBIDITY CURRENTS

To the knowledge of the authors, no studies have been published on the erosion of cohesive beds by (cohesive) turbidity currents. Yet, we expect that surface erosion by turbidity currents will be affected by a reduction in water available to fluidise the cohesive bed (i.e. a decrease in drainage). This phenomenon was observed during tests on sand-water mixtures by Winterwerp et al. (1992), and was referred to as "hindered erosion" in analogy to hindered settling.

Let us assume that the flow of water out of the turbidity current, necessary to replace particles eroding from the underlying bed, can be modelled as Poiseuille flow. Thus the flow rate is inversely proportional to the square of the pore diameter within the turbidity current, i.e. with $\left(1-\phi_f\right)^2$ and inversely proportional to the viscosity of the turbidity current, i.e. with $\left(1+2.5\phi_f\right)^{-1}$, e.g. Section 5.2.2. Hence, we anticipate that the hindered surface erosion rate E_{tc} by turbidity currents is affected by a volumetric effect and the mixture viscosity:

$$E_{tc} = E_0 \frac{\left(1-\phi_f\right)^2}{1+2.5\phi_f} \tag{9.37}$$

where E_0 is the surface erosion rate by clear water and ϕ_f is the volumetric concentrations of the flocs in the turbidity current (see also Section 5.2.2).

Note that (9.37) has not been validated at present. Moreover, a large unknown parameter in this formula is the bed shear stress, governing E_0, as friction between turbidity currents and an eroding bed is poorly understood.

Next to surface erosion, mass erosion is a likely mechanism, as flow velocity and fluid density are large in general (e.g. equ. (9.34)). Also the banks of a turbidity current are sensitive to erosion. This is observed frequently in the Yellow River, where the river's thalweg may change location during one flood only (Kriele et al., 1998).

In the literature, a number of models have been presented on turbidity currents, e.g. Parker et al. (1986, 1987), Zeng and Lowe (1997a,b); and Skene et al. (1997). The reader is also referred to Takahashi (1991) for an overview on debris flows. These authors have highlighted the importance of the vertical exchange processes, i.e. erosion, sedimentation c.q. consolidation and entrainment, on the behaviour of turbidity currents. Without going into much detail, it is interesting to contemplate on the basis of the ideas described in Chapter 6, 7 and 9 how turbidity currents would be affected by these processes.

Let us examine Fig. 6.5 again, which is repeated as Fig. 9.21 for convenience. The behaviour of turbidity currents is given in the right part of the graph. If the turbidity current erodes the bed (or the riverbank), its sediment concentration increases, and we move to the right of the saturation curve, i.e. the current becomes more stable, as Ri_f decreases. Moreover, as the current's total buoyancy increases (the sediment concentration increases more than the current's thickness), it starts to accelerate, eroding more sediment, etc. However, as bed erosion is progressively more hindered with increasing sediment concentration, the current starts to erode its banks severely. This is the reason that turbidity currents can be so persistent and devastating.

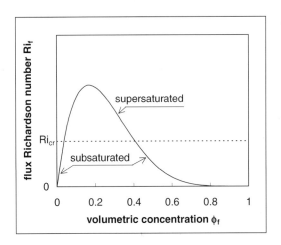

Fig. 9.21: Schematic variation of flux Richardson number with volumetric concentration.

Entrainment of ambient water by the turbidity current decreases the current's sediment concentration, and we move to the left of the saturation curve and Ri_f increases, and the flow becomes less stable. However, an increase in Ri_f yields a decrease in entrainment rate (e.g. Fig. 9.13). Hence, this entrainment effect is damped. Moreover, the total buoyancy of the current remains constant, at least initially: the current accelerates nor decelerates.

Sedimentation, c.q. consolidation also causes a decrease in the current's sediment concentration, again resulting in a shift to less stable conditions with higher Ri_f. Moreover, sedimentation and consolidation increase with increasing Ri_f, in the sense that an increase in Ri_f implies less kinetic energy for vertical mixing. Hence the damping effect, which occurs with entrainment, does not occur. Furthermore, the total buoyancy of the turbidity current decreases, as the current's sediment concentration decreases more than the current's thickness. As a result, the current decelerates, and net sedimentation (i.e. the difference between gross sedimentation and erosion) increases further. This is a snowball effect, and the turbidity current can come to a full stop in a remarkably short time. However, further research is required to validate these hypotheses.

10. BIOLOGICAL EFFECTS

Living organisms, their behaviour and secretions can have major effects on the stability of fine cohesive sediment deposits, in particular in intertidal areas (e.g. Paterson, 1997). De Wolf (1990) presented the schematised drawing in Fig. 10.1 of zoobenthos activity in the North Sea, showing a variety of species that rework the bed.

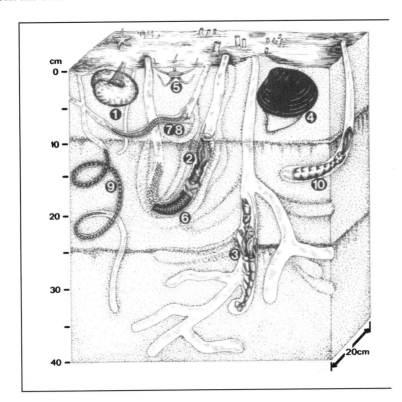

1 Sea potato (*Echinocardium cordatum*)
2 Parchment Worm (*Chaetopterus* sp.)
3 Mud Shrimp (*Calianassa sp.*)
4 Black Clam (*Arctica* sp.)
5 Brittle Star (*Amphiura* sp.)

6 Polychaete worm (*Gattyana* sp.).
7 Polychaete worm (*Glycera* sp.)
8 Rag worm (*Nereis* sp.)
9 Polychaete worm (*Notomastus* sp.)
10 Spoon worm (*Echiurus* sp.)

Fig. 10.1: Drawing of zoobenthos in North Sea (De Wolf, 1990).

Widdows and Brinsley (2002) distinguish between bio-stabilisers and bio-destabilisers, as shown in the schematised diagram of Fig. 10.2. Organisms that operate individually, like mysids (shrimp-like animals), hydrobia (small snails) and burrowing bivalves, like cockles destabilise the bed. Moreover, they may also rework and mix the sediment throughout the top layer of the bed. Other organisms stabilise the bed.

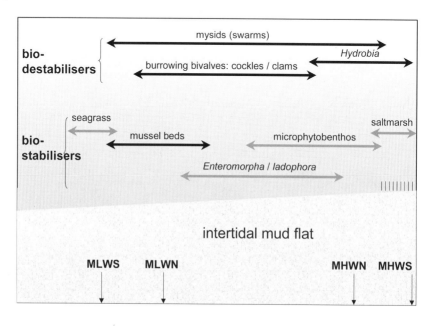

Fig. 10.2: Spatial distribution of some key biota acting as 'bio-engineers' on the intertidal shore: bio-stabilisers and bio-destabilisers (Widdows, 2002).

In this chapter we discuss some aspects of the role of biology on the dynamics of cohesive sediment in the marine environment. This chapter is not meant to present a full oversight of the extensive literature on these effects, but to give the reader a feeling of its importance and some literature for further information. Moreover, we limit ourselves to intertidal areas in the moderate climate areas of north-western Europe, and the reader should realise that most systems have their own specific eco-system interacting with the sediment dynamics (such as the role of mangroves in the stability and development of mud coasts, e.g. Robertson and Alongi, 1992).

In the following sections we discuss the effects of the following bio-engineers:
* vegetation,
* bio-deposition,

- bio-stabilisation and
- bio-destabilisation.

10.1 THE ROLE OF VEGETATION

Salt marshes are found on intertidal areas between MHWN and MHWS and are characterised by dense vegetation around a network of small channels and gullies, which drain the marsh. However, local scour is often observed at the borders of the dense vegetation. Salt marsh vegetation is zoned, depending on local salinity, inundation times and oxygen levels. At the higher bed elevations, which are flooded for less than 10 % of the time, macrophytes appear, such as *Puccinellia* and higher on the salt marsh *Aster, Limonium* and *Artemsia*.

The trapping efficiency on salt marshes is very large (e.g. Stumpf, 1983): almost all sediments transported over salt marshes are trapped. This sediment can become available to the system upon erosion of the salt marsh, for instance by wave attack during storm conditions.

Around MHWS the vegetation becomes less dense. Here we find for instance *Salicornia* and *Spartina*, which is often growing in patches. These patches also trap sediment efficiently, as a result of which they form small elevations in bed level (Pethick et al., 1990). These mounds in turn affect the water movement, causing local scour around the *Spartina* patches (see Fig. 10.3).

It should be noted that the entire intertidal area is strongly affected by mutual interactions between the hydrodynamics, sediment transport patterns and biology (e.g. Paterson and Black, 1999 and Widdows and Brinsley, 2002).

Mudflats are found in the zone between MLWN and MHWN, which is characterised by extreme dynamic conditions: they are inundated twice a day and exposed to solar heating, extreme temperatures, ice, rain, wind and waves. This area is largely unvegetated. Here however, fauna plays an important role.

A small number of macrophytes can live around MLWS and under subtidal conditions, amongst which the seagrass species *Zostera marina* and *Zostera noltii*. *Zostera marina* has 2 cm thick roots and long leaves, whereas *Zostera noltii* has much smaller roots. Both species can stabilise the bed by their root system (e.g. Fig. 10.2).

The seafloor vegetation is not only stabilising the bed, but also protects the bed from erosion and resuspension by reducing turbulence levels in the water column, damping waves and by increasing sedimentation rates. Bouma et al.

(2004) carried out experiments with *Zostera marina* and *Spartia* on a sediment bed in the laboratory at a population density of about 500 plants per m^2. These experiments showed that in the case of *Zostera marina* with their flexible stems, the wave heights and near bottom kinetic energy decreased by not more than a few %. However, in the case of *Spartia*, with their stiff stems, the wave heights and near bottom kinetic energy decreased by 5-10 % per m vegetation. In the latter case, also much fine sediment was trapped, whereas *Zostera marina* catches only very little sediment.

Fig. 10.3: Patches of Spartina at the edge of a salt marsh in the Paulina Polder, Western Scheldt, The Netherlands.

10.2 BIO-DEPOSITION

We distinguish between two effects of organic material on the sedimentation rate of cohesive sediment:
- effect of organic coatings on flocculation processes, and
- effect of pelletisation on particle aggregation.

Van Leussen (1988, 1994) presents a literature survey on the effects of organic coatings on the aggregation rate of cohesive sediment known at that time. This effect would be more or less maximal at organic matter concentrations of 1 to 2 mg/l. As such concentrations are always met in estuaries and coastal areas, it can be expected that organic coatings always play a role.

At larger organic matter concentrations floc sizes may increase rapidly if the environmental conditions are favourable. This may be the case in the deep sea, but most likely not in estuaries and coastal areas: turbulence-induced shear stresses are so large that the fragile flocs, bound together by organic filaments, will fall apart. This was also suggested by Van Leussen (1994) in an analysis of data from Uiterwijk Winkel (1975) and others, who observed that floc sizes in the Rhine-Meuse estuary decrease in the highly dynamic salinity intrusion zone in the estuary's mouth.

There are some examples of increases in floc size in coastal areas such as Cape Lookout Bight (USA, NC), where high sedimentation rates were found of marine "snow". This marine snow consists of flocs with a size of cm's, which are glued by bacterial mucilage, while suspended sediment concentrations are low and flow velocities are high (Wells, 1987).

Filter feeders in general, and suspension feeders, such as mussels in particular (Dittman et al., 1999) are very effective in cleansing the water column. Dankers and Koelemay (1989) estimate the filtering volume by mussels (*Mytilus edulis*) at about 1 m^3/hr per 10 kg of mussels. This would imply that the mussels in for instance the Wadden Sea, The Netherlands, would filter a volume equal to the entire Wadden Sea within 8 to 9 days in a rich mussel year, and within one month in a lean mussel year (Dankers and Koelemay, 1989).

Oost (1995) mapped the occurrence of mussel beds with the local sediment composition and concluded that the surface layer of the seabed is always very muddy in and around mussel beds, e.g. Fig. 10.4 for the eastern Wadden Sea.

Distinction has to be made between mussel cultures and mussel beds (Dankers, 2002):
- Mussel beds are natural fields with mussels, adhered to the seabed. They can become decades old. These beds filter fine sediment and stabilise the bed itself. The mud content around mussel beds is permanently high.
- Mussel cultures consist of mats with mussels, which are removed and restocked regularly (autumn and early spring). These mussel cultures contribute to filtering the seawater, but do not stabilise the bed from a morphological point of view.

The faecal pellets secreted by the mussels are very stable: they have been found in geological deposits. In contrast, pellets by clams, and by oysters in particular, are much less stable, hence do not have a large permanent impact on the sediment dynamics in marine systems.

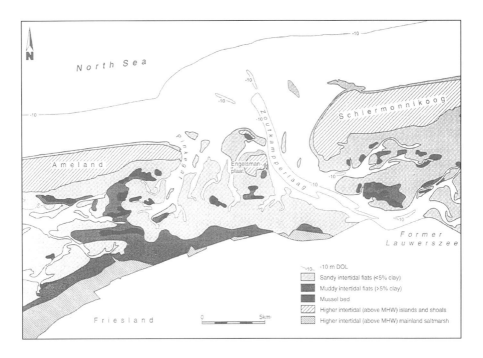

Fig. 10.4: Spatial distribution of mussel beds and mud content in eastern Wadden Sea (after Oost, 1995).

Andersen (2001) studied pelletisation effects by deposit feeders, such as the mud snail *Hydrobia ulvae* and the mud worm *Heteromastus filiformis* on a mudflat in the Lister Dyb tidal basin of the Danish Wadden Sea. The density of *Hydrobia ulvae* amounted up to 40,000 individuals per m^2. Andersen measured the settling velocity of the pelletised sediment and found settling velocities of 3 to 8 mm/s for the snail pellets and 12 to 33 mm/s for the worm pellets.

The pellets of these worms and snails on the seabed, though, are not very stable and can be mobilised and brought into suspension easily at high energetic conditions (spring tide, wave action), probably because of a lack of adhesion between the pellets and the mud bed, c.q. substrate. However, the pellets settle rapidly under calm weather conditions because of their large settling velocity, as a result of enclosed minerals and large size.

It is noted that Andersen (2001) also found that the settling velocity increased with increasing SPM-values. This however may be due to larger particles entrained at higher flow velocity, as a result of which also SPM concentrations increase.

In general faecal pellets consist for 70 to 90 % (by weight) of phylo-silicates (physils), mostly clay minerals and silt, and 4 to 8 % of organic carbon. By pelletisation a large amount of physils and organic matter are compacted and can be deposited in areas where normally the hydrodynamic conditions do not allow suspended fines to settle. The contribution of faecal pellets to the total sedimentation flux can be very large. Biggs and Howell (1984) reported sedimentation fluxes in Delaware Bay about 200 times larger than the fluvial sediment input. The deposition rate by pelletisation in the Wadden Sea was estimated at about 25,000 to 175,000 ton/yr (Verwey, 1952).

10.3 BIO-STABILISATION

As discussed in the preceding section, mussel beds have a large stabilising effect on the sediment and morphology of marine environments. Older mussel beds can even withstand severe storms and ice cover.

Widdows et al. (2002) established the stabilisation of sandy substrate by mussel beds (*Mytilus edulis*). Their results are presented in Fig. 10.5, showing that a small cover of the bed by mussels results in a large increase in erosion rates as a result of local scour around the beds. At larger cover percentages, the bed is stabilised by the mussels.

This is further substantiated by the correlation between mussel beds and mud content in the bed, as observed by Oost (1995) and depicted in Fig. 10.4.

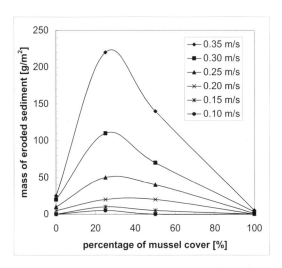

Fig. 10.5: Effect of mussel density on erodibility of sandy substrate (after Widdows et al., 2002).

If not already present in the form of a more or less permanent mudflat, muddy zones develop in March on the upper sandy parts of tidal flats, when microphytobenthos starts growing (De Deckere et al., 2002; De Beer and Kühl, 2001; and De Brouwer et al., 2000). This microphytobenthos is mainly composed of motile benthic diatoms, which may migrate over distances up to 0.5 cm to find nutrients in subsurface layers during the night, for primary production at the surface of the sediment during daylight and to protect themselves against intense light and/or strong tidal currents. The diatoms excrete large quantities of extracellular polymer substances (EPS, mainly polysaccharides), that form mucus covering the sediment particles (Herman et al., 2001; Staats et al. 2001; Riethmuller et al., 2000; Blanchard et al., 2000; Decho, 2000; Reise, 2002). The mucus facilitates the gliding migration of the diatoms.

EPS glues fine sediment particles together (see also Section 3.1.3) smoothing and strengthening the sediment surface structure. The top layer of cohesive sediment is stabilised by these changes, leading to net sediment accretion in a layer that may become several centimetres thick with a high water content (De Deckere et al., 2002). The mud content also builds up in the top 10 cm (Herman et al., 2001). The diatom biomass concentration may become very high, turning the top sediment layer into a coherent greenish mat. EPS is produced continuously as hydrolytic enzymes excreted by bacteria continuously decompose the mucus for their metabolisation of sugars (Herman et al., 2001). It is hypothesised that the amount of secreted polysaccharides is mainly the result of the "overflow metabolism" presumably due to nutrient limitation (Blanchard et al., 2000; Ruddy et al., 1998) and far in excess of the theoretical needs for locomotion only. This would imply that nutrient limitation stimulates sediment stabilisation by microphytobenthos. However, the factors that determine the production of EPS, its nature and its binding capacity, are yet largely unknown (Decho, 2000).
   De Brouwer (2002) distinguished between two types of EPS: the EDTA-extractable EPS appeared to be a very effective stabiliser. The benthic organisms form biofilms on intertidal mudflats, which become abundant in early spring (March/April) and disappear early summer (June/July).

Researchers have tried to link the stabilisation of muddy sediment by microphytobenthos to the chlorophyll *a* content of the top 1-2 mm sediment layer. In many cases this is a rather reliable measure for the biomass concentration. However, the chlorophyll *a* content alone cannot be considered to be a general index for sediment surface stabilisation. The actual stabilisation depends on various factors and is therefore site specific and time dependent

(Riethmuller et al., 2000; Decho, 2000). Blanchard et al. (2000) discuss the possibility of differences in the benthic diatom species composition that may be connected to differences in the physical environment.

Tolhurst et al. (2002) carried out laboratory experiments on the stabilising effect of EPS. They added various amounts of commercially available EPS (Xanthan gum) to dried and powdered samples of natural sediment. From these experiments it was concluded that the critical shear stress for erosion τ_e increased from about 0.4 - 0.6 Pa to about 1.8 - 2.1 Pa at a concentration of 10 ppt (EPS per dry weight of sediment). Similarly, the erosion rate was decreased by about 90 % over the same range of EPS-content.

De Brouwer (2000, 2002) concluded that the threshold of erosion increased by a factor of about 5 during the algae bloom period on a mudflat in the western Scheldt, whereas Kornman and De Deckere (1998) found a variation in τ_e from 0.2 Pa in March to 0.5 Pa in April and May to 0.1 Pa in June during measurements on the Heringplaat in the Dollart estuary. Simultaneously, SPM values in the Dollart channels decreased by a factor 5 (e.g. De Deckere et al., 2002).

Christie et al. (2000) correlated values of τ_e measured with the ISIS of HR Wallingford (e.g. Appendix C.7.2) on the Skeffling mudflats in the Humber estuary. They found an increase in τ_e from about 0.1 - 0.2 Pa to about 0.4 Pa, which would be described by $\tau_e = 0.025 \times CC^{0.6}$, (30 g/kg < CC < 120 g/kg), where CC is the colloidal carbohydrate content. This relation however is not very accurate, as the data show considerable scatter.

Riethmüller et al. (2000) carried out erosion experiments on tidal flats in the German and Danish Wadden Sea (e.g. Büsum flats and Sylt-Rømø Bight) and tried to correlate τ_e against the chlorophyll a content in the upper 1 mm of the bed. They concluded that microphytobenthos chlorophyll a content is not a general index for sediment surface stabilisation. Instead, the results were highly site-specific and influence of microphytobenthic composition, sediment reworking and of endobenthic macrofauna appeared to be important.

Austen et al. (1999) investigated the erodibility of mudflat surfaces in the Lister Dyb tidal area of the Danish Wadden Sea. The sediment is cohesive, consisting of about 90% clay. Austen found a significant positive correlation between chlorophyll a and erosion threshold (e.g. Fig. 10.6). The cross-shore variation of the chlorophyll a content was strongly related to the average length of the exposure time of the mudflats during each tidal period. The erosion threshold was negatively correlated to the number of the snail *Hydrobia ulvae*.

Andersen (2001) examined the erodibility of two microtidal areas in the Danish Wadden Sea, rich in organic matter. The erosion threshold appeared to be significantly correlated with the chlorophyll a content. Thresholds were low and fairly constant at high faecal pellet content mostly produced by *Hydrobia ulvae*. The correlation was even stronger for the erosion rate.

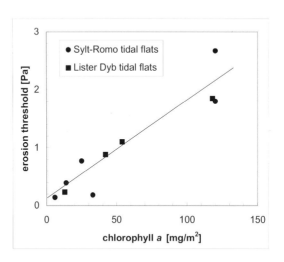

Fig. 10.6: Increase in erosion threshold as a function of chlorophyll a content (after Austen et al., 1999).

Staats et al. (2001) observed high concentrations of chlorophyll *a* in June in the top five mm of the sediment of a tidal flat in the Ems-Dollart estuary, particularly in a 100 m wide zone along the adjacent channel. Although the zone with high chlorophyll *a* concentration matches the zone with high mud content, indications for sediment stabilisation by microphytobenthos were not found. The mud content in the zone was rather constant throughout the year. In contrast, a broad range of mud content values was observed at low chlorophyll *a* concentrations in the area neighbouring the zone. The authors claim that the high mud content on the bordering intertidal areas is likely to be governed by hydrodynamic factors.

Contrary to these results, Michener and O'Brien (2001) found no relation between EPS and sediment stability on a mudflat in the macro-tidal Severn estuary. Apparently, the tidal conditions in this estuary are so dynamic that they dominate biological effects.

Patterson (1997) summarises the effect of bacteria, in particular filamentous cyanobacteria, which can increase the strength of the bed by a factor of 2 – 5, probably through the secretion of EPS.

10.4 BIO-DESTABILISATION

Hydrobia, Macoma and other macrofauna destabilise muddy sediments by means of bioturbation and grazing of diatoms, detritus and bacteria (Jørgensen, 2001; and Coull, 1999, see also Fig. 10.2). Particularly important for destabilisation is the surface burrowing activity of the *Hydrobia* and *Macoma* by reworking the top sediment by moving through the sediment and by ingestion/egestion. Abundant populations of *Hydrobia* cause an increasing grazing pressure, destroying the microphytobenthos mat in the course of June and subsequent destabilisation of the sediment. The effects of the grazing by *Macoma baltica* were described by Widdows et al. (2000) and Herman et al. (2001), those of *Hydrobia ulvae* by Austen et al. (1999) and Andersen (2001). The dominance by *Macoma* or *Hydrobia* might be dependent on the duration of air exposure. *Macoma baltica* avoids areas that are subjected to (enduring) air exposure (Widdows and Brinsley, 2002). The silt layer further erodes during the summer and the autumn. Stormy events may speed up the removal of the muddy top layer (De Deckere et al., 2002). A complete removal may take place during severe storms.

The effect of surface perforation by grazers and worms on erodibility is very large because of the triggering of mass erosion (Section 9.4). Also bioturbation and pelletisation of minerals and organic matter by these "tunneling" organisms results in changes of grain size distribution of the upper part of the bed. As a result, the sediment attains more granular ("sandy") properties (see Fig. 10.1).

Widdows et al. (2000) have studied the role of clams (*Macoma balthica*) on the stability of sediment deposits on the Skeffling mudflat in the Humber estuary, UK. They measured the erodibility of the mudflats with an annular flume and found a substantial increase in erodibility with increasing clam density (e.g. Fig. 10.7). The threshold for erosion decreased by a factor of 2 to 3 for a clam density increasing from 10 to about 1,000 - 10,000 individuals per m^2. The erosion rate increased even by a factor 10 over this range of clam densities.

Widdows et al. conclude that during the period of algae bloom the sediment on the mudflat was stabilised, but that after this bloom, bioturbation by clams promoted erosion of part of the sediments deposited during the bloom period.

Suspended sediment concentrations in the Dollart estuary, The Netherlands, have been found to decrease when microphytobenthos is abundant, i.e. during Spring season, with a second dip during Autumn (Coull, 1999; De Brouwer et al., 2000; De Deckere et al., 2003), as shown in Fig. 10.8. It was observed that meiobenthos on mudflats increased in numbers in June, accompanied by the decay of the microphytobenthos mat and the reduction of sediment stability (De Beer and Kühl, 2001). This is ascribed to the grazing of meiofauna on the microphytobenthos, which could imply that meiobenthos has an indirect effect on the stability of the sediment deposits. Their bioturbation (Aller, 2001) activity also plays a role, because protruding tubes cause micro-turbulence. The decrease of the meiofauna densities from August onwards is probably caused by grazing macrofauna.

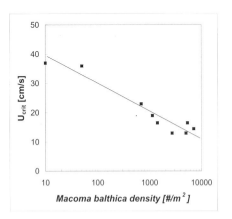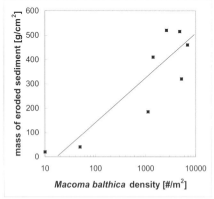

Fig. 10.7: Decrease in erosion threshold and increase in erodibility as a function of Macoma baltica density (after Widdows, 2000).

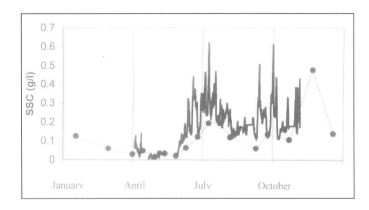

Fig. 10.8: Seasonal variation in suspended sediment concentration in the Dollart estuary ("Heringplaat" and "Groote Gat"), The Netherlands measured in 1996 (after De Deckere et al., 2002).

The combined effect of bio-stabilisation and –destabilisation was investigated by Mahatma et al. (2004, in prep.) in the East Frisian Wadden Sea. They sampled along a cross-shore transect of 1.5 km length starting at the saltmarsh transition region with high mud content (> 50 %) to regions with heterogeneously distributed mussel and cockle beds and mud contents of some 10 %. The inundation periods ranged from seven days during the spring-neap cycle to 4-5 hours per tide at the most outer stations. Bio-stabilisers present were diatoms and tube-building vagile and sessile worms with abundancies of more than 10^4 individuals per m^2; bio-destabilisers were mainly clams (*Macoma baltica*) reaching abundancies of more than 10^4 individuals per m^2. In the natural environment with both stabilisers and destabilisers present, the general increase of erosion threshold to maximal values of more the 3 Pa with EPS-surface concentrations clearly exhibits the impact of benthic diatoms present on the erosion thresholds. However, the large scatter in the data suggests that additional sources of stabilisation are present. The results are presented in Fig. 10.9, showing a bifurcation in the data. This behaviour can be explained fully by the density of tube-building worms: at abundancies above 2000 individuals per m^2 the increase of the erosion threshold with EPS is significantly increased. On the other hand, no impact of bio-destabilisers is detectable.

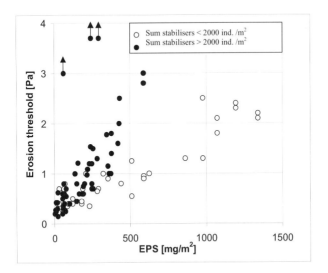

Fig. 10.9: Increase in erosion threshold as a function of EPS surface concentration for lower and higher densities of macrozoobenthos stabilisers; data with arrows indicate a lower limit of erosion threshold when the experiment was stopped (after Mahatma et al., 2004).

This brief overview shows that the effects of biology on cohesive sediment dynamics on intertidal areas in marine environments in north-west Europe can be large, both at short- and long-term time scales. Moreover, these effects cause significant inter-annular (seasonal) variations in sediment properties, such as deposition rates and erodibility.

The best way to account for these effects at present is to modify the parameters in the various sediment models on the basis of experimental data presented in the literature (such as settling velocity, threshold for erosion, etc.).

11. GAS IN COHESIVE SEDIMENTS

In the previous chapters we have argued that cohesive sediment contains various materials, such as sand, organic residue of plants, and sometimes contaminants. The organic matter is degraded under aerobic (top ~10 cm of the seabed) or anaerobic conditions (below ~10 cm within the seabed). A large number of bacteria species are involved in this degradation, resulting in the production of mainly two gasses: methane and carbon dioxide. After a time scale of weeks, years or centuries, depending on ambient conditions (temperature and pressure), the pore water can become saturated with dissolved gas, and a gas phase develops in the form of small bubbles. A cohesive sediment bed is an almost perfect trap for these gas bubbles, which behave as low density "particles". A large amount of gas can be stored in the sediment as bubbles, but when the sediment's bulk density becomes less than that of water, stability of the sediment bed is at risk. The presence of methane stored around the world in marine cohesive sediments is probably of the order of the presently known amounts of fossil fuel reserves. Releases of methane from marine sediments would form one of the major contributions to the global greenhouse effect and there are indications that such releases have played an important role in the cycles of ice-ages (Henriet et al., 1996). In this chapter, a concise description of the underlying processes is given, illustrated with a few case studies.

11.1 INTRODUCTION

In general, gas in sediment appears as a gas phase in unsaturated zones in the material. However gas can be present in sediment in other ways as well:
- as a gas phase within the pore system,
- as a gas phase in the form of bubbles surrounded by sediment,
- dissolved in the pore fluids of the sediment,
- adsorbed by the sediment particles, and/or
- in a solid phase as gas hydrates.

The composition of gas in sediment deposits depends on its source. Biogenic gas consists mainly of methane (CH_4) and carbon dioxide (CO_2) and is generated by bacterial decomposition and mineralization of organic matter. Thermogenic gas consists mainly of methane, ethane and propane, which has migrated from deep (oil) reservoir sources. In the marine environment gas from both origins may be found. In coastal regions, and in particular in estuaries and harbours it is well known that decomposition of organic matter in cohesive sediment results in accumulation of gas, mostly methane. Therefore most dredgers are provided with degassing equipment to prevent malfunctioning of pumps.

Thermogenic gas is also found in the upper layers of sediment deposits on the seafloor of the continental shelf, because of natural gas seeps (e.g. Judd et al., 1997), or because of human activity like oil and gas drilling. A potential new source by human activity is CO_2 sequestration in oil reservoirs, aimed at reducing the emission of green house gasses (Wildenborg et al., 2004). Natural gas accumulations from both thermogenic and biogenic origin are also found in the upper layers of sediment on continental slopes and abyssal planes.

The accumulation of gas in shallow seafloor sediments largely affects the stability of these sediment deposits when subject to mechanical impact (earthquakes, drilling), hydraulic impact (waves, tidal currents) or atmospheric impact (storm depression). Accumulating gas reduces the weight of the sediment and therefore creates under-consolidated conditions so that the soil strength remains low up to great depths. This results in meta-stable conditions of the seafloor, which can be disturbed by natural events and/or human activities, yielding the typical marine seafloor features as shown in Fig. 11.1 (Coleman et al., 1978).

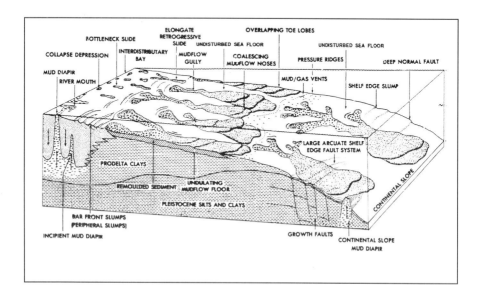

Fig. 11.1: Marine seafloor features (after Coleman et al., 1978).

Mud diapirs are formed by upward squeezing of under-consolidated mud by overburden in areas characterised by rapid deposition. Mud volcanoes are formed where mud is squeezed through the seafloor. This process is accompanied by vast amounts of methane release, which is known to occur in for instance the Caspian Sea (near Baku), in the Black Sea (Krim area), and in the Gulf of Mexico (Hovland and Judd, 1988).

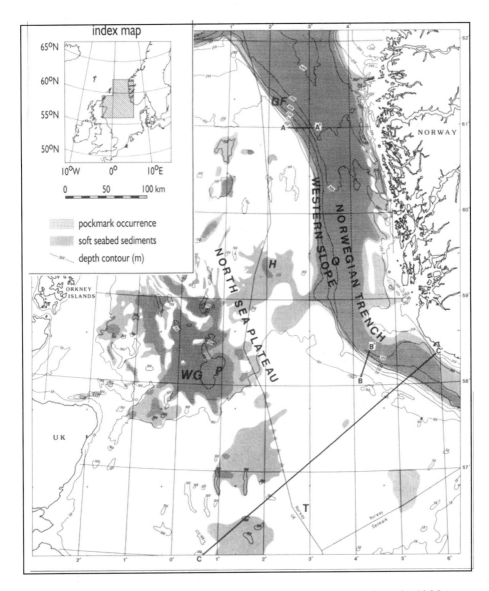

Fig. 11.2: Methane seeps in the North Sea (after Hovland et al., 1988.)

Marine landslides over the continental slopes are well-known events, which are also often related to the release of gas. The (3rd) Storegga slide on the continental slope near the West Coast of Norway at about 7000 years BP (Evans et al., 1996) is probably the most famous and largest slide ever. Such massive sub-aqueous ground displacements generate surface waves known as tsunamis. On somewhat smaller scale numerous marine landslides related to gas releases are encountered along the continental margins, e.g. near the East Coast of North America (Paull et al., 2001), causing cable failure and sometimes human casualties due to unexpected large single waves in coastal regions.

Continuous emissions through pockmarks of gas in the upper sediment layers of the North Sea (Fig. 11.2) seafloor are less catastrophic. The yearly amount of gas seepage can be very large, though observations are scarce. On the UK continental shelf gas seepage (mostly methane) is estimated to contribute to about 40 % of the total UK emissions (Judd et al., 1997). A well studied pockmark area is UK Block 15/25, which was formed approximately 13,000 years BP when gas trapped beneath the seabed permafrost was released during warming of the seawater. The seabed sediment consists of post-glacial soft, silty clays of the Witch Ground Formation. The underlying Coal Pit Formation (over-consolidated clays by ice-loading) is exposed in the pockmark. At present, gas is trapped at a depth of approximately 50 m below the seabed, but gas seepage has been reported during every survey between 1983 and 2000. The gas may have originated from the even deeper sediment deposits of the Kimmeridge Clay Formation (Jurassic) and/or Tertiary peat. Fig. 11.3 shows an example of a local methane release in that area (Judd et al., 1997).

Most gas emissions contain methane, but also pure carbon dioxide emissions may occur. An example of a catastrophic carbon dioxide release is the Lake Nyos (Cameroon) disaster in 1986 (Evans et al., 1993). In this 200 m deep lake in an extinct volcano crater, carbon dioxide from a deep reservoir accumulated in the lake's soft cohesive bed to super-saturated levels. The initiation of bubbles in the water caused a chain reaction resulting in a so-called limnic eruption: a buoyancy-driven flow of rising and expanding gas bubbles. The accelerating gas plume injected water and gas hundreds of meters into the atmosphere. The resulting carbon dioxide cloud, with a density larger than air, flowed down into the valleys and killed 1,800 people.

Other examples of carbon dioxide release can be found in brine lakes around marine mud volcanoes, such as the Napoli MV in the Mediterranean Sea at a depth of 2,000 m (Fig. 11.4, e.g. Medinaut/Medineth, 2000). At this large depth no gas bubbles can be formed, and liquid carbon dioxide, with a larger density than that of water, remains stable near the seafloor forming carbon dioxide lakes. At that depth hydrates may be formed on the water - liquid carbon dioxide interface, preventing carbon dioxide to dissolve into the water (Brewer et al., 1999).

However, marine examples of pure carbon dioxide releases are rare. In general natural gas releases in the marine environment contain mainly methane and about 5 to 10 % carbon dioxide.

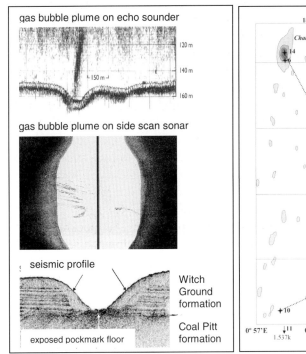

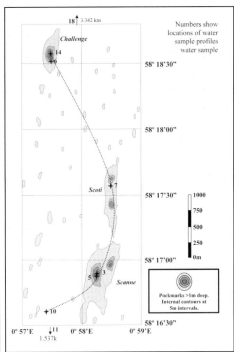

a) Acoustic images of gas release   b) Location of North Sea pockmarks

Fig. 11.3: Methane release from pockmark in the North Sea
(Judd et al., 1997).

Gas accumulation is not only important in natural deposits, but also in man-made deposits of (contaminated) mud from harbours, mine tailings and sewer sludge. Both confined disposal facilities (CDF) on land and sub-aqueous disposals (CAD = Contained Aquatic Disposal) can face capacity problems due to gas, as the mud remains under-consolidated (e.g. Wichman, 1999) and gas bubbles expand the mud volume (Fig.11.5 and Van Kessel et al., 2002). Gas release from a CAD may result in contamination of the aqueous and atmospheric environment. High gas concentrations in the water column may also alter the chemical equilibrium near the seafloor, resulting in major effects on the marine ecosystem.

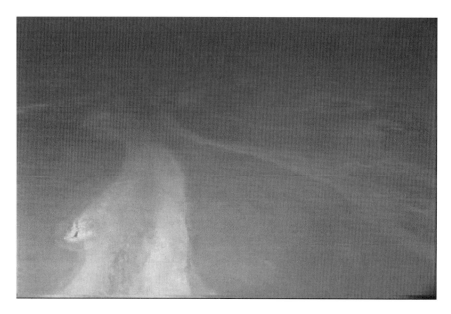

Fig. 11.4: Brine lake and rivers on Napoli mud volcano (depth 2000 m, Medinaut/Medineth, 2000).

Fig. 11.5: Mud expansion due to CO_2 production (Van Kesteren, 2004).

In order to assess the effects of gas on seafloor stability and the impact of gas releases on sediment transport and marine ecosystem, the (bio)chemical and physical processes involved must be known and quantified if possible. In this chapter the fundamental aspects of the processes involved in gas behaviour in cohesive sediments are treated and some applications are presented.

11.2 GAS RELATED PROCESSES IN SEDIMENT

A number of processes depending on the type of sediment and ambient conditions, such as temperature and pressure, determine how gas is generated, accumulated and transported through the seafloor sediment. An overview of these processes and their sequence is presented in Fig. 11.6 (Van Kessel et al., 2002 and Van Kesteren, 2004). The processes are marked bold and italic.

The flowchart in Fig. 11.6 starts with the two potential natural sources: biogenic and thermogenic gas. Biogenic gas originates from bacterial decomposition of organic matter within the sediment, while thermogenic gas reaching the sediment by upward seepage flow is generated by fractionating hydrocarbons at high temperature and pressure. Biogenic and thermogenic gasses in the sediment behave identical and may undergo the same processes sketched in the flowchart of Fig. 11.6.

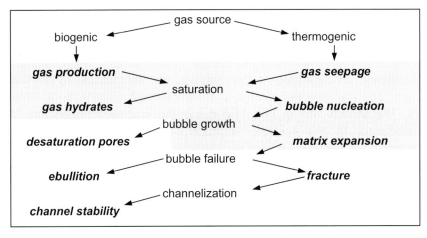

Fig. 11.6: Gas related processes in sediments.

Gas can accumulate in the pore fluid or can adsorb to the solid phase (precipitation of CO_2) of the sediment. In general the pore fluid is water and phase changes can occur from vapour to liquid to solid, depending on the temperature and pressure.

A solid phase, known as gas hydrates, may occur at temperatures above 0 °C. A gas hydrate is a kind of ice structure in which water molecules are wrapped around gas molecules. Gas hydrates can form an impermeable boundary in the seafloor blocking gas and pore water flow. Methane hydrates may form at water depths beyond 300 m, while carbon dioxide hydrates may form already at a water depth of 130 m.

Also a vapour phase can form in the pore water, when one of the dissolved gasses becomes saturated. In that case small bubbles are formed initially. These will grow when more gas is generated or supplied from below, until the bubble size becomes of the order of the pore size. Further growth of the bubble is possible by displacing pore water, desaturating (part of) the pore system. If capillary forces prevent such pore water flow, only deformation of the sediment matrix can occur and bubbles are formed in the sediment pores. The response of the sediment-water mixture to bubble growth is controlled by the pore size distribution: in coarse grained sediment, such as sand, pore water will be displaced, while in fine grained sediment, like most cohesive sediments, bubbles growth is accompanied by expansion of the sediment matrix.

The expansion of bubbles is governed by failure of the sediment matrix surrounding the bubbles. As discussed in Chapter 8, three failure mechanisms can be distinguished: tensile failure, shear failure and ductile failure. The first two mechanisms generate cracks around bubbles by hydraulic or ductile fracture, while the third mechanism allows the bubbles to grow. If the cracks coalesce and form channels to the surface, gas may escape to the sediment surface. The stability and erosion of these channels control the gas transport rates. The shaded area in Fig. 11.6 marks the processes responsible for gas accumulation in the seafloor sediment, while the other processes are responsible for gas releases from the seafloor.

The various processes are discussed in more detail in the next sections, which deal with biogenic gas production, gas phase equilibrium and bubble dynamics.

11.3 BIOGENIC GAS PRODUCTION

Bacteria within the sediment decompose organic matter depositing with the mineral fraction on the seabed. Close to the seabed oxygen is available and aerobic degradation takes place, but 5 to 10 cm below the seafloor surface all oxygen has been consumed and the degradation process becomes anaerobic (e.g. Section 3.1.3). Organic matter decay is the result of a complex synergy of multiple bacteria species, of which methanogenesis is the dominant anaerobic process, involving the consumption by methane producing bacteria of acetate formed in the final stage of the process. Besides methane also carbon dioxide is produced at the same quantities depending on the type of organic matter.

In general the decomposition rate is a function of the following parameters:
- temperature,
- concentration of organic matter, and
- the (initial) age of the deposit.

The latter parameter is important because of the dependency of the decomposition rate on the type of organic matter. Recent deposits have a much higher decomposition rate than old deposits, in which more resistant substances are found. A part of that organic material remains intact even after many thousands of years.

The decomposition rate is expressed with a second order decomposition model using a power-law decay function (see Fig. 11.7, Middelburg, 1989):

$$\frac{d\xi^{OM}}{dt} = -k(t)\xi^{OM} \tag{11.1}$$

where ξ^{OM} is the content of organic matter in the total solids [kg/kg], t is time [year] and k the time-dependent decomposition decay function, given by:

$$k(t) = f(T)\, k_1 \left(t_{age} + t\right)^{k_2} \quad [\text{year}^{-1}] \; ; \; f(T) = \frac{m^{T-T_c} - 1}{m^{T_r - T_c} - 1} \tag{11.2}$$

in which t [years] is time, k_1 and k_2 are coefficients, $f(T)$ is a correction function for temperature, m is a dimensionless temperature scale, T_c is the lowest temperature at which decomposition of organic matter occurs, T_r is a reference temperature, and T is the actual temperature.

The parameter t_{age} represents the initial age of the organic matter. For Dutch cohesive sediments $m \approx 1.1$ (Van Kesteren, 2002). T_r is the reference temperature at which the coefficients k_1 and k_2 are determined and T is the temperature for which k is to be assessed. The coefficients k_1 and k_2 can be obtained from Fig. 11.7, in which the decomposition rate $k(t)$ is given as a function of time at the reference temperature. Equ. (11.2) is a straight line on double log-scale when $t_{age} = 0$. The gradient of the line yields $k_2 = -0.95$, the point at $t = 1$ year yields $k_1 = 0.178$. These coefficients are determined on the basis of numerous experiments described in literature (Middelburg, 1989).

The power function (11.2) describes the decay of organic matter on a time-scale ranging from several hours to several hundred thousands of years. However, the error margin of the decomposition rate is a factor 10. Moreover, the age of the sediments is often uncertain. Therefore age and decomposition rates have to be obtained from experiments.

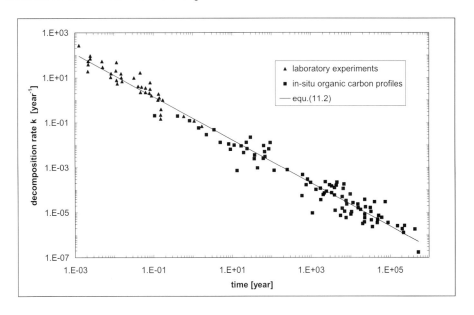

Fig. 11.7: Decomposition rate of organic matter in marine sediments (after Middelburg, 1989).

Depending on the composition of the organic matter the methanogenesis will produce a certain amount of methane and carbon dioxide. Ratio's of methane and carbon dioxide production varies between 1 for organic matter that is easy to decompose, such as sugars, to 5 for reduced organic matter (Dinel, 1988).

The gas production rate can be assessed on in-situ samples by measuring the methane and carbon dioxide concentration in a nitrogen or argon filled head space of a testing cell (Van Kesteren, 2002). Typical values of methane production rates found for cohesive sediments in Dutch harbours and lakes range from 1.5 g to 15 g methane per kg organic matter per year. With an organic content of 10% (by weight of the solids) and a typical void ratio of 5 (mud density 1250 kg/m^3), the methane production rate per m^3 mud $\dot{c}_g$ ranges from 2 μg/m^3/s (= 0.13 μmol/m^3/s) to 20 μg/m^3/s (= 1.3 μmol/m^3/s).

11.4 THERMODYNAMIC EQUILIBRIA OF GAS IN WATER

Possible thermodynamic states of gas in pore water or in the water column are:
* dissolved in the pore fluids,
* a liquid phase,
* vapor phase, or
* solid phase.

In general the pore fluid is water, but in areas of oil and gas exploration activities, the seabed may contain also some non-aqueous phase liquids (NAPL), such as diesel fuel, which was used in the past as a drilling fluid. In the following the thermodynamic states for methane and carbon dioxide are discussed in more detail as these gases are encountered predominantly in the marine environment.

11.4.1 SOLUBILITY

The solubility of gas in water depends on temperature and pore fluid pressure. Raoul's law defines the equilibrium between saturated dissolved gas and its partial vapour pressure. For low gas concentrations, Raoul's law reduces to the linear law of Henry (Section 8.1.3) for the solubility of gas in fluid:

$$p_i = H_i \, x_i \qquad\qquad (11.3)$$

where p_i is the partial vapour pressure of gas component i in the pore fluid, H_i is Henry's constant for this component and x_i is its mole fraction.

Equ. (11.3) shows that the vapour pressure is proportional to the concentration of dissolved gas. As soon as the total vapour pressure of all dissolved gasses exceeds the water pressure, gas bubbles can be formed below the critical point of water (374 ^{0}C, 21.9 MPa), if sufficient gas nuclei are present. Henry's constant is inversely related to the solubility of gas: larger H_i

implies smaller solubility (see also equ. (8.14). At high pressures, Henry's law is not valid anymore.

The solubility of methane and carbon dioxide in water can be obtained from measurements reported in the literature (e.g. Perry, 1984). For methane, Henry's constant can be approximated by:

$$H_{CH_4} = 76.34T + 2222 \qquad\qquad\qquad (11.4a)$$

in which T is temperature [^{0}C] and H is Henry's constant [MPa]. This approximation holds for temperatures from 0 to 40 ^{0}C and pressures up to 30 MPa, which are above the critical point of methane (-82 ^{0}C, 4.6 MPa).

For carbon dioxide the formulation for the solubility is more complex and below the critical point (31^{0}C, 7.3 MPa) equ. (11.3) can be approximated by:

$$c = a_1 p - a_2 p^2 \qquad\qquad [mol/mol]$$

$$\frac{1}{a_1} \equiv H = 64.11 + 3.786T + 0.01044T^2 \qquad [MPa] \qquad (11.4b)$$

$$\log a_2 = -2.927 - 0.02189T + 0.00007433T^2$$

This approximation holds also for temperatures above the critical temperature as long as the pressure is below critical. At supercritical pressures the saturation concentration is much less dependent on pressure and temperature, and its maximum varies between 25 and 33 mmol/mol, as shown in Fig. 11.8 (data from Perry, 1984). The black solid lines in Fig. 11.8 represent sub-seafloor conditions at water depths of 0, 100 and 200 m respectively, given a geothermal gradient of 3.2 ^{0}C per 100 m, showing that a maximum concentration in dissolved gas exists at depths of respectively 620 m, 430 m and 260 m below the seafloor.

Because of dissociation of CO_2 with H_2O, the solubility of CO_2 in water is about 20 times higher than that of methane and other gases, such as oxygen and nitrogen. As a consequence, gas generated by biogenic degradation of organic matter contains mainly methane (95%) and little carbon dioxide (5%), although their production rates differ much less, or are even equal.

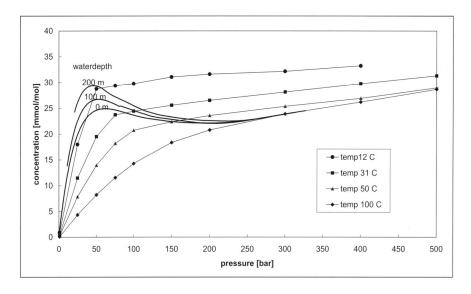

Fig. 11.8: Solubility CO_2 in water (after Perry, 1984).

The influence of fine pores on the solubility of gas is not accounted for in the analysis above. However, phase equilibriums may definitely change. In meso-pores (< 50 nm) gas molecules may adsorb to the surface of the pores, a process known as "pore condensation". Because of this, significantly more gas can be dissolved than given by equ. (11.3). This effect can be quantified with Kelvin's law (see e.g. Broekhoff, 1969), which gives the reduction of vapour pressure p_i as a function of the pore diameter filled with water (Van Kesteren et al., 2002):

$$p_i = p_{sat} \exp\left[-\frac{4\sigma_{st}V_{mol}^w}{RTd}\right] = Hx_i \exp\left[-\frac{4\sigma_{st}V_{mol}^w}{RTd}\right] \tag{11.5}$$

where p_{sat} is the saturation pressure given by equ. (11.3), σ_{st} is the surface tension, V_{mol}^w is the molar volume of water, r is the pore radius, R is the gas constant (= 8.2 J/mol/°K) and T is the temperature.

Pore condensation may be neglected if the volume of meso-pores is small, or if these are poorly accessible. In cohesive sediment the pore size depends on the particle size distribution and the way the particles are organised in the skeleton, which can be described as a fractal structure (see Section 4.2). Different pore size distributions are shown in Section 3.2.2 (Fig. 3.15) and both cohesive sediments with either a low volume of meso-pores or with a high volume of meso-pores exist. In natural deposits of non-cohesive sediments with particles in the range of silt, sand or gravel only macro pores will be present, and the effect of pore condensation can be neglected.

11.4.2 LIQUID PHASE

In general, ambient temperatures and pressures are above the critical point of methane. Therefore, liquid methane is generally not encountered in the marine environment. Carbon dioxide, however, with a critical point at 31 ^{0}C, 7.3 MPa, can become fluid at water depths of 350 m at temperatures close to 0 ^{0}C and beyond 730 m at temperatures of 31 ^{0}C. The density of liquid carbon dioxide is in general lower than that of water. However, at large water depths it can become larger: at 0 ^{0}C this occurs at a water depth of 1800 m and at 10 ^{0}C this occurs at a water depth of 3000 m. Therefore carbon dioxide fluxes through the seafloor of deep oceans may form liquid carbon dioxide lakes in depressions in the seabed. At these depths, also hydrates may be formed as discussed in the next section.

11.4.3 SOLID PHASE: GAS HYDRATES

A gas hydrate is a solid with an ice structure (clathrate) in which water molecules are wrapped around gas molecules at temperatures above 0 °C. The amount of water molecules in the hydrate for each gas molecule depends on the type of gas. For methane and carbon dioxide this ratio is about 6 (Sloan, 1990).

Equilibriums of gas hydrates of methane and carbon dioxide are shown in Fig. 11.9 (data from Sloan, 1990). The phase line "ice-water" (thick solid line) is the phase transition between solid and liquid water according to experimental data from Winterkorn (1953). It is noted that the phase transition is shifted to temperatures below 0 °C at increasing pressures.

The other two solid lines in Fig. 11.9 are the phase transitions between the hydrate and vapour phase for methane and carbon dioxide. To the right of the ice-water transition, these hydrates are formed in water, while on its left side the hydrates are formed within the solid water phase. For both methane and carbon dioxide a quadruplet (four phase transitions) exists at the crossing with the ice-water phase transition. For carbon dioxide a second quadruplet is found at the transition between its vapour and liquid phase (dashed line), where pressure and temperature are below critical (31 °C, 7.4 MPa). At this second quadruplet, the transition between the hydrate and vapour phase of carbon dioxide changes into a transition between the hydrate and liquid phase at increasing pressure and/or temperature. Methane does not have a second quadruplet, because the transition between the hydrate and vapour phase is above its critical temperature and pressure (-82.6 °C, 4.6 MPa).

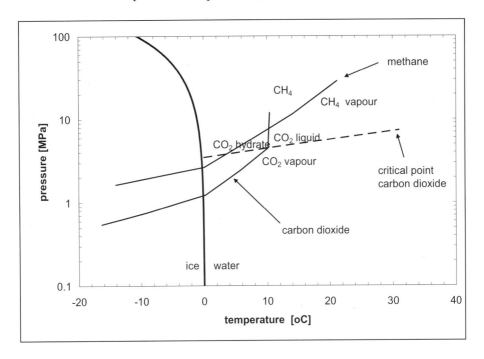

Fig. 11.9: Gas hydrate phase equilibriums
(data from Sloan, 1990 and Winterkorn, 1953).

From Fig. 11.9 it follows that methane hydrates are formed at a water depth of 280 m when the temperature is close to 0 °C. At higher temperatures the water depth required to form hydrates increases logarithmically; for instance at 10 °C the required water depth is 800 m. For carbon dioxide these water depths are much less: 130 m and 450 m, respectively. At 10 °C and 450 m also the second quadruplet exists. At water depths corresponding to pressures above the liquid-vapour transition, carbon dioxide hydrates are formed at the interface between liquid carbon dioxide and water. The hydrates prevent dissolving of liquid carbon dioxide in water and the two phases remain separated (e.g. Brewer, 1999).

The formation of gas hydrates in water requires ample availability of gas; in general the water must be saturated with dissolved gas. The molar concentration of gas in a hydrate is about 0.16 mol/mol. This is an order of magnitude higher than for dissolved gas, which is maximal 30 mmol/mol for carbon dioxide and 16 mmol/mol for methane at a water depth of 4 km (see Section 11.4.1).

However in the water column above the seafloor, it is difficult to keep the water saturated due to gas bubble formation, and to retain hydrates because the density of these hydrates is lower than that of water. The upward flow by bubble formation near the seafloor in water saturated with gas may induce catastrophic events, which are known as "limnic" eruptions, when occurring in lakes.

An exception is carbon dioxide at water depths of more than 2000 m, where the liquid phase and hydrate phase is more dense than water.

Dissolved gas can accumulate in the pore water of sediments in the seabed. Its transport is limited by diffusion. Therefore hydrates are generally formed within the seabed. The depth at which hydrates can form is limited by the increase of temperature with depth. The geothermal gradient amounts to about 3 ^{0}C per 100 m depth. As an example, the hydrate phase boundary and geothermal temperature profile in the methane stability diagram is shown as a function of depth in Fig. 11.10. Gas hydrates are only possible in the shaded area.

When gas hydrates are generated within the seabed, they can form an impermeable layer, blocking gas and pore fluid flow. Below the hydrates, gas can be trapped as a vapour phase. This vapour reflects acoustic signals efficiently, and gas is often well tractable on acoustic images. Because the lower boundary of the hydrate layer is fixed to a certain depth below the seafloor, these acoustic images have the same profile as the seabed bathymetry and are known as the "bottom simulating reflector" (BSR).

In cohesive sediment the pore size distribution can affect the phase equilibrium for hydrate formation. Due to adsorption pressures near clay particle surface, the phase equilibrium in small pores can be shifted to higher temperature or lower pressures (e.g. Cha, 1988). On the other hand, capillary forces can reduce the water pressure in small pores and the phase equilibrium is shifted to lower temperature or higher pressures (e.g. Yousif, 1989). It is noted that for temperature and pressure conditions in the hydrate zone, it is still difficult to predict the sediment conditions under which hydrates may be formed.

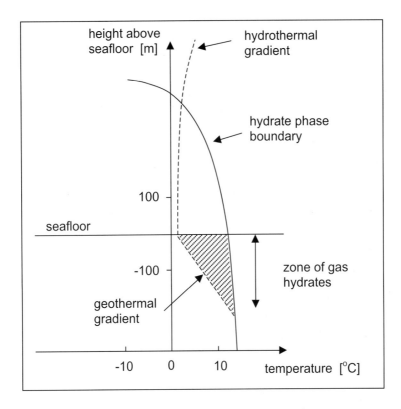

*Fig. 11.10: Methane hydrate stability zone in ocean sediment
(redrawn from Kvenvolden, 1988).*

11.5 BUBBLE MECHANICS

11.5.1 BUBBLE NUCLEATION

When pore water becomes over-saturated with gas, bubble nucleation may
occur, which may take place in a homogeneous or a heterogeneous way. Large
over-saturation may yield homogeneous nucleation, as a significant energy
barrier (activation energy) has to be overcome to generate micro bubbles
(Brennen, 1995). This is the result of the surface tension of water, which limits
nucleation. The excess gas pressure Δp_g to overcome the surface tension
amounts to:

$$\Delta p_g = \frac{2\sigma_{st}}{R_b} \qquad\qquad\qquad\qquad (11.6)$$

where R_b is the bubble radius and σ_{st} is the surface tension of water (0.074 N/m at 20 °C). From equ. (11.6) it follows that the excess pressure (and the over-saturation) increases with decreasing bubble radius.

However, nucleation generally takes place heterogeneously as a result of the energy barrier. Small dislocations, at which cavities are stable, are always present on solid surfaces. These are ideal locations for bubble nucleation, as initial cavities in the form of dislocations result in a strong decrease of the energy barrier. Heterogeneous nucleation can therefore occur at small over-saturation.

When both methane and carbon dioxide are present in the pore water, nucleation will be determined by saturation of methane, because it is less soluble in water than carbon dioxide (see Section 11.4.1).

As diffusive gas transport is small, except at very small length scales, a large number of bubbles per m³ sediment can be formed. As a result, the degree of over-saturation in pore water remains limited. The number of bubbles n_b and the average bubble distance l_b $\left(\sim \sqrt[3]{1/n_b}\right)$ depend on the rate of gas production $\dot{c}_g$. Assuming equilibrium between gas production and diffusive transport of gas towards the bubbles, a relation can be derived between bubble distance and gas production rate. For methane this relation can be approximated with $\log l_b \approx -0.33 \log \dot{c}_g - 3.41$ (Van Kesteren et al., 2002).

Van Kessel (1998) carried out a series of laboratory experiments. Pore water was saturated with CH_4 at high pressure, and subsequently the pressure was decreased to create over-saturation conditions. The number of bubbles observed amounted to about 10^7 per m³. At low production rates the number of bubbles is smaller; assuming $\dot{c}_g$ = 0.1 µmol/m³/s (see Section 11.3), the average bubble distance is calculated at about 0.1 m. In that case, only about 1,000 bubbles per m³ exist.

The number of nucleation sites in the seabed is generally unknown. Moreover, these are difficult to determine from bed samples taken at large water depth, hence large pressure. Because of the large pressure drop when the sample is brought to atmospheric conditions, many gas bubbles will be generated and the risk that the samples will "explode" is high. However, for modelling purposes it can be assumed that the total amount of gas in bubbles does not increase anymore when more than 10^5 nucleation sites per m^3 of sediment are available (Van Kessel et al., 2002).

11.5.2 BUBBLE GROWTH

The rate of bubble growth is proportional to the square root of time, as the amount of gas that has to diffuse towards the bubble increases with increasing bubble radius. For example, if it takes about one day to form a bubble with a diameter of 10^{-3} m, it will take another 30 years to form bubbles of 0.1 m size. However, the maximum bubble size is limited by the total amount of gas available. Bubbles larger than 0.1 m do generally not develop, but at the expense of other bubbles on a very long time scale.

Surface tension is important at small bubble radii. Because of this, the pressure within the bubbles decreases rapidly with increasing bubble size. The initial bubble growth rate is fairly large and the bubbles will attain a diameter equal to the pore diameter soon after their generation. This effect is important for bubbles smaller than 10^{-5} m. To grow further, the gas bubbles have to expel pore water from the pore system or the bubbles have to deform the sediment skeleton. In Fig. 11.11 these two modes are shown schematically. When the menisci of the gas-water interface remain stable (upper part in Fig. 11.11), the gas bubble pressure works directly on the sediment skeleton and deformation of the skeleton may occur. When the menisci are not stable, pore water may be expelled and replaced by gas (lower part in Fig. 11.11): gas enters the pore system. Which mode occurs depends on the shear strength of the particle matrix and the diameter of the pores:

- The excess gas pressure Δp_g in the bubbles at which gas may enter the pore system, the so-called "air-entry value", is the maximum pressure difference at which the menisci in the interface remain stable. This follows from equ. (11.6):

$$\Delta p_g > 2\sigma_{st}/R_{pore} \tag{11.7a}$$

- The excess gas pressure Δp_g in the bubbles should be smaller than the excess pressure that is required to deform the particle skeleton. This pressure is for undrained conditions given by (Vesic, 1972):

$$\Delta p_g < \frac{4}{3} c_u \left\{ 1 + \ln\left(\frac{G}{c_u}\right) + \ln\left(1 - \left[\frac{R_b}{R_{b,0}}\right]^{-3}\right)\right\}$$ (11.7b)

in which c_u is the undrained shear strength, G the shear modulus, R_b is the bubble radius and $R_{b,0}$ the initial radius.

Thus, gas may enter the pore system if:

$$\frac{4}{3} c_u \left\{ 1 + \ln\left(\frac{G}{c_u}\right) + \ln\left(1 - \left[\frac{R_b}{R_{b,0}}\right]^{-3}\right)\right\} > \Delta p_g > \frac{2\sigma_{st}}{R_{pore}}$$ (11.7c)

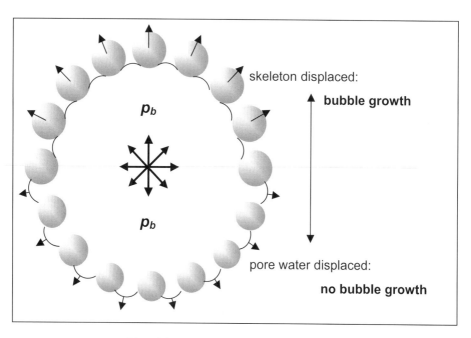

Fig. 11.11: Modes of bubble growth.

Typical values of the ratio of shear modulus and remoulded strength G/c_u are in the range of 10 to 100. As the surface tension between water and gas interface amounts to $\sigma_{st} = 0.074$ N·m^{-1}, equ. (11.6) yields a critical pore radius of 20 to 35 μm for sediment with a strength $c_u = 1$ kPa. For smaller pore radii the menisci remain stable and bubble expansion occurs. For larger pore radii, the menisci will fail, pore water will be displaced, and gas will enter the pore system. The critical pore radius is inversely proportional to the undrained shear strength: a ten times higher undrained shear strength will result in a ten times lower critical pore radius.

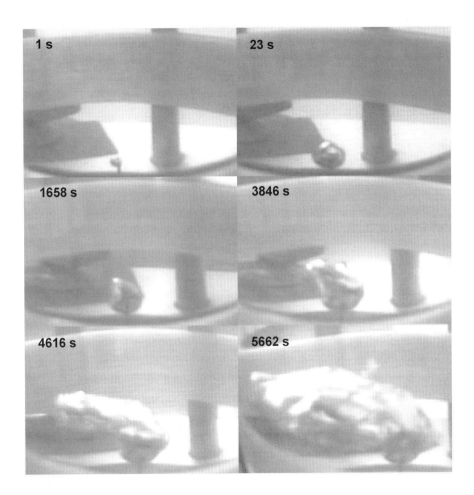

Fig. 11.12: Bubble growth in laponite clay (Van Kesteren et al., 2002).

Undrained conditions can occur only during rapid bubble growth. In general, bubble growth in cohesive sediment deposits is slow, because gas production generated by degradation of organic matter is a slow process. Therefore the right term of equ. (11.7b) must be replaced by a formulation that accounts for drainage (see Vesic, 1972; and Van Kesteren, 2004).

Pore water flowing in radial direction away from a bubble under drained conditions is limited in its movement by adjacent bubbles. The flow distance towards free boundaries, such as the seafloor, is often so large that undrained conditions apply, so that equ.(11.7b) can be used as an approximation.

Fig. 11.12 gives an example of bubble growth by deformation of a mud matrix, showing the time evolution of an air bubble in laponite, as measured by Van Kessel (Van Kesteren et al., 2002). Laponite is a transparent clay with a very high water content and very small pore sizes (e.g. Section 3.1.2).

11.5.3 EBULLITION OF BUBBLES

Because of the large difference between the density of a gas bubble ($\rho \approx 1$ kg/m^3, pressure dependent) and its surrounding medium (i.e. sediment $\rho \approx 1400$ kg/m^3), bubbles tend to rise (Dingemans, 1998). However, the particle skeleton can withstand these buoyancy-induced stresses as long as the yield stress is not exceeded. Hence, a criterion for the critical radius R_{crit} at which bubbles start to rise reads:

$$R_{crit} = f \frac{c_u}{\rho g} \tag{11.8}$$

From theoretical analyses, the dimensionless form factor f is computed at $f = 7.5$. Numerical Finite Element Model simulations (Cazemier et al. 1997) suggest that $f = 11.6$. In Section 11.4 we reasoned that the maximum bubble size amounts to about 0.1 m. Hence, with $f = 11.6$ and $\rho = 1400$ kg/m^3, and using equ. (8.12), it follows that bubbles can only rise in sediment with a strength $c_u \leq 120$ Pa. This is at least an order of magnitude smaller than the actual strength of marine sediment deposits, except for very fresh and soft deposits. The rise and escape of individual gas bubbles through the sediment matrix is therefore an unlikely mechanism for gas transport. Furthermore, sediment above a rising bubble "hardens", e.g. Chapter 8. Experimental results confirm the stationary position of bubbles in accordance with criterion equ. (11.8) (Van Kesteren, 2002; see also Fig. 11.12 and Fig. 11.15).

11.5.4 CRACK INITIATION AND PROPAGATION

During their growth, bubbles start to deviate from their initial spherical shape. Directly after nucleation, the pressure inside a bubble is completely determined by surface tension effects, as described by equ. (11.6). When bubbles start deforming the sediment matrix, the relation between bubble pressure and bubble radius is governed by equ. (11.7b) in case the menisci of the bubble wall interface are stable, transferring the gas pressure to the sediment skeleton.

However bubbles can become unstable, initiating cracks when they attain their critical size. As discussed in Section 11.5.2, two stress conditions at the bubble-sediment interface have to be considered (see Fig. 11.11):

A. bubble expansion by transferring excess gas pressures to the sediment skeleton through capillary forces: the lowest effective stresses are in tangential direction and occur at the transition from the plastic to the elastic zone,

B. excess gas pressure displaces pore water: in this case, the largest tensile stresses occur near the bubble surface.

Fig. 11.13 depicts the bubble-growth-induced failure process schematically. During bubble growth ductile failure (Section 8.3.1) in the plastic zone occurs, where the sediment is stretched in tangential direction. After a certain bubble growth cracks can be initiated in the area where tensile stresses occur, if enough energy can be released from the surrounding elastic zone to provide the required fracture energy. The required fracture energy in case A is much higher than in case B, because in case A crack propagation not only requires energy for creating new crack surface, but also requires unrecoverable energy for the additional plastic deformation in the ductile zone. This is not the case in case B, where cracks are initiated at the bubble surface. Therefore the critical bubble size at which failure occurs in case A is larger then in case B.

Crack initiation at the plastic-elastic zone interface is described in Section 8.3.1 as a shear-induced tensile crack. A shear failure plane bifurcates into a tensile crack (see also Fig. 11.13), which is referred to as ductile fracture of a bubble. Crack initiation at the bubble interface is referred to as hydraulic fracture. An example of hydraulic fracture is shown in Fig. 11.14, where blue dyed liquid is injected into laponite at a constant volume rate. Instead of bubble growth, as in Fig. 11.12, a planar crack is formed, propagating radially in the vertical plane. Section 8.3.3 (Fig. 8.28) describes how this injection method can be used to determine the fracture energy or fracture toughness of mud.

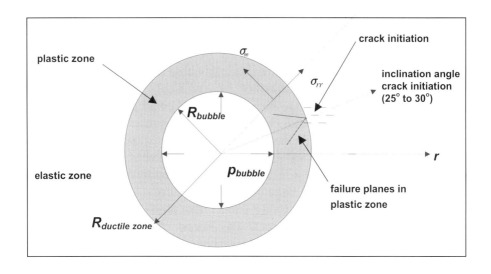

Fig. 11.13: Sketch of ductile fracture of a bubble.

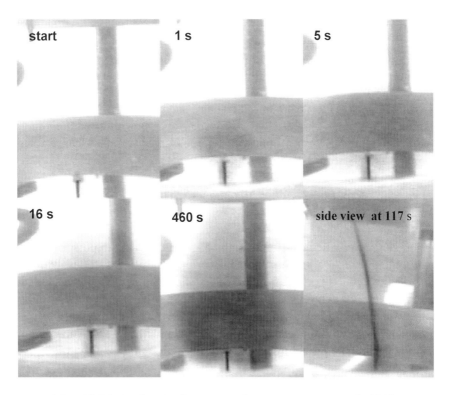

Fig. 11.14: Hydraulic fracture in laponite (Van Kessel, 1998).

During bubble expansion, the plastic zone around bubbles increases. The stresses remain constant along a radius normalized with the bubble radius. However due to localized plastic deformations in the plastic zone, localization of shear deformation occurs at the interface between the plastic and elastic zone when the tangential stress $\sigma_{\theta\theta}$ in the interface is tensile. This is the case when the isotropic effective stress far away from the bubble is smaller than the skeleton's cohesion. Micro-cracks will open and cracks are initiated when the amount of energy released from the elastic zone is sufficiently large. This process can be quantified by means of linear fracture mechanics applied to the elastic zone, including the stress state of the plastic zone.

The application of fracture mechanics for cohesive sediments is treated in Section 8.3.2. For convenience we recall the tensile stress concentration in the vicinity of a crack-tip in the plane of the crack (equ. (8.53a), and Broek, 1986):

$$\sigma_{\theta\theta} = \frac{K_I}{\sqrt{2\pi r}} \tag{11.9}$$

where r is radial distance form the crack tip, $\sigma_{\theta\theta}$ is the tangential stress, and K_I the stress concentration factor [Pa·m$^{0.5}$].

For the ductile fracture mode, the stress concentration for a "penny-shaped" crack is given by (see Murdoch, 1993 and Fig. 11.13):

$$K_I = \frac{2}{\pi}\sqrt{\pi R_p}\left[(p_b - p_\infty)\frac{2}{\pi}\arcsin\left(\frac{R_b}{R_p}\right)\right] +$$
$$+ (\sigma_{\theta\theta}^p - p_\infty)\left(1 - \frac{2}{\pi}\arcsin\left(\frac{R_b}{R_p}\right)\right) \tag{11.10}$$

where R_p the radius of the plastic zone, p_g the gas pressure in the bubble, p_∞ is the isotropic stress at a large distance from the bubble, R_b the radius of the bubble, and $\sigma_{\theta\theta}^p$ the tangential stress averaged over the plastic zone.

When the stress concentration factor K_I reaches a critical level, cracks are formed. The critical stress concentration $K_{I,c}$ is a material parameter for a given geometry and is known as the fracture toughness. As shown in Section 8.3.3

(equ. (8.57)), the fracture toughness of cohesive sediment can be related to the fracture energy G_{Ic}, which can be correlated to the undrained shear strength c_u. With equ. (8.58) and assuming a constant ratio of about 50 to 100 of Young's modulus E and the undrained shear strength c_u, a fracture toughness is found of:

$$K_{I,cr} = \alpha c_u \qquad \alpha \approx 0.13 \text{ to } 0.21 \tag{11.11}$$

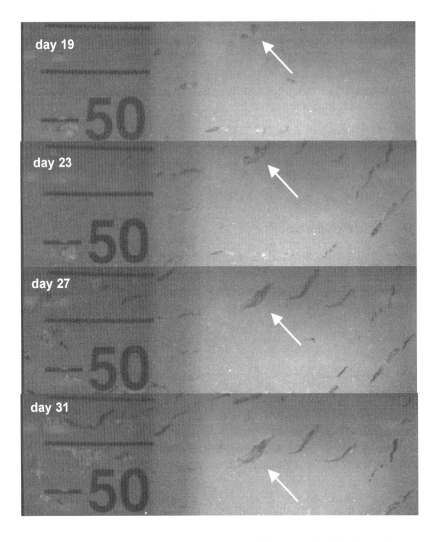

Fig.11.15: Observed fracture of bubbles in Lake Ketel mud.

Substituting from equ. (11.10) yields a critical bubble radius of 4 to 10 mm. This corresponds well with observed bubble sizes in the field and in large scale laboratory experiments (Van Kessel et al., 2002). Fig. 11.15 shows an example of observed bubble-induced fracturing in mud from Lake Ketel, The Netherlands. This figure shows that cracks are created slightly above the centre of the bubble, as indicated in Fig. 11.13. Also coalescence of two bubbles occurs as indicated with the white arrows.

11.6 CHANNEL FORMATION

Crack propagation itself is determined by the pore water flow rate from ambient sediment to the fracture. Neighbouring bubbles can connect by coalescence of fractures. Transport of gas bubbles through the fractures can occur only if the width of the crack openings exceeds the pore size for gas entry. When this opening of cracks is completed, the cavities are filled with both gas and water.

As discussed in the previous sections, gas accumulates in (the upper layers of) the sediment, unless channels to the bed surface are formed. It is therefore important to assess the conditions under which a connected channel system may be created, as this controls the outflow of gas, water and sediment (Crosato et al., 1998). Channels are created: (1) as a result of drainage of pore water during consolidation, and (2) by merging of cracks generated during bubble growth.

Channels induced by consolidation generally have vertical scales of a few decimetres and therefore cannot enhance transport of gas and water from deeper layers. However, by merging of bubble-created cracks, significantly longer channels may be created. Vertical channels can collapse in horizontal direction when the vertical stresses are sufficiently high. The channel radius R_{ch} after collapse can be found from cylindrical compression (Vesic, 1972):

$$\left(\frac{R_{ch}}{R_{ch,0}}\right)^2 = \frac{1}{1+\exp\left[-1-\ln\left(\dfrac{G}{2c_u}\right)-z\dfrac{K_0(\rho_s-\rho_w)g}{c_u(1+e)}\right]} \tag{11.12}$$

where z is the depth below bed surface, e is the voids ratio, K_0 the ratio between horizontal and vertical effective stress, ρ_s is the specific sediment density, ρ_w is the water density, $R_{ch,0}$ the channel radius before collapse, G is the shear modulus, and g is acceleration of gravity.

At its maximum depth, the channel diameter is reduced to the size of the largest pores in the sediment skeleton. Channels at a larger depth are unstable and will be closed.

Similar to pore water, gas can flow through drainage channels. Gas bubbles rising in channels have a fairly large velocity inducing fairly large wall shear stresses, especially in narrow channels. As a result, the drainage channel walls may erode. However, due to horizontal compaction by the vertical stress the channels often maintain their equilibrium radius, yielding a continuous horizontal supply of sediment, eroded from the channel wall. The erosion velocity of the channel can be determined with the erosion formulae presented in Section 9.5.1 (equ. (9.32)):

$$V_e = \frac{d\delta_e}{dt} = \frac{c_v \phi_{s,0}}{10 D_{50}} \frac{\tau_w - \tau_{cr}}{c_u} \tag{11.13}$$

in which c_v is the consolidation coefficient, $10 D_{50}$ is a measure for the floc diameter within sediment, τ_w the wall shear stress due to the gas-water flow through the channel and τ_{cr} the critical shear stress for erosion.

Laboratory observations (e.g. Fig. 11.18 and Van Kessel et al., 2002) show that the water-gas flow velocities can amount to several dm/s through channels with diameters up to a few dm. This implies that the flow through the channels is often turbulent (!) Hence, the wall shear stress becomes:

$$\tau_w = f \rho_{fl} \left(\frac{\Phi}{N_{ch} \pi R_{ch}^2} \right)^2 \tag{11.14}$$

in which ρ_{fl} is the density of gas-water mixture, N_{ch} the number of channels per m^2, Φ the bulk volume flux through the ssediment, and f the friction factor for turbulent channel flow ($f \approx 0.002$).

From these analyses the evolution of a channel can be assessed for a given gas and water fllux. An example of such a channel evolution is shown in Fig. 11.16, established for cohesive sediment from Stryker Bay (Minnesota, Van Kesteren, 2002). The channel evolution in the sediment with a thickness of 0.67

m is depicted for a time period of 47 years. The computed gas fluxes through the channels are compared with field measurements in Fig. 11.17. As temperatures during wintertime are too low for bacterial activity, gas flow occurs only in summertime.

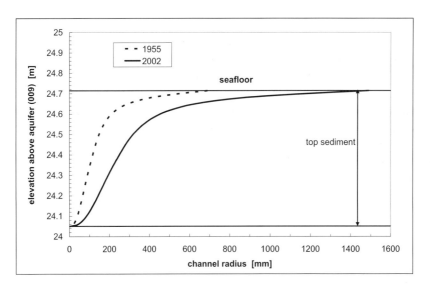

Fig. 11.16: Channel formation in Stryker Bay sediment.

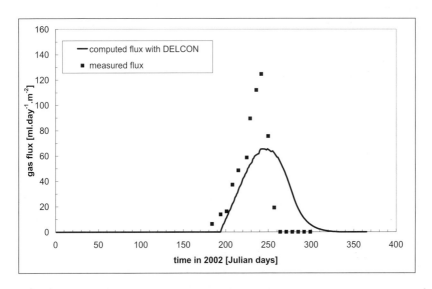

Fig. 11.17: Measured and computed gas flux through seafloor in Stryker Bay.

Gas-induced channels are found on the seabed as pockmarks (see Fig. 11.2 and Fig. 11.3). These channels can also be generated in the laboratory when length scales are large enough, i.e. of the order of meters, as shown in Fig. 11.18 (Cornelisse et al., 1997). Note that the channel radius in this experiment is much larger then depicted in Fig. 11.16, as the gas production rate was orders of magnitudes larger.

An important observation from field and laboratory experiments is that the ratio of gas and water fluxes through the channels remains constant in time (Van Kessel et al., 2002). Furthermore pore water flow through the channels results in rapid consolidation of the cohesive sediment at time scales much smaller than in the case without channels (e.g. "normal consolidation").

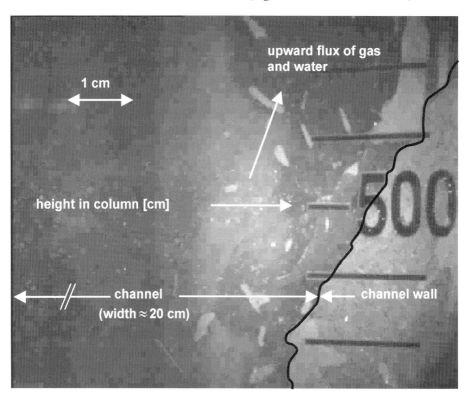

Fig. 11.18: Observed channel formation in laboratory test (Cornelisse et al. 1997).

It is concluded that under specific conditions gas may affect the behaviour of cohesive sediment (deposits) largely. These conditions may occur in the deep sea or in organic rich deposits. When gas forms bubbles, generating cracks, the sediment deposits may become unstable. Large amounts of gas may erupt and/or the sediment may start to flow when deposited on slopes. These processes are in particular of importance for deep sea engineering.

REFERENCES

Aitchison, G.D., 1961, Relationships of moisture stress and effective stress functions in unsaturated soils. Proceedings of the conference on pore pressure and suction in soils, Butterworths, London.

Agrawal, Y.C., and H.C. Pottsmith, 2000: Instruments for Particle Size and Settling Velocity Observations in Sediment Transport, Marine Geology, 168 (1-4) 89-114.

Agrawal, Y.C. and P. Traykovski, 2001, Particles in the bottom boundary layer: concentration and size dynamics through events, Journal of Coastal Research 106 (C5) 9533-9542.

Akers, R.J, A.G. Rushton and J.I.T. Stenhouse, 1987, Floc breakage: the dynamic response of particle size distribution in a flocculated suspension to a step change in turbulent energy dissipation, Chemical Engineering Sciences, 42, 787-798.

Al Tabbaa, A., 1984, Anisotropy of clays, MPhil thesis, Cambridge University, UK.

Alexis, A., P. Bassoulet, P. Le Hir and C. Teisson, 1992, Consolidation of soft marine soils: unifying theories, numerical modelling and in-situ experiments, Proceedings International Conference on Coastal Engineering, 226, 2949-2961.

Allen, G.P., J.C. Salomon, Y. Du Penhoat and C. de Grandpré, 1980, Effects of tides on mixing and suspended sediment transport in macro tidal estuaries, Sedimentary Geology, 26, 69-90.

Aller, R.C., 2001, Transport and reactions in the bio-irrigated zone, in: B.P. Boudreau and B.P. Jørgensen, The Benthic Boundary Layer- Transport Processes and Biochemistry, Oxford University Press, Oxford, 269-301.

Aller, R.C., 1982, The effect of macro benthos on chemical properties of marine sediment and overlying water, in: Animal-Sediment Relations; the biogenic alteration of sediments, ed. P.L. McCall and M.J.S. Tevesz, Plenum Press, New York, 53-104.

Allersma, E., 1988, Morphological studies for the Rhine-Meuse estuary – morphological modelling Part IV: composition and density of sediment, Delft Hydraulics, Report Z71-03 (in Dutch).

Ambrasys, N.N. and J.M. Menu, 1988, Earthquake induced ground displacements, Earthquake Engineering Structural Dynamics, 16 (7) 985-1006.

Amos, C.L., G.R. Daborn, H.A. Christian, A. Atkinson and A. Robertson, 1992, In situ erosion measurements on fine-grained sediments from the Bay of Fundy, Marine Geology, 108 (2) 175-196.

Andersen, T.J., 2001, The role of faecal pellets in sediment settling at an intertidal mudflat, the Danish Wadden Sea, in: Fine Sediment Dynamics in the Marine Environment, ed. W.H. McAnally and A.J. Mehta, Elsevier, Proceedings in Marine Science, 387-421.

Argamam, Y. and W.J. Kaufman, 1970, Turbulence and flocculation, ASCE, Journal of the Sanitary Engineering, 96 (SA2) 223-241.

Ariathurai, C.R., 1974, A finite element model for sediment transport in estuaries, PhD-thesis, University of California, Berkley, USA.

Ariathurai, C.R. and K. Arulanandan, 1978, Erosion rates of cohesive soils, ASCE, Journal of the Hydraulics Division, 104 (2) 279-282.

Arulanadan, K., P. Loganthan and R.B. Krone, 1975, Pore and eroding fluid influences on surface erosion of soil, ASCE, Journal of the Geotechnical Engineering Division, 101 (GT1) 51-66.

Arundale, A.M.W., E.J. Darbyshire, S.J. Hunt, K.G. Schmitz and J.R. West, 1997, Turbidity maxima formation in four estuaries, in Proceedings of the 4th Nearshore and Estuarine Cohesive Sediment Transport Conference, INTERCOH'94, Wallingford, UK, July 1994, ed. by N. Burt, R. Parker and J. Watts, John Wiley and Sons, 135-146.

ASTM, American Society for Testing Materials, West Conshohocken, Pennsylvania, USA , www.astm.org.

Austen I., T.J. Andersen and K. Edelvang, 1999, The Influence of Benthic Diatoms and Invertebrates on the Erodibility of an Intertidal Mudflat, the Danish Wadden Sea, Estuarine, Coastal and Shelf Science 49 (1) 99-111.

Ayesa, E., M.T. Margeli, J. Flórez and J.L. García-Heras, 1991, Estimation of breakup and aggregation coefficients in flocculation by a new adjustment algorithm, Chemical Engineering Science, 46 (1) 39-48.

Barends, F.B.J , 1980, Nonlinearity in groundwater flow, PhD Thesis, Delft University of Technology, Delft.

Barends, F.B.J. and E.O.F. Calle, 1985, A method to evaluate the geotechnical stability of offshore structures founded on a loosely packed seabed sand in a wave loading environment, Proceedings of the 4th Intern Conference on Behaviour of Offshore Structures, BOSS'85, Elsevier, 643-652.

Barenblatt, G.F., 1953, On the motion of suspended particles in a turbulent stream, Prikladnaja Matematika i Mekhanika, 17, 261-274 (English translation).

Bartholomeeusen, G., GC. Sills, D. Znidarcic, W.G.M. Van Kesteren, L.M. Merckelbach, R. Pyke, W.D. Carrier III, H. Lin, D. Penumada, J.C. Winterwerp, S. Masala and D. Chan, 2002, SIDERE, Numerical Prediction of Large Strain Consolidation, Géotechnique, 52 (9) 639-648.

Bastin, A., A. Caillot and B. Malherbe, 1982, Sediment transport measurements on and off the Belgium coast by means of tracers, Proceedings of the 8th International Harbour Congress, KVIV, Antwerp, 1982.

Batchelor, G.K., 1983, An introduction to fluid dynamics, Cambridge University Press.

Baumert, H. and G. Radach, 1992, Hysteresis of turbulent kinetic energy in non-rotational tidal flows: a model study, Journal of Geophysical Research, 97, (C3) 3669-3677.

Beardsley, R.C., J. Candela, R. Limeburner, W.R. Geyer, S.J. Lentz, B.M. Castro, D. Cacchione and N. Carneiro, 1995, The M_2 tide on the Amazon shelf, Journal of Geophysical Research, 100, 2283-2319.

Been, K., 1981, Non-destructive soil bulk density measurement using X-ray attenuation, Geotechnical Testing Journal, 4, 169-176.

Been, K. and G.C. Sills, 1981, Self-weight consolidation of soft soils: an experimental and theoretical study, Géotechnique, 31 (4) 519-535.

Been, K. and M.G. Jefferies. 1985, A state parameter for sand, Geotechnique, 35, No.3, pp. 99-112.

Berhane, I., R.W. Sternberg, G.C. Kineke, T.G. Milligan and K. Kranck, 1997, The variability of suspended aggregates on the Amazon Continental Shelf, Continental Shelf Research, 17 (3) 267-285.

Berlamont, J., M. Ockenden, E.A. Toorman and J.C. Winterwerp, 1993, The characterisation of Cohesive Sediment Properties, Coastal Engineering, 21 105-128.

Berner, R.A., 1980, Early diagenesis: a theoretical approach, Princeton University Press, New York.

Best, J.L. and M.R. Leeder, 1993, Drag reduction in turbulent muddy seawater flows and sedimentary consequences, Sedimentology, 40, 1129-1137.

Best, J.L., 1993, On the interaction between turbulent flow structure and bedform development: some considerations from recent experimental research, in: Turbulence – Perspectives on flow and sediment transport, ed. N.J. Clifford, J.R. French and J. Hardisty, John Wiley and Sons, 61-92.

Biggs, R.B. and B.A. Howell, 1984, The estuary as a sediment trap: alternate approaches to estimating its filtering efficiency, in: V.S. Kennedy, The Estuary as a Filter, Academic Press, 107-129.

Bird, R.B., W.E. Stewart and E.N. Lightfoot, 1960, Transport phenomena, Wiley, New York, Chapter 16.

Birdi, K.S., 2003, Handbook of Surface and Colloid Chemistry, Second Edition, CRC Press, Boca Raton, pp 765.

Biot, M.A., 1956, Theory of deformation of a porous visco-elastic anisotropic solid, Journal Applied Physics, 5 (27) 459-467.

Bishop, A.W. and V.K. Garga, 1969, Drained tension tests on London Clay, Geotechnique vol. 19, pp. 309-313.

Bjerrum, L., 1972, Embankments on soft ground, Proceedings ASCE, Special Conference on Performance of Earth and Earth-Supported Structures, II, 1-54.

Black, K.P. and M.A. Rosenberg, 1994. Suspended sand measurements in a turbulent environment: field comparison of optical and pump sampling techniques, Coastal Engineering, 24, 137–150.

Blanchard G.F., D.M. Paterson, L.J. Stal, P. Richard, R. Galois, V. Huet, J. Kelly, C. Honeywill, J. De Brouwer, K.R. Dyer, M. Christie and M. Seguignes, 2000, The effect of geomorphological structures on potential biostabilisation by microphytobenthos on intertidal mudflats, Continental Shelf Research 20 (10) 1243-1256.

Boadway, J.D., 1978, Dynamics of growth and breakage of alum flocs in presence of fluid shear, ASCE, Journal of the Environmental Engineering Division, 104 (EE5) 901-915.

Bolt, G.H. and M.G.M.Bruggenwert, 1976, Soil Chemistry, A Basic Elements Development in Soil Science 5A, Elsevier, Amsterdam.

Booij, R., P.J. Visser and H. Melis, 1993, Laser Doppler measurements in a rotating annular flume, Proceedings of the Fifth International Conference on Laser Anemometry Advances and Applications, SPIE – The International Society for Optical Engineering, Bellingham, Washington, 409-416.

Boudreau, B.P. and B.B. Jørgensen, 2001, The benthic boundary layer, Oxford University Press, 404 pp.

Bouma, T.J., M.B. de Vries, E. Low, G. Peralta, I.C. Tánczos and P.M.J. Herman, 2004, Plant traits and capacity for ecosystem-engineering: stiffness of emerging macrophytes, submitted to Ecology.

Bowden, R.K., 1988, Compression behaviour and shear strength characteristics of a natural silty clay sedimented in the laboratory, PhD-thesis, Oxford University, UK.

Brace, F.W. and E.G. Bombolakis, 1963, A note on brittle crack growth in compression, Journal of Geophysical Research, 68 (12) 3709-3713.

Bratby, J., 1980, Coagulation and flocculation, with an emphasis on water and waste water treatment, Uplands Press Ltd, Croydon, England.

Brennen, C.E., 1995, Cavitation and bubble dynamics. Oxford University Press, Oxford.

Brewer, P.G. et al. 1999, Direct experiments on the ocean disposal of fossil fuel CO_2. Science 284 (May 7): 943.

Briggs, K.B., P.D. Jackson, R.J. Holyer, R.C. Flint, J.C. Sandige, D.K. Young, 1998, Two-dimensional variability in porosity, density and electrical resistivity of Eckenforde Bay sediment, Continental Shelf Research, 18, 1939-1964.

British Standards Institution, British Standards Publishing Limited (BSPL), London, UK, www.bsi-lobal.com.

Broek, D., 1982, Elementary Engineering Fracture Mechanics, 3rd ed. Sijthoff & Noordhoff.

Broekhoff, J.C.P., 1969, Adsorption and Capillarity, PhD-thesis, Delft University of Technology, Faculty of Chemical Engineering, Waltman Delft.

Brown, W.D. and R.C. Ball, 1985, Computer simulation of chemically limited aggregation, Journal of Physics, Section A: Mathematics and General, 18, L517-L521.

Bruens, A.W., 2003, Entraining mud suspensions, PhD-thesis, Delft University of Technology, Department of Civil Engineering and Geosciences; also: Delft University of technology, Faculty of Civil Engineering and Geosciences, Communications on Hydraulic and Geotechnical Engineering, Report 03-1, ISSN 0169-6548.

Campbell, C.S., 1988, Boundary interactions for two-dimensional granular flows: asymmetric stresses and couple stresses, in: Micromechanics of Granular Materials, ed. M. Satake and J.T. Jenkins, Studies in Applied Mechanics 20, Elsevier, Amsterdam.

Cai, Z. and R.J. Bathurst, 1996, Deterministic sliding block methods for estimating seismic displacements of earth structures, Soil Dynamics and Earthquake Engineering, 15, 255-268

Carslaw, H.S. and J.C. Jeager, 1959, Conduction of heat in solids, 2nd ed., Oxford Science Publications.

Cazemier, W. and M.A.T. Visschedijk, 1997, Gas bubbles in sediments. TNO report no. 97-NM-R1066, DM11 (in Dutch).

Cha, S.B., 1988, Formation of hydrates in waterbased drilling muds and mud-like fluids, M.S. thesis, T-3506, Colorado School of Mines, Golden, CO.

Chen, S. and D. Eisma, 1995, Fractal geometry of in situ flocs in the estuarine and coastal environments, Netherlands Journal of Sea Research, 32 (2) 173-182.

Chen, W. and Saleeb, A.F., 1994, Constitutive equations for engineering materials, vol I, Elasticity and Modelling, Elsevier, Amsterdam.

Cellino, M. and W.H. Graf, 1999, Sediment-laden flow in open channels under non-capacity and capacity conditions, ASCE, Journal of Hydraulic Engineering, 125 (5) 456-462.

Cheng, N.-S., 1997, Effect of concentration on settling velocity of sediment particles, ASCE, Journal of Hydraulic Engineering, 123 (8) 728-731.

Chesner T.J. and M.C. Ockenden, 1997, Numerical modelling of mud and sand mixtures, in N. Burt, R. Parker and J. Watts, Cohesive Sediments, John Wiley & Sons, Chichester, 395-406.

Chou, H.-T., 1989, Rheological response of cohesive sediments in water waves, PhD-thesis University of California, Berkeley, USA.

Chou, H.-T, M.A. Foda and J.R. Hunt, 1993, Rheological response of cohesive sediment to oscillatory forcing, in: A.J. Mehta, Nearshore and Estuarine Cohesive Sediment Transport, American Geophysical Union, Coastal and Estuarine Studies, 42, 126-147.

Christensen, R.W. and B.M. Das, 1974, Hydraulic erosion of remoulded cohesive soils, in Special Report 135, Soil Erosion: causes and mechanisms, preventions and control, 8-19.

Christensen, B.A., 1965, Discussion of erosion and deposition of cohesive soils, ASCE, Journal of the Hydraulic Division, 91 (5) 301-308.

Christensen, B.A., 1972, Incipient motion on cohesionless channel banks, in: Sedimentation, ed. H.S. Shen, Colorado State University, Fort Collins, USA, 1-22.

Christie, M.C., K.R. Dyer, G. Blanchard, A. Cramp, H.J. Mitchener and D.M. Paterson, 2000, Temporal and spatial distributions of moisture and organic contents across a macro-tidal mudflat, Continental Shelf Research, 20, 1219-1241.

Claeys, S., G. Duman, J. Lancknens and K. Trouw, 2001, Mobile turbidity measurement as a tool for determining future volumes of dredged material in access channels to estuarine ports, Terra et Aqua, 84, 8-16.

Coleman, N.L., 1981, Velocity profiles with suspended sediment, Journal of Hydraulic Research, 19 (3) 211-229.

Coleman, N.L., 1986, Effects of suspended sediment on the open-channel velocity distribution, Water Resources Research, 22 (10) 1377-1384.

Coleman and Prior, 1978, Offshore Technology Conference Paper OTC 3170.

Concha, F. and M.C. Bustos, 1985, Theory of sedimentation of flocculated fine particles, in: Flocculation & Dewatering, ed. B.M. Moudgil and B.J. Scheiner, University of Florida, USA, 275-284.

Cools, P.M.C.B.M., 1990, Research activities on mechanical rock cutting and dredging at DELFT HYDRAULICS, Proceedings VIth Congress International Association of Engineering Geology, Amsterdam, August, 1990.

Coussot, P., 1997, Mudflow rheology and dynamics, IAHR monograph series, Balkema, Rotterdam, pp 255.

Cornelisse, J.M., W.G.M. van Kesteren and C. Kuijper, 1997, Evaluation of large-scale laboratory experiment on sediment from Lake Ketel with artificial gas production, Delft Hydraulics Report Z2111, DM14 (in Dutch).

Crickmore, M.J., 1982, Data collection - tides, tidal currents and suspended sediment, The Dock & Harbour Authority, LXIII (742) 183-186.

Crickmore, M.J., G.S. Tazioli, P.G. Appleby and F. Oldfield, 1990, The use of nuclear techniques in sediment transport and sedimentation problems, Tech. Doc. in Hydrology, IHP-III Project 5.2, Unesco, Paris, 169 pp.

Croad, R.N., 1981, Physics of erosion of cohesive soils, University of Auckland, School of Engineering, Department of Civil Engineering, Report 247, Auckland, New Zealand.

Crosato, A. and W.G.M. van Kesteren, 1998, Non-homogeneous gas escape and consolidation. Part 1: drainage system at consolidation, Delft Hydraulics Report Z2313, DM16.

Cross, M.M., 1965, Rheology of non-Newtonian fluids: a new flow equation for pseudo-plastic systems, Journal of Colloid Sciences, 20, 417-437.

Dalrymple, R.A. and P.L. Liu, 1978, Waves over soft mud: a two-layer fluid model, Jnl. of Physical Oceanography, 8, 1121-1131.

Dankers, N. and K. Koelemaij, 1989, Variations in the mussel population of the Dutch Wadden Sea in relation to monitoring of other ecological parameters, Helgoländer Meeresuntersuchungen, 43, 529-535.

Dankers, N., 2002, personal communication.

Dawson, T.H,. 1998, On Markov Theory and Wave-Crest Statistics, ASME Journal of Offshore Mechanics and Arctic Engineering, 120, 56-58.

De Beer, D. and Kühl, M., 2001, Interfacial microbial mats and biofilms, in: B.P. Boudreau and B.P. Jørgensen, The Benthic Boundary Layer- Transport Processes and Biochemistry, Oxford University Press, Oxford, 374-394.

De Boer, J., L.M. Merckelbach and J.C. Winterwerp, 2003, A parameterised consolidation model for cohesive sediment, Proceedings INTERCOH-2003, Elsevier, Proceedings in Marine Science.

De Brouwer, J.F.C., 2002, Dynamics in extracellular carbohydrate production by marine benthic diatoms, PhD-thesis, Katholieke Universiteit Nijmegen; Netherlands Institute of Ecology.

De Brouwer, J.F.C., S. Bjelic, E.M.G.T. De Deckere and L.J. Stal, 2000, Interplay between biology and sedimentology in a mudflat (Biezelingse Ham, Westerschelde, The Netherlands), Continental Shelf Research, 20, 1159-1177.

De Deckere E.M.G.T., B.A. Kornman, N. Staats, G.R. Termaat, B. De Winder, L.J. Stal and C.H.R. Heip, 2002, The seasonal dynamics of benthic (micro)organisms and extracellular carbohydrates in an intertidal mudflat and their effect on the concentration of suspended sediment, in: J.C. Winterwerp and C. Kranenburg, Fine Sediment Dynamics in the Marine Environment, Proceedings in Marine Science Vol 5, Elsevier, Amsterdam, 429-440.

De Vlieger, H. and J. De Cloedt, 1987, Navitracker: a giant step forward in tactics and economics of maintenance dredging, Terra et Aqua, 35, 2-18.

De Wit, P.J., 1995, Liquefaction of cohesive sediments caused by waves, PhD-thesis, Delft University of Technology, Faculty of Civil Engineering, also: Communications on Hydraulic and Geotechnical Engineering, Report 95-2.

De Wit, P.J. and C. Kranenburg, 1997, On the liquefaction and erosion of mud due to waves and currents, in: N. Burt, R. Parker and J. Watts, Cohesive Sediments, John Wiley & sons, 331-340.

De Wolf, P., 1990, The North Sea, Terra Publishers, Zutphen, The Netherlands.

Decho A.W., 2000, Microbial biofilms in intertidal systems: an overview. Continental Shelf Research, 20 (10) 1257-1273.

Desai, et. al., 1991, Implementation of hierarchical single surface δ_0 and δ_1 models in finite element procedures, International Journal of Numerical & Analytical Methods in Geomechanics, 15, 649-680.

Dinel, H., S.P. Mathur, A. Brown, and M. Levesque, 1988, A field study of the effect of depth on methane production in peatland water; equipment and preliminary results, The Journal of Ecology, 76 (4) 1083.

Dingemans, M.W., 1998, Movement of gas bubbles in artificial sludge depots, Delft Hydraulics Report Z2313, DM23.

Dittman, S., 1999, The Wadden Sea ecosystem – stability properties and mechanisms, Springer, Berlin, pp 305.

Dong, L., E. Wolanski and Y. Li, 1997, Field and modelling studies of fine sediment dynamics in the extremely turbid Jiaojianng River estuary, China, Journal of Coastal Research, 13 (4) 995-1003.

Dowling, J.J., 1990, Estimating porosity of partially saturated sediments, Engeneering Geology, Vol. 29, No. 2, July, pp. 139-147.

Downing, J.P. and R.A. Beach, 1989. Laboratory apparatus for calibrating optical suspended solids sensors. Marine Geology, 86, 243–249.

Downing, J. P., R.W. Sternberg and C. R. B. Lister, 1981, New instrumentation for the investigation of sediment suspension processes in the shallow marine environment, Journal of Marine Geology, 42, 19-34.

Doyle, H.A. (1995): Seismology, Wiley & Sons.

Dyer, K.R., 1989, Sediment processes in estuaries: future research requirements, Journal of Geophysical Research, 94 (C10) 14,327 - 14,339.

Einstein, H.A. and N.G. Chien, 1955, Effects of heavy sediment concentration near the bed on velocity and sediment distribution, MRD Sediment Series No 8, University of California, Berkeley.

Einstein, H.A. and R.B. Krone, 1962, Experiments to determine modes of cohesive sediment transport in salt water, Journal of Geophysical Research, 67 (4) 1451-1461.

Eisma, D. (ed.), 1996, In-situ measurements of suspended matter particle size and settling velocity, Journal of Sea Research, special issue, 36 (1-2).

Elata, E. and A.T. Ippen, 1961, The dynamics of open channel flow with suspensions of neutrally buoyant particles, MIT, Hydrodynamics Laboratory, Technical Report 45, Cambridge.

Evans, D., E.L. King, N.H. Kenyon, C. Brett, D. Wallis, 1996, Evidence for long-term instability in the Storegga Slide region off western Norway, Marine Geology 130, 281-292.

Evans, W.C., G.W. Kling, M.L. Tuttle, G. Tanyileke, and L.D. White, 1993, Gas buildup in Lake Nyos, Cameroon: The recharge process and its consequences: Applied Geochemistry, 8, 207-221.

Everett, D.H., 1988, Basic principles of colloid science, University of Bristol, Royal Society of Chemistry.

Eysink, W.D., 1989, Sedimentation in harbour basins – small density differences may cause serious effects, Delft Hydraulics publication 417.

Faas, R. and S. Wartel, 1985, Resuspension potential of fluid mud and its significance to sediment transport in the Western Scheldt estuary, Belgium, Estuaries, 8, 2b.

Felderhof, B.U. and G. Ooms, 1990, Effect of inertia, friction and hydrodynamic interactions on turbulent diffusion, European Journal of Mechanics, Part B/Fluids, 9 (4) 350-368.

Fenchel, T.M. and R.J. Riedl, 1970, The sulphide system: a new biogenic community underneath the oxidised layer of marine sand bottoms, Marine Biology, 7, 255-268.

Feng, J., 1993, Bottom mud transport due to water waves, Ph.D. Thesis, Report UFL/COEL-TR/090, University of Florida, Gainesville, USA.

Fennessy, M.J., K.R. Dyer and D.A. Huntley, 1994, INSSEV: an instrument to measure the size and settling velocity of flocs in situ, Marine Geology, 117, 107-117.

Flemming, B.W., 2000, A revised textural classification of gravel-free muddy sediments on the basis of ternary diagrams, Continental Shelf Research, 20, 1125-1137.

Floss, R., 1970, Vergleich der Verdichtungs- und Verformungseigenschaften unstetiger und stetiger Kiessanden hinsichtlich ihrer Eignung als ungebundenes Schuttmaterial im Strassenbau, Wissenschaftliche Berichtte der Bundesanstalt fur Strassenwesen, Heft 9.

Foda, M.A. and S.-Y. Tzang, 1994, Resonant waves of silty soil by water waves, Journal of Geophysical Research, 99 (C10) 20,463-20,475.

Ford, W.E., 1932, Dana's textbook of mineralogy, John Wiley & Sons.

Fredsøe, J. and R. Deigaard, 1992, Mechanics of coastal sediment transport, World Scientific, Advanced Series on Ocean Engineering, 3, 369.

Fredsøe, J., 1984, Turbulent boundary layer in wave-current motion, ASCE, Journal of Hydraulic Engineering, 110 (8) 1103-1120.

Friedlander, S.K., 1977, Smoke, dust and haze - fundamentals of aerosol behaviour, John Wiley and Sons, New York.

Fugate, D.C. and C.T. Friedrichs, 2002, Determining concentration and fall velocity of estuarine particle populations using ADV, OBS and LISST, Continental Shelf Research, 22, 1867-1886.

Fuller, W.B. and S.E. Thomson, 1907, The laws of proportioning concrete, Transactions of the American Society of Civil Engineering, 59.

Gabioux, M., S.B. Vinzon and A.M. Paiva, 2003, Tidal propagation over fluid mud layers in the Amazon Shelf, submitted to Continental Shelf Research

Galland, J.-C., D. Laurence and C. Teisson, 1997, Simulating turbulent vertical exchange of mud with a Reynolds stress model, Proceedings of the 4th Nearshore and Estuarine Cohesive Sediment Transport Conference

INTERCOH'94, July 1994, Wallingford, UK, ed. T. N. Burt, W.R. Parker and J. Watts, John Wiley & Sons, 439-448.

Gartner, J.W., R.T. Cheng, P.-F. Wang and K. Richter, 2001, Laboratory and field evaluations of the LISST-100 instrument for suspended particle size determinations, Marine Geology, 175, 199-219.

Gelfenbaum, G. and J.D. Smith, 1986, Experimental evaluation of a generalized suspended-sediment transport theory, in Shelf Sands and Sandstones, ed. R.J. Knight and J.R. McLean, Canadian Society of Petroleum Engineers, Calgary, 133-144.

Gibbs, R.J., 1985, Estuarine flocs: their size, settling velocity and density, Journal of Geophysical Research, 90 (C2) 3249-3251.

Gibson, R.E., G.L. England and M.J.L. Hussey, 1967, The theory of one-dimensional consolidation of saturated clays, Géotechnique, 17, 261-273.

Goldberg, U. and D. Apsley, 1997, A wall-distance-free low Re k-ε turbulence model, Computational Methods in Applied Mechanics and Engineering, 145, 227-238.

Graham, J., M.L. Loonan and K.V. Lew, 1983, Yield states and stress strain relationships in a natural plastic clay, Canadian Geotechnical Journal, 20, 502-516.

Gramberg, J., 1989, A non-conventional view on rock mechanics and fracture mechanics, Rotterdam.

Grant, W.D. and O.S. Madsen, 1979, Combined wave and current interaction with a rough bottom, Journal of Geophysical Research, 84 (C4) 1797-1808.

Gregory, J., 1997, The density of particle aggregates, Water Science Technology, 36 (4) 1-13.

Gregory, J., 1985, The action of polymeric flocculants, in: Flocculation, Sedimentation and Consolidation, ed. B.M. Moudgil and B.J. Scheiner, Engineering Publication Services, University of Florida, USA, 125-138.

Groen, A.E., R. de Borst and S.J. van Eekelen, 1995, An elasto-plastic model for clay: Formulation and algorithmic aspects, In: Proceedings 5th International Symposium Numerical Models Geomechanics, Rotterdam, Balkema, 27-32.

Groen, P. and R. Dorrestein, 1976, Sea waves, Royal Dutch Meteorological Institute, Reports on Oceanographic and Maritime Meteorology, No 11 (in Dutch).

Guan, W.B., E. Wolanski and L.X. Dong, 1998, Cohesive sediment transport in the Jiaojiang River Estuary, China, Estuarine, Coastal and Shelf Science, 46, 861-871.

Gularte, R.C., W.E. Kelly and V.A. Nacci, 1980, Erosion of cohesive sediments as a rate process, Ocean Engineering, 7, 539-551.

Guo, T and S. Prakash, 1999, Liquefaction of Silts and Silt-Clay Mixtures; Journal of Geotechnical and Geo-environmental Engineering, I, 706-710

Gust, G., 1976, Observations on turbulent-drag reduction in a dilute suspension of clay in sea water, Journal of Fluid Mechanics, 75 (1) 29-47.

Gust, G., 1984, discussion of N.L. Coleman's "Velocity profiles with suspended sediment", and reply by the author, IAHR, Journal of Hydraulic Research, 22 (4) 263-289.

Gust, G., 1990, Fluid velocity measuring instrument. US Patent No. 4,986. 122.

Gust, G. and V. Müller, 1997, Interfacial hydrodynamics and entrainment functions of currently used erosion devices, in: Cohesive Sediments, ed. N. Burt, R. Parker and J. Watts, Proceedings 4th Nearshore and Estuarine Cohesive Sediment Transport Conference INTERCOH-94, 49-174.

Harney, J.N. and D.M. Rubin, 2004, Spatial and temporal analysis of heterogeneous sediments on the Santa Cruz, CA Shelf, abstract AGU meeting, Ocean Sciences, paper OS41L-04.

Harrison, A.J.M. and M.W. Owen, 1971, Siltation of fine sediments in estuaries, Proceedings of the XIV IAHR Congress, Paris, Paper D1.

Haw, M.D., W.C.K. Poon and P.N. Pusey, 1997, Structure and arrangement of clusters in cluster aggregation, The American Physical Society, Physical Review E, 56 (2) 1918-1933.

Hayes, M.H.B. and R.S. Swift, 1990, Geneses, Isolation, Composition and Structure of humic substances, Soils Colloids and their associations in aggregates, NATO ASI series B215, 245-305, Plenum Press.

Hayter, E.J., 1983, Prediction of cohesive sediment movements in estuarial waters, PhD-thesis, University of Florida, Coastal and Engineering Department, Gainesville, USA.

Hawley, N., 1982, Settling velocity distribution of natural aggregates, Journal of Geophysical Research, 87 (C12) 9489-9498.

Head, K.H., 1986, Manual of Soil Laboratory Testing, Vol. 1, 2 and 3, Pentech Press London.

Headland, J.R., 1994, Application of an engineering model for harbour sedimentation, Proceedings 28th PIANC International Navigation Congress, Section II-4, 156-160.

Henriet, J.P., Mienert, J., 1996, Gas Hydrates, Relevance to World Margin Stability and Climate Change, Proceedings Workshop Gent.

Herman P.M.J., J.J. Middelburg and C.H.R. Heip, 2001, Benthic community structure and sediment processes on an intertidal flat: results from the ECOFLAT project, Continental Shelf Research, 21, 2055-2071.

Hill, P.S., G. Voulgaris and J.H. Trowbridge, 2001, Controls on floc size in a continental shelf bottom boundary layer, Journal of Geophysical Research, 106 (C5) 9543-9549.

Hill, D.C., S.E. Jones and D. Prandle, 2003, Derivation of sediment resuspension rates from acoustic back-scatter time-series in tidal water, Continental Shelf Research, 23, 19-40.

Hino, M., 1963, Turbulent flow with suspended particles, ASCE, Journal of the Hydraulics Division, 89 (HY4) 161-185.

Hinze, J.O., 1975, Turbulence, McGraw-Hill Book Company.

Hogg, R., R.C. Klimpel and D.T. Ray, 1985, Structural aspects of floc formation and growth, in: Flocculation, Sedimentation and Consolidation, ed. B.M. Moudgil and P. Somasundaran, University of Florida, Gainesville, USA, 217-227.

Hölscher, P., 1995, Dynamic response of saturated and dry soils, PhD-thesis Delft University of Technology, Delft.

Holzer, T.L., K. Hoeg and K. Arulanandan, 1973, Excess pore pressures during undrained clay creep, Canadian Geotechnical Journal, 10 (1) 12-24.

Hopfinger, E.J., 1987, Turbulence in stratified fluids: a review, Journal of Geophysical Research, 92 (C5) 5287-5303.

Houlsby, G.T. and A.M. Purzin, 2002, Rate-dependent plasticity models derived from potential functions, Journal of Rheology 46(1) 113-126.

Houwing, E.J. and L.C. van Rijn, 1997, In-situ Erosion Flume (ISEF): determination of bed-shear stress and erosion of a kaolinite bed, Journal of Sea Research, 39, 243-253.

Hovland, M. and A.G. Judd, 1988, Seabed Pockmarks and Seepages, Graham and Trotman, London.

Hu, S., 1990, Self-weight consolidation on impervious bases, MSc-thesis, IHE, The Netherlands.

Huisman, M. and W.G.M. Van Kesteren, 1998, Consolidation theory applied to the capillary suction time (CST) apparatus, Water Science and Technology, 37 (6-7) 117–124.

Hunter, R.J., 2001, Foundations of Colloid Science, Second Edition, Oxford University Press, pp 806.

Hydraulics Research Station, 1980, The Severn Estuary; Measurements of the vertical and lateral distribution of mud suspensions, Report No EX 965.

Inglis, C.C. and H.H. Allen, 1957, The regimen of the Thames Estuary as affected by currents, salinities and river flow, Proceedings of the Institution of Civil Engineers, Maritime and Waterways Engineering Division Meeting, 7, 827-879.

Ingraffea, A.R., 1987, Theory of crack initiation and propagation in rock. In: Fracture Mechanics of rock. B.K. Atkinson (ed.), London, 71-110.

IOS Taunton, 1977, Investigation of turbidity structures in the Severn Estuary and Inner Bristol Channel, Institute of Oceanographic Sciences, Taunton, Cruise Report No 92.

Itakura, T. and T. Kishi, 1980, Open channel flow with suspended sediments, ASCE, Journal of the Hydraulics Division, 106 (HY8) 1325-1343.

Jâki , J., 1944, The coefficient of earth pressure at rest, Journal of the Union of Hungarian Engineers and Architects, 355-8 (in Hungarian).

Jeffrey, D.J., 1982, Aggregation and break-up of clays in turbulent flow, Advances in Colloid and Interface Science, 17, 213-218.

Jespen, R., J. Roberts and W. Lick, 1997, Effects of bulk density on sediment erosion rates, Water, Air and Soil Pollution, 99, 21-31.

Jin, L. and G.S. Penny, 1994, Dimensionless method for the study of particle settling in non-Newtonian fluids, Paper SPE 28563, presented at SPE 69th Annual Technical Conference sand Exhibition, New Orleans, Louisiana, Sept. 25-28.

Jørgensen, B.B., 2001, Life in the diffusive boundary layer, in: B.P. Boudreau and B.P. Jørgensen, The Benthic Boundary Layer- Transport Processes and Biochemistry, Oxford University Press, Oxford, 348-373.

Jørgensen, B.B. and B.P Boudreau, 2001, Diagenesis and sediment-water exchange, in: B.P. Boudreau and B.P. Jørgensen, The Benthic Boundary Layer- Transport Processes and Biochemistry, Oxford University Press, Oxford, 211-244.

Judd, A., G. Davies, J. Wilson, R. Holmes, G. Baron and I. Bryden, 1997, Contributions to atmospheric methane by natural seepage on the UK continental shelf, Marine Geology, 137, 165-189.

Kachel, N.B. and J.D. Smith, 1989, Sediment transport and deposition on the Washington continental shelf, in Coastal Oceanography of Washington and Oregon, Elsevier Oceanography Series 47, ed. M.R. Landry and B.M. Hickey, 287-348.

Kajihara, M., 1971, Settling velocity and porosity of large suspended particles, Journal of the Oceanographic Society of Japan, 27 (4) 159-162.

Kandiah, 1974, Fundamental aspects of surface erosion of cohesive soils, PhD-thesis, University of California, USA.

Kantha, L.H., O.M. Philips and R.S. Azad, 1977, On turbulent entrainment at a stable density interface, Journal of Fluid Mechanics, 79 (4) 753-768.

Kao, J. and J.R. Bell, 1987, Compressibility and cementation characteristics of some calcareous sea-floor oozes, Proceedings International Symposium on Prediction and Performance in Geotechnical Engineering, Rotterdam, Balkema, 251-258.

Kappenberg, J., G. Schymura, H.-U. Fanger, 1995, Measurements of sediment dynamics and estuarine circulation in the turbidity maximum of the Elbe river, Netherlands Journal of Aquatic Ecology, 29, 229-238.

Karelse, M. and J.A.Th.M Van Kester, 1995, Validation of DELFT3D with tidal flume measurements, Delft Hydraulics, Report Z810.

Keedwell, M.J., 1984, Rheology and Soil Mechanics, Elsevier, Amsterdam.

Kelly, W.E. and R.C. Gularte, 1981, Erosion resistance of cohesive soils, ASCE, Journal of the Hydraulics Division, 107 (10) 1211-1224.

Kemeny, J. and N.G.W. Cook, Frictional stability of heterogeneous surfaces in contact, Department of Materials Science and Mineral Engineering University of California, Berkely, 1988.

Keurbis, R., D. Negussey, and Y.P. Void, 1988, Effect of gradation and fines on content on the undrained response of sand, Hydraulic Fill Structures, Special Conference Fort Collins, Geotechnical Special Publications no 21, ASCE New York, 330-345.

Kihara, T., Sasajima, H., Yoshinaga, K., Koizuka, T., Sasayama, H., Yoshinaga, H. and Fujimoto, T., 1994, Field surveys on siltation-prevention effects in waterways and anchorage by submerged walls, Hydro-Port'94, International Conference on Hydro-Technical Engineering for Port and Harbour Construction, October, Yokosuka, Japan, 1225-1242.

Kineke, G.C., 1993, Fluid muds on the Amazon continental shelf, PhD-thesis, University of Washington.

Kineke, G.C. and R.W. Sternberg, 1995, Distribution of fluid muds on the Amazon continental shelf, Marine Geology, 125, 193-233.

Kineke, G.C., R.W. Sternberg, J.H. Trowbridge and W.R. Geyer, 1996, Fluid mud processes on the Amazon continental shelf, Continental Shelf Research, 16 (5/6) 667-696.

Kirby, R. and W.R. Parker, 1980, Settled mud suspensions in Bridgewater Bay, Bristol Channel, Institute of Oceanographic Sciences, Taunton, IOS Report, No 107.

Kirby, R. and W.R. Parker, 1983, Distribution and behaviour of fine sediment in the Severn estuary and inner Bristol Channel, UK, Canadian Journal of Fisheries and Aquatic Science, 40 (suppl.1) 83-95.

Koenders, M.A, 1994, Least squares methods for the mechanics of non-homogeneous granular assemblies, Acta Mechanica, 106, 23-40.

Kolb, M. and H.J. Hermann, 1987, Surface fractals in irreversible aggregation, Physical Review Letters, 59 (4) 454-457.

Komamura, F. and R.J. Huang, 1974, A new rheological model for soil behaviour, ASCE, Journal of the Geotechnical Engineering Division, 100 (GT7) 807-824.

Kornman, B.A. and M.G.T. De Deckere, 1998, Temporal variation in sediment erodibility and suspended sediment dynamics in the Dollard estuary, in: Sedimentary Processes in the Intertidal Zone, e.d. K.S. Black, D.M. Paterson and A. Cramp, Special Publication, Geological Society, London, 231-241.

Kranck, K. and Milligan, T.G., 1992, Characteristics of suspended particles at an 11-hour anchor station in San Francisco Bay, California, Journal of Geophysical Research, 97 (C7) 11,373-11,382.

Kranenburg, C., 1992, Hindered settling and consolidation of mud – analytical results, Delft University of technology, Department of Civil Engineering, Hydromechanics Section, Report 11-92.

Kranenburg, C., 1994, An entrainment model for fluid mud, Communications on Hydraulic and Geotechnical Engineering, Faculty of Civil Engineering, Delft University of Technology, Report 94-10.

Kranenburg, C., 1994, On the fractal structure of cohesive sediment aggregates, Estuarine, Coastal and Shelf Science, 39, 451-460.

Kranenburg, C. and J.C. Winterwerp, 1997, Erosion of fluid mud layers – I: Entrainment model, ASCE, Journal of Hydraulic Engineering, 123 (6) 504-511.

Kriele, H.O., M. De Vries and Z.B. Wang, 1998, Morphological interaction between the Yellow River and its estuary, in: J. Dronkers and M. Scheffers (eds.), Physics of estuaries and coastal seas, A. A. Balkema, Rotterdam.

Krone, R.B., 1962, Flume studies of the transport of sediment in estuarial shoaling processes, Final Report Hydraulic Engineering Laboratory and Sanitary Engineering Research Laboratory, University of California, Berkeley, USA.

Krone R.B., 1963, A Study of Rheologic Properties of Estuarial Sediments. Report prepared for the Committee on Tidal Hydraulics under Contract DA-22-079-CIVENG-61-7 with the Waterways Experiment Station, Corps of Engineers, US Army.

Krone, R.B., 1986, The significance of aggregate properties to transport processes, Proceedings of a Workshop on Cohesive Sediment Dynamics with Special Reference to Physical Processes in Estuaries, Tampa, Florida, Springer Verlag, Lecture Notes on Coastal and Estuarine Studies, 14, Estuarine and Cohesive Sediment Dynamics, 66-84.

Krone, R.B., 1993, Sedimentation revisited, in Nearshore and Estuarine Cohesive Sediment Transport, ed. A.J. Mehta, American Geophysical Union, Coastal and Estuarine Studies, 108-125.

Kruijt, H.R., 1952, Colloid Science, Volume I: Irreversible systems, Elsevier, Amsterdam, pp 389

Krumbein, 1941, Measurement and geological significance of shape and roundness of sedimentary particles, Journal of Sedimentary Petrology, 11, Geotechnical Special Publications no 21, ASCE New York, 330-345.

Kuijper, C., J.M. Cornelisse and J.C. Winterwerp, 1989, The deposition of graded cohesive sediment, presented at the International Symposium on Sediment Transport Modelling, New Orleans.

Kuijper, C., J.M. Cornelisse, J.C. Winterwerp, 1989, Research on erosive properties of cohesive sediments, Journal of Geophysical Research, 94 (C10) 14,341-14,350.

Kuijper, C., J.M. Cornelisse and J.C. Winterwerp, 1990a, Erosion and deposition of natural muds – sediments from the Hollandsh Diep (north of Sassenplaat), The Netherlands, Rijkswaterstaat & Delft Hydraulics, Cohesive Sediments, Report 27.

Kuijper, C., J.M. Cornelisse and J.C. Winterwerp, 1990b, Erosion and deposition of natural muds – sediments from the Western Scheldt (Breskens), The Netherlands, Rijkswaterstaat & Delft Hydraulics, Cohesive Sediments, Report 29.

Kuijper, C., J.M. Cornelisse and J.C. Winterwerp, 1990c, Erosion and deposition characteristics of natural muds – Sediments from Ketelmeer (Netherlands), Delft Hydraulics and Rijkswaterstaat, Cohesive Sediments Report 30.

Kuijper, C. et al., 1993, On the methodology and accuracy of measuring physico-chemical properties to characterise cohesive sediments, European Commission, MAST-1 programme, G-6M coastal Morphodynamics.

Kusuda, T., T. Umita, K. Koga, T. Futawatari and Y. Awaya, 1984, Erosional processes of cohesive sediment, Water Science Engineering, Water Pollution Research Control, 117, 891-901.

Kumbhojkar, A.S. and P.K. Banerjee, 1993, An anisotropic hardening rule for saturated clays, International Journal of Plasticity, 9, 861-888.

Kynch, G.J., 1952, A theory of sedimentation, Transactions of the Faraday Society, 58, 166-176.

Kvenvolden, K.A., 1988, Methane hydrate – A major reservoir of carbon in the shallow geosphere, Chemical Geology, 71, 41.

Lambe, T.W. and R.V. Whitman, 1979, Soil Mechanics, SI Version, John Wiley and Sons, New York, pp 553.

Land, J.M., R.B. Kirby and J.B. Massey, 1997, Developments in the combined use of acoustic doppler current profilers and profiling silt meters for suspended solids monitoring", in Proceedings of the 4th Nearshore and Estuarine Cohesive Sediment Transport Conference, INTERCOH'94, ed. N. Burt, R. Parker and J. Watts, John Wiley & Sons, New York, 187-196.

Lau, Y.L and V.H. Chu, 1987, Suspended sediment effect on turbulent diffusion, Proceedings 22nd IAHR Congress, Lausanne, 221-226.

Launder, B.E. and D.B. Spalding, 1974, The numerical computation of turbulent flows, in: Computer Methods in Applied Mechanics and Engineering, 3, 269-289, North-Holland Publishing Company.

Lawn, B., Fracture of brittle solids, 2nd ed. 1993, Cambridge Solid State Science Series.

Le Hir, P. Bassoulet and H. Jetsin, H., 2001, Application of the continuous modelling concept to simulate high-concentrated suspended sediment in a macro-tidal estuary, in Coastal and estuarine Fine Sediment Processes, ed.

W.H. McAnally and A.J. Mehta, Elsevier, Proceedings in Marine Science, 3, 229-248.

Le Hir, P., 1997, Fluid and sediment "integrated" modelling application to fluid mud flows in estuaries, in: Proceedings of the 4th Nearshore and Estuarine Cohesive Sediment Transport Conference, INTERCOH'94, Wallingford, UK, July 1994, ed. by N. Burt, R. Parker and J. Watts, John Wiley & Sons, 417-428.

Lee, S.L., K.W. Lo and F.H. Lee, 1982, A numerical model for crack propagation in soils, Proceedings International Conference on Finite Element Methods, Shanghai, 412-418.

Lee, S.-I., I.-S. Seo and B. Koopman, 1994, Effect of mean velocity gradient and mixing time on particle removal in sea water induced flocculation, Water, Air and Soil Pollution, 78, 179-188.

Leentvaar, J. and M. Rebhun, 1983, Strength of ferric hydroxide flocs, Water Research, 17, 895-902.

Levich, V.G., 1962, Physico-chemical Hydrodynamics, Prentice Hall, Inc.

Li, M.Z. and G. Gust, 2000, Boundary layer dynamics and drag reduction in flows of high cohesive sediment suspensions, Sedimentology, 47, 71-86.

Lick, W. and J. Lick, 1988, Aggregation and disaggregation of fine-grained lake sediments, Journal of Great Lakes Research, 14 (4) 514-523.

Lick, W., 1982, The transport of contaminants in the Great Lakes, Annual Review of Earth and Planetary Science, 10, 327-353.

Lienhard, J.H. and C.W. Van Atta, 1990, The decay of turbulence in thermally stratified flow, Journal of Fluid Mechanics, 210, 57-112.

Lin, T.W. and R.A. Lohnes, 1984, Zone settling explained as self-weight consolidation, Journal of Powder and Bulk Solids Technology, 8 (2) 29-36.

Lin, P.C.-P. and A.J. Mehta, 1997, A study of fine sediment sedimentation in an elongated laboratory basin, Journal of Coastal Research, 25, 19-30.

Liu, J.C., and Znidarcic, D. 1991, Modelling one dimensional compression characteristics of soils, ASCE, Journal of Geotechnical Engineering, 117 (1) 162-169.

Liu, K. and C.C. Mei, 1989, Effects of wave-induced friction on a muddy seabed as a Bingham plastic fluid, Journal of Coastal Research, 5 (4) 777-789.

Logan, B.E. and J.R. Kilps, 1995, Fractal dimensions of aggregates formed in different fluid mechanical environments, Water Resources, 29 (2) 443-453.

Lubking, L., 1997, Soft Soil Correlation, Dredging Research Association, Report BAGT 569.

Lubking, P., 2004, Handbook Sandbook, CROW report 599 (in Dutch).

Lueck, R.G. and Y.Y. Lu, 1997, The logarithmic layer in a tidal channel, Continental Shelf Research, 17 (14) 1785-1801.

Lumley, J.L., 1969, Drag reduction by additives, Annual Review of Fluid Mechanics, 1, 367-384.

Lumley, J.L., 1973, Drag reduction in turbulent flow by polymer additives, Journal of Polymer Science, 7, 263-290.

Luther, G. W., P. J. Brendel, B. L. Lewis, B. Sundby, L. Lefrançois, and D. B. Nuzzio, 1998, Simultaneous measurement of O_2, Mn^{2+}, Fe^{2+}, I^- and S^{2+} in marine pore waters with a solid-state voltametric micro-electrode, Limnology and Oceanography, 43, 325-333.

Lyklema, J., 1989, The colloidal background of flocculation and dewatering, in: Flocculation and Dewatering, ed. B.M. Moudgil and B.J. Scheiner, Engineering Publication Services, University of Florida, USA, 1-20.

Lyn, D.A., 1986, Turbulence and turbulent transport in sediment-laden open-channel flows, W.M. Keck Laboratory of Hydraulics and Water Resources, Division of Engineering and Applied Science, California Institute of Technology, Report No KH-R-49.

Lynch, J.F., J.D. Irish, C.R. Sherwood and Y.C. Agrawal, 1994, Determining suspended sediment particle size information from acoustical and optical backscatter measurements, Continental Shelf Research 14 (10/11) 1139–1165.

Maa, J.P.-Y, L.D. Wright, C.-H. Lee and T.W. Shannon, 1993, VIMS Sea Carousel: a filed instrument for studying sediment transport, Marine Geology, 115, 271-287.

Maa, J.P.-Y. and A.J. Metha, 1987, Mud erosion by waves: a laboratory study, Continental Shelf Research, Vol. 7, Nos, 11/12, pp. 1269-1284.

Maa, J.P.-Y. and A.J. Metha, 1988, Soft mud properties: Voigt model, Journal of Waterway, Port, Coastal, and Ocean Engineering, 114 (6) 634-650.

Maa, J.P.-Y. and A.J. Metha, 1990, Soft mud response to water waves, Journal of Waterway, Port, Coastal, and Ocean Engineering, 116 (5) 765-770.

Magara, Y., S. Nambu and K. Utosawa, 1976, Biochemical and physical properties of an activated sludge on settling characteristics, Water Resources, 10, 71-77.

Mahatma, L., R. Riethmüller, C. Van Bernem, T.J. Andersen and K. Heymann, 2004, The combined effect of bio-stabilisers and bio-destabilisers on the erosion threshold in different of intertidal flat habitats (in preparation).

Malcherek, A., 1995, Mathematische Modellierung von Strömungen und Stofftransportprozessen in Ästuaren, Dissertation, Institut für Strömungsmechanik und Elektronisch Rechnen im Bauwesen der Universität Hannover, Bericht Nr. 44/1995 (in German).

Mallard, W.W. and R.A. Dalrymple, 1977, Water waves propagating over a deformable bottom, Proceedings of the 9th Annual Offshore Technology Conference, 3, 141-146.

Malvern, L.E., 1969, Introduction to the Mechanics of a Continuous Medium, Prentice-Hall, Series in Engineering of the Physical Sciences, New Jersey.

Manning, A.J., 2001, A study on the effect of turbulence on the properties of flocculated mud, PhD-thesis, Institute of Marine Science, University of Plymouth, UK.

Manning, A.J. and K.R. Dyer, 2002, A comparison of floc properties observed during neap and spring tidal conditions, in: J.C. Winterwerp and C. Kranenburg, Fine Sediment Dynamics in the marine Environment, Elsevier, Proceedings in Marine Science, 5, 233-250.

McAnally, W.H., 1999, Aggregation and deposition of estuarial fine sediment, PhD-thesis, University of Florida, Coastal and Oceanographic Engineering Department, USA.

McAnally, W.H. and A.J. Mehta, 2002, Significance of aggregation of fine sediment particles in their deposition, Estuarine, Coastal and Shelf Science, 54 643-653.

McCave, I.N., 1984, Size spectra and aggregation of suspended particles in the deep ocean, Deep Sea Research, 31 (4) 329-352.

McCave, I.N., R.J. Bryant, H.F. Cook, C.A. Coughanowr, 1986, Evaluation of a Laser-Diffraction-Size Analyzer for Use with Natural Sediments, Journal of Sedimentary Petrology, 56, 561-564.

McCutcheon, S.C., 1098, Vertical velocity profiles in stratified flows, ASCE, Journal of the Hydraulic Division, 107 (HY8) 973-988.

McKee S.L., Williams R.A., Boxman, A., 1996, Development of solid-liquid mixing models using tomographic techniques, International Journal of Multiphase Flow, 22 (suppl.1) 99.

McNeil, J., C. Taylor and W. Lick, 1996, Measurements of erosion of undisturbed bottom sediments with depth, ASCE, Journal of Hydraulic Engineering, 122 (6) 316-324.

Meakin, P., 1986, Computer simulations of growth and aggregation processes, in: On Growth and Form, Fractal and Non-fractal Patterns in Physics, ed. H.E. Stanley and N. Ostrowsky, Martinus Nijhoff Publishers, 111-135.

MEDINAUT/MEDINETH Shipboard Scientific Parties, 2000, (Aloisi, G., Asjes, S., Bakker, K., Bakker, M., Charlou, J.-L., De Lange, G., Donval, J.-P., Fiala-Medioni, A., Foucher, J.-P., Haanstra, R., Haese, R., Heijs, S., Henry, P., Huguen, C., Jelsma, B., De Lint, S., Van Der Maarel, M., Mascle, J., Muzet, S., Nobbe, G., Pancost, R., Pelle, H., Pierre, C., Polman, W., De Senerpont Domis, L., Sibuet, M., Van Wijk, T., Woodside, J., Zitter, T.), 2000, Linking Mediterranean brine pools and mud volcanism, EOS, Transactions, American Geophysical Union, 81 (51) 625-633.

Mei, C.C. and M.A. Foda, 1981, Wave-induced responses in a fluid-filled poro-elastic solid with a free surface - A boundary layer theory, Geophysical Journal Royal Astronomical Society, 66, 597-631.

Mehta, A.J., Depositional behaviour of cohesive sediments, 1973, PhD-thesis, University of Florida, Gainesville, USA.

Mehta, A.J. and E. Partheniades, 1975, An investigation of the deposition properties of flocculated fine sediment, IAHR, Journal of Hydraulic Research, 12 (4) 361-381.

Mehta, A.J. and E. Partheniades, 1979, Kaolinite resuspension properties, ASCE, Journal of the Hydraulic Division, Technical Note, 104 (HY4) 409-416.

Mehta, A.J., E. Partheniades, J.G. Dixit and W.H. McAnally, 1982, Properties of deposited kaolinite in a long flume, in: Proceedings of the ASCE Hydraulics Division Conference on Applied Research to Hydrodynamic Practice, 594-603.

Mehta, A.J., T.M. Parchure, J.G. Dixit and R. Ariathurai, 1982, Resuspension potential of deposited cohesive sediments, in: Estuarine Comparisons, ed. V.S. Kennedy, Academic Press, New York, 691-609.

Mehta, A.J. and P.-Y. Maa, 1985, Fine sedimentation in small harbour basins, in: Flocculation, Sedimentation and Consolidation, ed. B.M. Moudgil and P. Somasundaran, University of Florida, USA, 405-414.

Mehta, A.J., 1986, Characterisation of cohesive sediment properties and transport processes in estuaries, in: Lecture Notes on Coastal and Estuarine Studies, Vol 14, Estuarine Cohesive Sediment Dynamics, Proceedings of a Workshop on Cohesive Sediment Dynamics with Special Reference to Physical Processes in Estuaries, ed. A.J. Mehta, 290-325.

Mehta, A.J. and J.W. Lott, 1987, Sorting of fine sediment during deposition, in: Proceedings of the Conference on Advances in Understanding Coastal Sediment Processes, 1, 348-362.

Mehta, A.J., 1991, Review note on cohesive sediment erosion, Proceedings of Coastal Sediments Conference, I, 40-53

Mehta, A.J. and R. Srinivas, 1993, Observations on the entrainment of fluid mud by shear flow, in: A.J. Mehta, Coastal and Estuarine Studies, American Geophysical Union, 42, 224-246.

Mehta, A.J. and Y. Li, 1996, A PC-based short course on fine-grained sediment transport engineering, University of Florida,

Mehta, A.J., 2002, Mudshore dynamics and controls, in: Muddy Coasts of the World, ed. T. Healy, Y. Wang and J-A. Healy, Elsevier, Proceedings in Marine Science, 4, 19-60.

Mellor, G., 2002, Oscillatory bottom boundary layers, Journal of Physical Oceanography, 32, 3075-3088.

Merckelbach, L.M., 1996, Consolidation theory and rheology of mud – a literature survey, Delft University of Technology, Department of Civil Engineering, Hydromechanics Section, Report 9-96.

Merckelbach, L.M., 2000, Consolidation and strength evolution of soft mud layers, PhD-thesis, Delft University of Technology; also: Delft University of technology, Faculty of Civil Engineering and Geosciences, Communications on Hydraulic and Geotechnical Engineering, Report 00-2, ISSN 0169-6548.

Merckelbach, L.M. and C. Kranenburg, 2003a, New constitutive equations for soft mud-sand mixtures, submitted to Géotechnique.

Merckelbach, L.M. and C. Kranenburg, 2003b, Constitutive equations for soft mud determined from simple laboratory experiments, submitted to Géotechnique.

Metzner, A.B. and J.C. Reed, 1955, Flow of non-Newtonian fluids, correlation of the laminar transition and turbulent flow regions, American Institute for Chemical Engineering Journal, 1 (4) 434-440

Middelburg, J.J., 1989, A simple rate model for organic matter decomposition in marine sediments, Geochimica et Cosmochemica Acta, 53, 1577-1581.

Migniot, C., 1968, A study of the physical properties of various forms of very fine sediments and their behaviour under hydrodynamic action, Communications présenté au Comité Technique de la Société Hydrotechnique de France, La Houille Blanche, 23 (7) 591-620 (in French).

Mitchell, J.K., 1976, Fundamentals of soil behaviour, John Wiley & Sons, New York.

Mitchener, H.J. and D.J. O'Brien, 2001, Seasonal variability of sediment erodibility and properties on a macrotidal mudflat, Peterstone Wentlooge, Severn Estuary, UK, in: W.H. McAnally and A.J. Mehta, Fine Sediment Dynamics in the Marine Environment, Proceedings in Marine Science, Elsevier, 3, 301-321.

Mitchener, H.J., H. Torfs and R.J.S. Whitehouse, 1996, Erosion of mud/sand mixtures, Coastal Engineering, 29, 1-25 (errata, 1997, 30, 319).

Miyagi, N., M. Kimura, H. Shoji, A. Saima, C.-M. Ho, S. Tung and Y.-C. Tai, 2000, Statistical analysis on wall shear stress of turbulent boundary layer in a channel flow using micro-shear stress imager, International Journal of Heat and Fluid Flow, 21, 576-581.

Mogi, K., 1979, Flow and fracture of rocks under general triaxial compression, Third ISRM-Symposium, Montreux, 3, 123-130.

Moore, F., 1959, The rheology of ceramic slips and bodies, Transactions of the British Ceramic Society, 58, 470-494.

N.R. Morgenstern and J.S. Tchalenko, 1968, Microscope structure in Kaolin subjected to direct shear, Geotechnique, 17 (4) 309-328.

Moudgil, B.M. and T.V. Vasudevan, 1988, Characterisation of flocs, in: Flocculation & Dewatering, ed. B.M. Moudgil and B.J. Scheiner, University of Florida, USA, 167-178.

Murdoch, L.C., 1993, Hydralic fracturing of soil during laboratory experiments. Part 1 Methods and observations, Part 2 Propagation, Part 3 Theoretical analysis. Géotechnique v 43, no. 2 pp. 255-287.

Muste, M. and V.C. Patel, 1997, Velocity profiles for particles and liquid in open-channel flow with suspended sediment, ASCE, Journal of Hydraulic Engineering, 123 (9) 742-751.

NEN, Netherlands Normalisation Institute, Delft, The Netherlands, www.nen.nl.

Nezu, I and H. Nakagawa, 1993, Turbulence in open-channel flows, International Association for Hydraulic Research, Monograph Series, Balkema, Rotterdam.

Nezu, I. and W. Rodi, 1986, Open-channel flow measurements with a laser Doppler anemometer, ASCE, Journal of Hydraulic Engineering, 112 (5) 335-355.

Nielsen, P., 1992, Coastal bottom boundary layers and sediment transport, World Scientific, Advanced Series on Ocean Engineering, 4, 324.

Nieuwstadt, F.T.M. and J.M.J. Den Toonder, 2001, Drag reduction by additives: a review, in: Turbulence structure and modulation, ed. A. Soldati and R. Monti, International Centre for Mechanical Sciences, Springer Verlag, New York, CISM Lecture Series No 415, 269-316.

Nishi, K., and Kanatani, M., 1990, Constitutive relations for sand under cyclic loading based on elasto-plasticity theory, Soils and Foundations 30, 2, 43-59.

NNI, 1986, Dutch Standardisation Institute, 1989, Classification of unconsolidated soil samples.

Nur, A. and J.D. Beyerlee, 1971, An exact effective stress law for elastic deformation of rock with fluids, Journal of Geophysical Research, 76, 6414-6419.

Obi. S., K. Inoue, T. Furukawa and S. Masuda, 1996, Experimental study on the statistics of wall shear stress in turbulent channel flow, International Journal of Heat and Fluid Flow, 17, 187-192.

Odd, N.V.M., 1988, Mathematical modelling of mud transport in estuaries, in Physical Processes in Estuaries, ed. by J. Dronkers and W. van Leussen, Springer Verlag, 503-531.

Odd, N.V.M and A.J. Cooper, 1989, A two-dimensional model of the movement of fluid mud in a high energy turbid estuary, Journal of Coastal Research, Special Issue No 5, 175-184.

Odd, N.V.M, M.A. Bentley, C.B. Waters, 1993, Observations and analysis of the movement of fluid mud in an estuary, in Proceedings of the 3rd International Conference on Nearshore and Estuarine Cohesive Sediment

Transport, ed. A.J. Mehta, Coastal and Estuarine Studies, Vol. 42, American Geophysical Union, 430-446.

Oka, F., 1992, A cyclic elasto-viscoplastic constitutive model for clay based on the non-linear hardening rule, Proceedings of the 4th International Symposium on Numerical Models in Geomechanics NUMOG IV, 105-114.

Oka, F., 1988, The validity of the effective stress concept in soil mechanics, in Micromechanics of Granular Material, ed Satake and Jenkins, Elsevier Amsterdam, 207-214.

O'Melia, C.R., 1980, Aquasols: the behaviour of small particles in aquatic systems, Environmental Science and Technology, 14 (9) 1052-1060.

O'Melia, C.R., 1985, The influence of coagulation and sedimentation on the fate of particles, associated pollutants, and nutrients in lakes, Chemical Processes in Lakes, ed. Werner Stum, John Wiley and Sons, New York, 207-224.

Ooms, G. and W.M.M. Schinkel, 1998, Effect of particle properties on turbulence intensity of a suspension, Presented at the European Turbulence Conference, Cap Ferrat, France, 1-4.

Oost, A.P, 1995, Dynamics and sedimentary development of the Dutch Wadden Sea with emphasis on the Frisian inlet, PhD-thesis, University of Utrecht.

Owen, M.W., 1976, Determination of the settling velocities of cohesive muds, Hydraulic Research Station, Report IT 161.

Parchure, T.M. and A.J. Mehta, 1985, Erosion of soft cohesive sediment deposits, ASCE, Journal of Hydraulic Engineering, 111 (10) 1308-1326.

Parker, W.R. and P.M. Hooper, 1994, Criteria and methods to determine navigable depth in hyperconcentrated sediment layers, in: Proceedings of Hydro-Port'94, International Conference on Hydro-Technical Engineering for Port and Harbour Construction, October, Yokosuka, Japan, 1211-1224.

Parker, G., M. Garcia, Y. Fukushima and W. Yu, 1987, Experiments on turbidity currents over an erodible bed, IAHR, Journal of Hydraulic Research, 25, 123-147.

Parker, G., Y. Fukushima and H.M. Pantin, 1986, Self-accelerating turbidity currents, Journal of Fluid Mechanics, 171, 145-181.

Parker, D.S., W.J. Kaufman and D. Jenkins, 1972, Floc breakup in turbulent flocculation processes, ASCE, Journal of the Sanitary Engineering Division, 98 (SA1) 79 - 97.

Partheniades, E., 1962, A study of erosion and deposition of cohesive soils in salt water, PhD-thesis, University of California, Berkeley, California, USA.

Partheniades, E., 1965, Erosion and deposition of cohesive soils, ASCE, Journal of the Hydraulic Division, 91 (HY1) 105-139.

Partheniades, E., J. Kennedy, R.J. Etter and R.P. Hoyer, 1966, Investigations of the depositional behaviour of fine cohesive sediments in an annular rotating flume, Technical Report No 96, Hydromechanics laboratory, M.I.T., Cambridge, Massachusetts, USA.

Partheniades, E., R.H. Cross and A. Ayora, 1968, Further research on the deposition of cohesive sediments, in Proceedings of the 11th Conference on Coastal Engineering, 1, 723-72.

Partheniades, E. and R.E. Paaswell, 1970, Erodibility of channels with cohesive boundary, ASCE, Journal of the Hydraulics Division, 96 (HY3) 755-771.

Partheniades, E., 1971, Erosion and deposition of cohesive materials, in H.W. Chen, River Mechanics 2, Fort Collins, USA, Chapter 25, 1-91.

Partheniades, E., 1977, Unified view of wash load and bed material load, ASCE, Journal of the Hydraulics Division, 103 (HY9) 1037-1057.

Partheniades, E., 1986, The present state of knowledge and needs for future research on cohesive sediment dynamics, Proceedings of the 3rd International Symposium on River Sedimentation, 3-25.

Partheniades, E., 1986, A fundamental framework for cohesive sediment dynamics, in: A.J. Mehta, Lecture Notes on Coastal and Estuarine Studies, 14: Estuarine Sediment Dynamics, Chapter XII, 219-250.

Petit, H.A.H., 1999, Note on the use of non-symmetrical probability density functions in Van Rijn's stochastic transport formula, in: Stone stability - Annual Report, WL | Delft Hydraulics, Report Q2539.

Patel, V.C., W. Rodi and G. Scheuerer, 1984, Turbulence models for near-wall and low-Reynolds number flows: a review, AIAA Journal, 23 (9) 1308-1319.

Paterson, D.M., 1989, Short term changes in the erodibility of intertidal cohesive sediments related to the migratory behaviour of epipelic diatoms, Limnology and Oceanography, 34, 223-234.

Paterson, D.M., 1997, Biological mediation of sediment erodibility: ecology and physical dynamics, in: N. Burt, R. Parker and J. Watts, Cohesive Sediments, John Wiley & Sons, 215-229.

Paterson, D.M. and K.S. Black, 1999, Water flow, Sediment Dynamics and benthic biology, Advances in Ecological Research, 29, 155-193.

Paull, C. and W.P. Dillon, 2001, Natural gas hydrates : occurrence, distribution, and detection, Geophysical monograph, American Geophysical Union, Washington.

PDC, 1995, Regulation of the Yangtze Estuary, Volume B: Hydro-morphological assessment and mathematical models, Port and Delta Consortium and Ministry of Water Resources of the People's Republic of China, Shanghai Investigation Research and Design Institute (confidential).

Pedlosky, J., 1987, Geophysical Fluid Dynamics, Springer Verlag 2nd ed, New York, 710 pp.

Penhout, Y. du and J.C. Salomon, 1979, Numerical simulation of estuarine mud layers - application to the Gironde, Oceanologica Acta, 2 (3) 253-260 (in French).

Perry, R.H., 1984, Perry's Chemical Engineers' Handbook, McGraw Hill.

Pethick, J.S., D. Legget and L. Husain, 1990, Boundary layers under salt marsh vegetation developed in tidal currents, in: J.B. Thornes, Vegetation and erosion processes and environments, John Wiley & Sons, London, 113-124.

Peters Rit, A.W.P.G., H.G. Merkus and B Scarlett, 1987, Feasibility study for the development of reference materials and procedures for laser diffraction spectrometry. Part I, Literature review and proposals for further work, Technical University Delft, Department of Chemical Engineering.

PIANC (1990): Economic methods of channel maintenance, Supplement Bulletin of PIANC, n° 68/1990 (in French) and n° 67/1989.

Pope, S.B., 2000, Turbulent flows, Cambridge University Press, pp 771.

Ptasinski, P.K., 2002, Turbulent flow of polymer solutions near maximum drag reduction – experiments, simulations and mechanisms, PhD-thesis, Delft University of Technology, Faculty of Mechanical Engineering.

Ramsey J.D.F., 1986, Colloidal Properties of Synthetic Hectorite Dispersions: I Rheology, Journal of Colloid Interface Science, 109 (2) 441.

Raudkivi, A.J., 1990, Loose Boundary Hydraulics, Pergamon Press, Oxford, 3rd edition.

Rechlin, D., 1995, Definition of the nautical depth in the main muddy areas of the federal waterways board, PIANC bulletin 86, 18-30.

Reise, K., 2002, Sediment mediated species interactions in coastal waters, Journal of Sea Research, 48 (2) 127-141.

Riethmuller R., M. Heineke, H. Kuhl and R. Keuker-Rudiger, 2000, Chlorophyll a concentration as an index of sediment surface stabilisation by microphytobenthos?, Continental Shelf Research 20 (10) 1351-1372.

Riley, G.A., 1963, Organic aggregates in seawater and the dynamics of their formation and utilisation, Limnology and Oceanography, 8, 372-381.

Rice, J.R., 1968, A path dependent integral and the approximate analysis of strain concentration by notches and cracks, Journal of Applied Mechanics, 379-386.

Robertson, A.I. and D.M. Alongi, 1992, Tropical mangrove ecosystems, American Geophysical Union, Coastal and Estuarine Studies, 41.

Rodi, W., 1984, Turbulence models and their applications in hydraulics - a state-of-the-art review, IAHR monograph, Delft, The Netherlands.

Roscoe, K.H. and A.N.Schofield, 1963, Mechanics behaviour of a 'wet' clay, Proc. European Conference on Soil Mechanics and Foundation Engineering, Wiesbaden, Germany, 1, 47-54.

Ross, M.A. and A.J. Mehta, 1989, On the mechanics of lutoclines and fluid mud, Journal of Coastal Research, Special Issue 5: High Concentration Cohesive Sediment Transport, ed. A.J. Mehta and E.J. Hayter, 51-61.

Ross, M.A., 1988, Vertical structure of estuarine fine sediment suspensions, PhD-thesis, Coastal & Oceanographic Engineering Department, University of Florida, Gainesville, Florida, USA.

Rots, J.G., 1988, Computational modeling concrete fracture, PhD Thesis, Delft University of Technology.

Rots, J.G., 1991, Smeared and discrete representations of localized fracture, International Journal of Fracture, 51, 45-59.

Rowe, P.W., 1962, The stress-dilatancy relation for static equilibrium of an assembly of particles in contact, Proceedings of the Royal Society, A269, 500-527.

Rubin, D.M., 2004, A simple autocorrelation algorithm for determining grain size from digital images of sediment, Journal of Sedimentary Research, 74 (1) 160–165.

Ruddy G., C.M. Turley and T.E.R. Jones, 1998, Ecological interaction and sediment transport on an intertidal mudflat II. An experimental dynamic model of the sediment-water interface, in: K.S. Black, D.M. Paterson & A. Cramp, Sedimentary Processes in the Intertidal Zone, Geological Society Special Publication 139, 149-166.

Sanford, L.P. and J.P. Halka, 1993, Assessing the paradigm of mutually exclusive erosion and deposition of mud, with examples from upper Chesapeake Bay, Marine Geology, 114, 37-57.

Sanford, L.P. and J.P.-Y. Maa, 2001, A unified erosion formulation for fine sediments, Marine Geology, 179, 9-23.

Scarlatos, P.D. and A.J. Mehta, 1990, Some observations on erosion and entrainment of estuarine fluid muds, in: R.T. Cheng (ed.), Residual Currents and Long-term transport, Springer-Verlag, Coastal and Estuarine Studies, 38, 321-332.

Scarlatos P.D. and A.J. Mehta, 1988, Density stratification due to resuspension of fluid muds, in: Computer Modelling in Ocean Engineering, Schrefler & Zienkiewicz (eds.), Balkema, Rotterdam, 427-433.

Schatz, J.F., 1976, Models for inelastic volume deformation for porous geological materials, In: The effect of voids on material deformation, ASME, Vol. 16, Applied mechanics division meeting, Salt Lake City, Utah, U.S.A.

Y.P. Sheng, 1989, Consideration of flow in rotating annuli for sediment erosion and deposition studies, Journal of Coastal Research, Special Issue 5, 207-216.

Schiffman, R.L., V. Pane and V. Sunara, 1985, Sedimentation and Consolidation, in: Flocculation, Sedimentation and Consolidation, ed. B.M. Moudgil and P. Somasundaram, Engineering Publication Services, University of Florida, USA, 57-124.

Schofield, A.N. and C.P. Wroth, 1968, Critical State Soil Mechanics, Mc Graw Hill London.

Schwertmann, U and R.M. Taylor, 1989, Iron oxides, in: J.B. Dixon and S.B. Weeds, Minerals in Soil Environments, 2nd ed., Soil Science Society of America, Madison, Wisconsin, USA, 379-438.

Shen, H., J. Li, H. Zhu, M. Han and F. Zhou, 1993, Transport of suspended sediment in the Changjiang Estuary, International Journal of Sediment Research, 7 (3) 45-63.

Sheng, P.Y., 1986, Modelling bottom boundary layer and cohesive sediment dynamics in estuarine and coastal waters, in: A.J. Mehta, Lecture Notes on Coastal and Estuarine Studies, 14: Estuarine Sediment Dynamics, Chapter XVII, 360-400.

Sigg, L. and W. Stumm, 1994, Aquatic Chemistry: an introduction to the chemistry of dilute solutions and to the chemistry of natural water systems, Verlag der Fachvereine, ISBN 3-519-13651-1, Stuttgart (in German).

Sih, G.C., 1991, Mechanics of fracture initiation and propagation. Kluwer Academic Publishers, Dordrecht.

Simonin, O., R.E. Uittenbogaard, F. Baron and P.L. Viollet, 1989, Possibilities and limitations of the k-ε model to simulate turbulent fluxes of mass and momentum, measured in a steady, stratified mixing layer, Proceedings of the XXIII bi-annual Congress, Ottawa, A55-A62.

Singer, J.K., J.B. Anderson, M.T. Ledbetter, I.N. McCave, K.P.N. Jones and R. Wright, 1988, An assessment of analytical techniques for the size analysis of fine-grained sediments, Journal of Sedimentary Petrology, 58 (3) 534-543.

Skempton, A.W., 1961, Effective stress in soils, concrete and rocks, in: Pore pressure and suction in soils, Butterworths, London, 4-16.

Skempton, A.W., 1953, The colloidal activity of clay, Proceedings of the 3rd International Conference on Soil Mechanics and Foundation Engineering, I, 57-61.

Skene, I.K., T. Mulder and J.P.M. Syvitski, 1997, INFLO1: A model predicting the behaviour of turbidity currents generated at river mouths, Computers and Geosciences, 23 (9) 975-991.

Sloan, E.D., 1990, Clathrate hydrates of natural gases, Marcel Dekker Inc, New York

Smoluchowski, M., 1917, Versuch einer Mathematischen Theorie der Koagulations-kinetik Kolloid Lösungen, Zeitschrift für Physikalische Chemie, Leipzig, 92, 129-168 (in German).

Soulsby, R.L. and B.L.S.A.Wainwright, 1987, A criterion for the effect of suspended sediment on near-bottom velocity profiles, IAHR, Journal of Hydraulic Research, 25 (3) 341-356.

Soulsby, R.L., L. Hamm, G. Klopman, D. Myrhaug, R.R. Simons and G.P. Thomas, 1993, Wave-current interaction within and outside the bottom boundary layer, Coastal Engineering, 21, 41-69.

Soulsby, R., 1997, Dynamics of marine sands, Thomas Telford, pp 249.

Spicer, P.T. and S.E. Pratsinis, 1996, Coagulation and fragmentation: universal steady-state particle size distribution, American Institute of Chemical Engineers, AIChE Journal, 42 (6) 1612-1620.

Spierenburg, S.E.J., 1987, Seabed response to water waves, PhD-thesis Technical University Delft.

Staats N., E.M.G.T. De Deckere, B. De Winder B. and L.J. Stal, 2001, Spatial patterns of benthic diatoms, carbohydrates and mud on a tidal flat in the Ems-Dollard estuary, Hydrobiologia, 448 (1/3) 107-115.

Stein, R., 1985, Rapid grain-size analyses of clay and silt fraction by Sedigraph 5000D: comparison with Coulter Counter and Atterberg methods, Journal of Sedimentary Petrology, 55 (4) 590-615.

Stolzenbach, K.D. and M. Elimelich, 1994, The effect of density on collisions between sinking particles: implications for particle aggregation in the ocean, Journal of Deep Sea Research I, 41 (3) 469-483.

Stumm, W. and J.J. Morgan, 1981, Aquatic Chemistry, Chemical Equilibria and Rates in Natural Waters, 3rd ed., John Wiles & Sons, Inc.

Stumpf, R.P., 1983, The process of sedimentation on the surface of a salt marsh, Estuarine, Shelf and Coastal Science, 17, 495-508.

Sukhodolov, A., M. Thiele and H. Bungartz, 1998, Turbulence structure in a river reach with sand bed, Water Resources Research, 34 (5) 1317-1334.

Sutherland, T.F., P.M. Lane, C.L. Amos and J. Douwing, 2000, The calibration of optical backscatter sensors for suspended sediment of varying darkness levels, Marine Geology, 162, 587-795.

Takahashi, T., Debris Flow, 1991, IAHR Monograph Series, A.A. Balkema, Rotterdam.

Tambo, N. and Y. Watanabe, 1979, Physical characteristics of flocs - I. The floc density function and aluminium floc, Water Research, 13, 409-419.

Tambo, N. and H. Hozumi, 1979, Physical characteristics of flocs - II. Strength of floc, Water Research, 13, 421-427.

Tatsuoka,F. and K. Ishihara, 1974, Drained Deformation of sand under cyclic stresses reversing direction, Soils and Foundations, Journal of Japanese Society of Soil Mechanics and Foundation Engineering, 14 (3) 51-65.

Taylor, P.A. and K.R. Dyer, 1977, Theoretical models of flow near the bed and their implications for sediment transport, The Sea, 6, 579-601.

Teeter, A.M., 1986, Vertical transport in fine-grained suspension and newly deposited sediment, in Lecture Notes on Coastal and Estuarine Studies, 14, Estuarine Cohesive Sediment Dynamics, ed. A.J. Mehta, Springer-Verlag, Berlin, 170-191.

Teeter, A.M., 2001, Clay-silt sediment modelling using multiple grain classes. Part II: Application to shallow water resuspension and deposition, in:

Coastal and Estuarine Fine Sediment Processes, ed. W.H. McAnally and A.J. Mehta, Elsevier, Proceedings in Marine Science, 3, 173-185.

Teisson, C. and D. Fritsch, 1988, Numerical modelling of suspended sediment transport in the Loire estuary, in Proceedings of the 21st International Conference on Coastal Engineering, ICCE, Malaga, Spain, 2707-2722.

Tennekes, H. and J.L. Lumley, 1994, A first course in turbulence, The MIT press, Cambridge, Massachusetts.

Terzaghi, K., 1943, Theoretical Soil Mechanics, John Wiley and Sons, New York.

Thorn, M.F.C., 1981, Physical processes of siltation in tidal channels, Proceedings of Hydraulic Modelling applied to Maritime Engineering Problems, ICE, London, 47-55.

Thorne, P.D. and D. Hanes, 2002, A review of acoustic measurement of small-scale sediment processes, Continental Shelf Research, 22, 602-632.

Timoshenko, S.P., and J.N. Goodier, 1970, Theory of Elasticity, 3rd ed., McGraw-Hill.

Tipping, E., C. Woof and D. Cooke, 1981, Iron oxide from a seasonal anoxic lake, Geochimica et Cosmochimica Acta, 45, 1411-1419.

Tipping, E. and D. Cooke, 1982, The effects of adsorbed humic substance on the surface charge of goethite (a-FeOOH) fresh water, Geochimica et Cosmochimica Acta, 46, 75-80.

Tolhurst, T.J., K.S. Black, S.A. Shayler, S. Mather, I. Black, K. Baker and D.M. Paterson, 1999, Measuring the in-situ erosion shear stress of intertidal sediments with the Cohesive Strength Meter (CSM), Estuarine Coastal and Shelf Science, 49, 281-294.

Tolhurst, T.J., K.S. Black, D.M. Paterson, H.J. Mitchener, G.R. Termaat and S.A. Shayler, 2000, A comparison and measurement standardisation of four in-situ devices for determining the erosion shear stress of intertidal sediments, Continental Shelf Research, 20, 1397-1418.

Tolhurst, T.J., G. Gust and D.M. Paterson, 2002, The influence of an extracellular polymeric substance (EPS) on cohesive sediment stability, in: J.C. Winterwerp and C. Kranenburg, Fine Sediment Dynamics in the Marine Environment, Proceedings in Marine Science Vol 5, Elsevier, 409-425.

Toorman, E.A. and J.E. Berlamont, 1991, A hindered settling model for the prediction of settling and consolidation of cohesive sediment, Geo-Marine Letter, 11, 179-183.

Toorman, E.A., 1992, Modelling of fluid mud flow and consolidation, PhD-thesis, Katholieke Universiteit Leuven, Belgium.

Toorman, E.A. and J.E. Berlamont, 1993, Settling and consolidation of mixtures of cohesive and non-cohesive sediment, in: Advances in Hydro-Science and Engineering, ed. S. Wang, University of Mississippi, 606-613.

Toorman, E.A., 1996, Sedimentation and self-weight consolidation: general unifying theory, Géotechnique, 46 (1) 103-113.

Toorman, E.A., 1997, Modelling the thixotropic behaviour of dense cohesive sediment suspensions, Rheologica Acta, 36, 56-65.

Toorman, E.A. and H. Huysentruyt, 1997, Towards a new constitutive equation for effective stress in self-weight consolidation, in: Cohesive Sediments, ed. N. Burt, R. Parker and J. Watts, John Wiley & Sons, New York, 121-132.

Toorman, E.A., 1999, Sedimentation and self-weight consolidation: constitutive equations and numerical modelling, Géotechnique, 49 (6) 709-726.

Toorman, E.A., 2002, Modelling of turbulent flow with suspended cohesive sediment, in: Fine Sediment Dynamics in the Marine Environment, ed. J.C. Winterwerp and C. Kranenburg, Elsevier, Proceedings in Marine Sciences, 5, 155-170.

Torfs, H., 1995, Erosion of sand-mud mixtures, PhD-thesis, Katholieke Universiteit Leuven, Belgium.

Torfs, H., H. Mitchener, H. Huysentruyt and E.A. Toorman, 1996, Settling and consolidation of sand-mud mixtures, Coastal Engineering, 29, 27-45.

Torfs, H., J. Jiang and A.J. Mehta, 2001, Assessment of the erodibility of fine/coarse sediment mixtures, in W.H. McAnally and A.J. Mehta, Coastal and Estuarine Fine Sediment Processes, Proceedings in Marine Science, Vol 3, Elsevier, Amsterdam, 109-123.

Tovey, N.K., 1971, A selection of scanning electron micrographs of clays, University of Cambridge, Department of Engineering, Report CUED/C-SOILS/TR5a.

Townsend, F.C. and M.C. McVay, 1990, SOA: Large strain consolidation predictions, ASCE, Journal of Geotechnical Engineering, 116 (2) 222-243.

Trowbridge, J.H. and G.C. Kineke, 1994, Structure and dynamics of fluid muds on the Amazon continental shelf, Journal of Geophysical Research, C, Oceans, 99 (1) 865-874.

Truesdell, C. and R. Toupin, 1960, The Classical Field Theories, Handbuch der Physic III/1, ed. S. Flugge, Springer, Berlin.

Truesdell, C. and W. Noll, 1965, The Non-linear Field Theories in Mechanics, Handbuch der Physic III/1, ed. S.Flugge, Springer, Berlin.

Turner, J.S., 1973, Buoyancy effects in fluids, Cambridge University Press.

Tzang, S.Y., 1998, Unfluidized soil responses of a silty seabed to monochromatic waves, Coastal Engineering, 35, 283-301.

Uiterwijk Winkel, A.P.B., 1975, Micro-biological aspects and the sedimentation behaviour of riverine cohesive sediment, Rijkswaterstaat, Directie Waterhuishouding en Waterbeweging, District Zuid-West, Report 44.006.01, pp 60 (in Dutch).

Uittenbogaard, R.E., J.A.Th.M. Van Kester and G.S. Stelling, 1992, Implementation of three turbulence models in TRISULA for rectangular horizontal grids, including 2DV test cases, Delft Hydraulics, report Z162.

Uittenbogaard, R.E., 1994, Physics of turbulence; MAST-VERIPARSE technical report on subtask 5.2, Delft Hydraulics, Report Z649.

Uittenbogaard, R.E., 1995, Modelling seasonal temperature stratification with TRIWAQ, Delft Hydraulics, Report Z978

Uittenbogaard, R.E., 1995, The importance of Internal Waves for Mixing in a Stratified Estuarine Tidal Flow, PhD-thesis, Delft University of Technology, September 1995.

Uittenbogaard, R.E., J.C. Winterwerp, J.A.Th.M. van Kester and H. Leepel, 1996, 3D cohesive sediment transport. Part I and II. Delft Hydraulics, report Z1022, March.

UNESCO. 1981, The practical salinity scale 1978 and the international equation of state of seawater 1980. Tenth report of the Joint Panel on Oceanographic Tables and Standards (1981), (JPOTS), Canada, September 1980.

Valiani, A., 1988, An open question regarding shear flow with suspended sediments, Meccanica, 23, 36-43.

Van den Bosch, L. and J. Berlamont, 1991, Mud basin and capture reservoir experimental program. Column experiments, Hydraulics Laboratory - K.U. Leuven, August 1991 (in Dutch).

Van Craenenbroeck, K., M. Vantorre and P. De Wolf, 1991, Navigation in muddy areas – Establishing the navigable depth in the Port of Zeebrugge, Proceedings of the CEDA-PIANC Conference on Accessible Harbours, Amsterdam, 1991.

Van der Ham, R., 1999, Turbulent exchange of fine sediments in tidal flow, PhD-thesis, Delft University of Technology, Faculty of Civil Engineering and Geotechnical Sciences.

Van der Ham, R., C. Kranenburg and J.C. Winterwerp, 1998, Turbulent vertical exchange of fine sediments in stratified tidal flows, proceedings of the Conference on Physical Processes in Estuaries and Coastal Seas (PECS), ed. by J. Dronkers and M.B.A.M. Scheffers, The Hague, September 1997, 201-208.

Van Eekelen, H.A.M., 1980, Isotropic yield surface in three dimensions for use in soil mechanics, International Journal of Numerical and Analytical Methods in Geomechanics, 4, 89-101.

Van Eekelen, S.J. and P. Van den Berg, P., 1994, The Delft Egg model, a constitutive model for clay, In: Proc. 1st Int. DIANA Conference on Computational Mechanics, Delft, Dordrecht, Kluwer, pp. 103-116.

Van Kessel, T. van, 1997, Generation and transport of subaqueous fluid mud layers, PhD Thesis Delft University of Technology, Delft.

Van Kessel, T., 1997, Generation and transport of subaqueous fluid mud layers, PhD-thesis, Delft University of Technology, Department of Civil Engineering, The Netherlands.

Van Kessel, T., 1998, Bubble initiation and bubble growth in sediment layers. Delft Hydraulics, Report Z2314, DM18 (in Dutch).

Van Kessel, T. and W.G.M. van Kesteren, 2002, Gas production and transport in artificial sludge depots, Waste Management, 22 (1) 53-62

Van Kessel, T., 2003, Analysis of LISST-data, Delft Hydraulics, Report Z3671.

Van Kester, J.A.Th.M., 1994, Validation of DELFT3D against mixing layer experiment; Phase 1: improved implementation of k-ε model, Delft Hydraulics, Report Z810 (in Dutch).

Van Kester, J.A.Th.M., R.E. Uittenbogaard and G.S. Stelling, 1994, Sensitivity analysis of 3D-Noordwijkerraai-model, Delft Hydraulics, Report Z691 (in Dutch).

Van Kesteren, W.G.M., 1992, Porewater behaviour in dredging processes, XIIIth World Dredging Congress, Bombay, India.

Van Kesteren, W.G.M., 1995, Numerical simulations of crack bifurcation in the chip forming cutting process in rock, In: G. Baker and B.L. Karihaloo, Fracture of brittle and disordered materials, Proc. IUTAM, Brisbane Australia, pp 505-524.

Van Kesteren, W.G.M. and J.M. Cornelisse, 1995: Modelling the behaviour of natural mud beds under cyclical or wave forcing; Theoretical and experimental research, RIKZ/Delft Hydraulics - Cohesive Sediment Report 50, Z830 (in Dutch).

Van Kesteren, W.G.M, 1996, Slope stability analysis Orange River mouth, WL|Delft Hydraulics, report Z2739.

Van Kesteren, W.G.M., J.M. Cornelisse and C. Kuijper, 1997, DYNASTAR BED MODEL: bed strength, liquefaction and erosion, Rijkswaterstaat and WL | delft hydraulics, Cohesive Sediments, Report No 55.

Van Kesteren, W.G.M., 1998, Deposition and Consolidation of Titanium Tailings, WL | Delft Hydraulics report Z2462.

Van Kesteren, W.G.M., 2002, Consolidation/Permeability Modelling and flux management plan, Appendix GT3 in: Data Gap report St.Louis River/Interlake/ Duluth Tar Site, Duluth Minnesota, Service Engineering Group, St.Paul Minnesota, USA.

Van Kesteren, W.G.M. and T. van Kessel, 2002, Gas bubble nucleation and growth in cohesive sediments, in: J.C. Winterwerp and C. Kranenburg (Eds.), Fine Sediment Dynamics in the Marine Environment. Proc. in Marine Science 5, Elsevier, Amsterdam, 329–341.

Van Kesteren, W.G.M, 2004, Seafloor failure processes, PhD-thesis, Delft University of Technology.

Van Ledden, M., 2002, A process-based sand-mud model, in J.C. Winterwerp and C. Kranenburg, Fine Sediment Dynamics in the Marine Environment, Proceedings in Marine Science, Vol 5, Elsevier, Amsterdam, 577-594.

Van Ledden, M., 2003, Sand-mud segregation in estuaries and tidal basins, PhD-thesis, Delft University of Civil Engineering.

Van Ledden, M., W.G.M. Van Kesteren and J.C. Winterwerp, 2003, A classification for erosion behaviour of sand-mud mixtures, Continental Shelf Research, 24, 1-11.

Van Leussen, W., 1988, Aggregation of particles, settling velocity of mud flocs; a review, in Physical Processes in Estuaries, ed. JJ. Dronkers and W. van Leussen, Springer-Verlag, 347-403.

Van Leussen, W. and E. van Velzen, 1989, High concentration suspensions: their origin and importance in Dutch estuaries and coastal waters, Journal of Coastal Research, Special Issue No 5, 1-22.

Van Leussen, W., 1994, Estuarine macroflocs and their role in fine-grained sediment transport, PhD-thesis, University of Utrecht, The Netherlands.

Van Leussen, W., 1999b, personal communication.

Van Olphen, H., 1977, An introduction to Clay Colloid Chemistry, 2nd edition, John Wiley & Sons, New York, London, pp 301.

Os, A.G., van and W. van Leussen, 1987, Basic research on cutting forces in saturated sand. ASCE, Journal of Geotechnical Engineering, 113 (12) 1501-1516.

Van Rijn, L.C. and A.S. Schaafsma, 1986, Evaluation of measuring instruments for suspended sediment, Proc. Int. Conf. on Measuring Techniques of Hydraulic Phenomena in Offshore, Coastal and Inland Waters, London, 401-423.

Van Rijn, L.C., 1986, Manual Sediment Transport Measurements, Delft Hydraulics Laboratory, Special Publication, March 1986.

Van Rijn, L.C., 1990, Principles of fluid flow and surface waves in rivers, estuaries, seas, and oceans, AQUA Publications, I11, pp 335.

Van Rijn, L.C., 1993, Principles of sediment transport in rivers, estuaries and coastal seas, AQUA Publications, The Netherlands.

Van Rijn, L.C., 2004, Principles of Sedimentation and Erosion Engineering, in preparation.

Van Wijngaarden, L., 1972, One-dimensional flow of liquids containing small gas bubbles, Annual Review Fluid Mechanics, 4, 369-396.

Van Wijngaarden, M. and J.R. Roberti, 2002, In-situ measurements of settling velocity and particle size distribution with the LISST-ST, in: J.C. Winterwerp and C. Kranenburg, Fine Sediment Dynamics in the marine Environment, Elsevier, Proceedings in Marine Science, 5, 295-311.

Van Wijngaarden, M., L.B. Venema, R.J. De Meijer, J.J.G. Zwolsman, B. Van Os and J.M.J. Gieske, 2002a, Radiometric sand-mud characterisation in the Rhine-Meuse estuary; Part A. Fingerprinting, Geomorphology, 43, 87-101.

Van Wijngaarden, M., L.B. Venema, R.J. De Meijer, 2002b, Radiometric sand-mud characterisation in the Rhine-Meuse estuary; Part B. In-situ mapping, Geomorphology, 43, Special Issue No 5, 103-116.

Vane, L.M. en Zang, G.M., 1997, Effect of aqueous phase properties on clay particle zeta potential and electro-osmotic permeability: implications for electro-kinetic soil remediation processes, Journal of Hazardous Materials, 55, 1-22.

Vanoni, V.A., 1946, Transportation of suspended sediment by water, ASCE Transactions, 111 (2267) 67-133.

Venmans, A.A.M. and E.J. Den Haan, E.J., 1990, Classification of Dutch peats, Proceedings 6th International IAEG congress, 784-788.

Verbeek, H., C. Kuijper, J.M. Cornelisse and J.C. Winterwerp, 1993, Deposition of graded natural muds in The Netherlands, in: Nearshore and Estuarine Cohesive Sediment Transport, ed. A.J. Mehta, American Geophysical Union, Coastal and Estuarine Studies, 185-204.

Vermeer, P.A., 1980, Formulation and analysis of sand deformation problems, PhD Thesis, Delft University of Technology, Delft.

Verruijt, A., 1969, Elastic storage of aquifers, Chap. 8 in: Flow through porous media, de Wiest (ed.), Academic Press, London.

Verwey, J., 1952, On the ecology of distribution of cockle and mussel in the Dutch Wadden Sea, their role in sedimentation and the source of their food supply, with a short review of the feeding behaviour of bivalve mollusks, Archives Néerlandaise Zoologie, 10, 127-239.

Vesic, A.S., 1972, Expansion of cavities in infinite soil mass, J. Soil Mech. Found. Div., Proc. ASCE, Vol. 98, Nr. SM 3, p. 265 - 290.

Vicsek, T., 1992, Fractal growth phenomena, World Scientific, Singapore.

Vinzon, S.B. and A.J. Mehta, 2003, Lutoclines in high concentration estuaries: some observations at the mouth of the Amazon, Journal of Coastal Research, 19 (2) 243-253.

Virk, 1975, Drag reduction fundamental, Journal of the American Institute of Chemical Engineering, AIChE, 21, 625-656.

Wang, Z.Y., P. Larsen, F. Nestmann and A. Dittrich, 1998, Resistance and drag reduction of flows of clay suspensions, ASCE, Journal of Hydraulic Engineering, 124 (1) 41-49.

Wacholder, E. and N.F. Sather, 1974, The hydrodynamic interaction of two unequal spheres moving under gravity through quiescent viscous fluid, Journal of Fluid Mechanics, 5, 417-437.

Weaver, C.E., 1989, Clays, muds and shales, Elsevier, Developments in Sedimentology, 44, Amsterdam.

Weitz, D.A., J.S. Huang, M.Y. Lin and J. Sung, 1985, Limits of the fractal dimension for irreversible kinetic aggregation of gold colloids, Physical Review Letters, 54 (13) 1416-1419.

Wells, J.T., 1987, Entrapment of shelf-destined mud, Cape Lookout Bight, NC, South-eastern Section of the Geological Society of America, Abstracts with Progress, 135.

Wells, J.T., 1983, Dynamics of coastal fluid muds in low-, moderate- and high-tide-range environments, Canadian Journal of Fisheries and Aquatic Science, 40 (suppl 1) 130-142.

West, A.H.L., J.R. Melrose and R.C. Ball, 1994, Computer simulations of the break-up of colloid aggregates, Physical Review E, 49 (6) 4237-4249.

West, J.R. and K.O.K. Oduyemi, 1989, Turbulence measurements of suspended solids concentration in estuaries, ASCE, Journal of Hydraulic Engineering, 115 (4) 457-474.

Whitehouse, R., R. Soulsby, W., Roberts and H. Mitchener, 2000, Dynamics of estuarine muds, HR Wallingford, DETR, Thomas Telford, London, pp 210.

Wiberg, P.L., D.E. Drake and D.A. Cacchione, 1994, Sediment resuspension and bed armouring during high bottom stress events on the northern California inner continental shelf: measurements and predictions, Continental Shelf Research, 14 (10/11) 1191-1219.

Wichman, B.G.H.M., 1999, Consolidation behaviour of gassy mud: theory and experimental validation, PhD-thesis, Delft University of Technology.

Widdows, J., M.D. Brinsley, N. Bowley and C. Barrett, 1998, A Benthic Annular Flume for in-situ measurements of suspension feeding / biodeposition rates and erosion potential of intertidal cohesive sediments, Estuarine, Coastal and Shelf Science, 46, 27-38.

Widdows, J., S. Brown, M.D. Brinsley, P.N. Salkeld and M. Elliot, 2000, Temporal changes in intertidal sediment erodibility: influence of biological and climatic factors, Continental Shelf Research, 20, 1275-1289.

Widdows, J., J.S. Lucas, M.D. Brinsley, P.N. Salkfeld and F.J. Staff, 2002, Investigation of the effects of current velocity on mussel feeding and mussel bed stability using an annular flume, Helgoland Marine Research, 56, 3-12.

Widdows, J. and M. Brinsley, 2002, Impact of biotic and abiotic processes on sediment dynamics and the consequences to the structure and functioning of the intertidal zone, Journal of Sea Research, 48, 43-156.

Wiedemeyer, W.L. and R. Schwamborn, 1996, Detritus derived from eelgrass and macroalgae as potential carbon source for *Mytilus edulis* in Kiel Fjord, Germany: a preliminary carbon isotopic study, Helgolaender Meeresunter-suchungen, 50 (3) 409-413.

Wijdeveld, A.J., 1997, Experimental research on the remobilisation of heavy metals – Phase 2: Mobilisation experiments for various chemical and physical conditions, Delft Hydraulics, Report T2007-20.

Wildenborg, A.F.B., A.L. Leijnse, E. Kreft, A. Obdam, M. Nepveu, L. Wipfler, C. Hofstee, W. van Kesteren, I. Gaus, I. Czernichowski-Lauriol, P. Torfs, R. Wojcik, B. Orlic, 2004, CO_2 Capture Project - An Integrated, Collaborative Technology Development Project for Next Generation CO_2 Separation, Capture and Geologic Sequestration, Safety Assessment Methodology Assessment for CO_2 Sequestration (SAMCARDS).

Williams, P.R. and D.J.A. Williams, 1989, Rheometry for concentrated cohesive suspensions, Journal of Coastal Research, 5, 151-164.

Williamson, H.S. and M.C. Ockenden, 1996, ISIS: An instrument for measuring erosion shear stress in-situ, Estuarine Coastal and Shelf Science, 42, 1-18.

Winterkorn, H.F., 1953, The condition of water in porous systems, Soil Science 55, 109-115

Winterwerp, J.C., W.T. Bakker, D.R. Mastbergen and H. Van Rossum, 1992, Hyperconcentrated sand-water mixture flows over erodible bed, ASCE, Journal of Hydraulic Engineering, 118 (11) 1508-1525.

Winterwerp, J.C, 1993, Rheological experiments and thixotropic behaviour, presentation at the mid-term workshop of MAST G8M, Advances in Coastal Morphodynamics, Gregynog, Wales.

Winterwerp, J.C. and C. Kranenburg, 1997a, Erosion of fluid mud by entrainment, in: N. Burt, R. Parker and J. Watts, Cohesive Sediments, John Wiley & Sons, Chichester, 263-278.

Winterwerp, J.C. and C. Kranenburg, 1997b, Erosion of fluid mud layers – II: Experiments and model validation, ASCE, Journal of Hydraulic Engineering, 123 (6) 512-519.

Winterwerp, J.C., R.U. Uittenbogaard, W.G.M Van Kesteren, and Z.B. Wang, 1997, Dynastar Generic Mud Model; Physical design study, RIKZ/Delft Hydraulics - Cohesive Sediment Report 54, Z2176.

Winterwerp, J.C., 1998, A simple model for turbulence induced flocculation of cohesive sediment, IAHR, Journal of Hydraulic Engineering, 36 (3) 309-326.

Winterwerp, J.C., 1999, On the dynamics of high-concentrated mud suspensions, PhD thesis, Delft University of Technology, The Netherlands, also Delft University of Technology, Faculty of Civil Engineering and Geosciences, Communications on Hydraulics and Geotechnical Engineering, Report 99-3, ISSN 0169-6548.

Winterwerp, J.C., 2001, Stratification of mud suspensions by buoyancy and flocculation effects, Journal of Geophysical Research, 106 (10) 22,559-22,574.

Winterwerp, J.C., R.E. Uittenbogaard and J.M. de Kok, 2001, Rapid siltation from saturated mud suspensions, Proceedings of the 5th International Conference on Nearshore and Estuarine Cohesive Sediment Transport, INTERCOH'98,

Proceedings in Marine Science No 3, Coastal and Estuarine Fine Sediment Processes, ed. W.H. McAnally and A.J. Mehta, Elsevier, Amsterdam, 125-146.

Winterwerp, J.C., 2002, On the flocculation and settling velocity of estuarine mud, Continental Shelf Research, 22, 1339-1360.

Winterwerp, J.C., 2002, Scaling parameters for high-concentration mud suspensions in tidal flow, in: Elsevier, Proceedings in Marine Science, No 5; Proceedings of the 6th International Conference on Nearshore and Estuarine Cohesive Sediment Transport, INTERCOH-2000, ed. J.C. Winterwerp & C. Kranenburg, 171-186.

Winterwerp, J.C., A.J. Bale, M.C. Christie, K.R. Dyer, S. Jones, D.G. Lintern, A.J. Manning, W. Roberts, 2002, Flocculation and settling velocity of fine sediment, in: Elsevier, Proceedings in Marine Science, No 5; Proceedings of the 6th International Conference on Nearshore and Estuarine Cohesive Sediment Transport, INTERCOH-2000, ed. J.C. Winterwerp & C. Kranenburg, 25-40.

Winterwerp, J.C. and T. van Kessel, 2003, Sediment transport by sediment-induced density currents, Ocean Dynamics, 53 (3) 186-197.

Winterwerp, J.C., H.J. de Vriend and Z.B. Wang, 2003, Fluid-sediment interactions in sediment-laden flows, Proceedings of the International Yellow River Forum on River Basin Management, Zhengzhou, China, May, 2003.

Winterwerp, J.C., 2004, An integrated model for settling, consolidation and erosion of cohesive sediment – Part II: model application, submitted to ASCE, Journal of Hydraulic Engineering.

Wolanski, E., N.N. Huan, L.T. Dao, N.H. Nhan and N.N. Thuy, 1996, Fine-sediment dynamics in the Mekong River estuary, Vietnam, Estuarine, Coastal and Shelf Science, 43, 565-582.

Wolanski, E., R.J. Gibbs, Y. Mazda, A.J. Mehta and B. King, 1992, The role of turbulence in settling of mudflocs, Journal of Coastal Research, 8 (1) 35-46.

Wolanski, E., J. Chapell, P. Ridd and R. Vertessy, 1988, Fluidization of mud in estuaries, Journal of Geophysical Research, 93 (C3) 2351-2361.

Wood, D.J., 1997, Pressure-impulse impact problems and plunging wave jet impact, Ph.D. Thesis, University of Bristol, Faculty of Science, Bristol, U.K., 136 pp.

Wright, L.D., W.J. Wiseman, Z.-S. Yang, B.D. Bornhold, G.H. Keller, D.B. Prior and J.N. Suhayda, 1990, Processes of marine dispersal and deposition of suspended silts off the modern mouth of the Huanghe (Yellow River), Continental Shelf Research, 10 (1) 1-40.

Wright, L.D., S.-C. Kim and C.T. Friedrichs, 1999, Across-shelf variations in bed roughness, bed stress and sediment suspension on the northern California shelf, Marine Geology, 154, 99-115.

Yong, R.N. and C.E. Mc Keyes, 1971, Yield and failure of a clay under triaxial stresses, Proceedings of the Soil Mechanics and Foundation Division, 97 (SM1) 159-176.

Youd, T.L., 1973, Factors controlling maximum and minimum densities of sands, ASTM Special Technical Publications, no 523, 98-112.

Young, F. R., 1984, Cavitation, McGraw-Hill, London.

Yousif, M.H., P.M. Li, M.S. Selim and E.D. Sloan, 1989, Depressurization of natural gas hydrates in Berea sandstone cores, Journal Inclusion Phenomena and Molecular Recognition in Chemistry, D.W. Davidson Memorial Volume, 3, 71-88.

Zeng, J.-J. and D.R. Lowe, 1997a, Numerical simulation of turbidity current flow and sedimentation: I. Theory, Sedimentology, 44, 67-84.

Zeng, J.-J. and D.R. Lowe, 1997b, Numerical simulation of turbidity current flow and sedimentation: II, Results and geological applications, Sedimentology, 44, 85-104.

Zhou, D. and J.R. Ni, 1995, Effects of dynamic interaction on sediment-laden turbulent flows, Journal of Geophysical Research, 100 (C1) 981-996.

Zienkiewicz, O.C. and R.L. Taylor, 1989, The finite element method, Vol. I, McGraw-Hill Book Company ltd., London, UK.

A. NOMENCLATURE

A activity

A constant in log. velocity profile

A_s specific surface

a_{eb} floc break-up efficiency parameter in flocculation model

B buoyancy destruction in turbulence model

$\underline{\underline{C}}$ Green's deformation tensor

C_0 initial suspended sediment concentration, homogeneous over water depth

C_{aw} air-water compressibility

CEC   cation exchange capacity

CSL   critical state line

C_s solids compressibility

C_s depth-averaged saturation concentration

C_{sk} skeleton compressibility

C_w air-water compressibility

$\underline{\underline{c}}$ Cauchy's deformation tensor

c suspended sediment concentration by mass

c wave celirity

c_a reference sediment concentration

c_b (near) bed concentration

c_D drag coefficient

c_d drained shear strength

$\dot{c}_g$ gas production rate

c_g wave group velocity

c_{gel} gelling concentration

c_i consolidation coefficient

c_m concentration in fluid mud layer

c_q coefficient in integral entrainment model

c_s local saturation concentration

c_s coefficient in integral entrainment model

c_s' coefficient in integral entrainment model

c_u undrained shear strength

c_v vertical consolidation coefficient

c_w coefficient in integral entrainment model

c_y coefficient in integral entrainment model

$c_{1\varepsilon}$ coefficient in k-ε turbulence model

$c_{2\varepsilon}$ coefficient in k-ε turbulence model

$c_{3\varepsilon}$ coefficient in k-ε turbulence model

c_μ coefficient in k-ε turbulence model

c_σ coefficient in integral entrainment model

D deposition rate

D self-diffusion term in turbulence model

D_e equilibrium floc size

D_f floc size

D_{ij} deformation tensor

D_{ijkl}^e 4th order stiffness tensor

D_{max} maximal attainable floc size

D_p diameter primary mud particles

D_s molecular diffusion coefficient of mud flocs

D_0 initial floc size

D_{50} median floc size

$\underset{=}{d}$ deformation tensor

$\underset{=}{E}$ Green's strain tensor

E Young's modulus

E_b rate of sediment exchange between bed and water column

E_h redox potential

$\underset{=}{e}$ equivalent strain tensor

e total energy

e void ratio

e unity charge

e_c efficiency coefficient for coagulation in flocculation model

e_d efficiency coefficient for diffusion in flocculation model

$\underset{=}{F}$ deformation tensor

F_s settling flux

F_s sedimentation flux

$F_{s,T}$ sedimentation flux over tidal period

F_N flocculation and floc break-up function in flocculation model

Fo_w Fourier number pore water pressure dissipation

F_y yield strength of flocs

f Coriolis parameter

f yield surface function

f friction coefficient

f_c coefficient current-induced exchange flow rate

f_d coefficient density-induced exchange flow rate

$f_{t,e}$ coefficient tide-induced exchange flow rate

$f_{t,d}$ coefficient tide-induced exchange flow rate

f_s shape factor in flocculation model

f_w wave-induced bed friction coefficient

G shear rate parameter in flocculation model

G shear modulus

G elastic shear modulus

G^* complex shear modulus

G' storage modulus

G'' loss modulus

G_{Ic} failure energy

g acceleration of gravity

g plastic potential function

H wave height

H hardening parameter

H Henri's coefficient

He' Hedstrøm number

H_{rms} root mean square wave height

h water depth

h_{ij} hardening tensor

h_s sedimentation depth

I_i i^{th} invariant stress tensor

J Jacobi determinant

K isotropic compression modulus or bulk modulus

K_I stress intensity factor

K_k coefficient in power law description of permeability

K_p coefficient in power law description of effective stress

$K_{p,0}$ creep coefficient in effective stress function

K_0 coefficient of lateral stress

k turbulent kinetic energy

k permeability

k Boltzman constant

k_A aggregation coefficient in flocculation model

k'_A non-dimensional aggregation coefficient in flocculation model

k_B floc break-up coefficient in flocculation model

k'_B non-dimensional floc break-up coefficient in flocculation model

k_s Nikuradse's roughness height

$\underline{\underline{L}}$ spatial velocity gradient tensor

LI liquid index

LL liquid limit

L_{sl} Stokes' length

l turbulent length scale (mixing length)

ℓ Monin-Obokhov length scale

M erosion parameter in erosion law of Partheniades

M q/p ratio related to internal friction angle

M_E erodibility

M_{cyc} cylindrical constrained modulus

M_{ps} plane-strain constrained modulus

m exponent in hindered settling formula accounting for non-linear effects

NCL normal compression line

N number concentration of mud flocs

N Avogardo's number

N_0 initial number concentration of mud flocs

n effective stress ratio normally and over-consolidated mud

n exponent in hindered settling formula by Richardson and Zaki

n porosity

n_f fractal dimension of mud flocs

n_i normal vector

n_{max} maximum porosity

OVR over-consolidation ratio

P production term in turbulence model

Pe Péclet number

PI plasticity index

PL plastic limit

p exponent in flocculation model

p isotropic stress (pressure)

p_{atm} atmospheric pressure

p_e excess pore water pressure

p_e electron activity

pH acidity

p^{sk} effective stress

p_{vap} vapour pressure

p^w total water pressure

Q exchange flow rate

Q_e exchange flow rate by entrainment

Q_d exchange flow rate by slat-induced density current

Q_S exchange flow rate by sediment-induced density current

Q_T exchange flow rate by temperature-induced density current

Q_t exchange flow rate by tidal filling

q deviatoric stress

q exponent in flocculation model

$\underline{\underline{R}}$ rotation tensor

R gas constant

R yield surface shape

Re overall Reynolds number

Re' effective Reynolds number

Re_e effective Reynolds number

Re_f floc Reynolds number

Re_p particle Reynolds number

Re_T turbulent Reynolds number

Re_y yield stress Reynolds number

Re_w wave Reynolds number

Ri gradient Richardson number

Ri_f flux Richardson number

Ri_* bulk Richardson number

S salinity

SAR sodium adsorption ratio

SL swelling line

$\underline{\underline{T}}$ first Piola-Kirchoff stress tensor

$\underline{\underline{\hat{T}}}$ second Piola-Kirchoff stress tensor

T tidal period or accelerating or decelerating phase of tide

T absolute temperature

T' time parameter in flocculation model

T_c consolidation time

T_c' relative consolidation time

T_f flocculation or floc break-up time scale

T_m vertical mixing time

T_m' relative vertical mixing time

T_r residence time

T_{rel} relaxation time

T_s sedimentation time

T_s' relative sedimentation time

T_T turbulent time scale

t time

t_i deviatoric stress

U depth-averaged horizontal flow velocity

U_m amplitude tidal flow velocity

u horizontal flow velocity

u ore water pressure

u_i flow velocity in i-direction

$\hat{u}_{orb}$ amplitude orbital flow velocity near the bed

u_* shear velocity

u_{*f} flow-induced shear velocity

u_{*w} wave-induced shear velocity

V_e erosion velocity

V_f volume of mud flocs

VSL virgin compression line

v_f vertical fluid velocity in bed relative to fixed reference frame

v_s vertical velocity of sediment particles relative to fixed reference frame

W water content

W_s constant or characteristic settling velocity

$\underline{\underline{w}}$ Cauchy spin tensor

w vertical velocity

w_e entrainment velocity

w_s effective settling velocity, varying with depth and/or time

$w_{s,r}$ settling velocity of individual mud floc in still water

$w_{s,max}$ maximal settling velocity, limited by residence time

$\underline{\underline{X}}$ Lagrangean position tensor

$\underline{\underline{x}}$ Eulerian position tensor

x horizontal co-ordinate

x_i co-ordinate in i-direction

$y(\tau)$ shear stress probability distribution

Z_b bed level

Z_s level of water surface

z vertical co-ordinate

z_{wc} roughness height for waves and current

z_0 roughness height for current alone

α shape factor sediment

α yield surface coefficient for compression

α_c consolidation coefficient

α' shape parameter in settling velocity formula:
$$\alpha' \equiv \alpha/18\beta$$

β Rouse parameter

β dilatancy ratio

β shape factor sediment

γ_{ij} deviatoric strain

$\dot{\gamma}_{ij}$ deviatoric strain rate

Γ_c diffusion coefficient in consolidation formula

Δ relative sediment density:
$$\Delta \equiv (\rho_s - \rho_w)/\rho_s$$

Δt time step in 1DV POINT MODEL

Δz grid size in 1DV POINT MODEL

Δz_b size near bed grid in 1DV POINT MODEL

$\Delta \rho_f$ excess floc density:
$$\Delta \rho_f \equiv \rho_f - \rho_w$$

δ_m thickness fluid mud layer

δ_s swelling front

δ_w thickness wave boundary layer

ε turbulent energy dissipation

ε_v volume strain increment

$\dot{\varepsilon}_v^p$ hardening rule

ζ relative height above bed:
$$\zeta = z/h$$

ζ Gibson height

η stress ratio $\left(\eta = q/p^{sk}\right)$

η parameter in settling function

θ Lode angle

θ phase shift between stress and strain

θ_e non-dimensional threshold shear stress for erosion:
$$\theta_e = \tau_b/\tau_e$$

κ Von Kármàn constant

κ_s effective Von Kármàn constant

Λ critical state parameter

λ plastic multiplier

λ yield surface coefficient for shear

λ thickness diffusive double layer

λ friction coefficient

λ size floc forming eddies

λ_0 Kolmogorov micro-scale of turbulence

λ_L Lamé constant

λ_s structural parameter Moore model

μ dynamic viscosity

μ dielectric constant

μ_L Lamé constant

ν Poisson ratio

ν kinematic viscosity

ν_m kinematic viscosity of fluid mud

ν_T eddy viscosity

Ξ_s^m settling function for mud

Ξ_s^{sa} settling function for sand

ξ^i solids content

ξ_0 critical clay content

Π parameter wake function boundary layer

ρ bulk density of water-sediment suspension

ρ_{dry} dry bed density

ρ_f density of mud flocs

ρ_s density of primary sediment particles

ρ_w density of water

σ_k Prandtl-Schmidt number for the diffusion of k in k-ε turbulence model

σ_m critical stress for mass erosion

$\sigma_{m,w}$ critical stress for mass erosion by waves

σ_{st} surface tension

σ_T Prandtl-Schmidt number relating eddy diffusivity and eddy viscosity

σ_ε Prandtl-Schmidt number for the diffusion of ε in k-ε turbulence model

σ_{ij} stress tensor

σ_{ij}^w fluid-induced part of stress tensor

σ_{ij}^{sk} sediment-induced part of stress tensor (effective stress)

σ_v' effective stress

σ_{zz} total vertical stress

τ_B Bingham strength

τ_b bed shear stress (flow- and wave-induced contributions)

τ_c flow-induced bed shear stress

τ_{cr} (true) critical shear stress for erosion

τ_e apparent critical shear stress for erosion

τ_{ij} deviatoric stress tensor

τ_{ij}^{sk} skeleton shear stress tensor

$\tau_{m,ij}^w$ molecular shear stress tensor

τ_s water surface shear stress

τ_{xz} shear stress ($\tau_{xz} = \sigma_{ij}$)

τ_y yield stress

υ cation valence

Φ pore water flux

φ internal friction angle

ϕ volumetric concentration

ϕ finess factor

ϕ_e internal friction angle for compression

ϕ_c internal friction angle for expansion

ϕ'_g gas volume fraction

ϕ_* min $\{1, \phi\}$

ϕ_p volumetric concentration of primary particles

ξ^i solids content

ξ^{OM} solids content

ψ dilatancy angle

$\underline{\underline{\Omega}}$ angular velocity tensor

Ω solubility coefficient

$\overline{\omega}_i$ spin momentum

ω wave frequency

superscripts refer to size fraction:

$\blacksquare^{cl}$ clay fraction

$\blacksquare^{m}$ mud fraction (= clay + silt)

$\blacksquare^{sa}$ sand fraction

$\blacksquare^{si}$ silt fraction

superscripts refer to modes of deformation:

$\blacksquare^{cf}$ compressive failure

$\blacksquare^{e}$ elastic deformation

$\blacksquare^{p}$ plastic deformation

$\blacksquare^{tf}$ tensile failure

subscripts refer to phase (applied as superscripts in combination with stresses):

$\blacksquare_d$ drained behaviour

$\blacksquare_s$ solid phase

$\blacksquare_{sk}$ solids skeleton or fabric

$\blacksquare_w$ water/fluid phase

$\blacksquare_g$ gas phase

$\blacksquare_f$ flocculated phase

$\blacksquare_p$ primary particle

$\blacksquare_u$ undrained behaviour

operators:

D/Dt material derivative

δ_{ij} Kronecker delta

e_{ijk} permutator

x' turbulent fluctuating quantity

x_{rms} root mean square fluctuating quantity

$\overline{x}$ depth-averaged quantity

$\langle x \rangle$ tide-averaged quantity

$\underline{x}$ vector

$\underline{\underline{x}}$ tensor

$\underline{\underline{x}}^{-1}$ inverse of tensor

$\underline{\underline{x}}^{T}$ transpose of tensor

$\underline{\underline{\dot{x}}}$ time derivative of tensor

$\underline{\underline{\overset{o}{x}}}$ co-rotational Jaumann stress rate

B. DEFINITIONS AND USEFUL RELATIONS

This appendix summarises the definitions used in this book and provides some useful relations between the various parameters. Some of these relations were given before by Lambe and Whitman (1979).

GENERAL

First we define the masses and volumes of the solids, water and gas phase:

M_t	total mass of sediment-water sample [kg],
M_s	total mass of solids in sediment-water sample [kg],
M_w	total mass of water in sediment-water sample [kg],
M_g	total mass of gas in sediment-water sample [kg],
V_t	total volume of sediment-water sample [m³],
V_s	total volume of solids in sediment-water sample [m³],
V_e	total volume of voids in sediment-water sample [m³],
V_w	total volume of water in sediment-water sample [m³],
V_g	total volume of gas in sediment-water sample [m³],
LL	liquid limit [%],
PL	plastic limit [%].

These definitions have the following mutual relations:

$$M_t = M_s + M_w + M_g$$
$$V_t = V_s + V_e \qquad \text{(B.1)}$$
$$V_e = V_w + V_g$$

In shallow water the mass of gas is in general three orders of magnitude smaller than the mass of water and solids. However in deeper water, say below 100 m, the mass of gas cannot be neglected anymore. At such depths phase transitions to liquid gas and gas hydrates may occur. For example, methane hydrates are formed below 300 m depth at a temperature of 4 °C, and at 800 m water depth at 10 °C. Carbon-dioxide hydrates are formed below 150 m water depth at 4 °C, which become super-critical below 450 m depth at temperatures above 10 °C (see Chapter 11).

DEFINITIONS

bulk density ρ:

$$\rho = \frac{M_t}{V_t} \tag{B.2}$$

mass concentration c or dry bed density ρ_{dry}:

$$c \equiv \rho_{dry} = \frac{M_s}{V_t} \tag{B.3}$$

solids fraction ψ_s^i of component i:

$$\psi_s^i = \frac{V_s^i}{V_s} \tag{B.4}$$

solids fraction of clay fraction ψ_s^{cl} (grain size < 2 μm), silt fraction ψ_s^{si} (grain size > 2 μm and < 63 μm) and sand fraction ψ_s^{sa} (grain size > 63 μm and < 2 mm):

$$\psi_s \equiv 1 = \psi_s^{cl} + \psi_s^{si} + \psi_s^{sa} \tag{B.5}$$

solids fraction of mud fraction ψ_s^m

$$\psi_s^m = \psi_s^{cl} + \psi_s^{si} \tag{B.6}$$

total solids content S_c:

$$S_c = \frac{M_s}{M_t} = \frac{c}{\rho} \tag{B.7}$$

volume concentration phase ϕ_i of component i:

$$\phi_i = \frac{V_i}{V_t} \tag{B.8}$$

volume concentration solid phase ϕ_s, water phase ϕ_w and gas phase ϕ_g:

$$\phi \equiv 1 = \phi_s + \phi_w + \phi_g \tag{B.9}$$

volume concentration total solids ϕ_s:

$$\phi_s = \frac{V_s}{V_t} = \frac{c}{\rho_s} \tag{B.10}$$

volume concentration clay fraction ϕ_s^{cl}, silt fraction ϕ_s^{si} and sand fraction ϕ_s^{sa}:

$$\phi_s = \phi_s^{cl} + \phi_s^{si} + \phi_s^{sa} \tag{B.11}$$

volume concentration mud fraction ϕ_s^m:

$$\phi_s^m = \phi_s^{cl} + \phi_s^{si} \tag{B.12}$$

solids content of component ξ_s^i of component i:

$$\xi_s^i = \frac{M_s^i}{M_s} \tag{B.13}$$

solids content clay fraction ξ_s^{cl}, silt fraction ξ_s^{si} and sand fraction ξ_s^{sa}:

$$\xi_s \equiv 1 = \xi_s^{cl} + \xi_s^{si} + \xi_s^{sa} \tag{B.14}$$

number concentration N:

$$N = \frac{\text{number of particles}}{V_t} \tag{B.15}$$

void ratio e:

$$e = \frac{V_e}{V_s} = \frac{\phi_w + \phi_g}{\phi_s} = \frac{1 - \phi_s}{\phi_s} \tag{B.16}$$

porosity n:

$$n = \frac{V_e}{V_t} = \frac{e}{1+e} = \phi_w + \phi_g = 1 - \phi_s \qquad (B.17)$$

water content W:

$$W = \frac{M_w}{M_s} \qquad (B.18)$$

degree of saturation S:

$$S = \frac{V_w}{V_e} = \frac{\phi_w}{\phi_w + \phi_g} = \frac{\phi_w}{1 - \phi_s} \qquad (B.19)$$

plasticity index PI:

$$\text{PI} = \text{LL} - \text{PL} \qquad (B.20)$$

liquidity index LI:

$$\text{LI} = \frac{W - \text{PL}}{\text{PI}} = \frac{W - \text{PL}}{\text{LL} - \text{PL}} \qquad (B.21)$$

activity A:

$$A = \frac{\text{PI}}{\xi^{cl} - \xi_0} \qquad \left(\xi_0 = \text{critical } \xi^{cl} \right) \qquad (B.22)$$

USEFUL RELATIONS

Relations (B.23) through (B.29) are valid for unsaturated conditions.

$$\rho = c + \left(1 - \frac{c}{\rho_s}\right)S\rho_w + \left(1 - \frac{c}{\rho_s}\right)(1 - S)\rho_g$$

$$= \phi_s \rho_s + (1 - \phi_s)S\rho_w + (1 - \phi_s)(1 - S)\rho_g \tag{B.23}$$

$$= S\left[\rho_w + \left(\frac{\rho_s - \rho_w}{\rho_s}\right)c\right] + (1 - S)\left[\rho_g + \left(\frac{\rho_s - \rho_g}{\rho_s}\right)c\right]$$

$$c = \left(\frac{\rho - \rho_g - S(\rho_w - \rho_g)}{\rho_s - \rho_g - S(\rho_w - \rho_g)}\right)\rho_s \tag{B.24}$$

$$c \equiv \rho_{dry} = (1 - n)\rho_s = \frac{1}{1 + e}\rho_s \tag{B.25}$$

$$\phi_s = 1 - n = \frac{1}{1 + e} \tag{B.26}$$

$$\rho_b = n\rho_w + (1 - n)\rho_s \tag{B.27}$$

$$n = 1 - \phi_s = \frac{\rho_s - \rho}{\rho_s - \rho_w} = \frac{e}{1 + e} \tag{B.28}$$

$$e = \frac{n}{1 - n} = \frac{1 - \phi_s}{\phi_s} = \frac{\rho_s - c}{c} \tag{B.29}$$

$$W = \frac{1}{S_c} - 1 = \frac{\rho_w}{\rho_s} e = \frac{\rho_w}{\rho_s} \frac{n}{1-n} = \frac{\rho_w}{\rho_s} \frac{1-\phi_s}{\phi_s} \tag{B.30}$$

$$\Delta \rho_f = \rho_f - \rho_w = (\rho_s - \rho_w) \left[\frac{D_p}{D_f} \right]^{3-n_f} \tag{B.31}$$

$$\phi_f = \left(\frac{\rho_s - \rho_w}{\rho_f - \rho_w} \right) \frac{c}{\rho_s} = \frac{c}{\rho_s} \left[\frac{D_f}{D_p} \right]^{3-n_f} \tag{B.32}$$

$$\phi_f = f_s N D_f^3 \tag{B.33}$$

$$c_{gel} = \rho_s \left[\frac{D_p}{D_f} \right]^{3-n_f} \tag{B.34}$$

$$N = \frac{1}{f_s} \frac{c}{\rho_s} D_p^{n_f - 3} D_f^{-n_f} \tag{B.35}$$

C.   MEASURING TECHNIQUES

This appendix gives a brief summary on techniques available to measure the various physical and chemical sediment and water properties treated in this book. We will not consider measuring biological parameters. Where relevant, we will note whether the techniques are standardised. This appendix is partly based on a report drafted in commission of the European Union (Kuijper et al., 1993) and its summary by Berlamont et al. (1993), and additional literature.

Some of the measuring techniques described are applied in the field. Such applications require appropiate logistics; these are so site-specific in general that we will not discuss them. However, many measurements are carried out in the laboratory on sediment and/or water samples from the field.

Taking samples from the water column is not as trivial as it may seem. If one is interested in the concentration of the sediment suspension, its composition and/or its grain size distribution, so-called iso-kinetic sampling has to be applied, i.e. samples have to be siphoned or pumped at the local flow velocity so that representative samples are obtained. If not, the samples may be biased towards the larger or smaller fraction of the suspension. Sometimes sampling with bottles, which can be opened and closed at its two ends, can be deployed as well. Van Rijn (1986) gives an extensive overview of the various sampling techniques that existed at that time, summarising their advantages and disadvantages, and their applicability for various hydro-sedimentological conditions (e.g. flow velocity, water depth, sediment concentration, grain size distribution, etc.).

Taking appropriate bed samples is even more difficult, in particular when undisturbed samples with in-situ sediment density and structure are required. One problem is the definition of the bed, which may not be unambiguous in muddy environments, as discussed in Section 3.4.3 and C.3.

A commonly used instrument is the grab sampler. It is applicable for consolidated but not too stiff mud in not too deep water, when the exact bed structure is not relevant. Because of the consistency of the material, such samples can be used to determine in-situ bed density and sediment composition.

Undisturbed samples have to be obtained through cores taken from the bed. Corers exist in many configurations and dimensions, often designed for use under specific conditions (e.g. Van Rijn, 1986). More than ten different systems are in use in The Netherlands alone. Hence, world wide, possibly at least hundred different configurations or so may exist. Yet, a number of systems can be distinguished.

The gravity corer penetrates the bed by its own weight. For shallow water sampling, the core length is limited, depending on the sediment composition and stiffness. Sediment samples retrieved with gravity corers are generally deformed by compaction. Moreover, the upper (soft) part of the bed is often lost.

Piston and spring corers consist of a frame that is placed on the bed, from which a corer is (slowly) driven into the bed. Core lengths are limited, but the samples can be fairly undisturbed.

Longer samples are taken with boring techniques, such as box-corers and vibro-corers, either directly from a ship, or from an anchored frame. Samples taken with this technique are fairly undisturbed, and, also, soft sediments from the upper part of the bed can be retrieved.

The various corers often have devices such as an internal piston to assist in drawing the sample into the core tube, as well as core catchers to close their bottoms preventing loss of sediment. In the case of a stiff clay, such devices are not necessary, as the clay sticks to the wall of the corer, preventing its slippage.

While sampling, it is important to measure other hydro-sedimentological parameters, such as flow velocity, wave height, salinity and water temperature, as well as monitoring the weather conditions. Often, hydro-sedimentological conditions prior to sampling have also to be known.

The water and sediment samples should be stored away from sunlight, preferably in dark at low temperature to prevent algae growth. Note that organic substances are also easily decomposed by oxidation or reduction at a rate varying with different classes of organic compounds (e.g. polysaccharides, lipides, etc.), particle size, temperature and bacterial activity. Some examples of reduction rates are: faeces ~25 %/day, leaves ~5 %/day and organics in sediments ~3 %/year.

C.1   COMPOSITION AND PROPERTIES OF THE SEDIMENT-WATER MIXTURE

CEC: The standard method of determining the cation-exchange capacity CEC involves leaching the soil or sediment to remove adsorbed cations and replacing these with a standard cation (Ca, Ba, K, Na, etc.) until equilibrium is reached. Next, these cations are desorbed and the amount per unit weight of soil or sediment is determined. This value is expressed in milli-equivalents of the particular cation per 100 g dry soil or sediment. A more indicative test is the methylene blue test, which is described in standard NEN 933-9 and ASTM C837.

E_h: The redox-potential E_h is measured with a device, which consists of a platinum measuring electrode and an Ag/AgCl reference electrode. E_h follows from the reading by comparing the reference potential of the Ag/AgCl electrode against a normal hydrogen electrode. This latter potential varies with temperature, which is therefore to be measured simultaneously.

Measuring E_h in soils is described in the standards ISO 11271 and BS 1377-Part3-11. Measuring E_h in aqueous solutions is described in the standard IEC 60746-5.

gas content: Gas may be found in the unsaturated zone of the pore system or in bubbles (vacuoles) in the sediment matrix. The amount of gas can be determined by compressibility tests of the soil in closed containers. This measurement is part of the standard triaxial test procedure to determine the degree of saturation of a sample.

A direct measurement on in-situ samples of soft sediment can be done in special liners, which can be pressurised (e.g. mud sampler). Another in-situ instrument for unsaturated soils is the gamma-ray attenuation probe (NEN 5784). The gas content in sub-samples of cores can also be measured with a pycnometer, which is normally used for determining the density of solid particles (see NEN 5111 and ASTM D5550).

It must be understood that, while retrieving samples from the seabed, degassing of dissolved gas in the pore water and expansion of the gas will occur at atmospheric pressure. When samples are retrieved from deeper water, such degassing may result in mud foam or failure of the liner.

An indirect method to measure gas content is by determining the in-situ sample volume (NEN 5110), the pore water volume from the water content (NEN 5781 and ISO 16586), and the solids volume from the solid content (NEN 5781 and ISO 16586) and solid density (NEN 5111 and ASTM D5550). Small amounts of gas (less than 2 % by volume) can be determined with acoustic transmission or backscatter techniques (Van Kessel et al., 1996).

metal contents: Determination of metals in soil is done by dissolution of the soil sample with nitric acid and hydrochloric acid. The measurements are described in standard NEN 6961, which includes NEN-EN 13346 for mud and NEN-EN-ISO 15587-1 for water.

electrolyte concentration pore water: The electrolyte concentration in pore water can be determined by measuring the specific electrical conductivity in an aqueous extract of the soil sample. The measurement is described in standard NEN 5749 and ISO 11265.

mineralogical composition: The mineralogical composition of sediment is determined by a combination of X-ray diffraction measurements, related to the crystal lattices, and differential thermal analysis (DTA). Analysis of the measurements requires specific knowledge and is usually done by specialists. X-ray diffraction analysis is described in the standards NEN-EN 13925-1, 2, 3.

mineral density: the specific density of mineral particles is usually measured with a so-called pycnometer by weighing it empty, next with a specific amount of sediment and then fully filled with hexane, added onto the sediment.

Measuring mineral density is described in the standards ISO 11508, NEN 5111, DIN 18124 (Lubking, 2004) and BS1377-Part2-8.3 .

organic content: There exist different definitions and ways of determining organic content. Therefore an appropriate protocol is required.

Organic content is measured by weighing samples prior to and after oxidation of the organic substances, determining the amount of organic carbon. This can be done in a number of ways:

- Wet-oxidation: organic carbon is oxidised with for instance potassium bichromate in a sulphuric environment; silver sulphate can be added to reduce the effects of chlorides. By adsorbing the evolved CO_2, the fractions of carbonate and organic carbon can both be determined.
- Dry-oxidation: the sample is oven-dried at 105 °C, oxidised with hydrogen peroxide, redried and weighed prior to and after oxidation.
- Loss-on-ignition: the loss in weight by calcination of a sample dried at 80 °C; this includes loss of water in both sediment and organic matter. To determine the organic matter content, the remainder of the sample is heated at temperatures between 470 and 500 °C – but not higher to prevent decomposition of limestone.

The loss-on-ignition method is standardised in NEN 5756 and requires a preparation method as described in NEN 5751. The determination of total organic carbon (TOC) and dissolved organic carbon (DOC) in pore water is standardised in NEN-EN 1484.

pH: A pH-probe is a galvanic element, measuring the induced voltage between electrodes and electrolyte, the latter being a function of pH. Measuring pH is described in the standards BS 1377 Part 3-9.

salinity: Salinity S is defined as the mass of dissolved salts to the mass of sea water at $T = 15\ °C$ and is measured by determining water conductivity and temperature with (commercially available) so-called CTD-probes. The relation between salinity S [ppt], temperature T [°C] and the density ρ [kg/m³] for sea water is given in UNESCO (1981):

$$
\begin{aligned}
\rho &= \rho_0 + AS + BS^{3/2} + CS^2 ; \\
\rho_0 &= 999.84 + 6.79 \cdot 10^{-2}\,T - 9.10 \cdot 10^{-3}\,T^2 + \\
&\quad + 1.00 \cdot 10^{-4}\,T^3 - 1.12 \cdot 10^{-6}\,T^4 + 6.54 \cdot 10^{-9}\,T^5 \\
A &= 8.24 \cdot 10^{-1} - 4.09 \cdot 10^{-3}\,T + 7.64 \cdot 10^{-5}\,T^2 + \\
&\quad - 8.25 \cdot 10^{-7}\,T^3 + 5.39 \cdot 10^{-9}\,T^4 \\
B &= -5.72 \cdot 10^{-3} + 1.02 \cdot 10^{-4}\,T - 1.65 \cdot 10^{-6}\,T^2 \\
C &= 4.83 \cdot 10^{-4}
\end{aligned}
\tag{C.1}
$$

which is valid for $0\ °C < T < 40\ °C$ and $0.5\ ppt < S < 43\ ppt$.

The relation between chlorinity, density and salinity can be approximated by $Cl : \rho : S = 3 : 4 : 5$. The simplified equation of state reads: $\rho = \rho_0 + 0.75S$.

Note that salinity measurements in HCMS are often biased by the high sediment concentration, as the sediment affects the water conductivity as well.

SAR: The concentrations of Ca and Mg are determined by flame atomic absorption spectrophotometry, whereas Na (or K) is measured with flame emission spectrophotometry.

specific surface A_s: The specific surface can be measured directly with a laser sizer and analysed according to DIN-norm 6614. It can also be determined indirectly by adsorption techniques, e.g. measuring the amount of ethylene glycol, mono-ethyl ether, or nitrogen (BET-method) which can be adsorbed. The BET-method is standardised in NEN-ISO 9277. The specific surface can also be determined from the pore size distribution by means of capillary condensation of nitrogen (see Section C.2.4).

vertical gradients in physico-chemical parameters: (Large) vertical gradients in physico-chemical parameters are generally found in the upper decimetres (or smaller) of the sea bed (e.g. Fig. 3.9). These profiles can be measured with micro probes. Various probes are designed for measuring specific chemical components (e.g. oxygen) or physical parameters (e.g. permeability). Probes have also been developed to measure multiple components, such as the voltametric probe (Luther et al., 1998). This probe scans the redox-potential ranging from -100 to $+100$ mV, and at each potential the strength of electrical current is measured. Each redox-couple has its own characteristic fingerprint, so that the strength of current is a measure for the concentration. Chemical components that can be determined with voltametric probes are O_2, Fe^{2+}, Fe^{3+}, Mn^{2+}, NO^{-3}, SO^{-2} and S^{2+}.

ζ-**potential**: The ζ-potential reflects clay mineral activity as a function of pH, anion or cation concentration and absorbed surfactants (inorganic or organic). It is measured with standard equipment like the Malvern 3000. Generally, diluted samples are required. For suspended matter, in particular, the ζ-potential measurement can give information about flocculation in the water column. The method can also be used for the fines in the pores of a sand bed when only small amounts of solids are present. For larger samples the clay mineral activity can also be determined with standard geotechnical tests like the pore size distribution and the Atterberg limits (see section C.5.1).

C.2   PARTICLE SIZE DISTRIBUTION

Techniques to measure the particle size distribution of cohesive sediment have not been standardised. Moreover, many instruments and techniques are available. This explains amongst other things the large variety in particle size distributions reported in the literature. We distinguish between methods applied in the field and in the laboratory on field (or laboratory) samples.

C.2.1 IN-SITU PARTICLE SIZE DISTRIBUTION IN THE WATER COLUMN

Often, one is interested in the floc size (distribution) of the sediment. As flocs are fragile and their size and structure depend on their stress history, floc sizes should be measured in-situ, as discussed in Chapter 4.

   In-situ floc size measurements are often carried out with video camera systems and digital image processing techniques. Such measurements have the lowest chance of disrupting the flocs. Eisma (1996) has given an excellent review on the various instruments and methods available at that time.

Presnetly, the commercial LISST, c.q. LISST-100 instrument is also used frequently. The particle size distribution (and sediment concentration) is determined with the Lorenz-Mie theory for analysing the scatter and adsorption of a laser beam, measured with concentric diode rings. One should be aware of possible deviations, as the Lorenz-Mie theory is developed for massive, spherical particles. The settling velocity (see Section C.6) can be determined with the LISST-100 system, which contains a small settling chamber. For more information, the reader is referred to for instance Agrawal and Pottsmith (2000); Fugate and Friedrichs (2002); and Gartner et al. (2001).

C.2.2 IN-SITU PARTICLE SIZE DISTRIBUTION IN THE BED

To determine spatial distributions of the particle size in the field requires painstaking sampling and analysis of these samples in the laboratory (see C.2.3). Nowadays also a more direct method exists with the Medusa-system (Koomans et al., 1999), based on the radiometric properties of sediment. From the measured spectrum of the radio-nuclide distribution, fairly accurate site-covering pictures of the sand-mud composition of the upper 3 dm of the bed sediment can be obtained. This method has been applied a number of times in The Netherlands, for instance in the Rhine-Meuse estuary (e.g. Van Wijngaarden et al., 2002a, 2002b). The absolute total random error in the lithographic maps varied between 6 and 18 %.

Another method for rapid determining particle size distributions was developed by Rubin (2004) and Harney and Rubin (2004) using high-resolution optical cameras. Analysis of their images yields accurate data on the surface sediments.

C.2.3 PARTICLE SIZE DISTRIBUTION IN THE LABORATORY

Measuring the size distribution of flocculated material in the laboratory is very difficult, as it is almost impossible not to disturb or damage the flocs of the sediment sampled in the field. The results of such measurements are almost always biased by the methodology of sampling, storage, transport and further treatment of the sediment. We therefore recommend focusing on the particle size composition if the particle size distribution in the laboratory is to be measured.

Measurements of the distribution of the primary particles, as discussed in Section 3.1.1, must be carried out on deflocculated material. Deflocculation can be done by ultra-sonic stirring of the sample. However, this may not be sufficient to break down the primary flocs (e.g. Chapter 4), and further chemical treatment with a defloculent such as carboxyl methyl-cellulose (CMC) or poly-phosphate (e.g. sodium carbonate/oxalate, sodium hydroxide, ammonium hydroxide, sodium hexametaphosphate or tetra-sodium-diphosphate) may be required.

The particle size distribution is often measured with laser diffraction spectrometry; a large number of instruments is commercially available, and the underlying Lorenz-Mie theory is well developed. For larger particles $\left(2\pi d|m-1|/\lambda > 30\right)$, where d = particle size, m = relative refractive index and λ = laser wavelength, Fraunhofer's model is used to establish the particle size distribution.

The laser-diffraction technique is suitable for a wide range of particle sizes, and little material is required for accurate measurements. However, the lens-system should be selected carefully and tuned to the size range of the particles. At particle sizes of several μm, the method becomes unreliable. Further discussions on this technique for application on natural sediment is found in for instance McCave et al. (1986) and Singer et al. (1988).

For the determination of smaller particle sizes, use can be made of a Sedigraph, which is in effect a (small) settling column. The particle size is computed from Stokes' law. Organic material and carbonates should be removed from the sample with hydrogen peroxide and acetic acid, respectively. Some further information can be found in Stein (1985).

The particle size of coarser sediment fractions can be obtained from sieving, preferably after drying the sample.

The use of the co-called Coulter counter is not recommended for cohesive sediment, as this technique is mainly suited for solid, mono-dispersive material; see also Peters Rit et al. (1987).

C.2.4 PORE SIZE DISTRIBUTION

The pore size distribution of sediment is strongly related to its skeleton structure. It is therefore important that the skeleton structure is not damaged during measurements. Some tests must be executed on dry samples. Air-drying will damage the skeleton due to capillary surface tension. This damage can partly be avoided with freeze-drying using liquid nitrogen and evaporation in vacuum. An alternative but complicated method is drying above the critical point of water (374 °C, 220 bar). Several methods can be applied:

Mercury intrusion: a freeze-dried sample in vacuum is filled with mercury by increasing the mercury pressure and measuring the amount of mercury displaced; pore size range: 10 nm – 100 μm.

Capillary condensation of nitrogen: nitrogen is absorbed at -196 °C in a freeze-dried sample by increasing the nitrogen vapour pressure until saturation (Broekhoff, 1969). In the first stage of the test the adsorbed nitrogen volume increases linearly with vapour pressure (known as common *t*-curve), its rate of increase is a measure for the specific surface area. This test is similar to the BET-method. At increasing vapour pressure increasingly larger pores are filled by liquid nitrogen due to capillary surface tension. After saturation, desorption is applied and in general the process is reversible revealing the same pore size distribution during absorption and desorption; pore size range: 1 nm – 1 μm.

CT-scan: samples without treatment are scanned by X-ray tomography revealing water-saturated pores and gas-saturated pores. The resolution is about 5 μm, which limits its applicability to sand- and silt-dominated soils; pore size range: 10 μm – 10 mm.

Cryo-SEM: a sample is frozen in nitrogen and placed in a Cryo-SEM system (Scanning Electron Microscope). A cross section of the sample is examined during evaporation showing the skeleton structure. This technique can be applied for sediment with a fragile skeleton structure, such as laponite (e.g. Fig. C.2); pore size range: 1 μm – 100 μm.

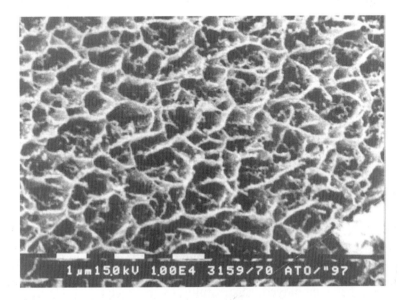

Fig. C.1: Cryo-SEM image of laponite (Van Kesteren, 2004).

C.3   SEDIMENT CONCENTRATION AND DENSITY

Sediment concentration in the water column and in the bed, c.q. the density or water content of the bed, can be measured with a number of instruments. All these instruments have to be calibrated against samples, which are analysed in the laboratory. None of the methods is standardised so that an appropriate protocol has to be developed.

It is further important to recognise that, when analysing and interpreting data, a large number of definitions on concentration and density exists (e.g. Chapter 3 and Appendix A and B).

C.3.1 IN-SITU SEDIMENT CONCENTRATION IN THE WATER COLUMN

Van Rijn and Schaafsma (1986) present a concise overview of the performance and accuracy of techniques to measure suspended sediment concentrations under field conditions. They distinguish between instruments suitable for sand and cohesive sediment suspensions.

Optical Back-Scatter sensors (OBS) are the most commonly used to measure suspended sediment concentration in the water column. The scatter of light is a function of the particle size (distribution) with respect to the light's wavelength, particle structure and sediment concentration (see for instance Downing et al., 1981; Lynch et al., 1994; or Black and Rosenberg, 1994). Sutherland et al. (2000) have discussed the effects of sediment colour on the back-scatter signal. Because of these dependencies, OBS-sensors have to be calibrated frequently, i.e. prior to, several times during and after their deployment against in-situ samples (e.g. Downing and Beach, 1989). Note that near the upper limit of measurable concentration, the OBS-signal saturates, depicted by a flat sensor response.

A persistent problem with the use of OBS sensors is contamination of the optics with algae or otherwise. The sensors have therefore to be cleaned regularly.

Suspended sediment concentrations can also be measured with the LISST-instrument (e.g. Appendix C.2.1).

A fairly new technique consists of the analysis of the energy in the back-scatter signal from a broadband ADCP (Acoustic Doppler Current Profiler). This method has been employed successfully in the waters of Hong Kong in conjunction with a profiling infrared siltmeter to monitor the fate of dredging spill in the early 1990's by Land et al. (1997). Later it was applied to measure the suspended sediment transport in the Marsdiep, a tidal inlet in the Dutch Wadden Sea, and in access channels to estuarine ports in Belgium (Claeys et al., 2001). More information can be found in Thorne and Hanes (2002) and Hill et al. (2003).

As with the OBS-system, this method has to be calibrated frequently and carefully, as the acoustic back-scatter is a function of particle size (distribution) and possibly composition. Note that the back-scatter signal may be disturbed severely by air bubbles in the water column.

The interface between the water column (LCMS) and fluid mud, c.q. soft mud beds is often assessed with dual frequency echo-sounders, generally operating at 33 kHz and 210 kHz. The higher frequency signal reflects on relatively small density gradients (gradients in acoustic impedance), generally attributed to the top of the fluid mud layer. The high-energy, low frequency signal can penetrate the soft mud and is only reflected at fairly large density gradients, i.e. the fluid mud – consolidated bed interface. Presently, this method is applied on an operational basis to determine the navigable depth in many fairways and harbour basins. Note that the acoustic signal does not reflect at a specific density, but at (steep) density gradients instead.

Calibration of the various methods should be done against samples taken from the water system at the time of deployment. It is important that these samples are representative for the water column at that moment, as discussed in the beginning of this appendix.

C.3.2 IN-SITU SEDIMENT CONCENTRATION (DENSITY) IN THE BED

The bulk density, the dry bed density, the water content of the bed and the sediment concentration in the bed are all related to each other (e.g. Chapter 3 and Appendix A and B). For convenience, we refer in this section to sediment concentration only.

The electrical conductivity probe is a classical instrument, deployed in soil mechanics for a long time (Dowling, 1990). As this instrument is based on the principle that sediment is a poorer conductor than water, the salinity of the pore water (determining the electrical conductivity of the pore water in the marine environment) should be known. This principle also implies that the method is unsuitable in areas where pore water salinity changes, such as in the brackish region of an estuary. The electrical conductivity of extracted pore water must be determined according to NEN 5749 or ISO 11265.

The conductivity probe is particularly suitable for use in the marine environment, as seawater is sufficiently conductive to attain an accuracy of about 10 % (e.g. Ariathurai and Arulanandan, 1986). Miniature probes are in use in several laboratories. Note that conductivity probes are not suitable in brackish water due to the varying salinity.

This measuring technique can be extended to 3D imaging with tomographic techniques (McKee et al., 1996; Briggs et al., 1998).

The gamma-densito meter is particularly suited for deployment in soft mud (Crickmore et al., 1990). A number of instruments are commercially available. Gamma-densito meters consist of a radioactive source and a detector. The source and detector can either be mounted in the same housing (back-scatter probe) or in separate housings, with the medium passing in between (transmission probe). One of the greatest advantages of the nuclear methods (γ-densitometer) is that measurements are independent of the diameter of sediment particles. The principle of these density measurements is based on the fact that materials absorb more gamma rays as their density increases.

The gamma-densito meter is used in many fairways to determine navigability (e.g. De Vlieger and De Cloedt, 1987; PIANC, 1990). This method is also used for non-destructive measurements in the laboratory, in particular for consolidation tests in large settling columns (Toorman, 1992).

Sediment density can also be measured with a tuning fork; the added mass, i.e. sediment density determines the eigen frequency of the fork. These instruments are commercially available. An advantage is that no pain-striking permission procedures have to followed, as in the case of using a γ-densitometer

The sediment concentration, c.q. bed density in stiff clays and sandy sediments has to be determined in the laboratory from samples – see page C.1.

C.3.3 SEDIMENT CONCENTRATION IN THE LABORATORY

Most of the instruments, when miniaturised (apart from the ADCP) are applicable in laboratory experiments in flumes and columns.

A handy instrument to measure water density, hence sediment concentration if the water's salinity is known, is the Paar densito-meter. This instrument measures the specific mass of a sample from its inertia during vibration.

Next to the nuclear densito meter to measure bed densities, X-ray sensors are in use in the laboratory (e.g. Been, 1981).

The sediment concentration in low-concentration water samples and in high-concentration bed samples are measured differently.

sediment concentration in low-concentrated water samples

The sediment concentration in low-concentration samples is usually determined by filtering. It is recommended to split the sand fraction from the fines, for instance with a 50-63 μm sieve. Various filters for the fine fractions are commercially available with mesh sizes of a few tenth of a μm. The following steps should be followed:

1. establish sample volume,
2. rinse filter with distilled or demineralised water to remove loose and soluble material,
3. dry filter in oven at a temperature for a specific period of time (of the order of hours) at a temperature of about 100 °C,
4. place filter in desiccator to cool down during a specific period of time preventing adsorption of water,
5. weigh the filter,
6. place filter on holder with vacuum and filter the water sample – rinse several times with distilled or demineralised water to remove loose and soluble material (in particular salt!),
7. dry and weigh filter with sediment again and determine concentration.

These steps are not standardised, so a protocol with respect to filter type, oven temperature, drying and cooling time is necessary. The accuracy of this method decreases with decreasing sediment concentration.

sediment concentration in high-concentration bed samples

The sediment concentration of high-concentration bed samples cannot be determined through filtering, as the permeability of the soil is too low. Therefore, the wet bulk density of the sample is determined. The easiest method is to use a container with known volume and weighing the container with and without the sample. When high accuracy is required, a pycnometer can be used (see Appendix C.1).

When the sample contains gas, the sample is placed in a chamber at reduced pressure for some time. Make sure that the pressure is reduced slowly to prevent the development of (large) bubbles.

C.4   RHEOLOGICAL PARAMETERS

Three sets of rheological tests may have to be executed to determine the rheological parameters that have been defined and used in this book:

- stationary rheometer tests to measure the flow curve,
- vane tests to measure peak and remoulded strength,
- oscillating rheometer tests to measure visco-elastic properties.

It is emphasised that the choice of the instrument and the measurements to be performed are largely governed by the rheological model applied. The latter is determined by the problem to be studied and its schematisation.

Many instruments are commercially available, and a careful evaluation is recommended before purchase. Some research institutes prefer to develop and deploy in-house designed dedicated tools.

A large number of books and papers have been published on rheometry, i.e. measuring rheological parameters. Some of these focus on measuring the parameters of cohesive sediment samples. The reader is referred to for instance Williams and Williams (1989) or Chou et al., (1993). We will not summarise these works in detail, but highlight a series of issues that should be addressed in determining rheological parameters.

stationary rheometer tests to measure the flow curve

The flow curve is measured in general with a Couette viscometer (also known as coaxial, concentric or rotational viscometer or rheometer). The driving of these devices can be either stress- or strain-controlled. The stress or strain is increased slowly, and often decreased again at the same rate. Sometimes this cycle is repeated a number of times to account for thixotropic effects. A typical flow curve is presented in Fig. C.2, showing a gradual increase in stress with increasing shear, until the sediment structure breaks down at the peak stress, reducing the stress level. After a further increase in shear the stress starts to build up again. Upon relaxation, lower stress levels are found because the sediment structure cannot recover fully (thixotropic effect).

The slope $d\tau/d\dot{\gamma}$ defines the sediment's (apparent) viscosity and the intercept with the ordinate is referred to as the yield strength. The yield and peak strength are poorly defined, as they depend on sample preparation and stress history, which include the effects of the acceleration of the viscometer.

The intercept of the tangent of the stress-strain curve is known as the Bingham strength, which also cannot be unambiguously defined in many cases.

Various measuring sensors can be used. Generally, either two coaxial cylinders (outer cylinder = cup, inner cylinder = bob), or a plate-cone configuration is deployed. The gap between the cylinders and plate and cone should be much larger than the maximum particle size in the sediment sample.

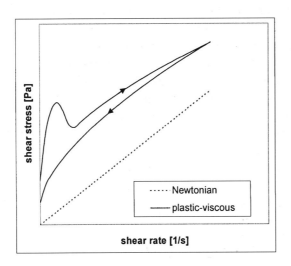

Fig. C.2: Sketch of flow curve measured with Couette viscometer.

Other important issues in sediment rheometry to be addressed are:
- wall slip: to prevent slippage, the wall may be roughened,
- end effects: the measured torque should be corrected for the upper and lower end (or side wall in case of plate-cone configuration) – this is often accounted for in the manufacturer's software,
- instability: the Reynolds number within the sensors is often well below its critical value; however Taylor vortices may develop resulting in an increase in apparent viscosity of the sample,
- segregation: at smaller concentrations, sediment particles can settle during the measurements - some devices have been tested with a counter flow to prevent settling,
- sample preparation: no optimal sampling procedure exists; however a standardised protocol is required to allow intercomparison of results from measurements.

The viscosity of a suspension can also be measured with a capillary viscometer. The sample flows through a capillary tube under the influence of a pressure gradient, and the viscosity can be computed from the (assumed) velocity profile

(Poiseuille flow) in the tube. If the suspension has yield strength, plug flow may result.

vane tests to measure peak and remoulded strength

Peak and remoulded strength are measured with a shear vane, which is a classical soil mechanical test (Bowles, 1979, see also NEN 5106). A typical result of a vane shear test is presented in Fig. C.3. The peak strength is a function of rotation speed, stress history and sample preparation. The undrained or remoulded shear strength, however, can be considered as a material property at the given water content of the sample.

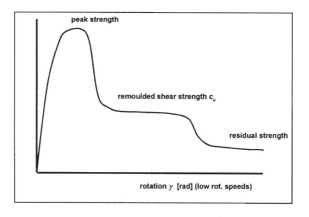

Fig. C.3: Sketch of stress response to shear vane test.

Special vane elements for rheometers are available in the size of 5, 25 and 40 mm diameter. The standard rotational speed is in the order of 1 rpm.

In Belgium and Germany, instruments have been developed to measure vane strength in-situ. The Belgian rheometer consists of a strain-controlled vane and a pressure gauge. In combination with in-situ density measurements, the navigability of soft mud layers is assessed (Van Craenenbroeck et al., 1991).

The in-situ rheometer developed in Germany consists of a strain-controlled vane; the results are correlated with rheological measurements in the laboratory to establish navigable depth (e.g. Rechlin, 1995).

oscillating rheometer tests to measure storage and loss modulus

The visco-elastic behaviour of soft (e.g. liquefied) mud layers can be established with oscillating rheometer tests, measuring the so-called storage (or elasticity or rigidity) modulus G' and the loss (or viscous) modulus G''. If the oscillating strain is given by:

$$\gamma = \gamma_0 \sin(\omega t) \qquad\qquad\qquad\qquad (C.2)$$

and the stress response by:

$$\tau = \tau_0 \sin(\omega t + \delta) \qquad\qquad\qquad\qquad (C.3)$$

the storage and loss moduli follow from:

$$\tau = \gamma_0 \left(G' \sin(\omega t) + G'' \cos(\omega t) \right) \qquad\qquad\qquad (C.4)$$

where $G' = G_0 \cos(\delta)$, $G'' = G_0 \sin(\delta)$ and $G_0 = \tau_0/\gamma_0$. Note that in the Kelvin-Voigt model, G'' is a function of the oscillation frequency ω, whereas both G' and G'' are a function of ω in the Maxwell model. Excellent reviews are given by Chou (1989), Chou et al. (1993) and Williams and Williams (1989).

   These oscillating rheometer tests can be carried out with the same equipment as the stationary tests described above, and the same remarks on gap size, end effects, etc. are applicable.

C.5   SOIL MECHANICAL PARAMETERS

Many textbooks are available on measuring soil mechanical parameters (e.g. Head, 1986; Whitman, 1964). Here we describe a number of methods briefly.

C.5.1 ATTERBERG LIMITS

The Atterberg or plasticity limits yield the water content of a cohesive sediment sample from two standardised soil mechanical tests. The water content W in soil mechanics is defined as the ratio between the weight of water and the weight of solids in a sample. To determine the liquid limit LL, a groove of specified dimensions is made in remoulded samples of different water contents. At the liquid limit, this groove becomes closed within a certain number of blows on the sample through (local) fluidisation. To determine the plastic limit PL, cylinder-shaped samples (sausages) at different water content are prepared. When deformed, these cylinders begin to crumble at the plastic limit. These tests require experience and skill and should be executed by specialised staff.

Currently, the Atterberg limits are also determined with more sophisticated tests, such as the Swedish fall-cone test (Head, 1986). In general it can be stated that the LL corresponds to an undrained shear strength of about 1 kPa, while the PL corresponds to an undrained shear strength of the order of 100 kPa. It is important to appreciate that the ratio between these two shear strengths classifies the behaviour of the sediment, more than their absolute values. More information about test procedures etc. can be found in many geotechnical textbooks, like Head (1986); Mitchell (1976); and Lambe and Whitman (1979); see also Chapter 3.

Measuring the Atterberg limits is described in the standards DIN 18122-1, ASTM D4318 and BS 1377 Part 2-4 and Part 2-5.

C.5.2 PERMEABILITY

The permeability coefficient k represents the resistance at which water can flow through a porous medium. It is assumed to be constant for a particular soil at a particular specific density and independent of the flow velocity and hydraulic gradient. The global permeability of mud can be obtained through Darcy's law from a settling or consolidation experiment:

$$k = \frac{Q}{A}\frac{L}{\Delta h} \qquad\qquad\qquad (C.5)$$

where k is the permeability, Q is the water flow rate (discharge) through the porous medium, A is the total cross-sectional area normal to the flow, and Δh is the head difference over the thickness L of the mud layer.

When the sample has approximately uniform density and structure, the flow rate can be estimated by observing the lowering of the mud-water interface during a consolidation experiment, and an average permeability can be calculated. The head difference is calculated from the difference between the water height in the column and the head of the lowest piezometer. From that head difference (Δh) and the height of the consolidating mud layer (L), the gradient is calculated. The greatest error is caused by calculating the flow rate, because of the low accuracy of time measurements at the beginning of the experiment and of height measurements after a few weeks.

Permeability is measured with piezometers mounted along a column drained at its bottom (Van den Bosch et al., 1991). This method is suitable to obtain values of permeability at higher densities. Water migrating through the filters is collected to establish the discharge. In the upper zone of the mud layer, however, water migrates upwards due to consolidation, whilst in the bottom zone water migrates downward and percolates through the filter. Thus, with this method only the permeability of the bottom zone is measured.

The permeability can also be determined directly with the so-called seepage tests on mud samples contained in a small tube closed with a membrane at its bottom and a capillary at its top. Water is driven through the sample by air pressure, and the flow rate can be read accurately from the water volume in the capillary. The device can be placed in a second container to carry out the experiments under pressure (see also Head, 1986, for more details).

Standard tests for permeability measurements on soils are the constant-head and falling-head method (see Head, 1986). The constant-head method is used in sandy soils with at most 10 % mud (< 63 μm), as described in the standards NEN 5123, ASTM D2434 and BS 1377–Part5-5. The falling-head method is used in cohesive soils with more than 10 % mud, as described in the standards NEN 5124. For soft cohesive soils the constant-head method is also applied by means of a hydraulic consolidation cell, standardised in BS 1377-Part6-4.

C.5.3 TOTAL AND PORE WATER STRESS

Total and pore water pressures in a consolidating suspension are measured with pressure transducers, which are commercially available. The accuracy of these transducers can be of the order of 10 Pa (i.e. 1 mm water pressure).

Pore pressure is measured through the use of a filter between the sediment-water mixture and the pressure transducer. A vyon plastic filter has been shown to be appropriate (Bowden, 1988), but also porous stone filters have been used successfully.

Some major problems may occur when measuring water pressure:
• the transducers should be calibrated accurately at the very low pressures at which they may be operated,
• small vibrations may disturb the readings, in particular at low stress levels,
• under unsaturated conditions probes become compressible, thus distorting the accuracy of the measurements significantly,

- if a reference pressure is obtained with a tube, condensation and blocking of the tube should be avoided; in long tubes extended measuring times are required to achieve equilibrium conditions.

C.5.4 CONSOLIDATION PARAMETERS

Oedometer test
The oedometer test is the most common test for determining consolidation parameters for cohesive sediments. The test determines the consolidation behaviour under confined conditions (horizontally) by applying progressively increasing vertical loading steps (often power of 2). The minimal initial loading that can be applied in standard equipment is 7 kPa. Therefore this test can be used for cohesive sediments with undrained shear strengths larger than 1 kPa. During each loading step the settlement is measured as a function of time. The consolidation coefficient c_v and the confined compressibility coefficient m_v are determined directly from the measured settling curve, while the permeability k is computed assuming linear consolidation during each step.

   The oedometer test is described in the standards NEN 5118, ASTM D2435 and BS 1377 Part 3.3.

Hydraulic consolidation test
This test is similar to the oedometer test, but in this case vertical loading is applied by hydraulic pressure on a closed top platen. This test is done in a so-called Rowe-cell, in which settling and pore water pressures are both measured. This test is described in the standards BS 1377 Part 6.3 and 6.4.

Seepage Induced Consolidation (SIC)
The seepage induced consolidation (SIC) test is a test in which sediment is consolidated by means of controlled pore water flow through the sediment. It is based on the seepage-induced consolidation method, developed by Znidarcic for very soft cohesive sediments (Liu and Znidarcic, 1991). A typical test set-up is shown in Fig. C.4 In this test, a constant discharge is applied through the sample by means of a displacement pump. Water pressure difference over the sample is measured until equilibrium is reached. The top of the sample remains close to its initial or in-situ void ratio, while in the lower part consolidation occurs. To prevent piping through the sample, a small weight is placed on its top. The permeability can be determined over the full range of void ratios, as the flow rate is constant. Multiple discharges can be applied to obtain a higher accuracy. At the end of the test an external surcharge is applied to squeeze out

pore water from the upper part of the sample to obtain a constant void ratio over the sample height. The permeability is determined by another seepage step. The SIC-test is also applicable for multi-phase materials, including desiccation tests.

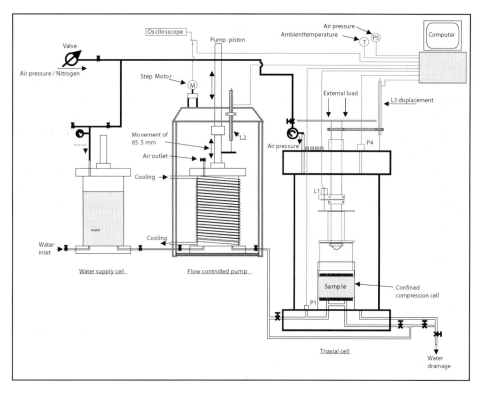

Fig. C.4: SIC-test set-up at Delft Hydraulics.

Capillary Suction Time (CST)

The Capillary Suction Time (CST) has been used since the 1970's as a quick and reliable method for characterising cohesive sediment filterability and hydraulic conductivity. The capillary suction pressure generated by standard filter paper is used to "suck" water from the sludge. The rate at which water permeates through the filter paper varies as a function of the condition of the sludge. The CST is obtained from two electrodes placed at a standard interval from the funnel of the instrument. The time necessary for the waterfront to pass between these two electrodes constitutes the CST (e.g. Huisman and Van Kesteren, 1998).

The force generated by capillary suction is much greater than the hydrostatic head within the funnel, so the measurement is independent of the amount of cohesive sediment tested, as long as there is sufficient to generate a CST. Each test can be completed within a few minutes.

CST instruments are commercially available, and are designed to comply with European and USA EMC standards.

C.5.5 STRESS-STRAIN RELATION

Stress-strain relationships for soils are commonly determined with triaxial tests. In the standard test set-up (see Fig. C.5) a cylindrical sample (h = 150 mm, Ø = 62 mm) is placed vertically in a cell which can be pressurised between two porous plates with filter paper. A rubber membrane encloses the sample. The lower porous plate is coupled to a drainage system to perform drained and undrained tests. The sample can be loaded vertically by a piston, which is normally done in a number of steps. The horizontal stress is applied by the cell pressure and is referred to as the confining pressure. A back-pressure is applied to saturate the sample and to retrieve the ambient pore pressure (e.g. water depth). By measuring vertical displacements and volume changes of the sample at each stress increment, the stress-strain relations are determined.

During the test softening of the sample may occur and therefore the loading should be displacement-controlled. Compression loading conditions are imposed by increasing the vertical stress, while extension loading conditions are imposed by increasing the horizontal stress. Three standard loading conditions exist:

- CU = consolidated undrained conditions,
- CD = consolidated drained conditions, and
- UU = unconsolidated undrained conditions.

Triaxial tests can be performed on very soft cohesive sediment (undrained shear strength less than 1 kPa) by using a mixture of glycerol and water as cell fluid at a density equal to the initial density of the sample (see Chapter 8).

The triaxial test is described in the standards NEN 5117, DIN 18137-Part2, ASTM D4767 and BS 1377 Part 7.9, 8.7 and 8.8.

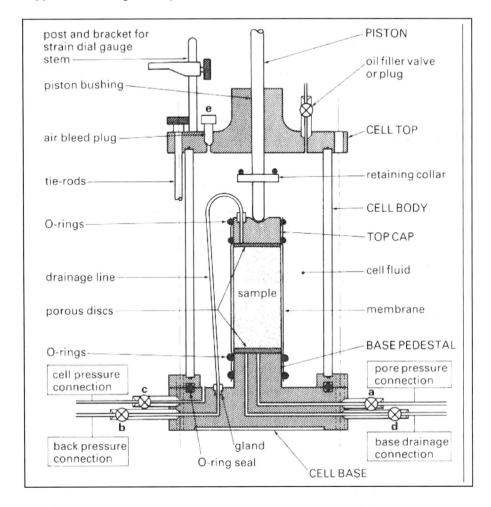

Fig. C.5: Standard triaxial-test set-up (after Head, 1986).

C.6 SETTLING VELOCITY

C.6.1 SETTLING VELOCITY IN THE LABORATORY

The settling velocity of mud flocs should preferably be measured in-situ, as these flocs are fragile and can be damaged easily during sampling, transport and storage (see also Section C.2). However, if settling velocities need to be measured in the laboratory, two techniques are suitable:

1. Sedimentation balance,

2. Optical techniques, discussed in Section C.6.2.

A sedimentation balance consists of a temperature-controlled settling tube generally with a length of a few dm. The sediment suspension is gently mixed throughout the tube, after which it is allowed to settle. The settling rate is measured with a balance mounted above the bottom of the tube, from which the distribution of the settling velocity is assessed, using the one-dimensional mass balance equation for the sediment in the tube (the so-called Oden equation).

Sedimentation balances are commercially available, and rather easy to operate. However, the user's main concern in its deployment is sample treatment and preparation. As this is not standardised, a protocol should be developed carefully, which includes:

*   sample collection,
*   sample transport and storage,
*   sample treatment (how to remix the sediment in the sampling container, measuring concentrations, use of flocculants or detergents, etc.),
*   filling of sedimentation balance and mixing of sediment.

C.6.2 IN-SITU SETTLING VELOCITY

Various instruments have been developed and deployed for in-situ measurements of the settling velocity of mud flocs. "In-situ" not only refers to deployment in the field, but also in laboratory facilities like flumes. Eisma (1996) gives an excellent review of the applicability and draw-backs of a number of these instruments and their intercomparison.

Basically, three methods can be distinguished:
1. Settling velocity determined from measuring gradients in sediment concentrations in a settling tube,
2. Settling velocity determined from measuring particle size distributions in a settling tube, and
3. Settling velocity measured from time-lapse optical images.

One of the first instruments for the in-situ measurement of settling velocity of mud flocs was the Owen Tube (Owen, 1976; Van Rijn, 1986). This instrument consists of a long slender tube, which is lowered in the water column in horizontal position with both ends open to allow through-flow of water. After closing both ends, the tube is brought into vertical direction and lifted out of the water. At specific time intervals sub-samples of the suspension are withdrawn from the bottom of the tube. From weighing the sediment collected in these sub-samples, the settling velocity (distribution) of the sediment is established.

Draw-backs of this method are the occurrence of large-scale circulations in the tube induced by its rotation and the settling sediment, and the adherence of sediment to the side walls. Other instruments were designed to circumvent these draw-backs, amongst which the so-called Field Pipette Withdrawal Tube (FPWT) (e.g. Van Rijn, 1986). With this instrument water samples are taken and lifted above water after closing both tube ends and rotating the tube in vertical direction. Sub-samples are taken halfway down the tube at fixed time intervals; sediment mass in these samples is used to establish the settling velocity distribution.

The settling velocity can also be measured with the LISST-ST system (e.g. Section C.2.1, Agrawal and Pottsmith (2000); Fugate and Friedrichs, 2002; Gartner et al, 2001; and Van Wijngaarden and Roberti, 2002). For data analysis it is assumed that the suspension is properly mixed over the LISST's settling tube and settles in still water. Then, the settling velocity distribution is found from the depletion of particles in various size classes, as measured with the LISST optical system.

At present, the authors prefer application of (in-situ) camera systems to monitor flocs settling in a settling chamber. Digital image analysis techniques have been developed to establish floc size and settling velocity distribution, and sometimes floc structure from the video recordings (see Eisma, 1996 for a summary on these systems and f.i. Manning and Dyer, 2002 for more recent results). The resolution that can be achieved depends on the lens and lighting system(s), the spread of the particle size distribution and the suspended sediment concentration (background turbidity).

Circulation in the settling tube, c.q. measuring chambers of the instruments is a major problem with all techniques for in-situ measurement of settling velocity. Circulation may be induced by the return flow due to settling of flocs, convection cells due to temperature gradients, eddies captured from the water column while sampling, etc. Circulation velocity may be of the same order or larger than the settling velocity of smaller flocs. Van Kessel (2003) for instance estimated that a 0.1 °C temperature gradient can induce convective currents as large as 2 mm/s.

Therefore, we advocate the use of small, closed still-water chambers, which separate a few flocs from the overall bulk suspension in which these flocs can settle in an undisturbed environment. Such a separation chamber has been developed for instance by Fennesy et al. (1994).

C.7   ERODIBILITY

In this section we discuss a number of devices that have and/or are being used to establish the erodibility, i.e. erosion threshold and erosion rates, of cohesive sediment in the marine environment.

   At present the various devices and methods yield very large differences in results, sometimes even by more than an order of magnitude (e.g. Tolhurst et al., 2000). This is only partly due to the scales of the inhomogeneities in the sediment bed (Cornelisse et al., 1997).

C.7.1 EROSION MEASURMENTS IN THE LABORATORY

Measurments to determine the erosion rate of cohesive sediment can be carried out in straight, recirculating flumes. The mean erosion rate can be obtained from the increase in suspended sediment concentration. A problem with this kind of experiment is that the flow in the flume is not uniform, and secondary currents cause a fairly non-uniform erosion pattern over the width of the flume (e.g. Kuijper et al., 1989). This non-uniformity can be overcome by placing relatively small containers with mud samples in a flume. Then, however, the problem of accurately measuring the suspended sediment concentration (or the erosion rate directly) arises.

Therefore, the so-called rotating annular flume was developed in the 1960's. This flume consists of a closed, circular channel, the fluid in which is driven by a rotating lid in contact with the fluid's surface. To minimise secondary currents, the flume itself is rotated in opposite direction. It was found that a fairly homogeneous bed shear stress could be obtained with this set-up. Another advantage of this flume is the absence of floc-breaking devices such as pumps and pipes. For further details, the reader is referred to one of the many detailed publications on the applicability of the annular rotating flume (e.g. Partheniades et al., 1966; Mehta, 1973; or Booij et al., 1993).

A problem in carrying out erosion experiments in the laboratory is sample preparation. Two methods have been described frequently in the literature:
1. Preparation of a placed bed, i.e. ("natural") sediment is remoulded and placed in the flume at in-situ bulk density.
2. Preparation of a deposited bed, i.e. a bed is formed by deposition from a suspension that is recirculated in the flume.
It is obvious that the soil structure is completely destroyed with method 1 and it will also be difficult to obtain a reasonable "natural" bed with method 2

(normally deposited beds will be too soft). These laboratory tools are therefore mainly suitable to study the erosion process in itself, but not to establish erosion properties from natural soils.

C.7.2 EROSION MEASURMENTS IN THE FIELD

A number of erosion devices for in-situ use have been described in the literature. A brief summary in alphabetical order is as follows:

Cohesive Strength Meter (CSM)

The CSM consists of a vertical jet confined within a small cylindrical chamber ($h = 6$ cm, $\varnothing = 2.8$ cm). The chamber is placed on the bed (dry intertidal flat), which is eroded by the jet. Sediment concentration within the chamber (actually in a return flow tube) is measured, from which the erosion rate follows. The CSM is calibrated against the initiation of movement of quartz particles varying between 130 and 1450 μm to relate jet pressure to bed shear stress.

Instrument and electronics are housed in a portable case to enable rapid and easy deployment, allowing densely spaced surveying. For more information, the reader is referred to Paterson (1989) and Tolhurst et al. (1999).

In-situ Erosion Flume (ISEF)

The ISEF is a vertically placed, semi-closed flume, the eroding flow in which is driven by a propeller. The flume has a length of 1.8 m, a height of 0.7 m and the width of its rectangular cross-section amounts to 0.1 m. It is deployable on dry intertidal areas. The open bottom of the flume is placed on the bed and leakage is prevented by slats that penetrate the bed at the sides of the exposed bed. The erosion rate as a function of flow velocity in the ISEF follows from measuring the suspended sediment concentration in the flume.

The ISEF has been calibrated against Shields' curve. For more information the reader is referred to Houwing and Van Rijn (1998).

Instrument for Measuring Erosion Shear Stress In-situ (ISIS)

In the ISIS a bed shear stress is generated by the flow through and around a specially shaped bell head, which draws water radially across the bed into the centre of the bell head. This bell is housed in a vertical cylinder ($h = 35$ cm, $\varnothing = 9$ cm), which is placed on the bed (dry intertidal area).

The ISIS is suited to study the onset of erosion. It was calibrated with hot film shear stress probes. For more information the reader is referred to Williamson and Ockenden (1996).

Sea Carousel or Benthic Annular Flume

The VIMS Sea Carousel consists of a semi-closed annular channel of 20 cm height and 15 cm width, with a radius of 1 m. Flow within the channel is driven by a rotating lid, and as such this set-up resembles the rotating annular flume described under C.7.1. The bed shear stress was established with a numerical and experimental study.

The Sea Carousel is used under water, deployed from a ship. For more information, the reader is referred to Maa et al. (1993).

The Benthic Annular Flume is very similar to, but much smaller than the Sea Carousel with a height of 35 cm, a channel width of 10 cm and a radius of 32 cm. Water flow is driven by a rotating lid. Erosion rates are determined from increases in suspended sediment concentration and are related to flow velocity instead of bed shear stress. For more information, the reader is referred to Widdows et al. (1998).

Sedflume

Sedflume is a mobile laboratory recirculating flume with a cross-section of 10 by 2 cm. Samples, taken in-situ, are mounted underneath the flume and are manually moved upward as erosion proceeds. The erosion rate is measured visually from the loss of material off the sample. The bed shear stress is obtained from relations generally known for pipe flow. For more information, the reader is referred to McNeil et al. (1996) or Jespen et al. (1997).

D.   TENSOR ANALYSIS

Cohesive sediment can deform largely at already moderate stress levels. This is discussed in Chapter 7, where we treat self-weight consolidation of cohesive sediment. Consolidation, however, is basically a one-dimensional process with deformations in the direction of gravity mainly.

In Chapter 8 we discuss the mechanical behaviour of cohesive sediment in a more general way, with stresses and deformations in directions other than that of gravity alone. This implies that one-dimensional analysis is not sufficient and three-dimensional analyses of stresses and deformations have to be considered. Therefore we need tensor analysis to describe and quantify the mechanical behaviour in terms of stresses and strains. In this appendix we recapitulate some basic tools with respect to tensor analysis.

The stresses and strains are second order tensors yielding relation between two vectors, which are generally linear. Besides the stress and strain tensors, also the permeability tensor relating the specific discharge and the hydraulic gradient is important in soil mechanics. More background can be found in Malvern (1969), Truesdell and Toupin (1960) and Truesdell and Noll (1965), and for engineering application in Chen and Saleeb (1994).

D.1   THE STRESS TENSOR

The behaviour of a medium in terms of deformations exerted by internal forces can be described as a continuum when the individual constituents (particles, molecules, etc.) of the medium are very small with respect to the scales of interest. For a continuum phase the actual forces between individual particles are averaged over a surface much larger than the size of the particles.

Fig. D.1 shows the resultant force $\underline{F}$ acting on a plane ABC in three dimensions. This force is balanced by normal and shear forces on the three orthogonal planes OAB, OAC and OBC, c.q. by the normal (σ_{ii}) and shear (σ_{ij}) stresses which are forces per unit the surface area. In Fig. D.1 these stresses are depicted as open arrows to distinguish them from the forces. The nine stress components form a second order tensor:

$$\underline{\underline{\sigma}} = \sigma_{ij}(\underline{e}_i\underline{e}_j) = \begin{bmatrix} \sigma_{11} & \sigma_{12} & \sigma_{13} \\ \sigma_{21} & \sigma_{22} & \sigma_{23} \\ \sigma_{31} & \sigma_{32} & \sigma_{33} \end{bmatrix} \tag{D.1}$$

We use both the tensor format ($\underline{\underline{\sigma}}$) and index format ($\sigma_{ij}$), with Einstein's summation-convention. The index format is coupled to the chosen reference system, while the tensor format is more generic and can be used for any reference system. The first index •$_i$ refers to the direction of the normal of the plane and the second index •$_j$ refers to the direction of the stress. The components of the stress tensor in a certain reference system can be represented by matrices like a 3×3 matrix or in case of a symmetric tensor by a 1×6 matrix.

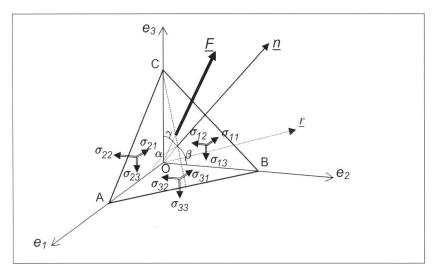

Fig. D.1: Internal forces and stresses on a soil element.

By changing the orientation of the plane ABC, the total force and the stresses on the orthogonal planes OAB, OAC and OBC remain the same, but the normal and shear stresses in the plane ABC will change. The orientation of the plane ABC is given by the normal vector $\underline{n}$ with components $n_1 = \cos\alpha$, $n_2 = \cos\beta$ and $n_3 = \cos\gamma$ (α, β, γ are defined in Fig. D.1).

From the equilibrium of the tetrahedron OABC the components of force $\underline{F}$ divided by the area ABC, denoted as f_i (i = 1, 2 , 3), can be expressed in the tensor components σ_{ij} and normal vector components n_j with:

$$f_i = \sigma_{ij} n_j \quad \text{or} \quad \underline{f} = \underline{\underline{\sigma}}\,\underline{n} \tag{D.2}$$

The torque equilibrium of the tetrahedron OABC requires that $\sigma_{ij} = \sigma_{ji}$ ($i \neq j$), which makes the stress tensor symmetric.

If we choose the plane's orientation such that $\underline{n}$ coincides with the direction of the internal force $\underline{F}$, no shear stresses are present and only the diagonal components of the tensor remain, the so-called principal stresses. The principal stresses and orientation can be determined by substituting $f_i = \sigma n_i$ in equ. (D.2):

$$(\sigma_{ij} - \delta_{ij}\sigma)n_j = 0 \tag{D.3}$$

where δ_{ij} is the unity tensor also known as the Kronecker delta defined by δ_{ij} = 1 for $i = j$ and $\delta_{ij} = 0$ for $i \neq j$.

The principal directions can be solved from equ. (D.3) if σ is chosen such that equ. (D.3) has non-trivial solutions, i.e. the determinant of the tensor $(\sigma_{ij} - \delta_{ij}\sigma)$ must equal zero:

$$
\begin{aligned}
0 = &-\sigma^3 + (\sigma_{11} + \sigma_{22} + \sigma_{33})\sigma^2 + \\
&-(\sigma_{11}\sigma_{22} + \sigma_{22}\sigma_{33} + \sigma_{33}\sigma_{11} - \sigma_{12}\sigma_{21} - \sigma_{13}\sigma_{31} - \sigma_{23}\sigma_{32})\sigma + \\
&+(\sigma_{11}\sigma_{22}\sigma_{33} - \sigma_{11}\sigma_{23}\sigma_{32} - \sigma_{22}\sigma_{13}\sigma_{31} - \sigma_{33}\sigma_{12}\sigma_{21} + \sigma_{12}\sigma_{23}\sigma_{31} + \sigma_{21}\sigma_{32}\sigma_{13}) \\
\Leftrightarrow\ & 0 = -\sigma^3 + I_1\sigma^2 - I_2\sigma + I_3
\end{aligned}
\tag{D.4}
$$

The three solutions for σ of this third order polynomial are the principal stresses σ_1, σ_2 and σ_3, also known as the eigenvalues of $\underline{\underline{\sigma}}$. Because these solutions are invariant for the orientation, the terms between the brackets in equ. (D.4) are also invariant and can therefore be expressed as a function of the principal stresses:

$$I_1 = \sigma_{ii} = (\sigma_{11} + \sigma_{22} + \sigma_{33}) = (\sigma_1 + \sigma_2 + \sigma_3)$$

$$I_2 = \frac{1}{2}\left[I_1^2 - \sigma_{ij}\sigma_{ji}\right] = (\sigma_{11}\sigma_{22} + \sigma_{22}\sigma_{33} + \sigma_{33}\sigma_{11} - \sigma_{12}\sigma_{21} - \sigma_{13}\sigma_{31} - \sigma_{23}\sigma_{32})$$

$$= (\sigma_1\sigma_2 + \sigma_2\sigma_3 + \sigma_3\sigma_1) \tag{D.5}$$

$$I_3 = \det \underline{\underline{\sigma}} = (\sigma_{11}\sigma_{22}\sigma_{33} - \sigma_{11}\sigma_{23}\sigma_{32} - \sigma_{22}\sigma_{13}\sigma_{31} - \sigma_{33}\sigma_{12}\sigma_{21} + \sigma_{12}\sigma_{23}\sigma_{31} + \sigma_{21}\sigma_{32}\sigma_{13})$$

$$= (\sigma_1\sigma_2\sigma_3)$$

The third invariant I_3 equals the determinant value of $\underline{\underline{\sigma}}$. It is shown in Chapter 8 that these invariants play an important role in the description of the mechanical behaviour of cohesive sediment.

The stress tensor can be represented graphically in the so-called principle stress space with the three orthogonal axes σ_1, σ_2 and σ_3 (see Fig. D.2).

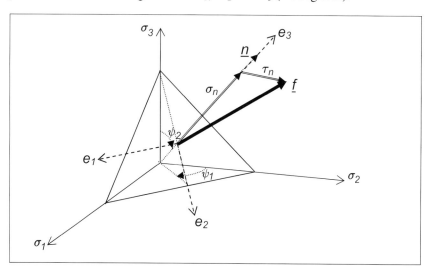

Fig. D.2: Principal stress space.

The stress tensor for an arbitrary plane in the principal stress space of the material can be computed with:

$$\sigma_{ij}^* = R_{ij}\sigma_{ji} R_{ij}^{-1} \quad \text{or} \quad \underline{\underline{\sigma}}^* = \underline{\underline{R}}\,\underline{\underline{\sigma}}\,\underline{\underline{R}}^{-1} \tag{D.6}$$

In which $\underline{\underline{R}}$ is the so-called rotation tensor. For a rotation α in anti-clockwise direction around an axis $\underline{r}$ the rotation tensor is given by:

$$R_{ij} = \delta_{ij}\cos\alpha + r_i r_j(1-\cos\alpha) - e_{ijk}r_k\sin\alpha \quad \Leftrightarrow$$

$$\underline{\underline{R}} = \begin{bmatrix} \cos\alpha + (1-\cos\alpha)r_1^2 & (1-\cos\alpha)r_1 r_2 + r_3\sin\alpha & (1-\cos\alpha)r_1 r_3 - r_2\sin\alpha \\ (1-\cos\alpha)r_2 r_1 - r_3\sin\alpha & \cos\alpha + (1-\cos\alpha)r_2^2 & (1-\cos\alpha)r_2 r_3 + r_1\sin\alpha \\ (1-\cos\alpha)r_3 r_1 + r_2\sin\alpha & (1-\cos\alpha)r_3 r_2 - r_1\sin\alpha & \cos\alpha + (1-\cos\alpha)r_3^2 \end{bmatrix} \quad (D.7)$$

in which e_{ijk} is the permutator (see equ. (2.10)). A special case is obtained for the two-dimensional case when $\underline{r}$ coincides with the σ_2-axis ($\underline{r} = (0,1,0)$). With equ. (D.6) and (D.7) the resulting stress tensor in the new orientation reads:

$$\sigma_{11}^* = \sigma_{11}\cos^2\alpha + \tfrac{1}{2}(\sigma_{13}+\sigma_{31})\sin 2\alpha + \sigma_{33}\sin^2\alpha$$

$$\sigma_{33}^* = \sigma_{11}\sin^2\alpha - \tfrac{1}{2}(\sigma_{13}+\sigma_{31})\sin 2\alpha + \sigma_{33}\cos^2\alpha$$

$$\sigma_{22}^* = \sigma_{22} \quad\quad\quad (D.8)$$

$$\sigma_{13}^* = \tfrac{1}{2}(\sigma_{33}-\sigma_{11})\sin 2\alpha + (\sigma_{13}-\sigma_{31})\cos 2\alpha$$

$$\sigma_{31}^* = \tfrac{1}{2}(\sigma_{33}-\sigma_{11})\sin 2\alpha + (\sigma_{31}-\sigma_{13})\cos 2\alpha$$

$$\sigma_{12}^* = \sigma_{21}^* = 0$$

$$\sigma_{23}^* = \sigma_{32}^* = 0$$

For a symmetric stress tensor $\sigma_{13} = \sigma_{31}$.

Also the stress tensor can be represented graphically as a vector in the principle stress space (see Fig. D.2). Two rotations are necessary to find the stress tensor in the orientation of $\underline{n}$ in Fig. D.2: first around the σ_3-axis over ψ_1 and than around the e_1-axis over ψ_2. The magnitude of these stresses can be determined graphically by means of the so-called Mohr circles (e.g. Malvern, 1969 and Fig. D.3). The normal stress is plotted on the horizontal axis, and the shear stress is plotted on the vertical axis; both stresses act on the plane of interest ABC (e.g. Fig. D.1). For the three-dimensional case three circles appear with the principal stress components on the horizontal axis. The outer stress circle, determined by the principle stresses σ_1 and σ_2, is found by varying ψ_1 and keeping $\psi_2 = \pi/2$. The stress circle determined by the principle stresses σ_1 and σ_3 is found by varying ψ_2 and keeping $\psi_1 = 0$, and the stress circle determined by the principle stresses σ_1 and σ_3, is found by varying ψ_2 and keeping $\psi_1 = \pi/2$. For all other values of ψ_2 and ψ_1 stress points within the three circles are obtained.

When ψ_2 is kept constant at a value between 0 and $\pi/2$ and ψ_1 is varied, stress points on a concentric circle within the circle determined by σ_1 and σ_3 are obtained (see Fig. D.3). The intersections of this concentric circle with the other two circles are found by drawing a line from σ_2 and σ_1 with an angle ψ_2 with the horizontal axis.

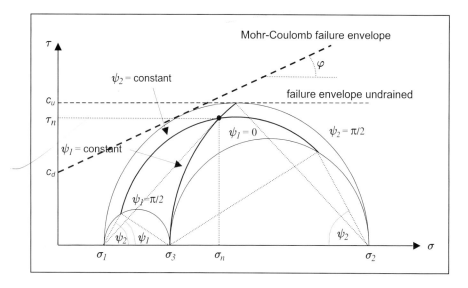

Fig. D.3: Stress circles of Mohr in three dimensions.

When ψ_1 is kept constant at a value between 0 and $\pi/2$, and ψ_2 is varied, stress points are obtained on a circle starting in σ_3 and intersecting the outer circle determined by σ_1 and σ_2 at a point on the line from σ_3 at angle ψ_1 (see Fig. D.3). The intersection of both circles for given ψ_2 and ψ_1 yields the normal stress σ_n and total shear stress τ_n on a plane rotated by ψ_2 and ψ_1 with respect to ABC.

However, it should be realised that this graphical approach does not yield the direction of the total shear stress. Therefore the Mohr circles are in general used to determine limiting stress states with respect to failure.

We have also drawn a (dashed) line which envelops other circles, which have not been drawn however (known as the Mohr-Coulomb failure envelope). This line intercepts the τ-axis at c_d, which is the drained shear strength, also known as the cohesion of the material. Its slope represents the internal angle of friction φ.

Under undrained conditions, the envelope of the Mohr circles is given by the horizontal dashed line in Fig. D.3, which intersects the τ-axis at the undrained shear strength c_u.

The principle stress space, shown in Fig. D.2, can also be used to visualise the stress state in terms of isotropic and deviatoric stresses as shown in Fig. D.4. The isotropic stress is defined as the average of the three principle stresses, and equals 1/3 of the first invariant I_1 of the stress tensor (see equ. (D.5)):

$$p = \frac{1}{3}\sigma_{ii} = \frac{1}{3}I_1 \qquad\qquad\qquad\qquad \text{(D.9a)}$$

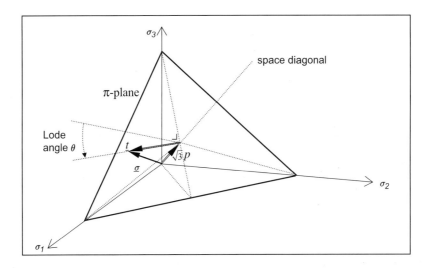

Fig. D.4: Principal stress space and π-plane.

The deviatoric stress tensor τ_{ij} is defined as the stress tensor minus the isotropic stress. The deviatoric stress tensor and its principal stress components t_i are given by:

$$\tau_{ij} = \sigma_{ij} - \delta_{ij}p \quad \text{and} \quad t_i = \sigma_i - p \qquad\qquad \text{(D.9b)}$$

In the principal stress space, the isotropic stress vector follows the space diagonal, while the deviatoric stress vector $\underline{t}$ with components t_i is located in a plane perpendicular to the space diagonal, the so-called deviatoric or π-plane. The invariants of the deviatoric stress tensor τ_{ij} can be expressed in the invariants of the stress tensor σ_{ij}:

$$J_1 = (t_1 + t_2 + t_3) = 0$$
$$J_2 = -(t_1 t_2 + t_2 t_3 + t_3 t_1) = \tfrac{1}{3} I_1^2 - I_2 \qquad\qquad \text{(D.10)}$$
$$J_3 = -t_1 t_2 t_3 = -I_3 + \tfrac{1}{3} I_1 I_2 - \tfrac{2}{27} I_1^3$$

The second invariant J_2 is a measure for the magnitude t of the deviatoric stress in the π-plane:

$$t = \sqrt{t_1^2 + t_2^2 + t_3^2} = \sqrt{\sigma_1^2 + \sigma_2^2 + \sigma_3^2 - 3p^2} = \sqrt{2}\sqrt{\tfrac{1}{3} I_1^2 - I_2} = \sqrt{2 J_2} \quad \text{(D.11)}$$

In soil mechanics it is more common to use a slightly different definition of the deviatoric stress, i.e. $q \equiv \sqrt{\tfrac{3}{2}} t = \sqrt{3 J_2}$, which is the notation used throughout this book. In case of triaxial loading q equals the stress difference between the vertical and horizontal radial stress ($q = \sigma_3 - \sigma_1$). Hence, the deviatoric stress at failure in cohesive sediment under undrained conditions equals twice the undrained shear strength, i.e. $q = 2c_u$ (e.g. Fig. D.3).

The third invariant J_3 is a measure for the angle θ between the deviatoric stress vector $\underline{t}$ and the normal of the projection of the principal axis on the π-plane, which is known as the Lode angle (see Fig. D.4). It can be expressed through the invariants of the deviatoric stress tensor:

$$\sin 3\theta = \frac{3}{2}\sqrt{3}\,\frac{J_3}{J_2^{3/2}} = 3\sqrt{6}\,\frac{J_3}{t^3} = \frac{27}{2}\frac{J_3}{q^3} \qquad -\frac{\pi}{6} \le \theta \le \frac{\pi}{6} \qquad \text{(D.12a)}$$

This results in an absolute value of the Lode angle with respect to the axis of the largest principal stress. A more direct way of computing the vector $\underline{t}$ in polar coordinates with respect to one of the principal axes is (e.g. the principal σ_1-axis):

$$\theta = \frac{\pi}{6} - \mathrm{atan}\left(\frac{\sqrt{3}\left(\sigma_3 - \sigma_2\right)/2}{\sigma_1 - \left(\sigma_3 + \sigma_2\right)/2}\right) \quad \left(-\pi \quad \text{if} \quad \sigma_1 < \left(\sigma_3 + \sigma_2\right)/2\right) \qquad \text{(D.12b)}$$

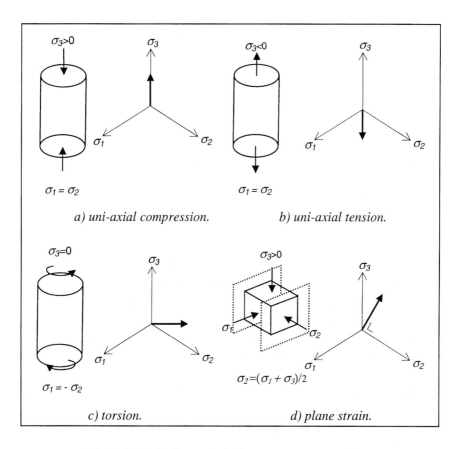

Fig. D.5: Projection principal stress space on π-plane.

When the soil is deformed, the stresses follow trajectories in the stress space of the soil, which are described by so-called stress paths. Such a stress path can be plotted in the three-dimensional principal stress space, but also as a projection on the π-plane. Typical stress conditions appear as specific patterns on the π-plane. In Fig. D.5 several examples are shown for uni-axial tension and compression, torsion, plane stress and plane strain conditions. A plane stress condition occurs in thin plates with only normal stresses on the plate faces. A plane strain condition occurs when the geometry is two dimensional and of infinite width in the third direction. Because of symmetry no shear stresses occur in a cross section perpendicular to the third direction. This condition is very common in seabed sediments, for instance when stresses build up by wave loading.

According to international standards, tensile stresses are defined as positive. This definition has been used in the discussions above. In soil mechanics, however, normal stresses are in general compressive, and therefore a compressive normal stress is defined positive. This does not affect the theory, but only the sign of the principle stress axis.

In this book we follow the soil mechanical definition: **compressive stresses are positive**.

D.2   THE STRAIN TENSOR

The internal stresses discussed in the preceding section induce deformations of the soil. These deformations can be expressed by means of changes in the length of line elements, the area of surface elements, or the volume of volume elements. Similar to the definition of stresses, where the actual inter-particle forces are averaged over a large area with respect to the particle size, it is assumed that deformations can be averaged over a length, surface area or volume scale, which is much larger than the particle scale. The consequence is that in a continuum description of deformations, the information on the actual particle movement is lost. As shown in Section D.4, failure mechanisms in granular materials, such as cohesive sediment, are determined by this particle movement, however. Yet, a continuum description for deformation is still applicable to establish the general response to changes in the stress field, as long as the particle scale is not important.

The position of a material point is given in Fig. D.6 in a coordinate system coupled to the material ($\underline{X}$), known as the Lagrangian description, and in a fixed reference system ($\underline{x}$), known as the Eulerian description. When we consider the deformation of a line element $d\underline{X}$ in the material coordinate system, the corresponding line element in the Eulerian reference system after deformation is given by:

$$dx_i = \frac{\partial x_i}{\partial X_j} dX_j \quad \text{or} \quad d\underline{x} = \underline{\underline{F}}\, d\underline{X} \tag{D.13}$$

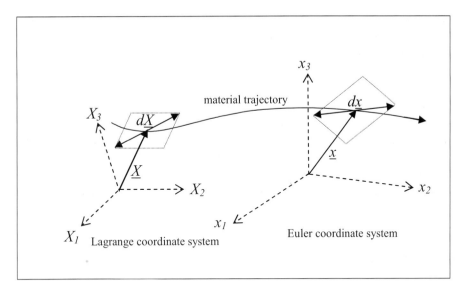

Fig. D.6: Deformation of line element.

The displacement gradients $\partial x_i / \partial X_j$ are components of the second order material deformation tensor $\underline{\underline{F}}$ in the Lagrangian reference system. The inverse tensor $\underline{\underline{F}}^{-1}$ is the displacement gradient tensor in the Eulerian reference system with components $\partial X_i / \partial x_j$. Note that $\underline{\underline{F}}^{-1}$ contains gradients with respect to a fixed reference system, while $\underline{\underline{F}}$ has a reference system that may move in time and space. The determinant of $\underline{\underline{F}}$ equals the volume ratio of a volume element after (dV) and before deformation (dV_0):

$$\frac{dV}{dV_0} = \det \underline{\underline{F}} = I_{3,F} = J \tag{D.14}$$

in which the determinant of $\underline{\underline{F}}$ is also known as Jacobi's determinant J.

The deformation tensor $\underline{\underline{F}}$ yields the total deformation, including possible rigid rotations, and is therefore referred to as the total deformation tensor. For the actual material deformations the change in length of line elements must be considered:

$$(dx)^2 = d\underline{x}\,d\underline{x} = d\underline{X}\,\underline{\underline{C}}\,d\underline{X} \ ; \quad \underline{\underline{C}} = \underline{\underline{F}}^T\underline{\underline{F}} \tag{D.15}$$

in which $\underline{\underline{F}}^T$ is the transposed tensor of $\underline{\underline{F}}$ $\left(F_{ij}^T = F_{ji}\right)$ and $\underline{\underline{C}}$ is known as the deformation tensor of Green. Green's tensor contains the actual deformations of the material.

The components of $\underline{\underline{C}}$ can be visualised by considering three orthogonal line elements in the initial Lagrangian reference system (see Fig. D.7). After deformation, the angle between line elements i and j has changed with Ω_{ij}, defined as positive when the angle becomes smaller. The quadratic length ratio before and after deformation of line element i results in the diagonal component C_{ii} of $\underline{\underline{C}}$; the non-diagonal components C_{ij} are related to changes in the angle between line element i and j:

$$C_{ij} = \sqrt{C_{ii}C_{jj}}\,\sin\Omega_{ij} \tag{D.16}$$

The principal directions of $\underline{\underline{C}}$ are found when $C_{ij} = 0$, which means that the orientation is such that three orthogonal line elements in the initial reference system remain orthogonal after deformation. The principal deformations of $\underline{\underline{C}}$ equal the ratio of the quadratic length of each orthogonal line element before and after deformation.

The relative volume change is given by:

$$\frac{dV - dV_0}{dV_0} = I_{3,F} - 1 = \sqrt{I_{3,C}} - 1 \ ; \quad I_{3,C} = \det\underline{\underline{C}} \tag{D.17}$$

A rigid rotation tensor $\underline{\underline{R}}$ can be determined from the total deformation tensor $\underline{\underline{F}}$ and Green's deformation tensor $\underline{\underline{C}}$:

$$\underline{\underline{F}} = \underline{\underline{R}}\,\underline{\underline{C}}^{\frac{1}{2}} \quad \Leftrightarrow \quad \underline{\underline{R}} = \underline{\underline{F}}\,\underline{\underline{C}}^{-\frac{1}{2}} \tag{D.18}$$

The square root of Green's tensor is obtained from the square root of the diagonal principal components of $\underline{\underline{C}}$. The principal direction of $\underline{\underline{C}}^{\frac{1}{2}}$ is coaxial with $\underline{\underline{C}}$.

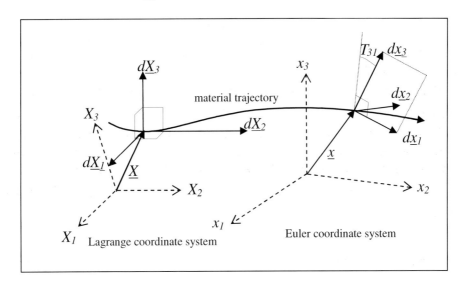

Fig. D.7: Deformation of orthogonal line elements.

The equivalent deformation tensor in Eulerian coordinates is given by:

$$(\mathrm{d}X)^2 = \mathrm{d}\underline{x}\,\underline{\underline{c}}\,\mathrm{d}\underline{x} \; ; \;\; \underline{\underline{c}} = (\underline{\underline{F}}^{-1})^T \underline{\underline{F}}^{-1} \tag{D.19}$$

The principal deformations of $\underline{\underline{c}}$ are reciprocal to the principal deformations of $\underline{\underline{C}}$, and the principal directions are coaxial. The inverse of the deformation tensor of Cauchy $\underline{\underline{c}}^{-1}$ is known as the Left-Cauchy-Green deformation tensor and $\underline{\underline{c}}^{-\frac{1}{2}}$ as the Left stretch tensor. The strain tensor of the deformation is obtained from the increase in length of the line elements:

$$(\mathrm{d}x)^2 - (\mathrm{d}X)^2 = 2\,\mathrm{d}\underline{X}\,\underline{\underline{E}}\,\mathrm{d}\underline{X} = 2\,\mathrm{d}\underline{x}\,\underline{\underline{e}}\,\mathrm{d}\underline{x}; \tag{D.20}$$
$$\underline{\underline{E}} = \tfrac{1}{2}(\underline{\underline{C}} - \underline{\underline{\delta}}); \;\; -\underline{\underline{e}} = \tfrac{1}{2}(\underline{\underline{c}} - \underline{\underline{\delta}})$$

in which $\underline{\underline{E}}$ is the strain tensor of Green, $\underline{\underline{e}}$ the equivalent strain tensor in the Eulerian coordinate system and $\underline{\underline{\delta}}$ is the Kronecker delta. $\underline{\underline{E}}$ and $\underline{\underline{e}}$ are coaxial with the deformation tensors $\underline{\underline{C}}$ and $\underline{\underline{c}}$. $\underline{\underline{E}}$ and $\underline{\underline{e}}$ are related by the total deformation tensor $\underline{\underline{F}}$ through equ. (D.15) and equ. (D.17):

$$\underline{\underline{E}} = \underline{\underline{F}}^T \underline{\underline{e}} \, \underline{\underline{F}}$$ (D.21)

The principal strains of $\underline{\underline{E}}$ and $\underline{\underline{e}}$ are obtained by substituting equ. (D.19) into equ. (D.15):

$$2E_i = C_i - 1; \quad 2e_i = 1 - c_i = 1 - C_i^{-1}$$ (D.22)

Because the principal deformations of $\underline{\underline{C}}$ and $\underline{\underline{c}}$ are reciprocal ($C_i = 1/c_i$), equ. (D.20) yields for the principal strains E_i and e_i :

$$2E_i = (1 - 2e_i)^{-1} - 1$$ (D.23)

Both principal strains E_i and e_i are equal for infinite small strains. However $\underline{\underline{E}}$ and $\underline{\underline{e}}$ are not equal, as the Lagrangian coordinate system can still be rotated with respect of the Eulerian reference system. The total deformation tensor $\underline{\underline{F}}$ for infinite small strains reduces with equ. (D.18) to the rigid rotation tensor $\underline{\underline{R}}$ and subsequently equ. (D.21) reduces to:

$$\underline{\underline{E}} = \underline{\underline{R}}^T \underline{\underline{e}} \, \underline{\underline{R}}$$ (D.24)

For the material strain tensor $\underline{\underline{E}}$ the more commonly used (small) strain tensor $\underline{\underline{\varepsilon}}$ is used, which can be expressed in the Green's tensor components:

$$\varepsilon_{ii} = \sqrt{C_{ii}} - 1 \; ; \; \varepsilon_{ij} = \Omega_{ij}/2$$ (D.25)

For infinitely small gradients of the displacements in the total deformation tensor (equ. (D.13)), the rotation tensor $\underline{\underline{R}}$ becomes also infinitively small and $\underline{\underline{E}}$ and $\underline{\underline{e}}$ become equal. In that case, there is no difference between the strain description in a Lagrangian or Eulerian reference system. For infinitely small displacements gradients, the total deformation tensor can be divided in the strain tensor $\underline{\underline{\varepsilon}}$, being its symmetric part, and the rotation tensor $\underline{\underline{\omega}}$, being its asymmetric part:

$$\varepsilon_{ii} = \frac{\partial s_i}{\partial x_i} \; ; \; \varepsilon_{ij} = \frac{1}{2}\left(\frac{\partial s_i}{\partial x_j} + \frac{\partial s_j}{\partial x_i} \right) \; ; \; \omega_{ii} = 0 \; ; \; \omega_{ij} = \frac{1}{2}\left(\frac{\partial s_i}{\partial x_j} - \frac{\partial s_j}{\partial x_i} \right) \quad (D.26)$$

deformation   rotation

in which s_i is the displacement in i-direction.

D.3   STRAIN-RATE TENSOR

The spatial velocity (strain rate) of a material line element $d\underline{x}$, which is displaced through the Eulerian coordinate system in time (see Fig. D.6), is given by:

$$\frac{D}{Dt}d\underline{x} = \frac{D}{Dt}\{\underline{\underline{F}}\,d\underline{X}\} = \frac{D}{Dt}\{\underline{\underline{F}}\}\underline{\underline{F}}^{-1}\,d\underline{x} \equiv \underline{\underline{L}}\,d\underline{x} \; ; \; \underline{\underline{L}} = \dot{\underline{\underline{F}}}\underline{\underline{F}}^{-1} \quad (D.27)$$

where $\underline{\underline{L}}$ is the spatial velocity gradient ($L_{ij} = \partial u_i/\partial x_j$) and the material time derivative operator D/Dt is denoted by a dot (the spatial time derivative operator is $\partial/\partial t$; see Chapter 2). The spatial velocity gradient tensor includes velocity gradients by rigid displacement and rigid rotation. In order to get the real deformation rate, the material derivative of the length of a line element $d\underline{x}$ should be considered:

$$\frac{D}{Dt}\{d\underline{x}.d\underline{x}\} = d\underline{x}.\left\{\frac{D}{Dt}d\underline{x}\right\} + \left\{\frac{D}{Dt}d\underline{x}\right\}d\underline{x} = 2d\underline{x}.\underline{\underline{d}}\,d\underline{x} \qquad (D.28)$$

in which $\underline{\underline{d}}$ is the deformation rate tensor of Euler, which equals the material time derivative operator of the strain tensor $\underline{\underline{e}}$ (see equ. (D.20)).

The deformation rate tensor can be written as the symmetric part of the spatial velocity gradient tensor:

$$\frac{D}{Dt}\underline{\underline{e}} = \dot{\underline{\underline{e}}} = \underline{\underline{d}} = \frac{1}{2}\left[\underline{\underline{L}} + \underline{\underline{L}}^T\right] \Leftrightarrow \dot{e}_{ij} = d_{ij} = \frac{1}{2}\left[\frac{\partial u_i}{\partial x_j} + \frac{\partial u_j}{\partial x_i}\right] \qquad (D.29)$$

The material time derivative of the strain tensor in material coordinates $\underline{\underline{E}}$, i.e. the strain rate tensor of Green, is related to the strain rate tensor $\underline{\underline{e}}$ with equ. (D.21):

$$\dot{\underline{\underline{E}}} = \underline{\underline{F}}^T \dot{\underline{\underline{e}}}\, \underline{\underline{F}} \qquad (D.30)$$

When there is no deformation, but only a rigid displacement and rotation, the deformation rate tensor equals zero ($\underline{\underline{d}} = \underline{\underline{0}}$). It follows from equ. (D.27) that the spatial velocity gradient tensor becomes anti-symmetric ($\underline{\underline{L}} = -\underline{\underline{L}}^T$) in this case.

The angular velocity of line element $d\underline{x}$ in Eulerian coordinates follows from the material derivative of the orientation of $d\underline{x}$, which is known as the spin tensor of Cauchy $\underline{\underline{w}}$ which equals the anti-symmetric or skew part of the spatial velocity gradient tensor:

$$\underline{\underline{w}} = \frac{1}{2}\left[\underline{\underline{L}} - \underline{\underline{L}}^T\right] \Leftrightarrow w_{ij} = \frac{1}{2}\left[\frac{\partial u_i}{\partial x_j} - \frac{\partial u_j}{\partial x_i}\right] \qquad (D.31)$$

The spin tensor $\underline{\underline{w}}$ can be related to the vorticity vector $\underline{w}$ defined by:

$$\underline{w} = \mathrm{rot}\,\underline{u} \equiv \underline{\nabla} \times \underline{u} \quad \text{or} \quad w_i = e_{ijk} \frac{\partial u_k}{\partial x_j} \tag{D.32}$$

in which e_{ijk} is the permutator (see equ. (2.10)).

The angular velocity of line element $d\underline{x}$ in Eulerian coordinates is half the vorticity vector $\underline{w}$, which can be related to the spin tensor by:

$$w_{jk} = -\frac{1}{2} e_{ijk} w_i \tag{D.33}$$

In general the spin tensor $\underline{\underline{w}}$ is not reference invariant[1], contrary to the deformation rate tensor $\underline{\underline{d}}$. Therefore, $\underline{\underline{w}}$ is not equal to the reference invariant material angular velocity tensor $\underline{\underline{\Omega}}$, which is obtained from the rigid rotational tensor $\underline{\underline{R}}$:

$$\underline{\underline{\Omega}} = \underline{\underline{\dot{R}}}\,\underline{\underline{R}}^T \tag{D.34}$$

For instance, when a moving orthogonal reference system is coupled to a certain material body deforming in time, the deformation rate tensor $\underline{\underline{d'}}$ within this body equals the deformation rate tensor $\underline{\underline{d}}$ in the fixed reference system, because $\underline{\underline{d}}$ is reference invariant.

The spin tensor $\underline{\underline{w}}$ with respect to a fixed reference system is the sum of the spin tensor $\underline{\underline{w'}}$ in the moving reference system and the angular velocity tensor $\underline{\underline{\Omega}}$ of the material body to which the moving reference system is coupled:

---

[1] A quantity is reference invariant when it is independent of the kinematics of the observer's orthogonal reference system.

$$\underline{\underline{w}} = \underline{\underline{w}}' + \underline{\underline{\Omega}} \qquad \text{(D.35)}$$

The only case where the spin tensor in the fixed reference system equals the angular velocity of the material is when the principal directions of the actual deformation with respect to the material, represented by Green's deformation tensor $\underline{\underline{C}}$, do not change in time, i.e. $\underline{\underline{w}}' = \underline{\underline{0}}$.

More general, the reference invariance of vector and tensor properties of the material are determined by the material time derivative D/Dt. This operator itself is reference invariant, but when it is applied to a vector or tensor, the material time derivative of the axis of the moving reference system must be taken into account as well. This yields for any vector $\underline{a}$:

$$\frac{D}{Dt}\underline{a} = \frac{D'}{Dt}\underline{a} + \underline{\underline{\Omega}}\,\underline{a} \qquad \text{(D.36)}$$

and for any second order tensor $\underline{\underline{a}}$:

$$\frac{D}{Dt}\underline{\underline{a}} = \frac{D'}{Dt}\underline{\underline{a}} + \underline{\underline{\Omega}}\,\underline{\underline{a}} - \underline{\underline{a}}\,\underline{\underline{\Omega}} \qquad \text{(D.37)}$$

where D'/Dt is the material time derivative in the moving reference system, also known as the Jaumann rate. When $\underline{\underline{a}}$ is symmetric the 2nd and 3rd term in equ. (D.37) are equal, and $\underline{\underline{a}}$ is reference invariant.

D.4   MOMENTUM AND ENERGY EQUATIONS

The first momentum equation of Cauchy in a fixed spatial orthogonal coordinate system is given by (see Chapter 2, equ. (2.2)):

$$\frac{\partial u_i}{\partial t} + \frac{\partial u_i u_j}{\partial x_j} = \frac{1}{\rho}\frac{\partial \sigma_{ij}}{\partial x_j} + f_i \qquad \text{(D.38}$$

where σ_{ij} is the stress tensor in spatial coordinates, known as the Cauchy stress tensor, f_i is an external mass force, for instance gravity. In order to write this balance equation in material coordinates, a material coupled stress tensor is defined, known as the first stress tensor of Piola-Kirchoff $\underline{\underline{T}}$. The resultant force on a certain plane and surface area must be equal in both spatial and material coordinates. This yields the following condition for $\underline{\underline{T}}$:

$$\underline{\underline{T}} = J\underline{\underline{F}}^{-1}\underline{\underline{\sigma}} \quad \Leftrightarrow \quad \underline{\underline{\sigma}} = J^{-1}\underline{\underline{F}}\,\underline{\underline{T}} \tag{D.39}$$

in which $\underline{\underline{F}}$ is the deformation gradient tensor (see equ. (D.13)), and J is the Jaccobi determinant of $\underline{\underline{F}}$ (see equ. (D.14)), which equals the ratio of the material density before and after deformation:

$$J \equiv \det\underline{\underline{F}} = \frac{\rho}{\rho_0} \tag{D.40}$$

Hence, the momentum equation in material coordinates reads:

$$\frac{\partial u_i}{\partial t} + \frac{\partial u_i u_j}{\partial X_j} = \frac{1}{\rho_0}\frac{\partial T_{ij}}{\partial X_j} + f_i \tag{D.41}$$

The second momentum equation of Cauchy yields the balance equation for spin momentum (Malvern, 1969):

$$\frac{\partial \varpi_i}{\partial t} + \frac{\partial \varpi_i \varpi_j}{\partial x_j} = \frac{1}{\rho}\frac{\partial m_{ij}}{\partial x_j} - \frac{1}{\rho}e_{ijk}\sigma_{kj} + l_i \tag{D.42}$$

in which ϖ_i is the intrinsic spin momentum per unit mass, l_i is the external spin couple and m_{ij} is the couple stress tensor. For non-polar materials, such as soil at macro-scale, ϖ_i, l_i and m_{ij} are all zero and the second momentum equation of Cauchy yields the symmetric Cauchy stress tensor:

$$e_{ijk}\sigma_{kj} = 0 \iff \sigma_{jk} = \sigma_{kj} \iff \underline{\underline{\sigma}} = \underline{\underline{\sigma}}^{T} \tag{D.43}$$

The couple stress tensor m_{ij} is not zero for a so-called Cosserat material, such as suspensions of granular material at high concentrations. Therefore the assumption of a symmetric stress tensor is not allowed, when the dynamic equations for flow of granular sediment are considered (see e.g. Campbell, 1988).

The total energy e of a unit mass of material is the sum of its thermodynamic internal energy and its kinetic energy. On macroscopic scale, the kinetic energy is determined by the material velocity field only. Therefore the kinetic energy related to the internal spin momentum is added as an intrinsic energy quantity per unit mass to the thermodynamic internal energy. The energy balance equation is then given by:

$$\rho \frac{De}{Dt} = \rho\varphi - \frac{\partial q_i}{\partial x_i} + \sigma_{ij}d_{ji} - \sigma_{ij}e_{kji}(w_k - v_k) + m_{ij}\frac{\partial v_i}{\partial x_j} \tag{D.44}$$

in which d_{ij} is the deformation rate tensor (equ. (D.29)), q_i is the heat flux, φ is the radiation heat, w_i the vorticity vector (equ. (D.31)) and v_i is the vorticity vector coupled to the intrinsic spin momentum ϖ_i.

For non-polar materials the energy balance equation reduces to:

$$\rho \frac{De}{Dt} = \rho\varphi - \frac{\partial q_i}{\partial x_i} + \sigma_{ij}d_{ji} \tag{D.45}$$

The third term of the right-hand side of equ. (D.44) and (D.45) is called the stress power, which represents the internal work done during deformation.

In order to write the energy balance equation in material coordinates, the second Piola-Kirchoff stress tensor $\hat{\underline{\underline{T}}}$ is introduced, defined by:

$$\hat{\underline{\underline{T}}} = J\underline{\underline{F}}^{-1}\underline{\underline{\sigma}}\,\underline{\underline{F}}^{-1T} \iff \underline{\underline{\sigma}} = J^{-1}\underline{\underline{F}}\hat{\underline{\underline{T}}}\underline{\underline{F}}^{T} \iff \underline{\underline{T}} = \hat{\underline{\underline{T}}}\underline{\underline{F}}^{T} \tag{D.46}$$

With equ. (D.46) and (D.29) it can be shown that the stress power term in spatial coordinates can be written as:

$$\sigma_{ij} d_{ji} = J^{-1} \hat{T}_{ij} \dot{E}_{ji} \tag{D.47}$$

With equ. (D.42) the energy balance equation in material coordinates becomes:

$$\rho_0 \frac{De}{Dt} = \rho_0 \varphi - \frac{\partial Q_i}{\partial X_i} + \hat{T}_{ij} \dot{E}_{ji} \tag{D.48}$$

Therefore the second Piola-Kirchoff stress tensor is an appropriate work-conjugate measure for the stresses in the material.

E. THE 1DV POINT MODEL

The 1DV POINT MODEL used in Chapter 4, 5, 6 and 7 was developed on the basis of DELFT3D-FLOW, the software system of WL | delft hydraulics to simulate the water movement and transport of matter in three-dimensions, by stripping all horizontal gradients, except the horizontal pressure gradient. This model was originally developed to study the implementation of the k-ε turbulence model in DELFT3D-FLOW by Uittenbogaard et al. (1992) and Van Kester (1994). Later, the model was extended (Uittenbogaard, 1995) by incorporating the effects of temperature-induced stratification. The version that is used as a basis to implement the various physical-mathematical formulations described in this book was developed by Uittenbogaard et al. (1996)[1]. For further details, the reader is referred to Winterwerp (1999, 2002).

In Section E.1 we present the various mathematical-physical formulations of the relevant processes, which are described in Chapters 2, 4, 5, 6 and 7. Section E.2 contains information on the numerical implementation of the equations, and in Section E.3 some numerical accuracy aspects are discussed.

E.1 The 1DV-equations

the water movement

The 1DV-equation for horizontal momentum reads:

$$\frac{\partial u}{\partial t}+\frac{1}{\rho}\frac{\partial p}{\partial x}=\frac{\partial}{\partial z}\left[(\nu+\nu_T)\frac{\partial u}{\partial z}+\frac{\tau_{xz}^{sk}}{\rho}\right]-\frac{1}{\rho}\frac{2\tau_{sf}}{b} \tag{E.1}$$

where:
b = flow (channel) width,
p = pressure,
t = time,

---

[1] Implementation in the 1DV POINT MODEL of the various formula described in this book has been carried out by several persons of WL|delft hydraulics, i.e. Dr. R.E. Uittenbogaard, Mrs. H.H. Leepel, Mr. J.Th.M. van Kester, Mr. J.M. Cornelisse and Mrs. P.M.C. Thoolen, whose work is gratefully acknowledged.

u = horizontal flow velocity, positive in x-direction

x = horizontal co-ordinate,

z = vertical co-ordinate,

ρ = bulk density of water-sediment mixture,

v = molecular viscosity,

v_T = eddy viscosity (see (E.12)),

τ_{xz}^{sk} = inter-particle stresses (see (E.36)), and

τ_{sf} = side wall friction.

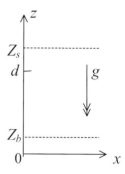

The last term in (E.1) is added to account for side wall friction effects in case of simulating the flow in a confined channel (laboratory flume) of width b. The orientation of the co-ordinate system is consistent with that of Chapter 2. The bed or lower boundary condition is defined at $z = Z_b$, and the water surface at $z = Z_s(t)$. Hence, the water depth $h = Z_s - Z_b$. Note that in DELFT3D the following convention is used: $Z_s = d + z(t)$; d and Z_b are given with respect to a horizontal reference plane.

We note that by omitting the horizontal advection terms, we implicitly assume that the Froude number of the flow is small, which is generally the case for the flow conditions studied in this book. We have also neglected the vertical velocity component w, which is small with respect to the horizontal component u. However, w may be of the same order of magnitude as the settling velocity in converging or diverging flow, for instance over a spatially varying bed. Hence, we should assume that the bed is flat and horizontal in our applications with the 1DV POINT MODEL. Vertical velocities induced by variations in water level with time, induced by tidal movements, for instance, are accounted for through the application of the σ-transformation (equ. (E.38)).

The pressure term in (E.1) is adjusted to maintain a given, time-dependent depth-averaged velocity:

$$\frac{1}{\rho}\frac{\partial p}{\partial x} = \frac{\tau_s - \tau_b}{\rho h} + \frac{U(t) - U_0(t)}{T_{rel}}, \qquad U(t) = \frac{1}{h}\int_{z_{bc}}^{Z_s} u(z',t)\,\mathrm{d}z' \qquad (E.2)$$

where:

T_{rel} = relaxation time (see also Appendix E.3),

U = actual computed depth-averaged flow velocity,

U_0 = desired depth-averaged flow velocity,

z_{bc} = apparent roughness height,

τ_b = bed shear stress, and

τ_s = surface shear stress.

A quadratic friction law, satisfying the log-law, is used:

$$\tau_b = \rho |u_{*b}| u_{*b} \ ; \ u_{*b} = \frac{\kappa u\left(z_{bc} + \tfrac{1}{2}\Delta z_b\right)}{\ln\left(1 + \tfrac{1}{2}\Delta z_b / z_{bc}\right)} \tag{E.3}$$

where:

κ = von Kármàn constant, and

Δz_b = defined in Appendix E.2.

The boundary conditions to (E.1) read:

$$\tau_b = \left\{ \rho(v + v_T)\frac{\partial u}{\partial z} + \tau_{xz}^{sk} \right\}\Bigg|_{z=z_{bc}} \ ; \ \tau_s = \left\{ \rho(v + v_T)\frac{\partial u}{\partial z} + \tau_{xz}^{sk} \right\}\Bigg|_{z=Z_s} \tag{E.4}$$

The second part of condition (E.4) describes stresses at the water surface and is used in case of wind shear stresses, or when the 1DV POINT MODEL is used to simulate the behaviour of sediment suspensions in an annular flume. The side wall friction is given by:

$$\tau_{sf} = \lambda_{sf} \rho |u| u \tag{E.5}$$

where:

λ_{sf} = friction coefficient.

the mass balance for suspended sediment

The transport of sediment is modelled with the advection-diffusion equation for various fractions numbered by the superscript (i):

$$\frac{\partial c^{(i)}}{\partial t} - \frac{\partial}{\partial z}\left\{w_s^{(i)}c^{(i)}\right\} - \frac{\partial}{\partial z}\left\{\left(D_s^{(i)} + \Gamma_T^{(i)}\right)\frac{\partial c^{(i)}}{\partial z}\right\} = 0 \qquad (E.6)$$

where:

$c^{(i)}$ = sediment concentration by mass for fraction (i),
D_s = molecular diffusion coefficient for sediment (E.10),
w_s = effective settling velocity – see also page E-11 and further,
$w_{s,r}$ = settling velocity of individual particle (see (E.24)), and
Γ_T = eddy diffusivity (see also (E.13)).

The volumetric concentration of mud flocs ϕ_f, and their primary particles ϕ_p are related to the mass concentration c and sediment density ρ_s through:

$$\phi_f = \frac{\sum_{(i)} c^{(i)}}{c_{gel}} \quad, \text{ and } \quad \phi_p = \frac{\sum_{(i)} c^{(i)}}{\rho_s} \qquad (E.7)$$

where

c_{gel} = gelling concentration (see also (E.23)), and
ρ_s = density of the sediment.

At the water surface and the bed the boundary conditions read:

$$\left\{w_s^{(i)}c^{(i)}\right\}\Big|_{z=Z_s} = 0 \quad ; \quad \left\{\left(D_s + \Gamma_T\right)\frac{\partial c^{(i)}}{\partial z}\right\}\Big|_{z=Z_s} = 0 \text{ and}$$

$$\left\{w_s^{(i)}c^{(i)}\right\}\Big|_{z=Z_b} = E_{b,c} \quad ; \quad \left\{\left(D_s + \Gamma_T\right)\frac{\partial c^{(i)}}{\partial z}\right\}\Big|_{z=Z_b} = 0 \qquad (E.8)$$

We have split the two boundary conditions (2.14) in four parts, which is allowed as the diffusion term at the bed and at the water surface is zero. At the rigid bed $z = Z_b$ we apply the classical formula of Partheniades:

$$E_{b,c} = -w_s^{(i)}\, c^{(i)} + M^{(i)} S\!\left(\theta_e^{(i)} - 1\right) \quad \text{at} \quad z = Z_b \tag{E.9}$$

in which

M = empirical erosion parameter,
θ_e = non-dimensional threshold shear stress for deposition: $\theta_e = \tau_b / \tau_e$,
τ_e = threshold shear stress for erosion, and
$S(x)$ = step function: $S = x$ for $x > 0$ and $S = 0$ for $x \leq 0$.

The molecular diffusion term D_s is given by:

$$D_s = \frac{kT}{6\pi\mu D} \tag{E.10}$$

in which

D = particle size,
k = Boltzman constant ($= 1.38 \cdot 10^{-23}$ J/K), and
T = absolute water temperature.

The influence of the suspended sediment concentration on the bulk fluid density is given by the equation of state:

$$\rho\!\left(S, c^{(i)}\right) = \rho_w(S) + \sum_{(i)} \left\{ \left(1 - \frac{\rho_w(S)}{\rho_s^{(i)}}\right) c^{(i)} \right\} \tag{E.11}$$

with

$\rho_w(S)$ = density of the water due to salinity only.

the k-ε turbulence model

The k-ε turbulence model consists of transport equations for the turbulent kinetic energy k and the turbulent dissipation ε, neglecting horizontal transport components:

$$\frac{\partial k}{\partial t} = \frac{\partial}{\partial z}\left\{(v+v_T)\frac{\partial k}{\partial z}\right\} + v_T\left(\frac{\partial u}{\partial z}\right)^2 + \frac{g}{\rho}\Gamma_T\frac{\partial \rho}{\partial z} - \varepsilon \qquad (E.12)$$

and

$$\frac{\partial \varepsilon}{\partial t} = \frac{\partial}{\partial z}\left\{\left(v+\frac{v_T}{\sigma_\varepsilon}\right)\frac{\partial \varepsilon}{\partial z}\right\} + c_{1\varepsilon}\frac{\varepsilon}{k}v_T\left(\frac{\partial u}{\partial z}\right)^2 + $$

$$+(1-c_{3\varepsilon})\frac{g}{\rho}\Gamma_T\frac{\varepsilon}{k}\frac{\partial \rho}{\partial z} - c_{2\varepsilon}\frac{\varepsilon^2}{k} \qquad (E.13)$$

where
v_T = eddy viscosity; $v_T = c_\mu k^2/\varepsilon$,
Γ_T = eddy diffusivity; $\Gamma_T = v_T/\sigma_T$, and
σ_T = turbulent Prandtl-Schmidt number.

We have carried out a few simulations for confined flows in laboratory flumes with an additional side-wall-friction induced turbulence production term (i.e. $\tau_{sf}\partial u/\partial z$ and $c_{1\varepsilon}\varepsilon/k\,\tau_{sf}\partial u/\partial z$ are added to the right hand side of (E.12) and (E.13) respectively).

The various coefficients in the standard k-ε turbulence model are summarised in Table E.1:

Table E.1: Coefficients in standard k-ε turbulence model.

						stable	unstable
c_μ	$c_{1\varepsilon}$	$c_{2\varepsilon}$	σ_T	σ_ε	κ	$c_{3\varepsilon}$	$c_{3\varepsilon}$
0.09	1.44	1.92	0.7	1.3	0.41	1	0

The model is subject to the following set of boundary conditions:

$$k\big|_{x_3=Z_b} = \frac{u_*^2}{\sqrt{c_\mu}} \quad , \quad \varepsilon\big|_{x_3=Z_b} = \frac{u_*^3}{\kappa z_{wc}} \quad , \quad k\big|_{x_3=Z_s} = \frac{u_{*,s}^2}{\sqrt{c_\mu}} \quad , \quad \varepsilon\big|_{x_3=Z_s} = \frac{u_{*,s}^3}{\kappa z_{wc}} \qquad (E.14)$$

in which the effects of waves may be included in the shear velocity u_* and where the roughness length z_{wc} for current and waves is given by (E.20); when no waves are present $z_{wc} = z_0$, which is the well-known roughness length for flow only.

the effect of surface waves

The effect of surface waves is modelled through the approach of Grant and Madsen, though we recommend the use of Van Kesteren and Bakker (1984) or Soulsby et al. (1993) in Section 2.3. The rms-value of the wave-induced bed shear stress is defined by:

$$\tau_w = \left\langle \tilde{\tau}_w^2 \right\rangle^{\frac{1}{2}} = \rho u_{*w}^2 \; ; \quad u_{*w}^2 = \frac{1}{2} f_w \hat{u}_{orb}^2 \qquad \text{(E.15)}$$

in which the near-bed horizontal orbital velocity amplitude $\hat{u}_{orb}$ is determined from the rms wave height H_{rms}. The friction coefficient f_w follows from Swart's formula:

$$\text{for } \frac{A}{k_s} > \frac{\pi}{2} : \; f_w = 0.00251 \cdot \exp\left\{ 5.21\left(\frac{A}{k_s} \right)^{-0.19} \right\} , \quad \text{and}$$

$$\text{for } \frac{A}{k_s} \le \frac{\pi}{2} : \; f_w = 0.3 \qquad \text{(E.16)}$$

where k_s is the Nikuradse roughness height related to the roughness length z_0 through $k_s \approx z_0/30$ and $A \equiv \hat{u}_{orb}/\omega$, with ω = wave frequency. The wave-affected boundary layer thickness δ_w is given by:

$$\delta_w = \frac{2\kappa}{\omega}\left| u_{*fw} \right| = \frac{2\kappa}{\omega}\sqrt{u_{*f}^2 + u_{*w}^2} \qquad \text{(E.17)}$$

where the subscript $\bullet_w$ reflects wave-related parameters, $\bullet_f$ flow-related parameters and $\bullet_{fw}$ the effects of current-wave interaction. Within the turbulent wave-boundary layer the eddy viscosity v_T reads:

$$z_0 \leq z - Z_b \leq \delta_w : \quad v_T = \kappa |u_{*fw}| (z - Z_b) = \kappa \sqrt{u_{*f}^2 + u_{*w}^2} (z - Z_b) \qquad \text{(E.18)}$$

Note that the near-bed shear stress $\tau_b = \rho_w u_{*f}^2$ is constant, but as yet unknown, throughout the wave-boundary and yields a logarithmic velocity profile based on $|u_{*fw}|$. Above the wave-boundary layer, wave-induced turbulence is not notable and Grant and Madsen assume:

$$z - Z_b > \delta_w : \quad v_T = \kappa |u_{*f}| (z - Z_b) \qquad \text{(E.19)}$$

which yields the usual logarithmic velocity profile with $|u_{*f}|$ as a parameter. The two velocity profiles are matched at $z - Z_b = \delta_w$. The wave-induced turbulence is assumed to contribute to the mean flow above the wave-boundary through an increase in effective roughness z_{wc}:

$$\delta_w \geq z_0 : \quad \frac{z_{wc}}{z_0} = \left(\frac{\delta_w}{z_0} \right)^\beta ; \quad \beta = 1 - \left(1 + \left(\frac{u_{*w}}{u_{*f}} \right)^2 \right)^{-\frac{1}{2}} \qquad \text{(E.20)}$$

The effective bed shear stress τ_b and shear velocity u_*, as applied in the preceding sections, follow from applying the logarithmic law of the wall using z_{wc} in stead of z_0. No other effects of wave-current interactions are accounted for at present.

the flocculation model

In the flocculation model (4.35) the geometric relation (4.23) is substituted, yielding an equation for the number and mass concentration N and c, but not for the floc diameter D_f. The flocculation model, as implemented in the 1DV POINT MODEL, contains the following set of equations:

$$\frac{\partial N}{\partial t} + \frac{\partial}{\partial z}\left(\frac{(1-\phi_*)(1-\phi_p)}{(1+2.5\phi_f)}w_{s,r}N\right) - \frac{\partial}{\partial z}\left(\Gamma_t\frac{\partial N}{\partial z}\right) = \tag{E.21}$$

$$-k'_A k_N^3(1-\phi_*)Gc^{\frac{3}{n_f}}N^{\frac{2n_f-3}{n_f}} + k'_B k_N^{2q}G^{q+1}\left(k_N c^{\frac{1}{n_f}}N^{\frac{-1}{n_f}} - D_p\right)^p c^{\frac{2q}{n_f}}N^{\frac{n_f-2q}{n_f}}$$

where

ae_b	= break-up efficiency parameter,
D_f	= diameter of mud flocs,
D_p	= diameter of primary particles,
D_s	= molecular diffusion coefficient for sediment,
$e_c e_d$	= flocculation efficiency parameter,
F_y	= yield strength of flocs,
G	= shear rate parameter; $G = \sqrt{(\varepsilon/v)}$
k'_A	= flocculation parameter; $k'_A = 1.5e_c\pi e_d$
k'_B	= floc breakup parameter; $k_B = ae_b D_p^{-p}(\mu/F_y)^q$
k_N	$= \left(D_p^{n_f-3}/f_s\rho_s\right)^{\frac{1}{n_f}}$
N	= number concentration of the mud flocs,
n_f	= fractal dimension,
p	= empirical coefficient; $p = 3 - n_f$,
q	= empirical coefficient; $q = 0.5$,
ϕ_*	= min $\{1, \phi\}$, and
μ	= dynamic viscosity of sediment suspension.

The relation between the number concentration N, the mass concentration c and the floc diameter D_f, and between the volumetric and mass concentration ϕ and c is given by simple algebraic relations:

$$N = \frac{1}{f_s}\frac{c}{\rho_s}D_p^{n_f-3}D_f^{-n_f} \tag{E.22}$$

$$\phi_f = \left(\frac{\rho_s - \rho_w}{\rho_f - \rho_w}\right)\frac{c}{\rho_s} = \frac{c}{\rho_s}\left[\frac{D_f}{D_p}\right]^{3-n_f} \qquad (E.23)$$

where

f_s = shape factor, and

ρ_f = floc density.

Note that the gelling concentration c_{gel} is obtained for unit volumetric concentration, i.e. $\phi_f = 1$, yielding:

$$c_{gel} = \rho_s\left[\frac{D_p}{D_f}\right]^{3-n_f} \qquad (E.24)$$

The relation between the floc size D_f and the settling velocity $w_{s,r}$ for a single particle in still water is given by:

$$w_{s,r} = \frac{\alpha}{18\beta}\frac{(\rho_s - \rho_w)g}{\mu}D_p^{3-n_f}\frac{D_f^{n_f-1}}{1+0.15Re_f^{0.687}} \qquad (E.25)$$

where

Re_f $\equiv$ floc particle Reynolds number; $Re_f = w_{s,r}D_f/\nu$.

The evolution of the settling velocity in the IDV POINT MODEL is obtained by a simultaneous solution of (E.21) and (E.5), from which the floc diameter D_f and the settling velocity $w_{s,r}$ are obtained from (E.22) and (E.25), using (E.23) to establish the fractal dimension n_f, together with the following boundary conditions, split again in 2×2 equations:

$$\{w_s N\}\big|_{z=Z_s} = 0 \quad ; \quad \left\{(D_s + \Gamma_T)\frac{\partial N}{\partial z}\right\}\bigg|_{z=Z_s} = 0 \quad \text{and}$$

$$\{w_s N\}\big|_{z=Z_b} = E_{b,N} \quad ; \quad \left\{(D_s + \Gamma_T)\frac{\partial N}{\partial z}\right\}\bigg|_{z=Z_b} = 0$$

(E.26)

and

$$E_{b,N} = -w_s N + \frac{M}{f_s D_e^3 \rho_{f,e}} S(\theta_e - 1) \quad \text{at} \quad z = Z_b$$

(E.27)

where the equilibrium floc diameter D_e and the equilibrium floc density $\rho_{f,e}$ are given by:

$$D_e = D_p + \frac{k_A c}{k_B \sqrt{G}} \quad \text{and}$$

(E.28)

$$\rho_{f,e} = \rho_w + (\rho_s - \rho_w)\left[\frac{D_p}{D_e}\right]^{3-n_f}$$

(E.29)

with:
$k_A \quad = 0.75 e_c \pi e_d / f_s \rho_s D_p$ and
$k_B \quad = k_B' / n_f$

The model contains the following empirical parameters: ae_b, $e_c e_d$, D_p, F_y, f_s, n_{f1}, n_{f2}, p, q, α and β, which have to be specified by the user through the input to the model.

the consolidation model

The consolidation model for sand-mud mixtures, as implemented in the 1DV POINT MODEL, contains the following set of equations:

$$\frac{\partial \phi_s^m}{\partial t} - \frac{\partial}{\partial z}\left(\Xi_s^m \phi_s^m\right) - \frac{\partial}{\partial z}\left(\left(D_s + \Gamma_T + \Gamma_c\right)\frac{\partial \phi_s^m}{\partial z}\right) =$$

$$= \frac{\partial}{\partial z}\left(\Gamma_c \frac{\phi_s^m}{1-\phi_s^{sa}}\frac{\partial \phi_s^{sa}}{\partial z}\right) \qquad (E.30a)$$

$$\frac{\partial \phi_s^{sa}}{\partial t} - \frac{\partial}{\partial z}\left(\Xi_s^s \phi_s^{sa}\right) - \frac{\partial}{\partial z}\left(\left(D_s + \Gamma_T + \Gamma_c \frac{\phi_s^{sa}}{1-\phi_s^{sa}}\right)\frac{\partial \phi_s^{sa}}{\partial z}\right) =$$

$$= \frac{\partial}{\partial z}\left(\Gamma_c \frac{\phi_s^{sa}}{\phi_s^m}\frac{\partial \phi_s^m}{\partial z}\right) \qquad (E.30b)$$

where superscript $\bullet^m$ refers to the mud fraction and superscript $\bullet^{sa}$ to the sand fraction, and:

$$\Xi_s^m = f_{hs}^m + \frac{f_c}{1+\eta f_c}, \qquad \text{with}$$

$$f_{hs}^m = \frac{\left(1-\phi_f^m - \phi_s^{sa}\right)\left(1-\phi_s^m - \phi_s^{sa}\right)}{1+2.5\phi_f^m} W_{s,r}^m, \quad \text{and} \qquad (E.31a)$$

$$f_c = \frac{\rho_s - \rho_w}{\rho_w} k\left(\phi_s^m + \phi_s^{sa}\right)$$

$$\Xi_s^{sa} = f_{hs}^{sa} + \frac{f_c}{1+\eta f_c}, \qquad \text{with}$$

$$f_{hs}^{sa} = \frac{\left(1-\phi_f^m - \phi_s^{sa}\right)}{\left(1-\phi_f^m\right)}\frac{\left(1-\phi_s^m - \phi_s^{sa}\right)}{1+2.5\phi_f^m}\left(W_{s,r}^{sa} - \phi_f^m W_{s,r}^m\right), \quad \text{and} \qquad (E.31b)$$

$$f_c = \frac{\rho_s - \rho_w}{\rho_w} k\left(\phi_s^m + \phi_s^{sa}\right)$$

$$\Gamma_c \equiv \frac{2K_k K_\sigma}{\left(3-n_f\right)g\rho_w} \qquad (E.32)$$

$$k = K_k \left(\frac{\phi_s^m}{1 - \phi_s^{sa}} \right)^{-\frac{2}{3 - n_f}} \qquad (E.33)$$

$$\sigma_{zz}^{sk} = K_p \left(\frac{\phi_s^m}{1 - \phi_s^{sa}} \right)^{\frac{2}{3 - n_f}} - K_{p,0} \qquad (E.34)$$

At the water surface and the bed, the boundary conditions read for both fractions:

$$\left. \{\Xi_s c\} \right|_{z=Z_s} = 0 \; ; \quad \left. \left\{ \left(D_s + \Gamma_T + \phi_*^2 \Gamma_c \right) \frac{\partial c}{\partial z} \right\} \right|_{z=Z_s} = 0 \text{ and}$$

$$\left. \{\Xi_s c\} \right|_{z=Z_b} = 0 \; ; \quad \left. \left\{ \left(D_s + \Gamma_T + \phi_*^2 \Gamma_c \right) \frac{\partial c}{\partial z} \right\} \right|_{z=Z_b} = 0 \qquad (E.35)$$

Again, the boundary conditions (5.16) and (5.17) may be split because the diffusion term is zero at the water surface and k is zero (for single-drained conditions) at the base of the consolidating bed.

The stress tensor τ_{xz}^{sk} for consolidating fluid mud in (E.1) is given by the following rheological model:

$$\tau_{xz}^{sk} = \mu_{mud} \frac{\partial u}{\partial z} \quad \text{with} \quad \mu_{mud} = \frac{a_y \tau_y}{1 + a_y |\partial u/\partial z|} + \mu^s \qquad (E.36)$$

in which:

a_y = coefficient ($a_y = 0.02$ implies $\tau_{xz}^{sk} = 0.95 \tau_y$ for $\partial u/\partial z = 10^{-3} \text{ s}^{-1}$).

$\mu^s = K_\mu \phi_p^n$

$\tau_y = K_y \phi_p^{\frac{2}{3-n_f}}$

The empirical parameters K_k, K_p, K_y, K_μ, n and η have to be specified by the user through the model input; n can vary between about 2 and 6 for various muds. Through the formulation of (E.36) for μ_{mud} we ascertain that τ_{xz}^{sk} is defined for all values of the shear rate.

The eddy diffusivity at the water-mud interface is set to zero when the yield strength of the mud exceeds the turbulent stress:

$$\Gamma_T = 0 \quad \text{if} \quad \tau_y > \tau_T \equiv \rho v_T \frac{\partial u}{\partial z} \tag{E.37}$$

E.2   Numerical implementation of the 1DV-equations

In the 1DV POINT MODEL we apply a simplified version of the so-called σ-coordinate transformation:

$$\frac{\partial}{\partial z} = \frac{1}{h} \frac{\partial}{\partial \sigma} \quad \text{with} \quad \sigma \equiv \frac{z - Z_s(t)}{h} \; ; \; h = Z_s(t) - Z_b \tag{E.38}$$

This implies that for a conserved constituent, like suspended sediment, the depth-mean sediment concentration remains constant with varying water level. However, the total mass, integrated over the water column, is not conserved.

The equations are solved on a staggered grid. The vertical grid size distribution does not have to be uniform. The sketch below shows where the various parameters are defined.

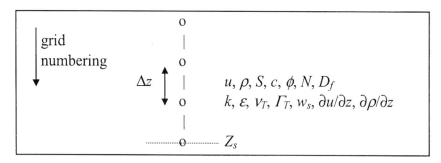

The time discretisation in the numerical solution technique is based on the θ-method; all simulations are carried out with $\theta = 1$, i.e. Euler-implicit time integration. In the vertical direction a first-order upwind scheme is used, together with a central difference scheme for the diffusion operator.

All source terms are modelled explicitly. The sink terms are modelled such that the relevant variables, i.e. k, ε, S, c, N and D_f never become negative; for details the reader is referred to Uittenbogaard et al. (1997).

A sub-time step can be set by the parameter n_{sub} through which the advection-diffusion equation for flocculation can be solved efficiently with a smaller effective time step than that for the overall process.

The communication between the various modules is given in the sketch below.

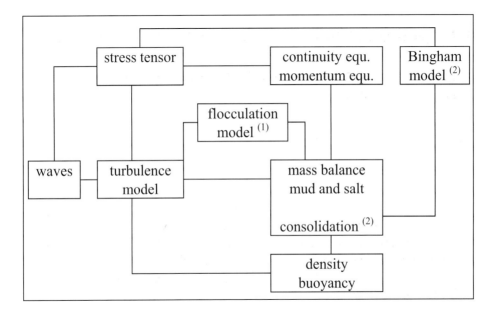

Fig. E.1: Set-up of and communication between modules; (1): only if floc evolution is simulated; (2): only if consolidation is simulated; (1) and (2): not together.

E.3   Requirements for numerical accuracy

Due to the implicit solvers and conservative form of the equations, the numerical scheme is stable and conserving for all numerical parameters. However, because of accuracy, we have the following requirements:

1. The grid-size may be chosen non-equidistant. However, it is recommended that the sizes of two neighbouring grid cells do not differ by more than a factor 1.5,
2. The size of the lowest grid cell should be smaller than the thickness of the fluid mud layer, that is when all sediment in the water column has settled,
3. From numerical experiments it appears that an optimal choice for the relaxation time is $T_{rel} = 2 \times \Delta t$,
4. The time step Δt should be small enough to accommodate for advective effects properly: $\Delta t < \Delta z / w_s$,
5. The numerical diffusivity D_{num} for the upwind scheme used amounts to $D_{num} = w_s \Delta z / 2$,
6. The parameter η in (E.31) is set at 10^5 s/m to obtain a smooth transition between consolidation and hindered settling.

SUBJECT INDEX

advection diffusion equation 9, 238
Atterberg limits 52, 70, 233, C-17
- A-line 54, 222
- activity 54, 70, 221, 351, B-3
- B-line 54, 222
- liquidity index 55, B-3
- liquid limit 52, 53, 68, 257, C-17
- plasticity index 54, 221, 350, B-3
- plasticity limits = Atterberg limits
- plastic limit 52, 53, 68, 257, C-17
- plasticity plot 54
bacteria 392, 393
bed forms 162, 193, 398
- brine lakes 400, 402
- channel 403, 423, 426
  - – collapse 423
  - – erosion 424, 426
  - – growth 403, 425
- diapirs 398, 399
- mud volcano 399, 400
- pockmarks 399, 400
- seeps 399
bed preparation
- deposited bed 347, C-26
- placed bed 347, C-26
bed shear stress / hydraulic resistance 14, 20, 27, 79
- drag reduction 171, 172, 175, 178, 179, 207
- Fanning friction coefficient 331

- flow-induced – 25
- form drag 14
- friction coefficient 25
- wave-related friction coefficient 25
- hydraulically rough 16, 17, 23, 168, 377
- hydraulically smooth 15, 16, 17, 23, 25, 377
- Moody diagram 331, 332
- Nikuradse roughness height 17, 25, 192
- probability density function of – 21, 22, 23, 370
  - kurtosis 21
  - mean 21
  - skewness 21
  - standard deviation 21
- roughness length 17, 27
- quadratic friction law 8, E-3
- shear velocity 14
- side wall friction 364, 367, 368, E-2
- skin friction 14
- wave-induced – 25, 378, 379
benthic boundary layer 5
biology 73, 383
- algae bloom 393, 394
- eco-system 384
- mudflats 385, 390
- salt marsh 384, 385
- seasonal effects 73, 394
bio-deposition 384, 386
- coatings 386
- deposition feeders 388

- filter feeders – see –
- organic filaments 387
- pelletisation 386, 388
- settling velocity 388
- vegetation 384
bio-destabilisation 384, 393
- bio-turbation 393, 395
- burrowing 393
- erodibility 393, 394
- grazing 393
- perforation 393
- tunnelling 393, 394
bio-engineering 383, 384
bio-stabilisation 384, 389
- chlorophyll *a* 390, 391, 392
- damping of turbulence 386
- damping of waves 386
- deposition feeders 388
- filter feeders – see –
- extracellular polymer substances (EPS) 390, 391, 392, 395
- erodibility 385, 389, 391, 392
- macrophytes – see –
- polysaccharides 390
- trapping efficiency 385, 386
bottom simulating reflector (BSR) 412
Boussinesq
- approximation 5
- viscosity concept 8
boundary conditions 8, 104, E-3
boundary layer 14
- benthic – 5
- buffer layer 15, 174, 207
- constant stress layer 18
- core region 15, 18, 171
- defect law 15, 165, 170
- inertial layer 15, 174

- law of the wall 208
- logarithmic layer 14, 15, 165, 171, 172, 174, 181, 209
- viscous sublayer 14, 20
- wake function 15, 166, 167
- wave-induced – 24, 25, 26, 105
buoyancy effects – see stratification
chemical environment 73, 349, 357
- acidity (pH) 37, 39, 40, 42, 43, 354, C-4
- cation (concentration) 44, C-3
- chemical profiles 43, C-4
- metal content C-3
- redox potential (E_h) 39, 40, 43
- salinity 10, 44, 193, 204, 354, C-4
- sodium adsorption ratio (SAR) 46, 351, 354, 356, C-4
Caland Canal 242, 244, 368
clay 3, 29
clay properties
- cation exchange capacity (CEC) 37, 47, 355, 356, C-2
- charge density 46
- electron activity 39, 40
- specific surface 37, 47, C-4
coagulation – see flocculation
cohesion
- bipolar 42
- diffusive double layer 44, 45, 48
- energy barrier 45
- hydrogen bonding 42
- van der Waals forces 42, 44
- ζ-potential 47, 48, 49, 50, 54, 102, C-6

cohesive sediment appearances
- consolidated bed 74, 75
- consolidating bed 74, 75
- debris flow 73, 75
- fluid mud 72, 81, 84, 177, 180, 199, 200
  - – concentration 82
  - mobile fluid mud 75
  - stationary fluid mud 75
- high concentration mud suspensions (HCMS) 73, 75, 81, 96, 175, 177, 200
- low concentration mud suspensions (LCMS) 73, 75, 81, 96
- mud banks 191
- turbidity currents 74, 75
- turbidity maximum 192
cohesive sediment paradigm 141, 142, 150
concentration 196
- bed – see density dry bed
- equilibrium – 180
- gelling – 75, 94, 111, 112, 134, 153, E-4
- gelling point 6
- mass – 29, 214, B-2, C-10, C-11, C-13
- near-bed – 143, 152
- number – 94, 96, B-3, E-4
- reference – 152
- saturation – 183
- volumetric – 29, 214, B-2, B-3, E-4
consolidation 50, 71, C-20
- aging 333
- bed phase 213
- coefficient of earth pressure at rest (K_0) 281, 314
- – coefficient 67, 77, 217, 219, 237, C-20

- isotropic – coefficient 259
- compressibility coefficient 219, 256, 268, C-20
  - skeleton – 256, 257, 259, 268
  - solids – 256, 257, 259
  - water – 257, 259
  - water-gas – 259, 261, 262
- effective stress – see stress
- equation/model
  - constitutive – 211
  - Eulerian – 224, 233, 238, E-12
  - Gibson – 211, 214, 217, 224, 225, 243, 246
- gelling point – see concentration
- Gibson height 241, 242, 248
- material co-ordinate 216, 217
- material functions 217, 221, 225, 229, 239, 242, 318
- normal – 277, 278, 313, 321, 322, 332, 333
- over – 278, 321, 322, 358, 360
- over – ratio (OCR) 278, 322
- permeability – see permeability
- – phase/regime 134, 211, 234
- point of contraction 137
- primary – 332, 333
- secondary – 333
- self-weight – 128, 211, 236, 281
- structural density 6
- under- – 399, 401

- virgin compression (line)
  (VCL) 277, 278, 315, 358
- swelling (line) 277, 278,
  315, 358, 361, 371, 372
constitutive relations 264, 267,
  271, 311
- see also non-Newtonian
- see also principal stress
  space
- see also yield surface
- associated flow 273, 274
- cyclical 310, 319
- consistency conditions 271
- flow rule 271, 277
- hardening
  - double – 273
  - isotropic – 271
  - kinematic – 271, 282,
    283, 314, 317
  - modulus 273, 317
  - parameter 273
  - rule 271, 317
  - tensor 271
- linear elastic fracture
  mechanics (LEFM) 291, 421
- modified Griffith theory
  298
- non-associated flow 277
- plastic multiplier 271
- plastic potential 271, 315,
  330
- softening
- stiffness tensor 267, 270
- stress-strain relations 267,
  C-23
continuity equation – see mass
  balance
Darcy 215, 220
density
- bulk – 29, B-2

- dry bed – 29, 66, 350, B-2,
  C-10, C-11, C-13
- excess floc – 29, 87, 94,
  124
- specific – 29
- structural – see
  consolidation
density currents – see sediment-
  fluid interaction
deposition – see sedimentation
diffusion / diffusivity
- eddy diffusion / -ivity 9, 12,
  15, 99, 169, 181, 197, 238,
  E-6
- molecular – 9, 238
- numerical – E-16
double drainage 238
drag – see bed shear stress
drained – see pore water flow
entrainment – see erosion
equation of state 6, 9, 10, 181
(Ems)-Dollart 110, 111, 193, 391,
  393
erosion 47, 58, 79, 141, 192, 343
- see also bio-engineering
- apparent critical shear stress
  – 47, 79, 347, 348, 359, E-5
- bulk erosion – see mass
  erosion
- classification diagram 357,
  361
- entrainment 77, 343, 344,
  359, 362
  - Case I – 363, 366
  - Case II – 363, 368
- erosion rate 144, 346, 348,
  373, C-26
- floc erosion 104, 343, 344,
  349, 359, 370, 371
- flow-induced – 370
- hindered erosion 380

- mass erosion 343, 344, 361, 276, 378
- mass failure 361
- non-cohesive – 353
- Partheniades – see –
- potential erosion rate 146
- resuspension 77, 139
- stochastic erosion formula 144
- surface erosion 343, 344, 359, 371, 374
- true critical shear stress for – 359
- wave-induced – 380

external forcing – see loading

failure mechanisms
- bifurcation 286, 287
- cracks 286, 375, 419
  - brittle crack 289, 290, 296, 297
  - coalesce of – 286, 289, 290
  - – drainage 292, 293, 294
  - – initiation 287, 293
  - – growth 287, 290, 291
  - – tip opening 301, 302
  - ductile crack 291, 292, 296
  - en echelon – 286
  - macro – 287
  - micro – 286
  - planar (shear) – 286, 307
  - penny-shaped – 291, 293, 421
  - shear – 286, 287, 309
  - wing-shaped – 286, 307
- deformation paradox 288
- ductile failure 254, 284, 285, 296, 404

- ductile fracture 419, 420
- hydraulic fracture 302, 303, 419, 420
- liquefaction – see pore water flow
- localisation 288
- shear failure 254, 284, 285, 303, 404
- tensile failure 254, 284, 285, 289, 404
  - direct – 285, 286
  - splitting – 285, 286
- stress intensity factor 294, 295, 307, 309, 421

Fick 7

filter feeders 357
- mussels 387, 388, 389
- suspension feeders 387

floc 87
- equilibrium floc size 104, 106, 198, E-11
- (excess) density – see density
- floc structure 88, 89, 229
- fractal structure – see self-similarity
- floc strength 102, 109
- flocculi 88
- primary particle 88, 94, 104
- floc size 94, 110, 124, 138, 141, 197, 200, E-9, E-10
- maximal floc size 92, 114, 117, 118
- space filling network 94, 112, 128, 135, 137

flocculation 44, 87, 109, 122, 138, 145, 153, 193, 195, 203
- aggregation / - parameter 87, 88, 96, 103, 109
- aggregation
  - cluster-cluster (CCA) 95

  - diffusion limited (DLA)
    95
  - diffusion limited cluster-
    cluster (DLCCA) 95, 125
  - high-concentrated cluster
    cluster (HCCCA) 95
  - reaction limited cluster-
    cluster (RLCCA) 95
  - reaction limited (RLA) 95
- break-up / - parameter 87,
   99, 101, 103, 109
- Brownian motion 88, 89
- coagulation 87
- differential settling 88, 89,
   126
- flocculation equation
     - Eulerian – 103, 109, 153
     - Lagrangean – 105, 106,
       113
- flocculation time 107, 108,
   113, 114, 116, 154, 198
- collision rate 89
- dispergents 41
- flocculants 41
- number equation 98, 99
- residence time 92, 114
- shear-induced – 88
- shear rate parameter 90, 91,
   102, 198
- Smulochowski equation 99,
   100
fluid mud – see cohesive sediment
   appearances
fluidisation – see pore water flow
Fourier number 293, 294
fractal dimension – see self-
   similarity
gelling concentration – see
   concentration
gas
   - bubbles 397, 403

   - ebullition 403, 418
   - failure 403, 419
   - growth 403, 415
   - nucleation 403, 413, 414,
     415, 416
- biogenic – 397, 403
- carbon dioxide 397
- – content 30, 407, C-3
- decomposition – see organic
   matter
- methane 397
- methanogenesis 405
- – phase
     - critical point 407, 408,
       410, 411
     - hydrates (clathrate) 401,
       403, 404, 410, 411
     - liquid 401, 410
     - quadruplet, 410, 411
     - vapour 407, 409, 411
- production 403, 404, 407
- sequestration 398
- thermogenic – 397, 403
Gibson – see consolidation
granular skeleton 34, 44, 50
- clay skeleton 60, 64
- critical porosity 60, 71
- density index 59, 60
- granular network / skeleton
   44
- fabric 4
- maximum porosity 59, 62,
   63, 68, 69
- minimum porosity 59, 62,
   63
- packing 59
- porosity 29, 59, 60
- roundness coefficient 64
- sand skeleton 60, 61
- silt skeleton 61

- uniformity coefficient 64, 68

harbour exchange flow 155
- density currents by salinity 156, 157, 158
- density currents by temperature 156, 157, 158
- density currents by sediment 156, 157, 158
- horizontal entrainment 156, 157, 158
- tidal filling 156, 157, 158

harbour siltation 155
Hedstrøm number 331
hindered settling – see settling velocity
hydraulic resistance – see bed shear stress
hysteresis 83, C-15
Kelvin-Voigt – see non-Newtonian
Kembs reservoir 362, 374
Knapp-Bagnold 184
Kolmogorov scale – see turbulence
Krone 128, 143, 146, 148, 374
Kynch 136, 219, 220
Lake Ketel 362, 422
Lake Nyos 400
laminar flow – see turbulence
limnic eruption 400, 412
loading (external)
- cyclical – 310, 312, 358
- torsion – D-9
- plane strain – D-9
- plane stress – D-9
- stress path
  - drained –
    - K_0 – 281, 283
  - undrained –
- shear (mode II) – 295, 307
- simple shear – 312, 319
- tear (mode III) – 295
- tensile (mode I) – 295, 307
- (uni-)axial compression 275, D-9
- (uni-)axial tension D-9
- radial compression (extension) 275

liquefaction – see pore water flow
macro-fauna 835, 392
- cockles 393
- clams 393
- meio-benthos 393
- snails 388, 391
- worms 388, 393, 395

macrophytes 385, 390, 391, 394
- diatoms 390, 391, 395

mass balance 5, 9, 98, 136, 151, 154, 155, 214, E-4
marine snow 387
material properties
- compression index 279, 317
- compression modulus 268, 269, 281
- critical porosity
- critical state parameter 280
- cylindrical constrained modulus 268, 269
- fracture energy 287, 299, 419
- fracture toughness 295, 310, 421, 422
- J-integral 300, 301
- loss modulus – see non-Newtonian
- phase transformation point 298
- plane strain constrained modulus 268, 269
- storage modulus – see Non-Newtonian
- shear modulus 267, 281

-    swell index 279, 317
-    Young's modulus 267, 268
Maxwell model – see non-
  Newtonian
measuring instruments/techniques
-    acoustic Doppler current
     profiler (ADCP) C-10
-    capillary suction time (CST)
     C-21
-    Coulter Counter C-8
-    CTD-probe C-5
-    dual frequency echo-
     sounder C11
-    extension test 298, 299
-    gamma-densitometer C-12
-    LISST C-7, C-25
-    micro-probe 301, C-5, C-12
-    oedometer 223, 317, C-20
-    optical back-scatter probe
     C-8
-    Owen tube C-24
-    rheometer 337, C-14
-    rotating annular flume C-24
-    sedigraph C-8
-    sedimentation balance C-23
-    seepage induced
     consolidation (SIC) 317,
     C-20
-    shear vane C-16
-    simple shear test 312
-    tensile test 289, 290, 298,
     300
-    tuning fork C-12
-    triaxial tests C-22
-    video camera C-6, C-7,
     C-25
Mediterranean Sea 400, 402
minerals – see sediment-clay
  minerals
Mississippi 36

momentum equation 5, 6, E-1,
  D-18
-    relaxation time E-2, E-16
Monin-Obukhov scale 166, 167,
  168
mud 29, 31
mutually exclusive erosion and
  deposition – see cohesive
  sediment paradigm
non-Newtonian 81, 206
-    see also constitutive
     relations
-    Bingham plastic 76, 327,
     328, E-13
-    creep – see strain
-    dashpot 328
-    dilatant 327
-    elasto-plastic 264, 327,
     330, 336
-    flow curve 327
-    generalised plastic 82
-    Herschel-Bulkley 82, 329
-    Kelvin-Voigt 6, 328, 337,
     C-17
-    Maxwell 265, 328, 337,
     C-17
-    Moore thixotropy 329
       - structure parameter 329,
         330
-    pseudo plastic 327
-    shear thinning 82, 83, 327
-    slider 328
-    spring 328
-    stress relaxation 333
-    thixotropy 83, 327
-    visco-elastic 336, C-14,
     C-17
       - loss modulus 337, C-17
       - storage modulus 337,
         C-17
-    visco-plastic 327, 358

North Sea 399, 400, 401
1DV model 105, 109, 153, 170,
    183, 186, 188, 191, 195, 199,
    203, 243, E-1
Orange River 72
organic matter 34, 41
    - see also bio engineering
    - extra cellular substance
      (EPS) 41
    - decomposition 398, 403,
      405, 406
    - humic acids 41
    - lipides 41
    - oxidised / aerobic 41, 42
    - polymers 42, 173
    - polysaccharides 41, 96,
      172, 175
    - reduced / anaerobic 43
Partheniades 142, 144, 149, 346,
    371
particle size
    - clay 31
    - colloidal 31
    - – distribution 88, 140, 150
    - finesse factor 31
    - mud 31
    - sand 31
    - silt 31
    - sand-silt-clay triangle – see
      sediment-phase diagram
Peclet number 75, 77, 257, 359,
    259, 325, 326
permeability 47, 55, 56, 163, 175,
    211, 215, 221, 231, 233, 241,
    356, 260, C-18, C-24
plasticity limits – see Atterberg
pore water flow
    - cavitation 261
    - drained – 345, 360, 257,
      260, 325, 332
    - fluidisation 358

    - liquefaction 77, 82, 191,
      320, 323, 340, 358, 375
    - undrained – 257, 260, 318,
      325, 332
polymers – see organic matter
pore size distribution 50, 125
pore water pressure – see stress -
    pressure
porosity 29, 30, B-3
    - critical – 320
    - maximum – 61, 63
    - minimum – 61, 63
    - flow through flocs 124, 125
    - void ratio – see –
Prandtl-Schmidt number 12, 76,
    210
pressure – see stress
principal stress space 272, 271,
    274 D-5, D-7
    - see also strength
    - deviatoric stress D-7
    - Lode angle 275, 311, 312
      D-8,
    - π-plane 266, 274, D-7, D-8
    - principal stress D-3
    - p-q diagram 274, 277, 285
    - stress invariants 312, D-3,
      D-4, D-8
    - yield criteria 270
    - yield surfaces – see –
rapid settling – see settling
resuspension – see water-bed
    exchange
Reynolds number
    - effective – 75, 76
    - generalised – 331
    - overall – 75, 76
    - particle – 124, 130
    - turbulence – 10, 22, 326,
      331
    - wave – 26

rheology – see non-Newtonian
Richardson number 12, 75
-    bulk – 76, 162, 189
-    critical – 161, 181, 182
-    gradient – 161, 176, 208
-    flux – 161, 176, 181, 182
Richardson and Zaki – see settling
rotating annular flume 140
Rotterdam 84, 200, 206
Rouse
-    – number 75, 76, 152, 162,
     189
-    – profile 151, 152, 184
sampling C-1
-    box-corer C-2
-    grab samples C-1
-    gravity-corer C-2
-    iso-kinetic sampling C-1
-    piston corer C-2
-    sample storage C-2
-    spring-corer C-2
-    vibro-corer C-2
sand 29, 31
sand-mud mixtures – see sediment
    mixtures
San Francisco Bay 333, 334, 345
saturation – see sediment-fluid
    interaction
sediment availability 73, 146
sediment composition
-    granular 30, 34, C-7
-    mineralogical 35, C-4
-    organic 34, 41, C-4
-    total organic content 34,
     C-4
sediment-fluid interaction 161, 207
-    buoyancy – see
     stratification
-    saturation 180, 184, 186,
     191, 209
       - auto-saturation 184, 192,

     210
   - capacity conditions 150,
     169, 180, 188
   - capacity flow 186
   - equilibrium concentration
     179
   - non-capacity conditions
     169
   - starved bed 169
   - sub-saturated 182
   - super-saturated 182, 185,
     189, 381
-    stratification 11, 13, 161,
     166, 192, 195, 197, 198,
     206, 209, 366
   - sediment-induced density
     current 123, 193
   - interface 136, 137, 180,
     213, 250
   - isolutals 195
   - lutocline 177
-    Toms effect 171, 173, 175
sediment phase
-    flocculated 30
-    gas 30
-    liquid 30
-    solid 30
sediment-phase diagram 67, 68,
    70, 72
-    cohesive discriminator 64,
     79
-    classification 66
-    granular discriminator 61,
     64, 69
-    modes of sediment
     behaviour 66
-    sand-slit clay triangle 32,
     33, 65, 69
sediment-clay minerals
-    bentonite 353
-    Boom clay 296, 325

-   carbonates 35
-   chlorite 35, 36, 39
-   clays 35
-   feldspar 35
-   Haaften clay 325
-   illite 35, 36, 38, 54
-   kaolinite 35, 35, 39, 51, 54, 352, 355
-   laponite 39, 302, 417, 420
-   montmorillonite 35, 36, 38, 54, 355
-   quartz 35
-   smectite 35

sediment mixtures
-   clay content/fraction 30, 34, 221, B-2
-   consolidation of – 30, 31
-   composition/ratio 79, 244
-   critical clay content 54, 64, 170
-   hindered settling of – 133
-   mud content/fraction 30, 237, B-2
-   equilibrium mud content 80
-   sand content/fraction 30, 237, B-2
-   sand-mud mixtures 78, 132, 229, 236, E-12
-   segregation 245
-   selling velocity 133
-   silt content/fraction 30, B-2
-   solids content/fraction 29, B-2
-   sorting 122

sedimentation
-   see also bio-engineering
-   accretion 121
-   deposition 77, 121, 128, 142, 143, 146
-   sedimentation 121, 128, 151

-   shoaling 121
-   siltation 121, 155

seepage 218

self similarity
-   fractal dimension 51, 93, 95, 96, 106, 112, 119, 127, 230, 241, 246
-   fractal structure 87, 93, 124, 228
-   theory 50, 93, 222, 231
-   power law behaviour 93

settling 77, 121, 203
-   settling flux 80
-   hindered – 127, 130, 153, 192, 203, 205
    - cloud formation 129
    - – formula 134, 135, 243
    - reduced gravity 129, 132
    - return flow 128, 132, 133
    - Richardson & Zaki 130
    - thermal 129
    - viscosity 129, 132
    - wake effect 128
-   rapid settling 84, 195, 200, 205
-   settling phase/regime 134, 136, 211, 233
-   settling time 114, 189
-   Stokes' – velocity 127, 131, 153, 200
-   – velocity 9, 76, 90, 114, 121, 123, 124, 127, 197, C-23
    - drag coefficient
    - effective – 123, 130
    - equilibrium – 114
    - maximum – 114
    - reference – 90

shear rate / shear rate tensor 82

shear rate parameter (G) 19

shear velocity 14, 15, 26, 28
silt 29, 31
siltation – see water-bed exchange
single-phase fluid 162, 163
Smoluchowski – see flocculation
solids content/fraction – see
   sediment mixtures
solubility (gas) 407, 408, 409
-    air entry value 415
-    capillary force 404
-    coefficient 261
-    desaturation 404
-    Henry's coefficient 261,
     407, 408
-    Henry's law 407
-    Kelvin's law 409
-    pore condensation 409
-    Raoul's law 407
-    saturation 407, 409, 414
-    surface tension 261, 409,
     414, 415
-    vapour pressure 261
Storegga slide 400
strain / strain tensor D-10
-    see also stress
-    angular frequency 336, 338
-    creep 326, 332
      - deviatoric 334
      - isotropic 334
      - – phases 335
-    dilatancy
      - – angle 276, 303, 316
      - – ratio 266, 281, 316
-    deviatoric (shear) strain
     (rate) 266, 282, 327, D-12
-    elongation/stretching 307,
     327
-    irreversible strain 265
-    Lamé constant 267
-    localisation of deformation
-    Poisson ratio 267

-    reversible strain 265
-    strain invariant D-12
-    stress dilatancy theory 276,
     316
-    volume – (rate) 257, 282,
     D-12
-    zero extension line 304
stratification – see sediment-fluid
   interaction
strength
-    Bingham strength 76, C-14
-    cohesion – see cohesion
-    drained shear strength 298,
     299, 313, C-14
-    peak strength 57, C-14
-    remoulded strength 53, 54
-    residual strength 57
-    undrained shear strength
     55, 57, 58, 67, 223, 233,
     296, 297, 298, 361, C-14,
     C-16, D-7
-    undrained tensile strength
     298
-    vane strength C-14
-    yield strength 75, 82, 230,
     C-14
-    yield velocity 366
stress / stress tensor 6, D-1
-    see also strain
-    critical state 313, 358, 371
-    deviatoric stress 7, 212,
     255, 327, D-7,
-    effective stress 7, 211, 212,
     216, 221, 228, 231, 242,
     245, 247, 255, 312, 358
-    fluid-induced stresses 6
-    history 73
-    hydraulic pressure 5, 8
-    interparticle stresses 6, 7
-    isotropic stress 7, 255, D-7

- Jaumann stress rate 268, D-18
- pressure
  - barotropic – gradient 205
  - (excess) pore water – 75, 77, 215, 247, 358, C-19
  - isotropic – 75, 230
  - pore water 212, 215, 256, C-19
  - – dissipation 257, 259, 293, 325
  - stagnation – 358, 361, 376, 378
  - water (over) – 75, 77
- Reynolds stresses – see turbulence
- skeleton stress 255, 256, 311
- stress ratio 281
- stress-strain relations – see constitutive relations
- total stress 212, 255
- wave-induced stresses 191

Stryker Bay 425
temperature 126, 355
- geothermal gradient 412, 413
- hydrothermal gradient 412, 413

tensor
- coaxial – 271
- eigen values D-3
- Jacobi determinant – 270, D-11
- total deformation – D-11
- deformation rate – 266, D-16
- deformation – of Green 264, 305, 306, D-12
- deviatoric – 255, 327

- displacement gradient – 265, D-11
- hardening – 271, 273, 280
- Kronecker delta 255, D-3, D-14
- Left Cauch-Green – D13
- Left stretch– D-13
- material angular velocity – D-18
- Mohr circle D-5, D-6
- permutator D-5, D-17
- Piola-Kirchoff
  - first stress – D-19
  - second stress – 269, D-20, D-21
- principal directions 271, D-3
- rotation – D-5, D12, D-15
- stress – 255, D-1
- strain – 264 D-10
- strain rate – D-16
- spatial velocity gradient – 265, D-15
- spin – of Cauchy 266, D16
- strain – of Green D-14

Terzaghi 218, 219, 248
Toms effect – see sediment-fluid interaction
turbidity current 323, 379, 381
turbidity maximum 193, 199
turbulence/turbulent 7
- buoyancy destruction 12, 166, 181, 208
- buoyancy effects 165, 169, 175, 192, 195, 200, 205
- bursts 20, 21, 174
- coherent structures 20, 23
  - ejections 20, 21
  - streaks 20
  - sweeps 20, 21

- – collapse 180, 183, 188, 192, 204
- – damping 165, 180, 183
- – damping functions 208, 210
- – dissipation (spectrum) 11, 12, 17, 18, 90, 208
- – energy (spectrum) 8, 11, 17, 18, 19, 102
- eddy diffusivity – see diffusivity
- eddy viscosity – see viscosity
- isotropy 11
- k-ε turbulence model 12, 13, 15, 169, 171, E-6
- Kolmogorov scale 18, 19, 21, 92, 99, 106
- laminar flow 76, 82, 83
- low-Reynolds effects 13
- (vertical) mixing 26, 177, 203, 209
- – mixing time 189
- model 11, 12
- – production 12, 17, 18, 208
- Reynolds stresses 6, 17, 18, E-2
- stirring (entrainment) 366
- turbulent stresses – see Reynolds stresses
undrained – see pore water flow
Van der Waals – see cohesion
viscosity
- apparent viscosity 327, C-15
- eddy viscosity 8, 12, 26, 27, E-6
- effective viscosity 129
- kinematic viscosity 7

- dynamic viscosity 7, 337, 368, C-15
void ratio 29, 60, 213, 221, B-3
Von Kármàn constant 15, 171, 172, 174, 207
- effective – 165, 209
Wadden Sea 387, 391, 395
wash load 72, 150
water-bed exchange processes 5, 43, 153, 343
- accretion – see sedimentation
- deposition – see sedimentation
- entrainment – see erosion
- erosion – see erosion
- erosion-deposition cycle 177, 195
- rapid settling – see sedimentation
- resuspension 77, 139
- sedimentation – see sedimentation
- shoaling – see sedimentation
- siltation – see sedimentation
water content 29, 52, 87, 373, B-3
waves
- wave damping 340, 377
- wave energy 340
- internal waves 11, 176
- surface waves 24, 178, 191, E-7
  - bed shear stress 25, 378, 379
  - boundary layer 24, 25
  - boundary layer thickness 24, 25, 26, 105, E-7
  - friction coefficient 25
  - height 25
  - orbital velocity 25

- rms-wave height 200,
        E-7
-   wave-induced erosion 377
Western Scheldt 65, 67, 81, 347
IJmuiden 32, 49, 53, 55, 58, 325,
    327, 374
yield surface 270, 272
-   brittle-ductile transition
    stress 2 85, 287
-   Cam Clay model 274, 277,
    278
        - critical state – see –
        - wet side 277, 279
        - dry side 277, 279
-   Critical State (Line) 276,
    278, 287, 311
-   Delft-Egg model 282
-   Desai model 284, 311, 313
-   Double Cap model 273
-   Drucker-Prager model 271,
    272
-   failure envelope 298, D-6
-   Modified Cam Clay model
    277, 278, 311
-   Mohr-Coulomb 273, D-6
-   over-consolidation ratio
    (OCR) – see consolidation
-   normal consolidation – see
    consolidation
-   swelling line (SL) – see
    consolidation
-   tension cut-off 285
-   virgin compression line
    (VCL) – see consolidation
-   Von Mises – 271, 272